Physics of Shock and Impact

Volume 1: Fundamentals and dynamic failure

Physics of Shock and Impact

Volume 1: Fundamentals and dynamic failure

Dennis Grady
Applied Research Associates, Albuquerque, New Mexico

IOP Publishing, Bristol, UK

ISBN 978-0-7503-1254-7 (ebook)
ISBN 978-0-7503-1255-4 (print)
ISBN 978-0-7503-1256-1 (mobi)

DOI 10.1088/978-0-7503-1254-7

Version: 20171201

IOP Expanding Physics
ISSN 2053-2563 (online)
ISSN 2054-7315 (print)

British Library Cataloguing-in-Publication Data: A catalogue record for this book is available from the British Library.

Published by IOP Publishing, wholly owned by The Institute of Physics, London

IOP Publishing, Temple Circus, Temple Way, Bristol, BS1 6HG, UK

US Office: IOP Publishing, Inc., 190 North Independence Mall West, Suite 601, Philadelphia, PA 19106, USA

Contents

Preface

This set of two books deals with the shock physics of solids, with an emphasis on impact applications that are brought about by any of a variety of intense-energy stimulations. A number of topics in shock physics are addressed in some depth. This book does not come close to exhausting the topic of shock physics, however. Rather, emphasis falls heavily on a selection of topics that have been the focus of my attention over the past several decades. The book covers the physical, material, mechanical and thermodynamic fundamentals of shock physics where appropriate. Also addressed are current research topics in shock physics, a number of which have yet to achieve a mature status of understanding. The various chapters also include a range of analytic solutions that address topics in the many applications involving shock physics. The solutions are tutorial in nature but are, in many cases, useful complements to comparable computational solutions, in that the underlying shock physics is usually more clearly revealed. In a number of cases unique shock wave and impact data are used to supplement the presented topic. The source of these data spans the three to four decades over which substantial strides in the development of shock wave instrumentation have been made. The shock data presented reflect such advances.

At this stage the book is a work in progress. The wonders of present day computer word processing allow, with some effort, completion of the text at intermediate points along the journey. My intent is to eventually supplement current chapters with further topics that have as yet to be prepared, and to add several further chapters. At periodic later stages the plan is to update the book when sufficient further progress has been made. Should personal energy and motivation stall along the path, the present organization will leave some product of the efforts expended.

Introduction

The title of this set of two books—*Physics of Shock and Impact*—has a noticeable redundancy. After all, impacts of bodies onto other bodies produce shock waves. And it is the physics of shock waves and the consequences of shock waves that are of paramount interest in this text. Other intense-energy impulsive loading events are the cause of similar shock waves. The sudden deposition of high-energy radiation, or the rapid release of chemical energy in the detonation of an explosive, are equally common events leading to strong shock waves.

Nevertheless, it is the various impact applications in practice over the past few decades which have provided the motivation for much of the analysis and physics modeling that is offered in the following chapters. Many shock physics issues are raised by the numerous terminal ballistics effects encountered in conventional armor and anti-armor applications. Most of the efforts on the strength and equation-of-state (EOS) modeling of ceramics have been driven by armor applications. Adiabatic shear failure in metals is clearly a consequence of conventional terminal ballistics. Hypervelocity impacts related to space station protection against meteoroid and orbital debris opens a broader arena of impact shock physics. Intense-energy fragmentation, a multiphase EOS and shock-induced vaporization, are all consequences of a hypervelocity impact event.

Other program applications, however, also provided impetus for the topics addressed here. The brief energy shortage in the early 1970s motivated efforts to find methods for effectively fracturing oil shale with explosives for the purposes of *in situ* retorting and oil extraction. Within the same time frame, concerns with the non-nuclear detonation of nuclear weapons due to accident or terrorist circumstances led to studies of the intense-energy breakup and dispersal of component nuclear materials. Both programs stimulated more fundamental investigations of dynamic fragmentation—the first focusing on more brittle materials such as hard rock and concrete, and the second on the spall and instability breakup of ductile solids and liquid metals.

Efforts to more accurately characterize the shock EOS and the strength properties of silicate and carbonate rocks for the purposes of nuclear and conventional explosive energy coupling and ground shock analysis, uncovered evidence for intense heterogeneous deformation on the microscale in the shock compression event. These observations led to some of the first endeavors to assess the characteristics of thermoplastic shear localization in the dynamic compression of solids. A similar program motivation drove more detailed investigations of the EOS and kinetics features of polymorphic phase transformation under strong shock compression. Both silicate and carbonate rocks and minerals were found to exhibit a rich range of phase transitions over the shock pressure ranges of interest. The rapidly improving time-resolved shock wave diagnostics over the same period, including piezo-resistance pressure gauges, electromagnetic particle-velocity gauges, and laser-velocity interferometry, allowed these shock wave effects to be examined in ever increasing detail.

The topics addressed in the present text have been partitioned over two volumes totaling eleven chapters. Briefly, these topics include the fundamentals of the shock EOS and shock wave propagation, fracture and fragmentation in shock and impact environments, the shock EOS and strength properties of brittle solids including ceramics and hard minerals, viscosity and structured shock waves, shock compaction of distended solids and adiabatic shear localization under high-strain-rate deformation.

Chapter 1, although far from exhaustive, covers a number of the fundamentals of shock wave physics. Acoustic wave propagation and the transition to large-amplitude nonlinear wave propagation is addressed. A description is provided of the respective Lagrangian, Eulerian and inertial representation of material position and motion. Classic nonlinear waves in solids, along with applications, are described, including the shock wave, simple waves, centered waves and steady structured waves. The governing equations of motion including the Hugoniot relations are developed, and solution techniques through the method of characteristics are described. The influences of material strength and viscosity are considered as they influence the structuring of shocks and large-amplitude waves. The spall wave with application to a number of material types is explored in some detail.

The dynamic fragmentation of condensed media subjected to intense-energy loading is covered in chapters 2–4. Chapter 2 focuses principally on the fragment size statistics of the resulting fragment distribution, and methods for predicting the governing distribution length scales, or equivalently, the characteristic fragment size. These topics are examined from both a historic perspective, and from that of more recent advances in theories of dynamic fragmentation. A wider range of applications in dynamic fragmentation is pursued in chapters 3 and 4, with the former leaning towards issues in physics while the latter focused on problems in engineering. Fragmentation due to both shock-induced spall and impact shatter is examined, along with specific spall and fragmentation criteria that have found application in shock and impact science. Instability fragmentation in ductile and liquid bodies subjected to intense distortional loading is pursued. Fragmentation in some special cases is described, including nuclear spallation, molecular dynamics simulation of intense-energy fragmentation and domain structure fragmentation in shocked ferromagnetic materials. Engineering models such as the Gurney theory and the equivalent Hugoniot theory have found profitable application in fragmentation. The models are relevant to space debris, exploding casings and ballistic impact.

Chapter 5 addresses the topic of thermoplastic shear localization in solids. This unique deformation phenomenon is of interest to the strong shock environment, as well as to the high-rate deformations achieved in ballistic impact and explosive loading. The chapter first explores the underlying physics of adiabatic shear and constraints imposed by the dimensional properties of the governing physics. Some calculations through stability analysis of the linearized equations of thermoplastic flow are described. The preponderance of attention, however, is focused on a two-temperature boundary layer model of the adiabatic shear localization process that finds inspiration from the seminal studies of Semenov, concerning thermal explosions in reactive materials, and of Mott, concerning plastic wave propagation in metals. The model identifies well-defined length scales associated with both the adiabatic shear-band width and the characteristic spacing of multiple shear bands. The time to localization and dissipation energy characteristic of the shear band also follow from the model. Topics briefly explored include turbulence in the wake of the shear band, dissipation contributions from elastic strain energy and concepts of an adiabatic shear-band toughness. Analysis and modeling are applied to a suite of metals for which appropriate thermomechanical properties are available.

Chapter 6, on the shock EOS and strength properties of ceramics, was motivated by programmatic opportunities to conduct extensive time-resolved shock profile measurements on this interesting class of materials in the early 1990s. The aim of the chapter is to pursue physics-based modeling and analysis for the purpose of ferreting out the complexities of the strong shock response of ceramics. The rich base of shock Hugoniot and shock wave profile data generated both by myself, and others, provides the backdrop for this effort. The lattice EOS, solid state phase transformation and high crystal strength intermix in ways that markedly complicate the interpretation of the measured data and the ability to achieve a clear physical understanding of the shock and high-rate response of ceramics. Boron carbide, despite several decades of continued study, still remains enigmatic.

The shock compaction of porous solids has received considerable attention since the seminal computational modeling efforts of Herrmann, and the resulting P-alpha crush model. Also notable are the even earlier geoscience-based analyses of Kompaneets on the propagation and attenuation of strong shocks through initially porous media. The self-similar features of shock propagation in crushable material that approximates extended geologic media parallels the better known solutions of Sedov and Taylor on the propagation of blast waves in air. In chapter 7 the solutions of Kompaneets are used as the starting point for exploring the analysis and modeling of shock compaction with an emphasis on brittle granular media. Both experimental data and crush models are examined that address the dynamic compaction of ceramic materials. The P-alpha model is discussed, along with extenuations introduced by Carroll and Holt. An alternative engineering crush model, the P-lambda model, which allows for the introduction of intermediate compliance materials such as water or polymeric media within the markedly stiffer porous material free space, is also explored.

The emergence of increasingly higher resolution shock wave profile measurement techniques during the 1970s revealed a rich time-resolved structure to the shock wave. The idealization assumed by many of a discontinuous shock wave was not observed. The assumption of such an idealized shock wave, in fact, confuses the physics of the material change of state through the shock wave. The material strength and phase transformation, of course, contribute a unique structure to the shock wave profile. Of further interest is the measurement of finite risetime through steady structured shock waves and consistent changes in that risetime with corresponding changes in the amplitude of the shock wave. This structure of the shock wave has been attributed to some form of solid state shock viscosity, although the responsible physics has not achieved consensus. In the early 1980s a systematic trend in this behavior over a reasonably wide range of materials was noted. This trend has subsequently become known as the fourth-power law. The history, and some of the developments, of the fourth-power law are described in chapter 8. Less well known was the proposition, during the same time frame, of a shock adiabatic invariant relating to the same shock wave structure. Whereas this shock invariant implies the fourth-power law of steady structured waves, the invariant has wider application to unsteady shock waves. The shock invariant, as well as issues of solid state shock viscosity, are also addressed in chapter 8.

In chapter 9 shock compression features unique to brittle solids are described, which have been garnered, after considerable thought and analysis, from time-resolved shock profile data. The chapter in part echoes some of the observations discussed in chapter 6 on ceramics. Other materials, analysis and modeling efforts enrich the description, however. The materials include natural minerals (quartz, calcite), rocks (dolomite, novaculite, limestone) and glass, among others. Again, the dynamic strength and phase transformation are observed to profoundly influence the shock wave structure in these materials. Dolomite, a rock formed of a natural carbonate mineral of the same name, has perhaps been the most extensively studied brittle material in terms of variations in both the impact amplitude and spatial propagation distance of experimentally measured time-resolved shock and release wave profiles. These data reveal unique kinetics characteristics in the dynamic strength of brittle solids that strongly hint at the fracture failure wave behavior later reported for similar brittle materials. The fundamentals of the fracture mechanics related to the shock failure of brittle solids are discussed in this chapter, as are descriptive models based on the governing fracture mechanics. Underlying causes of time and strain-rate-dependent strength and failure in brittle solids are revealed through such physics-based models.

Chapter 10 provides a more extensive exploration of the thermodynamic EOS of matter in the shock compression environment. The EOS of gases are briefly examined with emphasis on the shock Hugoniot and wave propagation. Several EOS, including the Mie–Gruneisen EOS, which have found frequent use in shock wave applications are covered. Also described is the less commonly used enthalpy-based Rice–Walsh EOS and its application in the EOS description of intense shocks in highly distended media.

The concept of a universal compression EOS commonly used as the reference compression path for more complete EOS descriptions of solids is examined. EOSs that have been identified with Murnaghan, Birch, Vinet, Holzapfel and others are contrasted and compared with material compression extracted from shock Hugoniot data. The applications in chapter 10 also explore the extreme shock compression of highly distended solids and the EOS of mixtures.

Lastly, chapter 11 focuses on methods of shock wave analysis. Useful tools for the interpretation of wave interaction in experiments and applications, including pressure–particle-velocity plots and distance–time plots are surveyed. Methods of Lagrangian analysis are described that were developed to extract dynamic constitutive properties from time-resolved unsteady experimental shock wave profile data. The shock wave analysis tools described are applied to several specific applications and identified with related experimental shock wave data.

Author biography

Dennis Grady

Dennis Grady is currently an Associate and Principal Scientist with the South West Division of Applied Research Associates headquartered in Albuquerque, New Mexico. He performed Military service from 1958 to 1963, in the U.S. Navy Submarine Service. He received his BS in physics and mathematics from Lewis and Clark College in 1967, and a PhD in physics from Washington State University in 1971. Following three years on the research staff of Poulter Laboratory at SRI International he joined Sandia National Laboratories. He retired from Sandia Laboratories as a Distinguished Member of the Technical Staff in 1996 and joined the South West Division of Applied Research Associates the same year. He has been involved with the experimental measurement and theoretical description of condensed matter under the extreme pressure and temperature stimulus of shock and high-velocity impact for over 40 years. He has published over 300 technical papers and reports on a range of materials and applications issues in the intense shock environment including experimental methods, electric and magnetic effects, phase transformation, high-pressure equation of state, transient strength, energetic reaction, hypervelocity impact, and dynamic fragmentation. He is a member or past member of the American Physical Society, International Hypervelocity Society, American Geophysical Union, International DYMAT Association, and has received honors from, and served various official roles in, these societies.

IOP Publishing

Physics of Shock and Impact
Volume 1: Fundamentals and dynamic failure
Dennis Grady

Chapter 1

Shocks and structured waves

The present chapter describes the motivation for, and analysis of, shock and large-amplitude wave propagation in condensed matter. Impact loading, or any intense-energy loading for that matter, is a localized event. The consequences of the event, including change of state, failure, extent of crush, fragmentation, or any of a number of other physical changes brought about by the impact, are dependent on the details of the transport of energy and momentum. This redistribution of the energy and momentum initially imparted by the impulsive loading is largely performed by nonlinear wave propagation. The shock and structured wave concepts that are addressed will principally focus on the response of solids, although some of the discussions are also applicable to liquids. A cursory description of sonic waves is pursued for the purpose of motivating considerations of the propagation of large-amplitude waves and the determining of material nonlinearities. The governing equations of large-amplitude wave propagation are developed and characteristic solution methods are described. Some classic waveforms are explored, including structured steady waves, simple waves and centered waves. Applications include among others, vaporization on the emergence of a strong shock at a free surface, elastic precursor waves in uranium alloys, ramp wave evolution through the pressure-induced silicon phase transformation and shock wave structure brought about by viscosity in the shock compression event. Spall waves and the processes of spall failure further contribute to the study of shock waves in condensed matter.

1.1 Introduction

This chapter provides the basics of the large-amplitude nonlinear wave propagation necessary to the interpretation and physical understanding of shock effects brought about by explosion or impact loading of condensed matter. The emphasis is on the analysis and interpretation of shock effects in controlled experiments and, where possible, in material and engineering applications.

doi:10.1088/978-0-7503-1254-7ch1

Large-amplitude waves in solids are intimately related to the constitutive responses of the material brought into play during the dynamic motions imparted by extreme shock wave loading. Such constitutive features include elasticity, thermodynamics, change of phase, inelastic deformation, and material strength and failure, including adiabatic shear, volumetric crush and dynamic spall. In turn, the creation, measurement and interpretation of large-amplitude waves in materials offer perhaps the most favorable methods for achieving the knowledge necessary for the understanding and prediction of material response in the extreme shock environment.

This chapter addresses much of the basic shock physics necessary for both the elucidation and the application of large-amplitude waves. The simple sonic wave in solids and liquids provides the starting point. Coordinate structures necessary to the description of large-amplitude waves are identified, including the Lagrangian, Eulerian and inertial frames of reference. Shock wave fundamentals are addressed, including the Hugoniot conservation equations, the Hugoniot equation-of-state (EOS), and the effects of strength and phase transformation on the Hugoniot relations. The intrinsic nonlinearity of condensed matter, both in lattice compressibility and in dynamic inelastic deformation including shear strength and viscosity, is intrinsic to the propagation, structuring and evolution of large-amplitude waves. There are special classic waveforms for which instructive analytic solutions can be derived. Such waves include the steady structured wave, the simple wave and the centered simple wave, among others. The solutions of such waves are discussed where appropriate. The evolution of ramp waves in nonlinear materials and the application of ramp waves in experiments receive additional attention.

Representative shock wave experimental data accompany the selected discourses. The one-dimensional governing equations of mass, momentum and energy are put forth and characteristic solution methods touched on. Topics include rarefaction waves and rarefaction shocks, shock structure evolution, elastic precursor structure and wave splitting with representative experimental shock wave data. A fairly extensive treatment is put forth for spall waves and the spall failure phenomena with application to a number of material types.

1.2 Waves in solids

The propagation of waves is fundamental to the dynamics of solids. Any solid structure will respond to a transient disturbance through the propagation of stress waves. The disturbance can result from the slightest of impulses to the most intense impact, explosion, or radiation loading. Calculation of wave propagation is critical to assessing the response of solid bodies to transient loading. The measurement of the wave propagation features of a solid can reveal many aspects of the material EOS, failure characteristics and deformation properties of a solid.

1.2.1 Sonic wave propagation

A body subjected to a modest impulsive loading such as that illustrated in figure 1.1 will propagate as an elastic pressure wave at a velocity determined by properties of

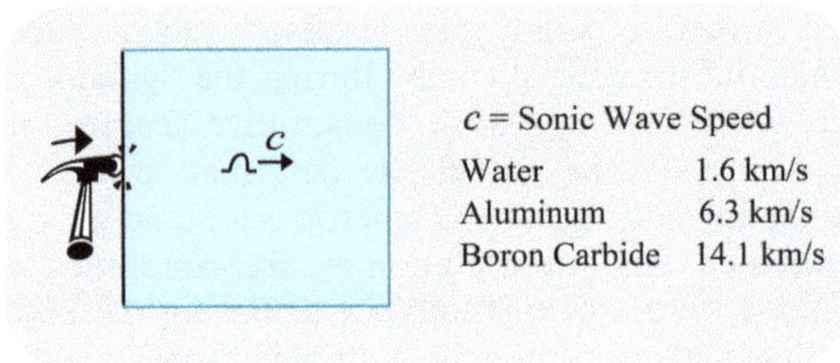

Figure 1.1. Sonic wave propagation in a material.

the material. A more detailed description of elastic wave propagation will be reserved until a later chapter. For the moment, the elastic pulse is considered to propagate at a velocity determined by the elastic stiffness (or compliance) of the material along the direction of wave propagation. Sonic wave velocities for several diverse materials are provided in the figure.

Both the compliance and the mass density of the material determine sonic speeds. The concept is illustrated here by first considering a simple spring–mass system such as is shown in figure 1.2. The spring has a linear spring constant k such that the force δP is necessary to displace the mass a distance δa from the equilibrium length a of the spring. That is $\delta P = k\delta a$. When set into axial motion the spring–mass system has a natural frequency of

$$\omega \sim \sqrt{k/m}. \tag{1.1}$$

In figure 1.3 a system of masses and springs is joined to provide a linear chain. The linear chain could be considered as, for example, a simple model of the intermolecular lattice in a condensed material. A stimulus is introduced at one end which results in the propagation of a sonic pulse along the linear chain. The wavelength λ of the pulse is considered to be long compared to the lattice spacing a between masses. Dispersion, or the dependence of wave velocity on wavelength, which occurs when wavelengths become comparable to the lattice spacing, is not considered here.

The sonic wave speed is related to the natural spring–mass frequency through the relation,

$$c \sim a\omega \sim a\sqrt{\frac{k}{m}}. \tag{1.2}$$

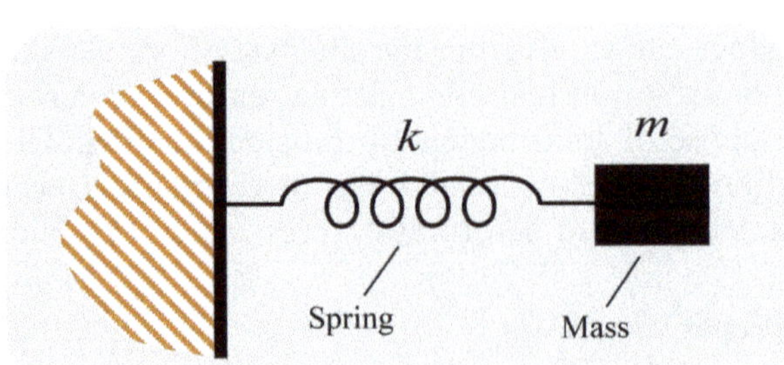

Figure 1.2. A linear spring–mass system.

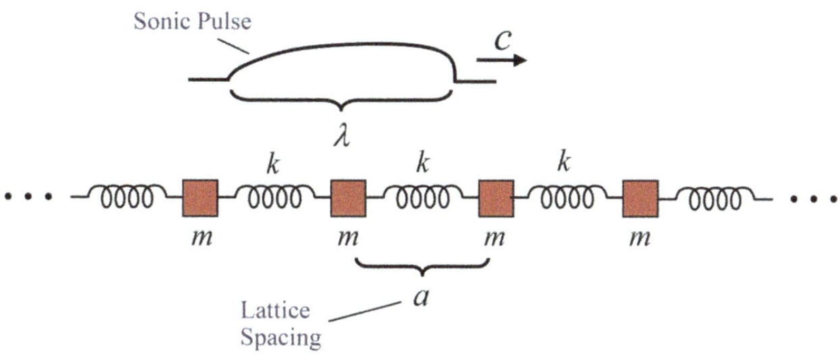

Figure 1.3. Extended linear spring–mass system.

Translation of the sonic speed in equation (1.2) to a continuum description of the linear chain is apparent. The density (mass per unit length) is provided through $\rho = m/a$. From the previous force displacement relation with $\delta P = k\delta a = ka\ \delta a/a$, an elastic modulus of $E = ka$ and compression strain $\delta\varepsilon = \delta a/a$ are identified, and $\delta P = E\delta\varepsilon$ provides the normal continuum pressure increment (force per unit area) in the propagating wave.

(Note: in the present development through much of this section, the pressure increment δP refers to stress or pressure on planes orthogonal to the direction of wave propagation. It will not be necessary to distinguish between stress and pressure. When such a distinction becomes necessary, appropriate notice is provided and a change in notation is introduced.)

Substitution into equation (1.2) provides for the sonic wave speed in terms of the elastic modulus and density of the material,

$$c = \sqrt{E/\rho}. \tag{1.3}$$

1.2.2 Inertial, Lagrangian and Eulerian wave speeds

The natural frequency of the unit spring and mass $\omega \sim \sqrt{k/m}$ provides the rate of energy propagation through the spring–mass system. The frequency ω provides the number of mass elements encompassed by the wave per unit time. Similar to equation (1.2), a velocity defined as

$$c_m \sim m\omega = m\sqrt{\frac{k}{m}} = \sqrt{mk}, \tag{1.4}$$

provides the speed of sonic wave propagation in units of mass per unit time. c_m is identified with the inertial wave speed of the material and $Z = mk$ with the inertial modulus. With the previous continuum relations for $\rho = m/a$ and $E = ka$, equation (1.4) becomes

$$c_m = \sqrt{\rho E} = \sqrt{Z} = z, \tag{1.5}$$

where z is the acoustic impedance of the material. Thus, the acoustic impedance is the sonic velocity in mass per unit time. (Actually mass per unit area per unit time.) The inertial wave speed is independent of the cell spacing a. It provides the speed that an acoustic disturbance spans the mass of the solid body.

The inertial velocity defined in equation (1.5) and identified with the acoustic impedance can be written as,

$$c_m = \frac{dM}{dt}, \tag{1.6}$$

where M is the mass per unit area subsumed by the propagating stress wave. Lagrangian and Eulerian coordinates are introduced in a later section. However, if a coordinate h (the Lagrangian or material coordinate) is identified as a distance in the direction of wave propagation before the material is compressed and still at the initial mass density ρ_o (a reference configuration), while x (the Eulerian or spatial coordinate) is the corresponding distance when the material is at a compressed density ρ, then the same mass increment can be represented as $dM = \rho_o dh = \rho dx$. It follows that

$$c_m = z = \frac{dM}{dt} = \rho_o \frac{dh}{dt} = \rho \frac{dx}{dt}, \tag{1.7}$$

and

$$z = \rho_o c_L = \rho c, \tag{1.8}$$

relating the inertial velocity, or acoustic impedance to the Lagrangian and Eulerian sonic velocity of the material. More encompassing discussions are provided in both chapters 10 and 11.

1.2.3 Nonlinear compressibility

If the spring constant in figure 1.2 or figure 1.3 is nonlinear (linear implies proportionality between the force and the spacing through a constant value of k) with a dependence $k(a)$ on the inter-molecular spacing, then a force (or pressure) versus compression relation will result such as that illustrated in figure 1.4. A dependence is shown for the pressure on either the spacing a or the density $\rho = m/a$.

A representative functional dependence is that of the Lennard-Jones 6-12 intermolecular potential, which has the functional form for the energy (more on Lennard-Jones and other universal potentials in chapter 10)

$$U(a) = \varepsilon\left[\left(\frac{\sigma}{a}\right)^{12} - 2\left(\frac{\sigma}{a}\right)^{6}\right], \tag{1.9}$$

and pressure

$$P(a) = \frac{12\varepsilon}{a}\left[\left(\frac{\sigma}{a}\right)^{12} - \left(\frac{\sigma}{a}\right)^{6}\right]. \tag{1.10}$$

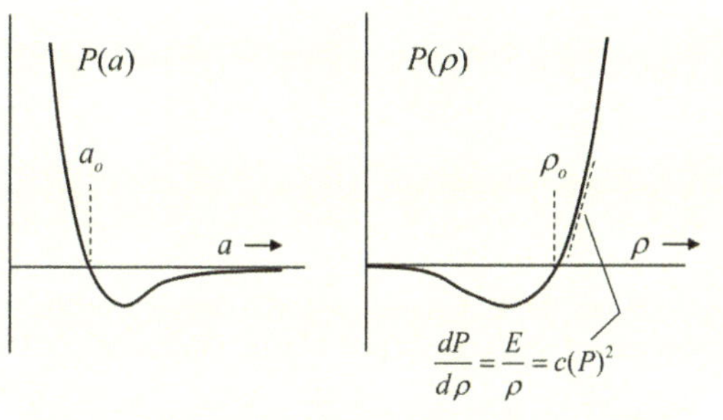

Figure 1.4. Representation of the dependence of pressure on the intermolecular spacing or, alternatively, the material density.

The axial elastic modulus of the linear system, as previously related to the spring constant through $E(a) = k(a)a$, correspondingly depends on the spacing a. At any pressure P the magnitude of the modulus $E(P)$ can be determined and the acoustic velocity at that pressure amplitude follows from $c(P) = \sqrt{E(P)/\rho}$. Namely, a small pressure increment δP about the pressure amplitude P will propagate through the system with an acoustic velocity $c(P)$.

Conversely, identify the spacing per unit mass $v = a/m$ (inverse of the density). The derivative at the same pressure yields $Z(P) = z(P)^2 = -dP/dv$. Namely, the acoustic impedance and inertial wave speed at pressure P.

1.2.4 Compression relations for solids

Various compressibility relations have been found to be both convenient and useful in the analysis of large-amplitude wave propagation in solids. These relations are usually referred to as EOSs. This identification is not entirely appropriate, as such relations are not usually intended to address the full thermodynamic description of the material commonly ascribed to an EOS. A reference EOS is perhaps a more appropriate identifying term. Frequently, however, these relations are joined with other thermodynamic models to construct a complete EOS. Several of these compression relations, or reference EOSs, are described here. A number of others are discussed more fully in chapter 10.

Perhaps the most common is the Murnaghan equation (Murnaghan 1944), which follows from the observation that the ultrasonic bulk modulus for many solids and liquids exhibits a near linear dependence on pressure, as illustrated in figure 1.5. The linear relation $K(P) = K_o + K_o'P$ provides the differential relation,

$$\frac{d\rho}{\rho} = \frac{dP}{K_o + K_o'P}, \tag{1.11}$$

for the compressibility of a Murnaghan material. The bulk modulus K replaces the modulus E used previously, as the present discussion transitions from the compression of a linear spring to that of a bulk solid (or liquid). Material strength issues will

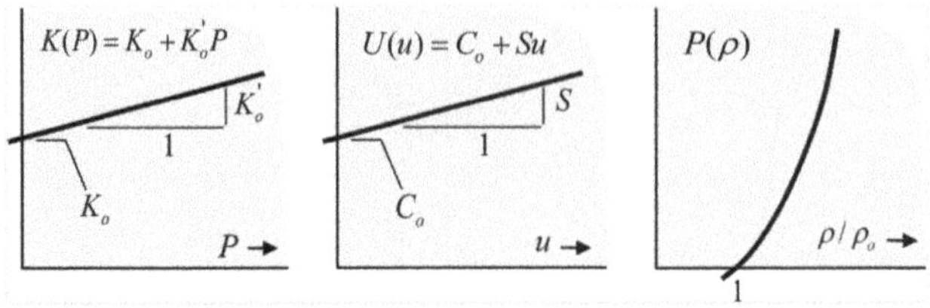

Figure 1.5. Linear behavior in either the bulk modulus versus pressure or the shock velocity versus particle velocity leads to familiar relations for the large-amplitude compression of a solid.

continue to be ignored. Equation (1.11) integrates to the well-known Murnaghan EOS for the material compressibility,

$$P(\rho) = \frac{K_o}{K_o'}\left[\left(\frac{\rho}{\rho_o}\right)^{K_o'} - 1\right]. \tag{1.12}$$

In a later chapter addressing EOS issues in the shock wave environment in substantially more detail, difficulties with the Murnaghan EOS under large compressions are pointed out. For the present purposes these difficulties will be ignored. For most materials the Murnaghan EOS is a quite adequate representation to volume strains of about 20%–25%.

A further familiar compressibility relation follows from the observation of linearity in the plot of experimentally measured shock velocity versus particle velocity, as is also illustrated in figure 1.5. Joining this linear shock-velocity versus particle-velocity relation $U = C_o + Su$ with the shock wave conservation relations for mass $u = (1 - \rho_o/\rho)U$ and momentum $P = \rho_o uU$ results in the relation

$$P(\rho) = \frac{\rho_o C_o^2(1 - \rho_o/\rho)}{\left(1 - S(1 - \rho_o/\rho)\right)^2}. \tag{1.13}$$

A power series expansion of the two compression relations provides, respectively,

$$P(\varepsilon)/K_o = \varepsilon + \tfrac{1}{2}\left(K_o' + 1\right)\varepsilon^2 + \tfrac{1}{6}\left(K_o' + 2\right)\left(K_o' + 1\right)\varepsilon^3 + \cdots, \tag{1.14}$$

and

$$P(\varepsilon)/\rho_o C_o^2 = \varepsilon + 2S\varepsilon^2 + 3S^2\varepsilon^3 + \cdots, \tag{1.15}$$

where $\varepsilon = 1 - \rho_o/\rho$ provides a measure of the material compression. Equating the first- and second-order terms yields the relations,

$$K_o = \rho_o C_o^2,\ K_o' + 1 = 4S. \tag{1.16}$$

A comparison of third-order terms reveals that the Murnaghan and the linear shock-velocity versus particle-velocity equations when related through equation (1.16) are tangent to second-order but diverge in the third-order term.

There are a number of materials that exhibit sensible agreement between the second-order Murnaghan equation and the linear shock-velocity versus particle-velocity equation, usually over a restricted compression range. It is more interesting to make the same comparisons for a material that is not as nicely behaved. Gieske and Barsch (1968) (see also Li *et al* 1998) measured the ultrasonic elastic constants of single-crystal aluminum oxide and observed a linear dependence of the bulk modulus on the hydrostatic pressure to 10 GPa. Their measured values are $K_o = 257$ GPa and $K_o' = 4.3$. McQueen (1968) (see Marsh 1980) determined the shock Hugoniot properties of a near-full-density polycrystalline aluminum oxide ($\rho_o = 3977$ kg m^{-3}) over the pressure range of 20–145 GPa. A linear shock-velocity versus particle-velocity behavior is observed with $C_o = 8.6$ km s^{-1} and $S = 0.98$. Through the relations provided in equation (1.16), both the shock and the ultrasonic results can be plotted in either the bulk modulus versus pressure or the shock-velocity versus particle-velocity plane. These plots are shown in figure 1.6 and illustrate the level of disagreement between the two sets of compression data.

Discrepancy between the two experimental studies could certainly be due to other than linear behavior of either of the sets of experimental data. Such nonlinearity is not, however, observed within the experimental accuracy in either set of data. In the present case a more likely explanation rests with the shock data acquired on a high-strength ceramic. The dynamic strength of the aluminum oxide polycrystalline ceramic tested by McQueen (1968) would lead to an elastic precursor traveling at speeds in excess of the bulk elastic wave speed of the solid and the Hugoniot elastic limit (HEL) amplitude of order 6–10 GPa. Dynamic strength markedly complicates the interpretation of the shock wave transit time and the corresponding shock wave velocity. Possible errors in shock velocity are more pronounced at lower impact pressures and diminish with increasing shock amplitude. The character of the shock data in the shock-velocity versus particle-velocity plot shown in figure 1.6, in contrast to the ultrasonic prediction, is consistent with this explanation. Difficulties with linear shock-velocity versus particle-velocity representations of material compression are pursued further in chapter 10.

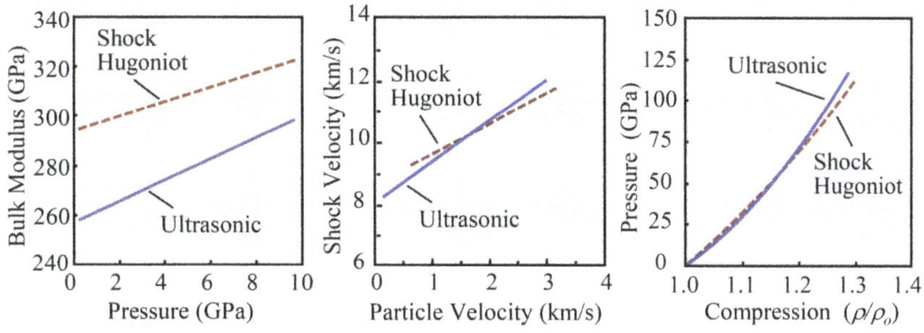

Figure 1.6. The left plot compares the experimental ultrasonic bulk modulus (Gieske and Barsch 1968) with values determined from the shock Hugoniot data (McQueen 1968) for aluminum oxide. The center plot shows the experimental Hugoniot and the U versus u curve determined from the ultrasonic data. Experimental Hugoniot and extrapolated Hugoniot compression from ultrasonic data are compared on the right.

The experimental Hugoniot and extrapolated Hugoniot compression based on the ultrasonic data are compared on the right in figure 1.6. Whether a Murnaghan or a linear shock-velocity versus particle-velocity equation is more appropriate to represent either set of experimental data cannot be discerned at the highest pressures. There is less than a 2 GPa difference between the two compression curves at 100 GPa pressure.

To summarize, for certain materials, the compression parameters obtained by different experimental methods can differ markedly. In the present case differences in compressibility (and moduli) do not indicate difficulties with the application of simple empirical compression relations such as the Murnaghan or the linear shock-velocity versus particle-velocity equations. For the shock pressures achieved on aluminum oxide, the two relations have reasonable overlay in both the compression and the bulk modulus plots over the full range of Hugoniot data. For the present application to aluminum oxide ceramic the ultrasonic modulus is more believable at lower pressures because of the known shock wave experimental difficulty associated with the strength of the material. The shock Hugoniot data are, on the other hand, more believable at higher pressures than the extrapolation of lower-pressure ultrasonic data.

The Murnaghan and the linear shock-velocity versus particle-velocity compression relations are compared in figure 1.7 using equivalent parameters provided through the expressions in equation (1.16). Values of S ranging from 1 to 2 (K_o' from 3 to 7) span the range of most materials, and are used in the comparisons. Either relation is appropriate in both the pressure and the bulk modulus, provided pressures do not exceed the initial bulk modulus, and probably more reasonably, are about one-half the bulk modulus. Note that this criterion spans the compression of aluminum oxide well into the several hundred gigapascal range.

When warranted, however, more discerning empirical compression relations, including the Birch–Murnaghan finite strain relation (Birch 1978), as well as those of Rose *et al* (1984) and Holzapfel (1991), can be used (e.g. Ahrens 1992, Sikka *et al* 1997). An empirical relation due to Vinet *et al* (1986) has been suggested for selected metals in the study of Akobuije *et al* (2005). The relation is perhaps more

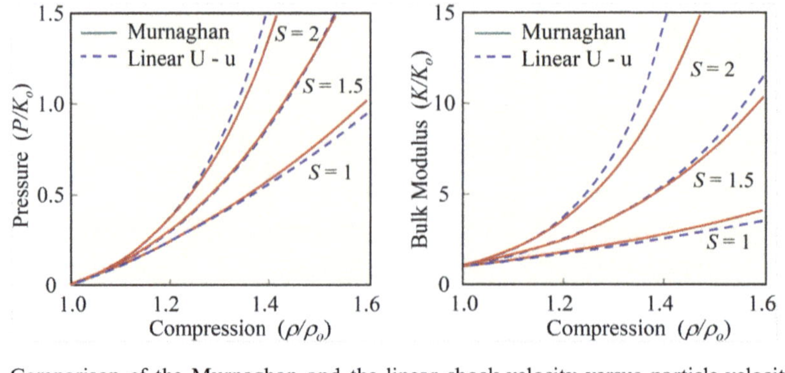

Figure 1.7. Comparison of the Murnaghan and the linear shock-velocity versus particle-velocity empirical compression relations.

appropriately associated with Rose *et al* (1984), or even Rydberg (1931). Jeanloz (1988) has noted that the latter compression relation is numerically equivalent in the leading terms to the Birch–Murnaghan finite strain relation, and that no two-parameter EOS is in full accord with all higher-order elastic moduli. These relations are discussed in further detail in chapter 10 on EOS issues of solids.

Further, these empirical models are at best only appropriate within limited ranges of the lattice compression of solids. When lattice instabilities are encountered, resulting in reorganization of the lattice structure (phase transitions), such models are not appropriate. Kerley (2006) has also pointed out that the linear shock-velocity versus particle-velocity equation is not appropriate for many materials.

1.2.5 Further expressions for the elastic stiffness of solids

The bulk modulus of a material, defined as,

$$K = -v\frac{dP}{dv} = \rho\frac{dP}{d\rho}, \tag{1.17}$$

is the common measure of the volumetric stiffness of a material. The differential is considered to be under either isothermal or isentropic constraints. The specific volume $v = 1/\rho$ is the inverse of the material density.

Two other expressions for the elastic bulk stiffness of a material are found useful. They are the sonic modulus,

$$C = \frac{dP}{d\rho}, \tag{1.18}$$

and the inertial modulus,

$$Z = -\frac{dP}{dv}. \tag{1.19}$$

Only two of the three moduli are independent. It is readily shown that

$$K = \sqrt{ZC}. \tag{1.20}$$

The sonic wave speed of the material in units of distance per unit time is $c = C^{1/2}$. The inertial wave speed is $z = Z^{1/2}$. Units of z are the mass per unit time over a unit area in the direction of wave propagation. The inertial wave speed is identical to the bulk elastic impedance of the material. From equation (1.20) the bulk modulus is $K = zc$.

The experimental observation, or assumption, that,

$$K(P) = K_o + K_o'P, \tag{1.21}$$

leads to the familiar Murnaghan reference EOS. Both the inertial modulus and the sonic modulus can be similarly expressed as a linear function of the pressure. In most cases, low-pressure ultrasonic data are equally well represented. Different high-pressure EOS relations result, however. In chapter 10 the reference EOS that follows from a linear pressure description of the inertial modulus is shown to provide a better representation of the high-pressure compression of a number of materials than does either the Murnaghan or the Birch–Murnaghan EOS.

1.3 Large-amplitude waves

The linear spring–mass system is also instructive in the topic of large-amplitude wave propagation. The wave is a motion of the mass elements, and a corresponding compression and expansion of the springs, that propagates through the system of otherwise stationary mass elements and uncompressed springs. Such a wave in the spring–mass system is illustrated in figure 1.8.

The wave is characterized by a quantitative description of the pressure or the velocity as it moves through the system. (The velocity is commonly referred to as the particle velocity or mass velocity; particles would be the mass elements in the present discrete system.) The description could be, for example, for the velocity, the speed of each mass element through the region in which the motion is disturbed or, for the pressure, the supporting force of the compressed springs within the same region. The location of each moving mass element provides the particle-velocity, or pressure (through compression of the spring), versus position description of the wave as it progresses through the system.

1.3.1 Lagrangian and Eulerian coordinates

Identification of a set of coordinates to quantify the location of a large-amplitude wave introduces an important feature in the characterization of wave propagation. This issue is particularly transparent within the present linear spring–mass system. The first is the Lagrangian, or material, set of coordinates. The second is the Eulerian, or spatial, or laboratory, set of coordinates.

A Lagrangian coordinate system can, for example, be identified with the initial position of each of the mass elements before motion is imparted by the wave—the set of discrete h_i coordinates is illustrated in figure 1.9. A set of appropriate Lagrangian coordinates need not, however, have the dimensions of length. Alternative Lagrangian coordinates could, for example, be the cumulative count of mass elements (dimensionless), or the cumulative mass of the elements

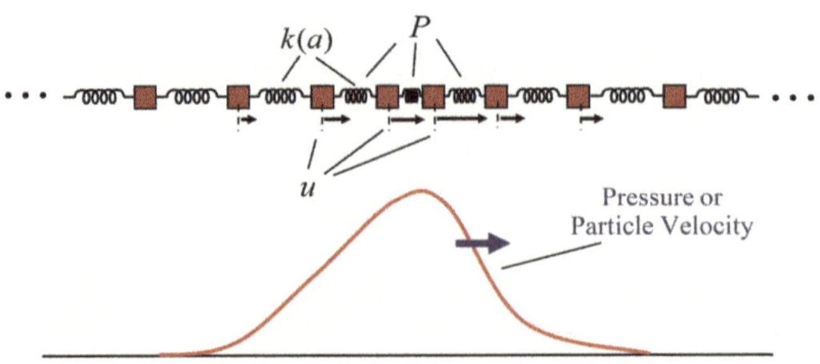

Figure 1.8. Large-amplitude wave of pressure or particle velocity in a linear spring–mass system.

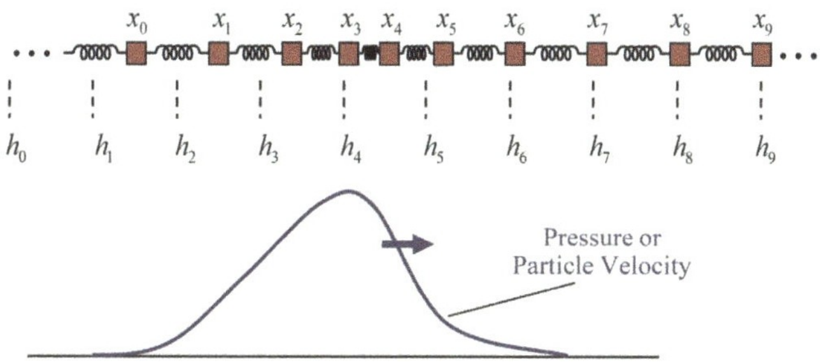

Figure 1.9. Identification of the discrete Lagrangian (h_i) and Eulerian (x_i) coordinates in a linear spring–mass system.

(dimensions of mass). The latter is identified as the inertial coordinate system, which is used frequently throughout the text.

The Eulerian coordinate identifies the current spatial position of each mass element relative to a stationary, or laboratory, measurement reference frame. In figure 1.9 each mass element is originally positioned at coordinated h_i (the Lagrangian coordinate). Following motion imparted by the wave up to a prescribed time, the corresponding mass element is at the position x_i (the Eulerian coordinate).

Some basic features of the motion and the deformation imparted by the large-amplitude wave are readily extracted. The separation of two adjacent mass elements at any later time, when specified as a ratio with the corresponding initial separation, identifies the compression deformation at that location,

$$\frac{x_{i+1} - x_i}{h_{i+1} - x_i} \simeq \left. \frac{\partial x(h,t)}{\partial h} \right|_t = \frac{\rho_o}{\rho}. \tag{1.22}$$

In the continuum limit the large-amplitude wave imparted to the mass and spring system is appropriately described by the functional dependence of the Eulerian position on both the Lagrangian coordinate and the time $x(h,t)$. A useful measure of the linear strain is provided by

$$\varepsilon(h,t) = 1 - \left. \frac{\partial x(h,t)}{\partial h} \right|_t. \tag{1.23}$$

The particle velocity at the corresponding material position is given by,

$$u(h,t) = \left. \frac{\partial x(h,t)}{\partial t} \right|_h. \tag{1.24}$$

1.3.2 Variable wave speed

The large-amplitude wave does not remain unchanged in form as it propagates through the medium. The behavior contrasts with the limiting small amplitude wave in which a linear compressibility results in waves that propagate unchanged in shape. Finite compression of the transmitting material usually requires a nonlinear description of the pressure versus compression. If the media can be reasonably described by a pressure versus density relation of the form $P(\rho)$, as illustrated in figure 1.10, then the speed of the wave at a pressure P is determined through the relation

$$c^2(P) = dP/d\rho. \tag{1.25}$$

Thus, a large-amplitude wave introduced into the material with a finite rise time, as illustrated in figure 1.10, will steepen with increasing propagation distance because the speed of the wave at increasingly higher pressure is correspondingly faster.

An illustration of the change in the form of the wave as it propagates through the material is provided by the successive profiles of the pressure (or the particle velocity) at increasing distance into the material. Increasing wave speed with pressure amplitude leads to an increasingly precipitous wave front. The predicted behavior eventually passes through a vertical singularity and becomes multivalued, as illustrated by the final two waveforms in figure 1.10. Such behavior is unphysical.

1.3.3 Viscosity and the large-amplitude wave

The physics that emerges, and prohibits the formation of multivalued wave fronts, is that of a supplemental pressure component in addition to that provided by the lattice compressibility of the solid and characterized by the EOS $P(\rho)$ in figure 1.10. It is a pressure component that becomes appreciable when gradients in the wave (the steepness of the wave) become large. If the profiles in figure 1.10 are that of the particle velocity u, then the added pressure component becomes prominent when the gradient $\partial u/\partial h$ becomes appreciable. That this derivative is also the deformation strain rate follows directly from the relations in equations (1.23) and (1.24).

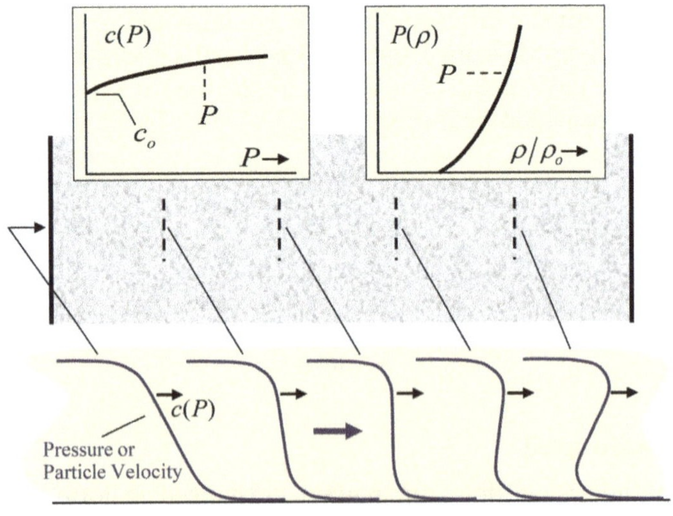

Figure 1.10. Wave propagation in a material for which wave speed increases with pressure. Consecutive wave profiles represent spatial snapshots of pressure or particle velocity at increasing times at the approximate positions shown.

The added pressure component is commonly identified as the viscous pressure. (Or viscous stress—recall that at this point in the discussion, pressure normal to the plane of the wave front is of concern and distinctions between stress and pressure in a solid medium are not addressed.) More common than not, the underlying physics responsible for the viscous pressure is not known. The rate dependence associated with deformation dislocation plasticity has received considerable attention as a possible viscous mechanism. Thermal conduction, elastic scattering and turbulence, among others, are also possible contributors to the viscous pressure. The issue of viscosity and structured shock waves is pursued in added depth in a later chapter.

Lacking the physical understanding, empirical expressions for the viscous pressure are introduced to provide the physics necessary to obtain properly behaved large-amplitude wave propagation. A power-law dependence of the viscous pressure on the velocity gradient is appropriate,

$$Q(\rho) = \eta\left(\frac{\partial u}{\partial h}\right)^m. \tag{1.26}$$

A power of $m = 1$ reduces the expression to linear, or Newtonian, viscosity. A power of $m = 1/2$ leads to structured steady wave fronts in which the maximum velocity gradient is proportional to the fourth power of the pressure change through passage of the wave. Such behavior has been observed experimentally for a number of materials (Grady 1981, 2010, Swegle and Grady 1985).

In physics-based computational wave propagation codes the same wave moderating effect is affected through an artificial viscosity term commonly provided in the form

$$Q(\rho) = \alpha\rho c\ \delta h\frac{\partial u}{\partial h} + \beta\rho\ \delta h\left(\frac{\partial u}{\partial h}\right)^2. \tag{1.27}$$

The artificial viscosity constants are α and β, while ρc is the wave impedance, and length scale δh is the computational cell size.

The contribution of the viscous pressure and its influence on the wave evolution is illustrated in figure 1.11. Three profiles of a propagating large-amplitude pressure wave that is loading the material from ambient pressure to some elevated pressure are shown in the figure. A correspondence is shown between each profile and the pressure (EOS plus viscous) versus compression strain path through the wave. In the first wave profile, viscous pressure is negligible and the marked variation of the wave speed over the pressure range causes rapid steepening of the wave, as is depicted by the variation in arrow length in the figure. The added contribution of viscous pressure for the second profile reduces the extent of curvature in the pressure versus volume path, the variation in wave speed through the wave reduces and the rate of steepening of the wave front correspondingly lessens. Wave speed is constant and independent of pressure in the final wave profile. The wave has achieved a steady structured form that is dictated by both the nonlinearity of the lattice compressibility and the nature of the viscous pressure law. This final wave structure will continue to propagate unchanged in form. The pressure versus strain follows a straight line path, called the Rayleigh line, which is supported by appropriate measures of both lattice and viscous pressure.

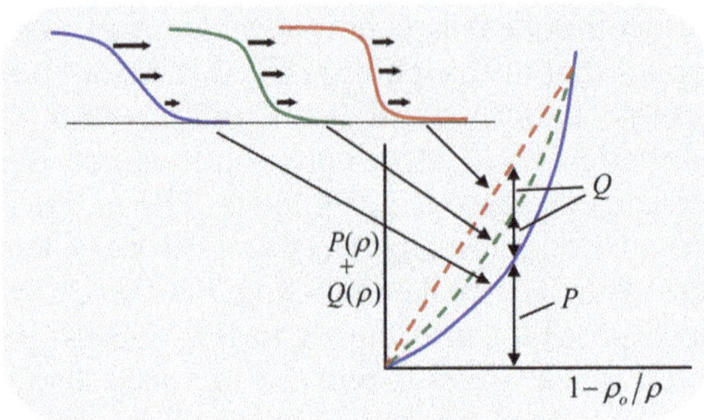

Figure 1.11. Contribution of the viscous stress in determining the wave velocity dependence and wave structure on pressure amplitude.

1.3.4 The steady structured waves

The closing paragraphs of the previous subsection allude to the formation and stability of steady structured large-amplitude waves. The structured steady wave is a class of large-amplitude waves that is commonly encountered in applications of shock and impact. The formation of steady waves on compression roughly requires only that the compressibility of the material decreases (increasing stiffness) with increasing compression. That is, the lattice compressibility exhibits positive curvature, as illustrated in figure 1.11. Structured steady decompression waves are less common but do occur.

For a set of material conditions, quantitative analysis of both the formation and the structure of steady waves can be pursued. For materials for which pressure in the large-amplitude wave depends only on the extent of compression, and not the rate of compression, the total pressure can be written as the sum of lattice and viscous terms,

$$\sigma(\varepsilon) = P(\varepsilon) + Q(\dot{\varepsilon}), \tag{1.28}$$

where $\varepsilon = 1 - \rho_0/\rho$ is an appropriate measure of the volume strain. Several examples are explored that illustrate the character of structured steady waves under common material response conditions.

The form used for the lattice compression will be the polynomial expression,

$$P(\varepsilon) = E\varepsilon + h_1\varepsilon^2 - b\varepsilon^3 + h_2\varepsilon^4. \tag{1.29}$$

The parameter E provides the initial stiffness of the material. The term h_1 is a hardening (stiffening) parameter. The property b captures the onset and extent of softening due to phase transformation or perhaps plastic yielding. The term h_2 determines subsequent second phase hardening.

A relation for the viscous pressure that is dependent on strain rate only, previously introduced in equation (1.26),

$$Q(\varepsilon) = \eta\dot{\varepsilon}^m, \tag{1.30}$$

completes the constitutive description.

1.3.5 Square-root viscosity and fourth-power steady waves

As one example, structured steady-shock waves in aluminum metal are considered. The lattice compressibility of aluminum is described through equation (1.29) by specification of E and h_1, setting the cubic and quadratic terms to zero. From the experimental linear shock-velocity versus particle-velocity relation, $U = C_o + Su$, sensible estimates of $E = \rho_o C_o^2$ and $h_1 = 2SE$ are provided (see equation (1.16)). Equation (1.30) is used for the viscous stress component with $m = 1/2$. Equation (1.28) then yields

$$\sigma(\varepsilon) = E\varepsilon + h_1\varepsilon^2 + \eta\dot{\varepsilon}^{1/2}. \tag{1.31}$$

Within the assumed material response, equation (1.31) is a generally applicable constitutive description for the purpose of describing large-amplitude wave propagation. Reasonable Hugoniot properties for aluminum of $\rho_o = 2700\ \mathrm{kgm^{-3}}$, $C_o = 5.38\ \mathrm{km\ s^{-1}}$ and $S = 1.34$ (Steinberg 1996) yield $E = 78$ GPa and $h_1 = 170$ GPa. The viscosity coefficient η will be assessed later. When limited to steady structured waves the stress versus strain path follows the straight Rayleigh line from the ambient state in front of the wave to the final (Hugoniot) state behind.

The total pressure and strain are then related through $\sigma(\varepsilon) = \rho_o U^2 \varepsilon$ as the pressure–strain state traverses the Rayleigh line. The property U is the velocity of the steady wave and equation (1.31) specializes to

$$\rho_o U^2 \varepsilon = E\varepsilon + h_1\varepsilon^2 + \eta\dot{\varepsilon}^{1/2}. \tag{1.32}$$

The difference modulus is defined as $M = \rho_o U^2 - E$ and equation (1.31) is rewritten,

$$\frac{d\varepsilon}{dt} = \frac{1}{\eta^2}(M\varepsilon - h_1\varepsilon^2)^2. \tag{1.33}$$

Before proceeding with the solution, several convenient changes of variables are introduced. First, a steady structured wave propagating to the right is self-similar with respect to the similarity parameter $\xi = t - h/U$, with h the Lagrangian position coordinate. Differentiation by time will be replaced with differentiation by the similarity parameter in equation (1.33). Second, strain will be replaced by the parameter $x = h_1\varepsilon/M$. Equation (1.33) transforms to

$$\frac{dx}{d\xi} = \frac{h_1^2}{\eta^2}\frac{M^3}{h_1^3}x^2(1 - x)^2 = \alpha x^2(1 - x)^2. \tag{1.34}$$

Note from equation (1.33) that the final strain state (or Hugoniot strain) is achieved when the right-hand side is zero and is given by $\varepsilon_h = M/h_1$. The range $0 \leqslant x \leqslant 1$ is the domain of interest. Further, a characteristic strain rate is provided by the ratio $\dot{\varepsilon}_c = h_1^2/\eta^2$. The constant in the differential equation is then $\alpha = \dot{\varepsilon}_c\varepsilon_h^3$. The differential equation is analytically solved, yielding

$$\xi(x) = \frac{1}{\alpha}\left[\frac{1}{1 - x} - \frac{1}{x} - 2\ln\left(\frac{1 - x}{x}\right)\right]. \tag{1.35}$$

With the steady wave solution for aluminum provided by equation (1.35), either time history profiles at a specified Lagrangian position, or spatial wave profiles at specific times can be plotted. Calculated spatial wave profiles for the present aluminum model are shown on the left in figure 1.12 after a one-microsecond propagation time. The increasingly steeper structured wave profiles correspond to successive shock velocities of 5.5, 5.6, 5.7 and 5.8 km s^{-1}, respectively. The corresponding Hugoniot pressures are 1.7, 3.2, 4.9 and 6.8 GPa.

From equation (1.33) a plot of the strain rate through the wave profile is provided, as shown on the right. The maximum strain rate $\dot{\varepsilon}_{max}$ is shown to occur at a compression strain of $\varepsilon = \varepsilon_h/2$ in the wave profile. A relationship between the maximum strain rate in the wave and the Hugoniot strain amplitude is

$$\dot{\varepsilon}_{max} = \frac{1}{16}\frac{h_1^2}{\eta^2}\varepsilon_h^4. \tag{1.36}$$

A law of viscosity of the square-root form leads to a fourth-power dependence between the steady wave maximum strain rate and the corresponding Hugoniot strain amplitude (Grady 1981, 2010, Swegle and Grady 1985).

The analytic wave profiles provided in figure 1.12 can be compared with the seminal time-resolved interface velocity measurements of Barker (1968) on 6061-T6 aluminum, shown in figure 1.13, using the then newly developed velocity interferometer shock wave diagnostic. Velocities are those of the interface between the aluminum and a quartz window with the exception of the highest shock pressure 9.0 GPa experiment, which is that of the free surface corrected for quartz impedance.

In the present steady structured wave solution, a viscosity coefficient of $\eta = 0.8 \times 10^5$ Pa · $s^{1/2}$ provides reasonable agreement with experimental structured wave data for aluminum. Despite efforts by a number of workers over a substantial span of time, the underlying physics that determines the magnitude of η, and the

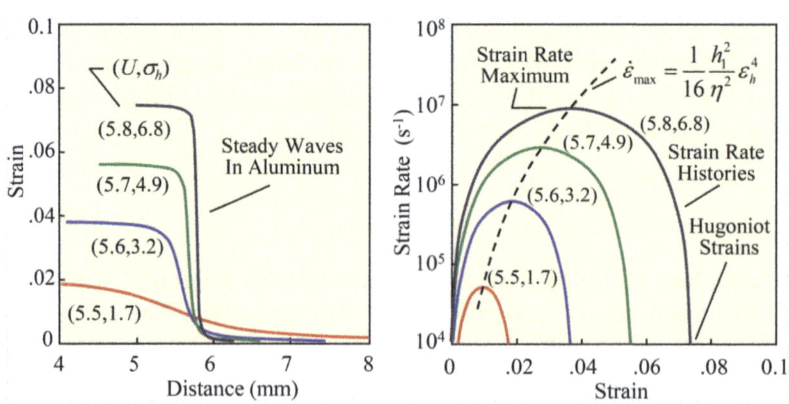

Figure 1.12. Wave profiles in aluminum based on a steady wave analysis with square-root viscosity. Strain-rate histories through pressure wave loading are shown on the right. Each profile is identified by shock velocity and Hugoniot pressure in parentheses. A fourth-power dependence of the maximum strain rate on the Hugoniot strain amplitude is a consequence of the square-root viscosity model.

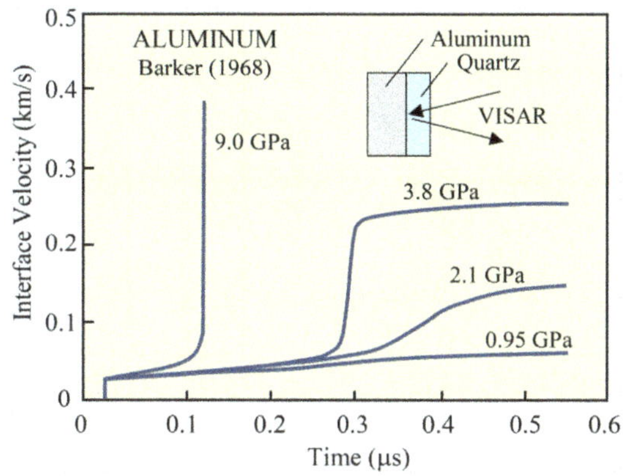

Figure 1.13. Time-resolved interface wave profiles for 6061-T6 aluminum with a quartz window using velocity interferometer diagnostics (Barker 1968).

form of the viscosity law, remain uncertain (Grady 1998, 2010). The present topic is pursued further in chapter 8.

1.3.6 Anomalous compressibility and steady structured waves

The nature of steady structured waves in materials exhibiting anomalous compressibility (phase transformation or plastic yielding, for example) offers an instructive contrast to the previous example. It has been speculated that the structured wave will exhibit a signature of the anomaly (a phase transformation will be considered) as a consequence of the character of the viscous stress, even when the transformation is overdriven and a single steady structured wave emerges. The merits of this concept can be examined with the analysis methods already outlined.

As a specific example, shock and structured wave data for the carbonate mineral calcite ($CaCO_3$) is used in the analysis. Hugoniot data for calcite are shown on the left in figure 1.14. On the right are shown shock and release wave particle-velocity profiles that correspond to the 18 GPa Hugoniot state. The profiles exhibit one of the most striking examples of a rarefaction shock wave resulting from the reversion of a high-pressure phase during decompression from a shock state, where transformation to the high-pressure phase occurs during the initial shock compression process. Diamond anvil cell measurements identify calcite VI as the likely high-pressure phase achieved above 15 GPa shock pressure (Merlini *et al* 2012). Reverse transformation on decompression through the rarefaction shock indicates return to the calcite III phase or transition to the aragonite phase. Further shock EOS data on carbonate rocks and minerals, including extensive efforts by Soviet scientists in the 1960s and 1970s, are included in a later chapter on EOS issues.

For the compression Hugoniot shown on the left, the fourth-order polynomial compression relation provided in equation (1.29) is used to describe the data. Polynomial parameters are provided in the figure. Note that the first-order modulus is set to $E = 0$. This ploy provides a not unreasonable description of the lower pressure compression that is dominated by an earlier phase transformation in the

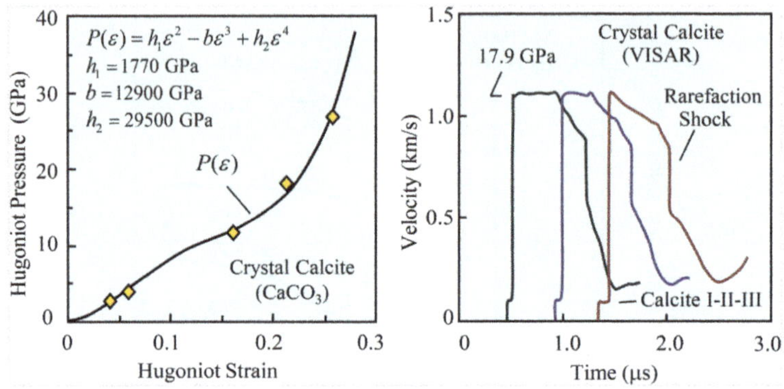

Figure 1.14. Shock Hugoniot data, and shock wave compression and release wave VISAR particle-velocity profiles (lithium fluoride window) for crystalline calcite (Grady and Moody 1985, Grady 1986a).

first 1–2 GPa pressure range (Grady 1979, Grady *et al* 1978). (This calcite I to calcite II and III is identified on the right in figure 1.14.) The analytic polynomial compression relation provides an adequate description of both the experimental Hugoniot compression and the decompression behavior revealed by the rarefaction shock wave.

The relation for the steady wave compression of calcite comparable to equation (1.32) is written as

$$\rho_o U^2\varepsilon = h_1\varepsilon^2 - b\varepsilon^3 + h_2\varepsilon^4 + \eta\dot{\varepsilon}, \tag{1.37}$$

where a viscous stress linear in the strain rate is selected for the present analysis. The differential equation for the strain through the steady wave is then

$$\frac{d\varepsilon}{dt} = \frac{1}{\eta}\left[\rho_o U^2\varepsilon - (h_1\varepsilon^2 - b\varepsilon^3 + h_2\varepsilon^4)\right]. \tag{1.38}$$

Solutions of equation (1.38) for structured steady waves encompassing the phase transformation region of anomalous compressibility are shown in figure 1.15. Solutions are provided for three different pressure amplitudes. On the left are shown the Rayleigh line pressure–strain paths experienced by the material through the steady wave to the final pressure amplitude (the Hugoniot pressure). The center plot shows the strain rate in the material through the steady waves (a direct plot of strain rate from equation (1.38)). On the right are the corresponding structured steady waves (the solution of equation (1.38)). As suggested, the waves do provide structure that is indicative of the phase transformation anomaly in the calcite mineral. It is observed, however, that the indicating structure becomes rapidly less apparent as the final Hugoniot pressure increases.

The viscosity coefficient η is not specified in the plots shown. Rather, both time and strain rate are normalized by the viscosity. The experimental profiles shown in figure 1.14 provide no indication of the phase transformation anomaly through the compression wave. The possible reasons for this are several. The material viscosity of crystal calcite may be too low to structure the wave within the current

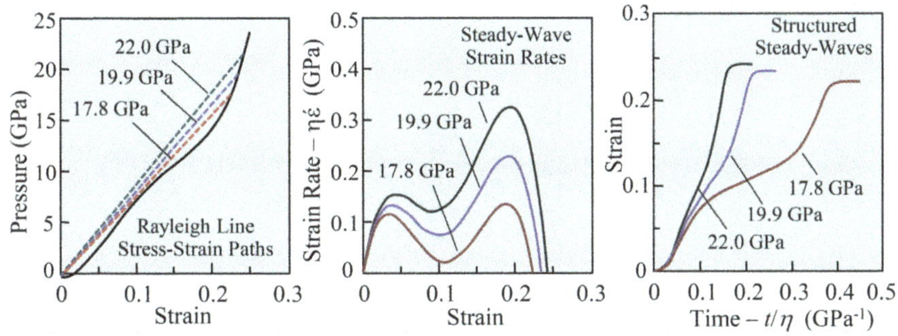

Figure 1.15. Structured steady waves based on a material response model including phase transformation and linear viscous stress for calcite mineral.

instrumentation resolution. The shock Hugoniot state of this one set of data may not be appropriate to reveal the compression wave structure. Alternatively, the physics of viscosity required to structure the steady wave may be markedly more complex than the linear viscosity assumed.

1.4 The shock wave

When a structured wave becomes sufficiently rapid in its rise time that the thickness of the wave can be ignored, or is chosen to be ignored, or is not measurable, then the wave is viewed as a discontinuous jump in the principal mechanical and thermodynamic parameters (e.g. pressure, strain, velocity, temperature). The wave is a classic shock wave. A rigorous definition of a shock wave is problematic. The Hugoniot jump relations, the analytic thermomechanical fundamentals of a shock wave, are precisely applicable through a structured steady wave, whether or not it is a discontinuity. Shock waves are not always steady waves, however. For example, an attenuating shock wave, as illustrated in figure 1.16, has a finite wave structure whether or not it can be measured. Since the structure (thickness) of a steady wave is dependent on the final amplitude of the wave, as revealed by the several previous examples of viscosity-structured waves, the thickness of an attenuating shock wave will change as it propagates, and is not steady. Nonetheless, when gradients of the thermomechanical parameters in front of, and behind, the shock wave are small compared to gradients in the shock, then the Hugoniot jump relations are applicable to a sensible degree of accuracy. The possibility of exceptions should nevertheless be recognized.

1.4.1 The Hugoniot jump relations

There are a number of methods for arriving at the Hugoniot jump conditions that provide relations among the parameters that undergo discontinuous changes through passage of a shock wave. A common heuristic approach is described here. Other methods are identified in later sections.

An illustration of a propagating shock wave is shown in figure 1.17. From some stationary reference plane the shock wave has propagated a distance $U\delta t$, while the

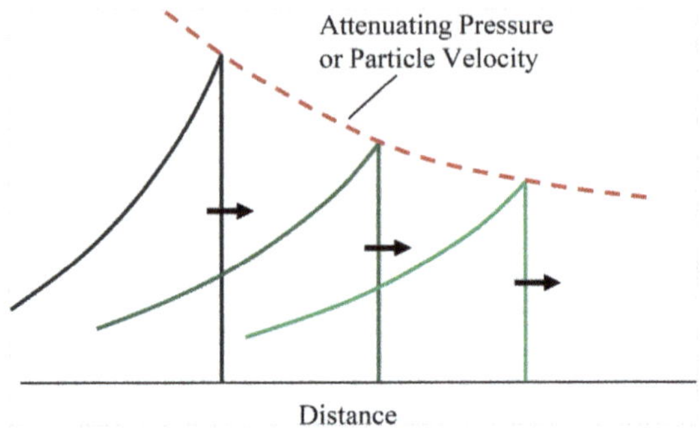

Figure 1.16. Shock wave propagation and attenuation due to an overtaking release wave.

material initially at that same plane has traveled a distance $u\delta t$, where U and u are the shock velocity and particle velocity, respectively. Behind the shock wave the constant density, pressure, specific energy and particle velocity are, respectively, ρ, P, E and u. In front of the shock wave the initial density is ρ_o, whereas initial conditions $P_o = 0$, $E_o = 0$ and $u_o = 0$ are assumed. The Hugoniot jump relations that are derived can later be transformed to more general initial conditions when necessary.

The first jump relation that is derived is a compatibility condition often identified with conservation of mass. The initial volume of material $U\delta t$ (a unit cross section is assumed) is reduced in volume by the amount $u\delta t$ following passage of the shock wave. Thus, the compression strain behind the shock wave (the Hugoniot strain) is

$$\varepsilon = \frac{u\delta t}{U\delta t} = \frac{u}{U}, \tag{1.39}$$

where $\varepsilon = 1 - v/v_o = 1 - \rho_o/\rho$.

The same material element is subjected to an accelerating pressure over the time δt, and conservation of momentum requires

$$\delta mu = \left(\rho_o U\delta t\right)u = P\delta t, \tag{1.40}$$

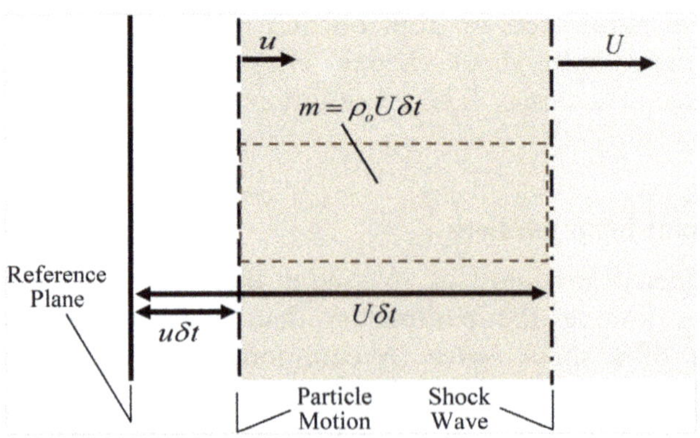

Figure 1.17. Motion and deformation of a mass element subject to passage of a shock wave.

or

$$P = \rho_o U u. \tag{1.41}$$

Lastly, work on the material element over the same time increment increases both kinetic energy and internal energy,

$$\frac{1}{2}(\rho_o U \,\delta t)u^2 + (\rho_o U \,\delta t)E = Pu \,\delta t, \tag{1.42}$$

or

$$\rho_o U(u^2/2 + E) = Pu. \tag{1.43}$$

Using the previous Hugoniot relations yields

$$\rho_o E = \frac{1}{2}\rho_o u^2 = \frac{1}{2}P\varepsilon. \tag{1.44}$$

Note that work expended by the shock wave is equally partitioned into kinetic energy and internal energy. Equations (1.39), (1.41) and (1.44) are, respectively, the Hugoniot jump relations governing conservation of mass, momentum and energy through passage of the shock wave. They provide three algebraic equations among the four Hugoniot state properties. Closure of the system of equations requires a fourth relation, the constitutive description of the material.

1.4.2 The Hugoniot equation of state

The Hugoniot jump relations governing the shock compression are what make the extreme pressure and temperature EOS of matter accessible through the shock wave experiment. The Hugoniot equations relate the mechanical properties of mass velocity and shock velocity, u and U, to the EOS properties P, ρ and E. The mass and shock velocities are readily measured in appropriate shock wave experiments. Pressure, density and specific energy states are calculated through the Hugoniot equations.

Efforts have been pursued to directly measure EOS properties in the shock wave experiment. Piezoresistive and piezoelectric effects have been exploited to measure shock pressure. Measurements of material density in the shock state with x-ray techniques have also been attempted. Even temperature under shock conditions has been measured.

Nevertheless, the overwhelming preponderance of EOS information determined through shock compression methods relies on precision measurement of the shock velocity and the mass velocity at the shock state. Even at present, when high precision time-resolved measurements of complex large-amplitude wave propagation are routinely conducted, it is through measurement of mass and wave velocities that the constitutive properties of materials are experimentally assessed.

Closure with the Hugoniot relations is achieved through specification of the Hugoniot of the material—a pressure versus density relation for the material that determines the span of pressure–density end states achieved in a shock compression event. More generally, closure is achieved through specification of a material EOS that determines the dependence of pressure on both the density and the specific energy $P(\rho,E)$. The $P(\rho,E)$ relation is not a complete EOS (the specific heat of the material is not provided). It is, however, sufficient to achieve closure with the Hugoniot conservation equations, and predict the response of a material to intense shock wave compression.

The principal Hugoniot (that Hugoniot originating from ambient conditions of the material) and the $P(\rho,E)$ EOS are illustrated in figure 1.18. The principal Hugoniot is one path on the surface of the $P(\rho,E)$ EOS. States on the principal Hugoniot are accessed through single-shock compression continuously traversing non-equilibrium states along the Rayleigh line from the ambient to the Hugoniot end state.

Since shock velocity U and particle velocity u are the commonly measured shock compression properties, the shock Hugoniot is specified through a functional relation of the form

$$U = f(u). \tag{1.45}$$

There are many materials for which the relation in equation (1.45) is linear over a limited span of shock compression measurements and is written as

$$U = C_o + Su. \tag{1.46}$$

The underlying physics of such linearity under certain conditions has been explored by others (e.g. Jeanloz 1988). Further discussions are provided in chapter 10. When more complex physics governs the Hugoniot compression process (polymorphic phase transformation for one) linearity between shock velocity and particle velocity is generally not observed and more involved function forms for equation (1.45) must be applied.

Plots are provided in figures 1.19 and 1.20 that are representative of experimental measurements of Hugoniot properties of solids. The Hugoniot measurements for copper metal shown in figure 1.19 are illustrative of the seminal shock EOS studies conducted at Los Alamos National Laboratory (LANL) in the early 1960s. The plots in this figure show 22 Hugoniot data points for copper of the approximately

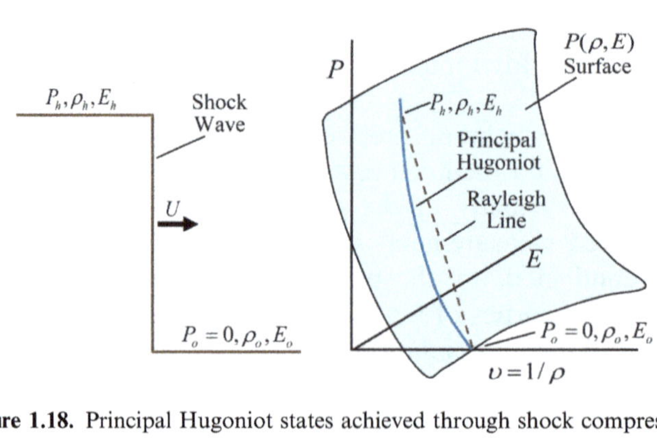

Figure 1.18. Principal Hugoniot states achieved through shock compression.

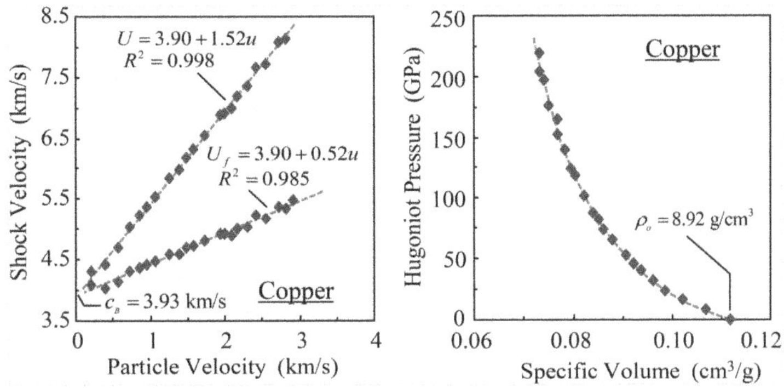

Figure 1.19. Shock wave Hugoniot data for full density copper metal from the LANL compendium (Marsh 1980).

140 experimental measurements provided in the LANL Shock Hugoniot Handbook (Marsh 1980) for this metal. Hugoniot states were determined through the measurement of shock velocity using flash gaps and smear camera techniques. The method requires an impedance match analysis with a base plate material for which the EOS is known. The LANL handbook Hugoniot data were later updated (in the late 1970s) as better EOSs for the base plate materials became available. As observed from the plot in figure 1.19, copper metal exhibits a strikingly linear dependence of shock velocity versus particle velocity over the measured pressure range of approximately 8–220 GPa with a high degree of correlation. The bulk sound speed identified in the plot is calculated from the acoustic longitudinal and shear wave velocities through the relation $c_B^2 = c_L^2 - 4c_S^2/3$.

Kerley (2006) has emphasized that the conventional plot of the experimental shock velocity and particle velocity may not be the most effective in illuminating EOS anomalies in the shock Hugoniot data. He suggests that the alternative measure of shock velocity $U_f = U - u$ plotted against particle velocity may in fact be more appropriate. The velocity U_f is the shock speed relative to the compressed state behind the propagating shock wave. The two shock velocities can alternatively be related to the compression with application of the Hugoniot relations (1.39) through $\rho U_f = \rho_o U$. (Shock impedance is invariant with respect to the reference frame.) This alternative plot of the data is also shown in figure 1.19 and, at least for this copper metal, is equally well represented with a straight line. The linear least squares fit and the corresponding correlation coefficient is shown for each representation in the figure.

The Hugoniot pressure versus specific volume states shown on the right in figure 1.19 are calculated from the shock-velocity versus particle-velocity data through application of the mass and momentum Hugoniot relations from equations (1.39) and (1.41).

Hugoniot measurements on a material exhibiting a somewhat more complex shock wave response are shown in figure 1.20. These data were obtained on a zeolitic tuff from the Nevada Test Site (NTS) and are representative of efforts in the early 1970s to acquire shock EOS properties for a range of continental rock and earth

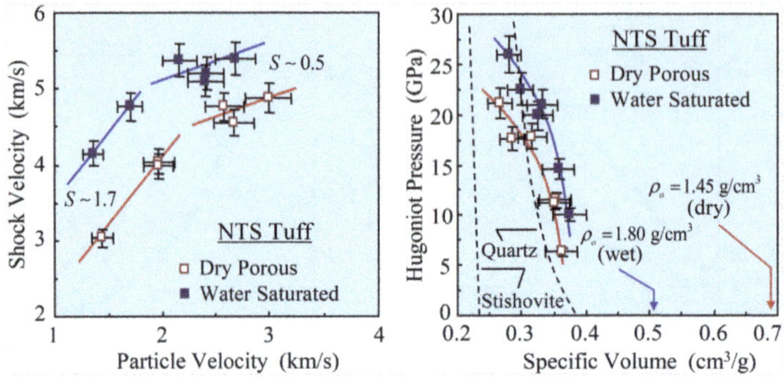

Figure 1.20. Experimental shock wave Hugoniot data for dry porous and water-saturated Mount Helen tuff (Murri *et al* 1975).

materials (Murri *et al* 1975). Hugoniot states were measured for Mount Helen tuff in both the dry porous and water-saturated conditions. The experimental method made use of multiple magneto-inductive transducers that provide measurement of both shock velocity and particle velocity.

Plots of both shock velocity versus particle velocity and Hugoniot pressure versus specific volume exhibit complexities involved with compaction of the distended material and molecular phase transformation. The hydrated zeolite that constitutes the rock component of the tuff is approximately 70% silica (SiO_2) by weight. Silica experiences phase transformation in the shock pressure range of 15–25 GPa with significant densification of the silica molecule. This shock densification and softening of the compressibility is revealed in the higher-pressure data for both the dry porous and the water-saturated tuff in figure 1.20. Shock Hugoniot curves for the NTS tuff are contrasted with compression curves for the low-pressure (α-quartz) and the high-pressure (stishovite) phases of crystalline SiO_2.

1.4.3 Hugoniot properties of solids

Theoretical attempts have been made to provide Hugoniot EOSs for simple materials from first-principles physics. Practical applications, however, must rely on a body of experimental Hugoniot data that is available for a surprisingly large number of materials. Obtaining a reasonably accurate Hugoniot over a useful range of shock pressures and energies is a fairly onerous, expensive and time-consuming task. The field of shock compression science is fortunate that in an earlier and more benevolent era both funding and encouragement was provided to the National Laboratories, and to a few research institutes and universities to pursue the acquisition of Hugoniot EOS measurements for a wide range of materials.

Probably the most monumental of these efforts was put forth by workers at the LANL in the collection and compilation of Hugoniot data for solids. This compilation of Hugoniot measurements has been made available in a compendium edited by Marsh (1980). It is a national treasure. The wide range of materials tested is quite remarkable. They include metals, ceramics, plastics and rocks, among

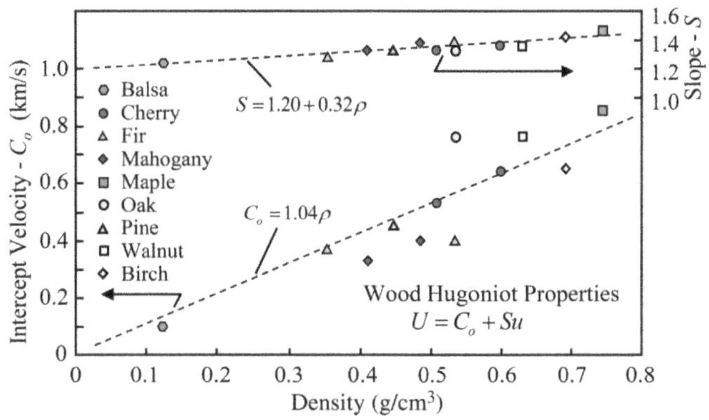

Figure 1.21. Linear shock-velocity versus particle-velocity properties for wood (Marsh 1980).

a number of others. Many of these materials reveal the linear $U = C_o + Su$ relation between the measured shock velocity and particle velocity with surprising accuracy. Where sensible, the experimental C_o and S shock properties are provided along with the applicable range. Further extensive compilations of shock Hugoniot data are provided in the compendiums of van Thiel *et al* (1977) and Trunin *et al* (2001).

The LANL compendium is readily accessible, and has been profitably used by researchers and engineers across the globe. A number of examples and assessments of the shock wave response of materials in the present text rely on the data made available in this compendium. As one example among the many possible, Hugoniot data for wood and the systematics emerging from this data, taken from the LANL compendium, are shown in figure 1.21.

The experimental shock Hugoniot linear shock-velocity versus particle-velocity EOS parameters are plotted against the initial density of the various woods. Taken together they provide a fairly satisfactory systematics for a first-order calculation of the Hugoniot EOS and shock properties of any wood material.

Applications of the Hugoniot measurements from the LANL compendium, and data from other sources of the same era, should be undertaken with a degree of prudence. Recognition of the substantial shock wave structure exhibited by many solids was only beginning to emerge at that time, and most of the shock measurement techniques were not developed to resolve such structure. Shock wave structure, of course, depends on both the material type and the intensity of the shock wave. Ceramics and high-strength rocks, for example, exhibit a strong two-wave structure due to their large HEL. Thus, the lower-amplitude Hugoniot data can be compromised by this complication. The possible complications of this nature on the shock Hugoniot for aluminum oxide ceramic were noted in a previous section. Shock-induced phase transformations can similarly contribute to shock wave structure and complicate interpretation of Hugoniot measurements.

1.4.4 Solid strength

In discussions to this point, pressure in the large-amplitude wave has been identified as that component of stress collinear with the direction of wave propagation. Tacit in this caveat is the possibility that the magnitude of stress components in directions other than that of the propagating wave may differ. In solids with strength, stress differences occur in the large-amplitude wave propagation process. In the propagation of one-dimensional waves in isotropic solids this issue reduces to the interpretation of two components of stress—the component along the direction of the propagating wave, or the longitudinal stress σ, and the component transverse to the direction of propagation σ_t. A difference in the two stress components in the wave propagation event reflects the strength of the solid material.

A fluid subjected to shock compression can exhibit transient strength (viscosity). Behind the shock wave, however, or within a structured wave at sufficiently low rates, stress in a fluid in any orientation is the same, $\sigma = \sigma_t$, and is identified with the pressure p of previous discussions. In solids subjected to sufficiently intense shock compression, such as the preponderance of data in the LANL shock compression handbook (Marsh 1980), the stress difference $\sigma - \sigma_t$ is probably nonzero, but is sensibly small compared to the shock stress σ and strength is ignored. Again, σ is identified with the pressure p. Both the stress state and the strength of material at the shock Hugoniot state continue to be ongoing topics of research, however, and have not been resolved for many materials.

When strength in the one-dimensional wave cannot be ignored, relations encompassing both stress components must be specified. The constitutive description of a material with the strength necessary to capture large-amplitude wave propagation will incorporate both stress components. Commonly used in the description of isotropic materials with strength are the resolved shear stress and average stress (mean pressure) defined as

$$\tau = \frac{1}{2}(\sigma - \sigma_t), \qquad p = \frac{1}{3}(\sigma + 2\sigma_t). \tag{1.47}$$

Conversely,

$$\sigma = p + \frac{4}{3}\tau, \qquad \sigma_t = p - \frac{2}{3}\tau. \tag{1.48}$$

Restricted, but very practical and useful models incorporate the EOS pressure dependence on the density and specific energy through the mean stress $p(\rho,E)$. Strength is associated with the resolved shear stress τ. Linear elastic strength, expressed through a constant Poisson's ratio ν, relates τ and p in the one-dimensional shock compression through the expression

$$\tau = \frac{3}{2}\frac{(1 - 2\nu)}{(1 + \nu)}p. \tag{1.49}$$

A strength-limiting model bounds τ through a yield criterion. Through either a von Mises or a Tresca yield criterion the magnitude of τ is limited by $|\tau| = Y/2$ where Y is the yield stress in simple tension. Although the terminology hints at metal plasticity, at this juncture no physical mechanism of failure is implied other than some process that bounds the elastic shear stress that the material can support

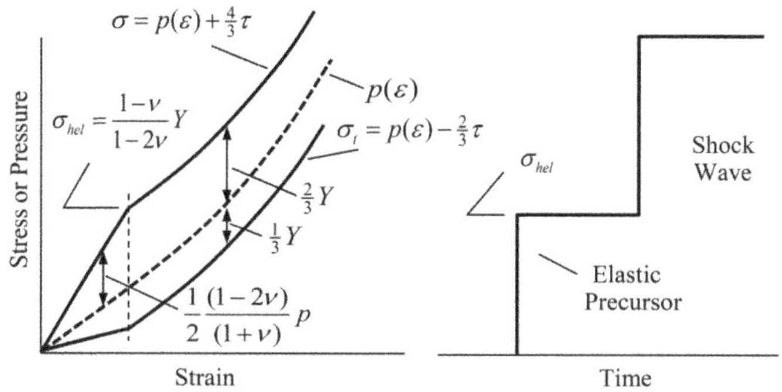

Figure 1.22. EOS and strength features of the elastic–plastic model governing a one-dimensional shock wave. The HEL and the resulting elastic precursor wave structure are illustrated.

without onset of inelastic deformation. This model of shock compression with strength is illustrated through the relevant stress and strain curves in figure 1.22.

The three curves in the figure are, respectively, the longitudinal (shock) stress σ, the mean pressure p and the transverse stress σ_t. Notable is the onset of dynamic yield on the shock stress curve, and its relationship to the yield stress Y. This dynamic yield point, and the elastic precursor wave that can emerge in the structured wave due to the stress path inflection, has been identified with the HEL of the material. The HEL of a material has become the iconic measure of dynamic strength relating to large-amplitude wave propagation in solids. Structured waves including both the HEL-governed precursor wave and the following shock wave are illustrated on the right in figure 1.22.

The HEL reflects the dynamic strength of the material. Observations of the signature precursor wave that is a consequence of the strength have revealed a rich display of structure over the many materials that have been investigated. Precursor wave structure can be a key indicator of the strength-limiting physics governing failure in the various material classes. The nature of the stress–strain path in materials with strength subject to shock was described at least as early as the work of Wood (1952). One of the earliest observations of an elastic precursor wave and reporting of a HEL is the work of Goranson *et al* (1955) using piezoelectric tourmaline as a shock wave diagnostic. The authors provided shock pressure measurements up to 32 GPa in iron and an HEL of 1.6 GPa. A thorough outline of elastic–plastic shock compression with experimental application to 2024 aluminum is provided in the study of Fowles (1961).

Representative experimental precursor wave profiles for several materials are shown in figure 1.23. Shock waves in single-crystal magnesium oxide, alpha-uranium metal and aluminum-oxide ceramic were produced by the planar impact of flat plates onto the test material. The observed wave structure evolves following propagation over distances of several millimeters.

A precursor wave shown for single-crystal magnesium oxide is illustrative of the limiting elastic wave structure in crystalline solids. The peak precursor stress is observed to decay with propagation distance. Both the peak attenuation and

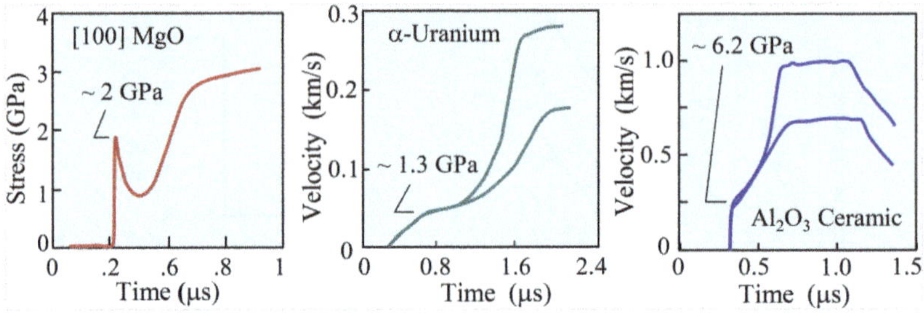

Figure 1.23. Representative elastic precursor waves in selected materials. A quartz gage stress profile for single-crystal magnesium oxide (Grady 1977a). Velocity interferometer (VISAR) profiles for alpha-uranium (Grady 1986b) and aluminum oxide ceramic Coors AD-995 (Grady 1995b).

subsequent stress relaxation behind the peak are believed to be a consequence of the kinetics of dislocation plasticity. The precursor wave observed in alpha-uranium is non-steady with the initial ramp widening with propagation distance. The unique precursor structure of this metal results from both twinning and pervasive dislocation plasticity. The shock compression response of uranium alloys is considered further in chapter 11. Aluminum-oxide ceramic is representative of the high-strength metal–nonmetal compounds. The failure processes governing precursor structure in ceramics are still open to conjecture. Some mix, dependent on the material, of localized plasticity and fine-scale fracture is probably responsible. Chapters 7 and 9 explore the strength properties of ceramics in the shock environment.

1.4.5 Governing equations

The differential equations governing large-amplitude wave propagation follow from conservation of mass, momentum and energy. Here the equations for wave propagation in one dimension are emphasized. In many applications a single-valued and invertible relation between stress and strain adequately describes the material response. In such cases the equation governing conservation of energy is not required. This simplification is equivalent, for example, to the assumption of isentropic flow in the gas dynamic equations such that the pressure becomes a single-valued function of the specific volume. When required by the material behavior all three equations governing mass, momentum and energy are used.

Let the coordinate h identify the location of a Lagrangian material point in the initially undeformed configuration of the one-dimensional body. Further, let $x(h,t)$ provide the position of the point initially at h after some later time t and such that $x(h,0) = h$. If the motion described by the coordinate x is coaxial with the Lagrangian coordinate h, then at any time t the derivative $(\partial x/\partial h)_t$ provides a measure of the axial compression (or extension). A measure of the axial strain is given by the relation

$$\varepsilon = \frac{\partial}{\partial h}(h - x(h,t))_t = 1 - \left(\frac{\partial x}{\partial h}\right)_t, \tag{1.50}$$

where the convention of positive strain in compression is customary. A time derivative of equation (1.50) yields

$$\left(\frac{\partial \varepsilon}{\partial t}\right)_h + \left(\frac{\partial u}{\partial h}\right)_t = 0, \tag{1.51}$$

where the mass velocity $u = (\partial x/\partial t)_h$. Equation (1.51) provides the relation governing conservation of mass in one-dimensional wave propagation.

If the axial motion is one of uniaxial strain such that no lateral motion of material elements occurs, then the relations provide the volume compression directly through

$$\frac{\rho_o}{\rho} = \left(\frac{\partial x}{\partial h}\right)_t = 1 - \varepsilon. \tag{1.52}$$

If the motion x is transverse to the coordinate h then the deformation is volume conserving and equation (1.51) is equally appropriate for transverse shear wave propagation where ε is interpreted as a measure of the shear strain.

Momentum balance in the same Lagrangian framework is also readily apparent. If σ is the axial stress collinear with the coordinate h then the force imbalance on a mass element $\rho_o\, dh$ is just

$$d\sigma = -\left(\frac{\partial \sigma}{\partial h}\right)_t dh. \tag{1.53}$$

The rate of change of the momentum of the element is then

$$\left(\frac{\partial(\rho_o u\, dh)}{\partial t}\right)_h = -\left(\frac{\partial \sigma}{\partial h}\right)_t dh, \tag{1.54}$$

and

$$\rho_o\left(\frac{\partial u}{\partial t}\right)_h + \left(\frac{\partial \sigma}{\partial h}\right)_t = 0 \tag{1.55}$$

provides the governing equation for momentum balance in the one-dimensional Lagrangian coordinate system. Equation (1.55) is equally appropriate for shear deformation where σ is interpreted as a shear stress τ, and u is the corresponding transverse mass velocity.

A similar development for the energy notes that the rate of work on the mass element is provided by

$$dw = -\left(\frac{\partial \sigma u}{\partial h}\right)_t dh, \tag{1.56}$$

and requires the corresponding rate of change in internal and kinetic energy,

$$\left(\frac{\partial}{\partial t}\left(\rho_o\, dh\left(E + \tfrac{1}{2}u^2\right)\right)\right)_h = -\left(\frac{\partial \sigma u}{\partial h}\right)_t dh, \tag{1.57}$$

or

$$\rho_0\left(\frac{\partial E}{\partial t}\right)_h + \rho_0 u\left(\frac{\partial u}{\partial t}\right)_h + \left(\frac{\partial \sigma u}{\partial h}\right)_t = 0. \tag{1.58}$$

Introducing the mass and momentum relations into equation (1.58) yields

$$\rho_0\left(\frac{\partial E}{\partial t}\right)_h - \sigma\left(\frac{\partial \varepsilon}{\partial t}\right)_h = 0. \tag{1.59}$$

Equations (1.51), (1.55), and either (1.58) or (1.59), provide the continuum mass, momentum and energy conservation equations in Lagrangian coordinates governing large-amplitude wave propagation in one dimension. The three differential equations relate the four material mechanical and state parameters u, σ, ε and E.

Lagrangian and Eulerian coordinates are related through

$$\left(\frac{\partial}{\partial t}\right)_h = \left(\frac{\partial}{\partial t}\right)_x + u\left(\frac{\partial}{\partial x}\right)_t, \quad \text{and} \quad \left(\frac{\partial}{\partial h}\right)_t = \frac{\rho_0}{\rho}\left(\frac{\partial}{\partial x}\right)_t, \tag{1.60}$$

providing the same governing equations in Eulerian coordinates,

$$\left(\frac{\partial \rho}{\partial t}\right)_x + u\left(\frac{\partial \rho}{\partial x}\right)_t + \rho\left(\frac{\partial u}{\partial x}\right)_t = 0, \tag{1.61}$$

$$\rho\left(\left(\frac{\partial u}{\partial t}\right)_x + u\left(\frac{\partial u}{\partial x}\right)_t\right) + \left(\frac{\partial \sigma}{\partial x}\right)_t = 0, \tag{1.62}$$

$$\left(\frac{\partial E}{\partial t}\right)_x + u\left(\frac{\partial E}{\partial x}\right)_t - \frac{\sigma}{\rho^2}\left(\left(\frac{\partial \rho}{\partial t}\right)_x + u\left(\frac{\partial \rho}{\partial x}\right)_t\right) = 0. \tag{1.63}$$

1.4.6 Characteristic equations

Characteristic solution methods of the one-dimensional continuum equations are particularly convenient for describing features of nonlinear wave propagation. Descriptions of the method of characteristics are commonly couched in the Eulerian framework of gas dynamics (e.g. Courant and Hilbert 1962, Zeldovich and Raiser 1966). These methods are essentially the same when couched in the Lagrangian reference frame commonly used for the characterization of wave propagation in solid matter. Interpretation is actually somewhat easier. The Lagrangian description of nonlinear wave motion in solids through characteristic solution methods is described here.

The one-dimensional equations governing mass and momentum are written in terms of the time t and the Lagrangian coordinate h as

$$\left(\frac{\partial \varepsilon}{\partial t}\right)_h + \left(\frac{\partial u}{\partial h}\right)_t = 0, \tag{1.64}$$

$$\rho_o\left(\frac{\partial u}{\partial t}\right)_h + \left(\frac{\partial \sigma}{\partial h}\right)_t = 0. \tag{1.65}$$

Closure in the system of equations is achieved through a constitutive relation between the stress and strain. Writing this relation,

$$\sigma = f(\varepsilon), \tag{1.66}$$

and identifying the Lagrangian characteristic wave speed (the Lagrangian sound speed is used exclusively through the subsequent sections and the subscripted symbol c_L is not used),

$$c(\varepsilon) = \sqrt{f'(\varepsilon)/\rho_o}, \tag{1.67}$$

the system of governing equations becomes

$$\left(\frac{\partial \varepsilon}{\partial t}\right)_h + \left(\frac{\partial u}{\partial h}\right)_t = 0, \tag{1.68}$$

$$\left(\frac{\partial u}{\partial t}\right)_h + c^2\left(\frac{\partial \varepsilon}{\partial h}\right)_t = 0. \tag{1.69}$$

The stress–strain relation in equation (1.66) is assumed to be invertible. Expressing the functional dependence of the wave speed on the stress, $c = c(\sigma)$, equations (1.68) and (1.69) can just as readily, and perhaps more commonly, be written in terms of u and σ.

It is convenient to define the variable,

$$\lambda = \int c\, d\varepsilon = \int \frac{d\sigma}{\rho_o c}. \tag{1.70}$$

By adding and subtracting equations (1.68) and (1.69), after some manipulation, the following alternative forms are obtained,

$$\left(\frac{\partial}{\partial t} + c\frac{\partial}{\partial h}\right)u + \left(\frac{\partial}{\partial t} + c\frac{\partial}{\partial h}\right)\lambda = 0, \tag{1.71}$$

$$\left(\frac{\partial}{\partial t} - c\frac{\partial}{\partial h}\right)u - \left(\frac{\partial}{\partial t} - c\frac{\partial}{\partial h}\right)\lambda = 0. \tag{1.72}$$

Characteristic directions in the $h - t$ plane are specified by the differential relations,

$$\frac{dh}{dt} = +c\,, \qquad \frac{dh}{dt} = -c\,, \tag{1.73}$$

and determine, respectively, forward-facing, or C_+, characteristics and backward-facing, or C_-, characteristics. Equations (1.71) and (1.72) are then recognized as directional derivatives of the flow variables along the C_+ and C_- characteristics, respectively.

It is useful to introduce new independent variables α and β determined by the directional derivatives and the characteristics paths as illustrated in figure 1.24. Equations (1.71) and (1.72) then simplify to

$$\frac{\partial}{\partial\alpha}(u + \lambda) = 0, \tag{1.74}$$

$$\frac{\partial}{\partial\beta}(u - \lambda) = 0. \tag{1.75}$$

The two additional differential expressions

$$\frac{\partial}{\partial\alpha}\left(h - \int c\, dt\right) = 0, \tag{1.76}$$

$$\frac{\partial}{\partial\beta}\left(h + \int c\, dt\right) = 0, \tag{1.77}$$

are then required to determine the characteristic paths.

Equations (1.74) and (1.75) integrate to provide the Riemann invariants,

$$J_+(\beta) = u + \lambda, \tag{1.78}$$

$$J_-(\alpha) = u - \lambda, \tag{1.79}$$

with $J_+(\beta)$ equal to a constant on C_+ characteristics while $J_-(\alpha)$ is correspondingly constant on C_- characteristics. When the relation between stress and strain is linear and the wave speed c is constant, the Riemann invariants J_+ and J_- reduce to the d'Alembert solutions of the linear elastic wave equation (e.g. Courant and Hilbert 1962).

Generally, the characteristic methods are not particularly suited to analytic solutions of wave interaction problems. There are several special classes of wave

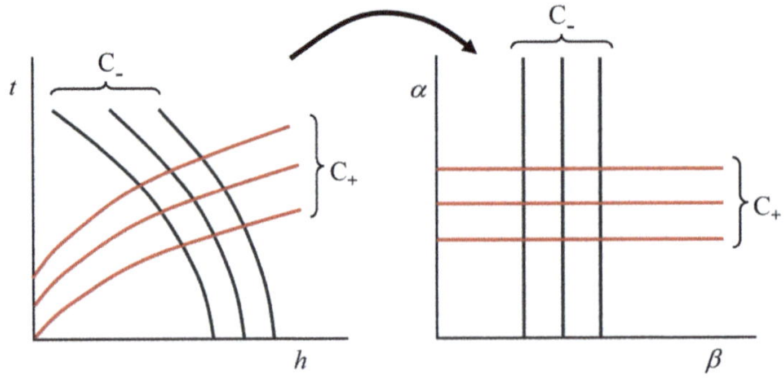

Figure 1.24. Forward C_+ and backward C_- facing characteristics in a plot of distance versus time and their mapping into the α versus β plane.

propagation problems that are nicely treated with the characteristic method, however, and these are discussed in subsequent sections. The system of characteristic equations (equations (1.74)–(1.77)) is generally amenable to numerical solution methods, and many characteristics-based computer codes have been developed over the years.

One elementary wave interaction problem is addressed here for illustrative purposes. Consider a forward-facing wave as illustrated in figure 1.25 that carries the material to stresses σ_1 and then σ_2 through two successive shock waves. Let the wave be incident on a rigid wall such that reflection of the first shock doubles the stress to $2\sigma_1$. Focus on the forward-facing C_+ characteristic through which the second shock achieves stress σ_2, and the backward-facing C_- characteristic of the reflected shock wave achieves stress $2\sigma_1$. The objective is to calculate the stress and particle-velocity amplitudes σ_3 and u_3 after wave interaction through a characteristic solution.

On the C_+ characteristic the Riemann invariant,

$$J_+ = u + \frac{\sigma}{\rho_o c}, \tag{1.80}$$

with wave speed c constant, is invariant though the wave interaction. Consequently,

$$u_2 + \frac{\sigma_2}{\rho_o c} = u_3 + \frac{\sigma_3}{\rho_o c}. \tag{1.81}$$

Similarly, invariance of

$$J_- = u - \frac{\sigma}{\rho_o c} \tag{1.82}$$

on the C_- characteristic requires

$$-\frac{2\sigma_1}{\rho_o c} = u_3 - \frac{\sigma_3}{\rho_o c}. \tag{1.83}$$

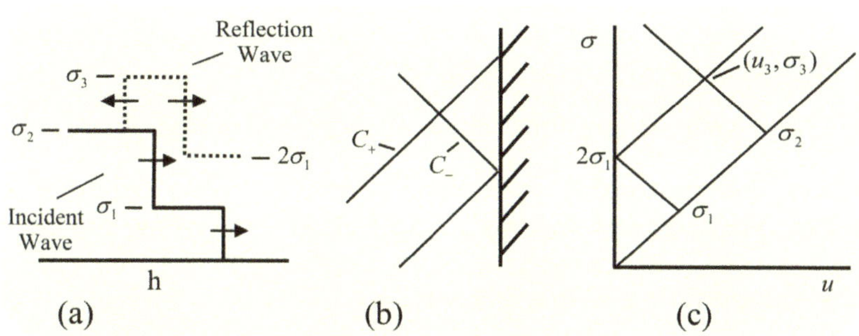

Figure 1.25. (a) A simple two-shock structure incident on and reflected from (b) a rigid wall leading to interaction of a second shock with the reflected shock achieving particle velocity and stress states u_3 and σ_3 shown in the (c) stress versus particle-velocity plane.

Solving equations (1.81) and (1.83) readily yields

$$u_3 = \frac{1}{\rho_o c}(\sigma_2 - \sigma_1), \qquad \sigma_3 = \sigma_1 + \sigma_2. \tag{1.84}$$

Features of the solution are illustrated in figure 1.25 through a stress versus particle-velocity graph of the wave interactions.

1.5 Waves of restricted form

Characteristic solution of the nonlinear wave equations has led to special forms of waves that have been widely studied and applied to many applications. These special wave solutions are particularly illuminating of the underlying physics and material properties governing the wave propagation. These waves include the simple wave, the centered wave and the steady structured wave, among others.

1.5.1 Simple waves

A simplification of the characteristic solution comes about when the application concerns one-dimensional waves propagating solely in one direction. Consider wave propagation in the forward C_+ direction with the stress and particle velocity ahead of the wave at a uniform state. For such a flow the Riemann invariant $J_-(\alpha)$, in addition to being constant on C_- characteristics, is also constant throughout the flow field (or independent of α), reflecting the constant uniform state ahead of the wave. Consequently, throughout the field,

$$J_-(\alpha) = u - \lambda \; = \; \text{constant}. \tag{1.85}$$

Since, on C_+ characteristics,

$$J_+(\beta) = u + \lambda \; = \; \text{constant}, \tag{1.86}$$

Equations (1.85) and (1.86) require that u and λ are independently constant on C_+ characteristics. In turn, since λ is a single-valued function of ε, or σ, both stress and strain, and also particle velocity, are individually constant on C_+ characteristics. Consider the wave speed c to functionally depend on the stress $c = c(\sigma)$, or alternatively on particle velocity $c = c(u)$, and is hence also constant on C_+ characteristics, equation (1.76) then yields

$$h - c(\sigma)t \; = \text{constant} \quad \text{or} \quad h + c(\sigma)t \; = \text{constant}. \tag{1.87}$$

Circumstances that reasonably approximate simple wave conditions occur frequently under intense-energy loading in both materials testing and applications. Some solutions are pursued here to gain appreciation for the nature of the propagation and evolution of simple waves.

1.5.1.1 Simple wave evolution from a ramp wave

Simple wave propagation evolving from an initial ramp wave input is informative. Initial ramp loading is often sought in materials testing. Subsequent evolution of the wave can be analyzed to determine the dynamic strength and EOS response of the material. Impact experiments using selected materials with gradually increasing density in the impacting projectile can be tailored to provide a ramp wave input. Pulsed magnetic loading methods also lead to near initial ramp wave loading. Further, consideration of the simple wave propagation of an initial ramp wave is instructive for evaluating more complex simple wave propagation.

A pressure ramp that at time $t = 0$ characterizes the initial conditions of a large-amplitude simple wave is shown in figure 1.26. This initial pressure ramp is expressed analytically as

$$\sigma(h) = \sigma_o(1 - h/h_o) = \sigma_o + \sigma_{ho}h, \tag{1.88}$$

where $\sigma_{ho} = -\sigma_o/h_o$ is the magnitude of the initial slope of the ramp. From the earlier simple wave solution,

$$\sigma(h,t) = \frac{2}{\rho c}J_+(\xi), \tag{1.89}$$

with $\xi = h - c(\sigma)t$. The initial ramp pressure condition provides the solution for the pressure,

$$\sigma(h,t) = \sigma_o + \sigma_{ho}(h - c(\sigma)t). \tag{1.90}$$

The gradient of the pressure in the wave front at any later time is then given by,

$$\left.\frac{\partial\sigma}{\partial h}\right|_t = \sigma_h = \sigma_{ho}(1 - c'\sigma_h t), \tag{1.91}$$

where $c' = dc/d\sigma$ is determined by the pressure dependence of the sound velocity. Equation (1.91) provides for the pressure gradient,

$$\sigma_h = \frac{\sigma_{ho}}{1 + c'\sigma_{ho}t}. \tag{1.92}$$

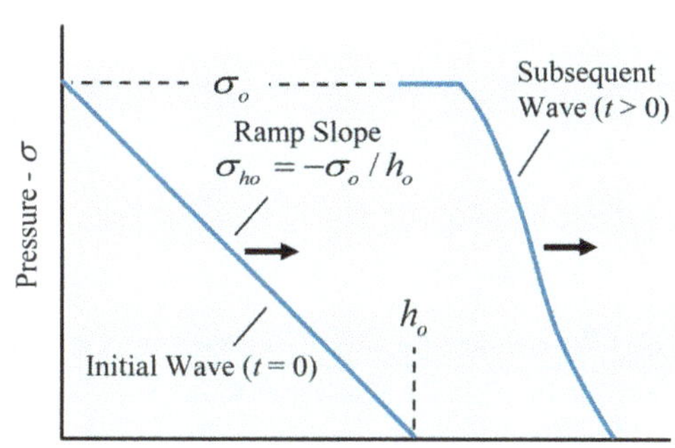

Figure 1.26. Initial pressure ramp initiating simple wave propagation in the forward h direction.

The time to break-over (verticality) of the wave at pressure amplitude σ occurs when the denominator achieves zero, or

$$t_b = \left| \frac{1}{c' \sigma_{ho}} \right|. \tag{1.93}$$

Consider a quadratic dependence of the sonic velocity on the pressure for purposes of exploring features of the simple wave propagation process,

$$c(\sigma) = c_o + c_o' \sigma + \frac{1}{2} c_o'' \sigma^2. \tag{1.94}$$

If pressure dependence is linear ($c_o'' = 0$), then all pressure amplitudes in the wave achieve break-over at the same time,

$$t_b = \frac{-1}{c_o' \sigma_{ho}}. \tag{1.95}$$

The position of the wave at break-over is provided by

$$h_b = h(\sigma) + c(\sigma) t_b, \tag{1.96}$$

where $h(\sigma)$ is the position of the initial pressure wave provided by equation (1.88). Again, if $c_o'' = 0$ simultaneous break-over of the wave occurs at a position,

$$h_b = h_o \left(1 + \frac{1}{\sigma_o} \frac{c_o}{c_o'} \right). \tag{1.97}$$

Earlier in this chapter it was noted that the EOS properties of many materials are reasonably approximated by the sound-velocity dependence provided in equation (1.94) with $c_o'' = 0$. This dependence leads to the Murnaghan EOS. Figure 1.27 illustrates simple wave propagation from an initial ramp when $c_o'' = 0$ as well as for when c_o'' is either negative or positive. The shape of the simple wave is markedly influenced by the sign and magnitude of curvature in the dependence of sound velocity on the pressure. The final wave profile in each example calculated from the

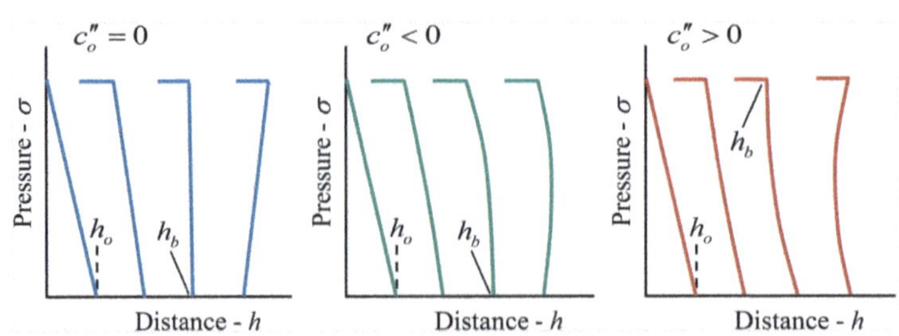

Figure 1.27. Simple wave propagation from an initial ramp wave. Dependence on curvature in the sound-velocity versus pressure relation is illustrated. The point at which the wave first achieves verticality is identified in each example.

simple wave characteristic solution is not of course physically achievable. The viscosity mechanisms considered in section 1.3.3 will come into play, leading to the formation of a structured steady wave.

1.5.1.2 Application to simple wave propagation in silicon

A practical example of simple one-dimensional wave propagation is provided by the experimental study of solids using ramp wave generators. Experimentally the method requires some preparatory technique so that stress loading on the input side of the test sample is spread over time and distance more or less uniformly (a ramp), as illustrated in figure 1.28. The resulting large-amplitude wave is then monitored at some material plane downstream from the input plane. VISAR (Barker and Hollenbach 1972), as indicated in the figure, is one such diagnostic. Ramp wave generators include materials in which the compression modulus softens with stress amplitude, such as fused silica (Grady 1977b), impact discs prepared from graded density material (Barker 1984) or layered material (Chhabildas *et al* 1993), and pulsed power systems in which rapid rising current pulses are used to provide continuously rising magnetic pressure loading (Hall *et al* 2001). Experimental ramp waves have commonly been referred to as ICE waves, alluding to both the quasi-isentropic character of the compression and to the markedly lower final temperature achieved at the compressed state in contrast to a comparable shock wave.

Experimental ramp wave methods are unique in their sensitivity to material compression features during the continuous rise of the stress wave. The evolution of such experimentally measured waves is ideally suited to simple wave analysis.

To explore the implications of simple wave behavior in the ramp wave event, an initial linear ramp loading expressed as a function of the particle velocity,

$$u(h) = -\alpha h, \tag{1.98}$$

is assumed. The initial uniform strain rate impressed on the material is provided by α. Since u is constant on any C_+ characteristic, the spatial position h in equation (1.98) can be replaced by $h - c(u)t$ corresponding to any later time and,

$$u(h,t) = -\alpha(h - c(u)t), \tag{1.99}$$

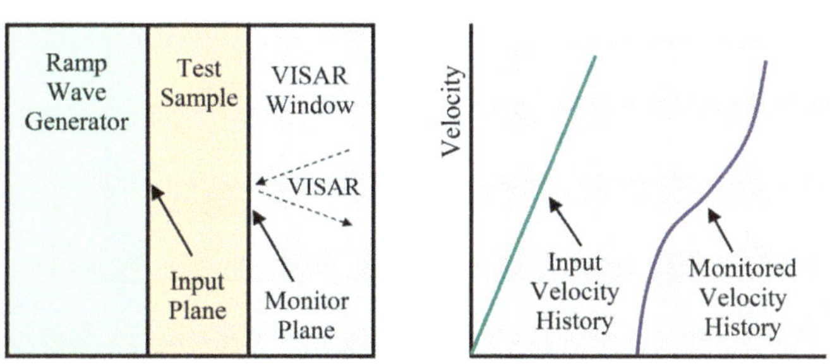

Figure 1.28. Ramp wave experiment to test the dynamic quasi-isentropic properties of matter.

provides the mass-velocity flow field throughout the forward-facing wave propagation event.

The Hugoniot compression of single-crystal silicon is used to exemplify the nature of the evolution of an initial ramp wave into a solid material. The measured Hugoniot states for silicon from experiments of Gust and Royce (1971), and later Goto *et al* (1982), are shown in figure 1.29. These compression data are adequately fit to the cubic relation,

$$c(u) = c_o + s_1 u + s_2 u^2 + s_3 u^3. \tag{1.100}$$

The constants are $c_o = 8.0$, $s_1 = 2.4$, $s_2 = -5.7$ and $s_3 = 2.0$, where $c(u)$ and u are expressed in units of km s^{-1}. The density of silicon is 2330 kg m^{-3}. The resulting analytic stress versus compression curve is compared with the experimental data in figure 1.29, and is probably adequate to about 60–70 GPa. Crystalline silicon exhibits an HEL in the neighborhood of 6–9 GPa, depending on crystal orientation (see also chapter 6), and undergoes a polymorphic phase transformation with substantial volume change in the 15–35 GPa shock pressure range. Physics-based functional forms would better model the phase transition region and extrapolation to higher shock pressures. The polynomial form in equation (1.100) is convenient for the present analysis. Differences in the quasi-isentropic and Hugoniot (shock) compression are ignored in the present analysis. Equivalently the Gruneisen parameter is assumed to be equal to zero.

Subsequent wave profiles in silicon evolving from the initial ramp through the formal solution provided from equation (1.99) are shown on the left in figure 1.30. An initial uniform strain rate of $\alpha = 10^6\ s^{-1}$ is used in the calculation. Ramp loading to an amplitude of 2.5 km s^{-1} (approximately 40 GPa) provides an initial ramp extent h_o of 2.5 mm and is consistent with current gas-gun ramp wave technology.

Wave profiles at successively later times show regions in which the particle-velocity profile is both spreading and steepening. The wave profiles become

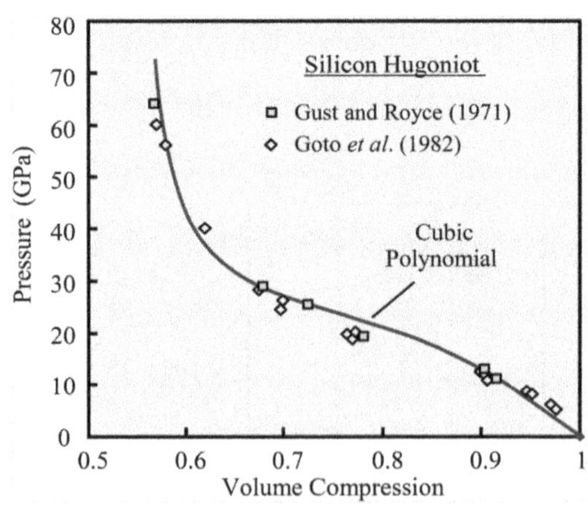

Figure 1.29. Hugoniot data for crystalline silicon (Gust and Royce 1971, Goto *et al* 1982) and an analytic fit to the data with a cubic polynomial.

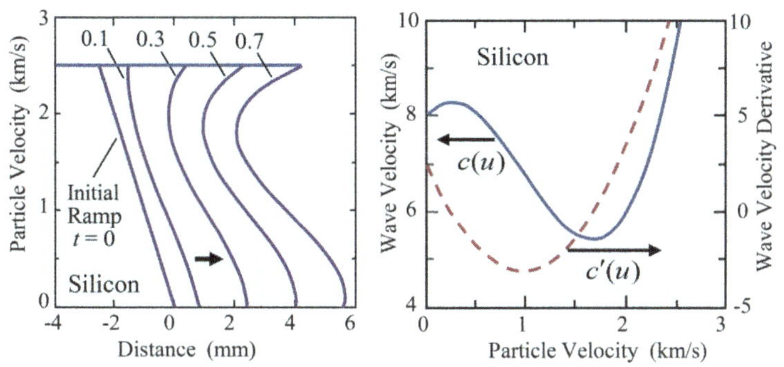

Figure 1.30. On the left are calculated simple wave particle-velocity profiles to an amplitude of 2.5 km s^{-1} from an initial ramp input based on a cubic polynomial representation of the compression of crystalline silicon. Wave arrival times are in microseconds. The characteristic velocity and first derivative as a function of particle-velocity amplitude are shown on the right.

multivalued at later times and are non-physical. The plot of both $c(u)$ and its first derivative $c'(u)$ on the right in figure 1.30 provide insight into the nature of this nonlinear wave propagation behavior. Using equation (1.99), the velocity gradient u_h (or strain rate) at any later time dependence on particle-velocity amplitude is given by

$$u_h(u,t) = \frac{-\alpha}{1 - \alpha c'(u)t}. \tag{1.101}$$

Forward-facing C_+ characteristics diverge and the gradient lessens (the wave becomes less steep) for all u for which $c'(u) < 0$. C_+ characteristics converge for $c'(u) > 0$ and intersect at a time $t = 1/\alpha c'(u)$ at which amplitude the gradient becomes locally vertical. The subsequent rate independent wave solution is not physical in regions where the velocity gradient achieves, and then exceeds, verticality.

Shocks originate when critical high-strain-rate conditions are achieved in the flow. Physically, when velocity gradients (strain rates) achieve some characteristic amplitude determined by intrinsic viscous properties of the material, additional stress arises to support the structured shock. Commonly this additional viscous stress is given a constitutive relation of the power-law form $\sigma_v = (\dot{\epsilon}/\dot{\epsilon}_*)^m$, as was described in previous sections. Shocks originate when the velocity gradient achieves amplitudes comparable to the characteristic strain rate $\dot{\epsilon}_*$ in the viscosity relation. For the present purpose, it is sufficient to equate the shock formation transition to the time of local verticality of the velocity gradient. In the present example of ramp compression of crystalline silicon, local verticality of the velocity gradient is first achieved on the peak stress ($u = 2.5$ km s^{-1}) C_+ characteristic for which $c'(u)$ is the largest, as is shown in figure 1.30.

The time and position in the $h - t$ plane at which verticality is achieved on the 2.5 km s^{-1} C_+ characteristic, is calculated from equation (1.101). The calculated critical point and corresponding characteristic is identified in the distance versus time plot shown in figure 1.31. Verticality is achieved and then exceeded on successively lower-amplitude characteristics as the upper portion of the wave folds

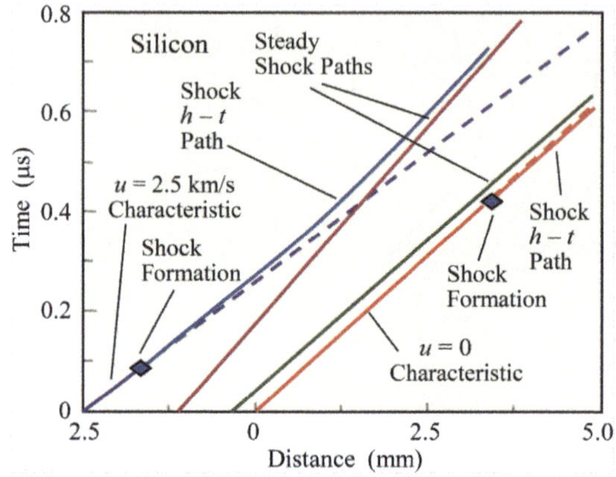

Figure 1.31. Selected characteristics of the ramp wave solution. Onset of upper and lower shock waves are identified by the diamond symbols. Shown also are the time dependent path of the transient shock and the characteristic towards which steady shock propagation is tending.

over. This character of the wave is illustrated by the profiles in figure 1.30. Similarly, at a later time, critical verticality is achieved on the $u = 0$ km s^{-1} C_+ characteristic also identified in figure 1.31.

Shocks initiate at critical points on both the $u = 2.5$ and $u = 0$ km s^{-1} characteristics, and grow in amplitude as the wave continues to propagate. The path of either shock is illustrated in figure 1.31. This path of the shock wave can be calculated from the wave solution and the EOS information.

Here the focus will be on the shock originating on the $u = 2.5$ km s^{-1} characteristic at $t_1 = 0.088$ μs. The path of the subsequent transient shock follows a curved trajectory as the shock grows and the corresponding transient shock velocity is time dependent. The $h - t$ path of the shock is calculated from the differential relation,

$$h(t) = h_1 + \int_{t_1}^{t} U(u)dt, \tag{1.102}$$

where (h_1, t_1) is the Lagrangian location and time of shock initiation. The time-dependent shock velocity in equation (1.102) is provided by

$$U(u) = \rho_o \frac{\sigma_a - \sigma(u)}{u_a - u}, \tag{1.103}$$

where σ_a and u_a are the peak amplitude of the stress wave on the 2.5 km s^{-1} characteristic, while $\sigma(u)$ and u are the current stress and velocity at the foot of the transient shock, as illustrated in figure 1.32. Since $u(h,t)$ is provided by equation (1.99), the differential expression represented by the integral relation in equation (1.102) can be explicitly calculated for the transient path of the shock. This transient shock will transition asymptotically to a velocity corresponding to the characteristic uniquely determined by the Rayleigh line that is tangent to the compression EOS at its lower end and intersects the point (σ_a, u_a) at the upper end, as illustrated on the

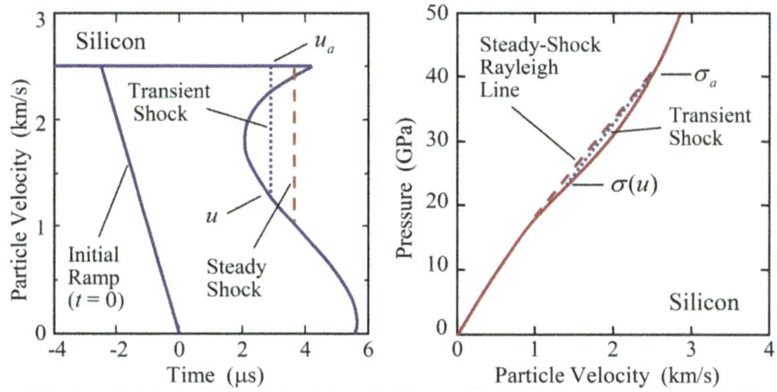

Figure 1.32. Evolving transient shock and steady shock amplitudes (left) and corresponding Rayleigh lines in the pressure versus particle-velocity plot (right).

right in figure 1.32. This steady shock characteristic is identified by the particle-velocity amplitude that satisfies

$$c(u) = U(u) = \rho_o \frac{\sigma_a - \sigma(u)}{u_a - u}. \tag{1.104}$$

Solving equation (1.104) for the present silicon ramp wave loading example provides a particle-velocity amplitude of 1.12 km s^{-1} for the foot of the shock and a characteristic velocity of 6.35 km s^{-1}.

1.5.2 Centered waves

A further special case is that of a simple wave in which all C_+ characteristics originate from a single point in the $h - t$ plane. Without loss of generality such a centered wave can be considered to emerge for the $h = 0$ and $t = 0$ origin. Since characteristics are straight lines for simple waves, any characteristic starting from the origin and passing through the point (h,t) has slope, and hence wave speed,

$$c = h/t. \tag{1.105}$$

Since c is a known function of either σ, ε or u, the solution is known immediately. For example, if $c(\sigma)$ is known for the material, then $\sigma = f(h/t)$ where $f()$ is the inverse of $c()$.

Alternatively, assume ε and u in the previous set of governing equations (1.68) and (1.69) depend on h and t only though the parameter $\xi = h/t$. These same equations then transform to

$$-\xi \frac{d\varepsilon}{d\xi} + \frac{du}{d\xi} = 0 \tag{1.106}$$

and

$$-\xi \frac{du}{d\xi} + c^2 \frac{d\varepsilon}{d\xi} = 0. \tag{1.107}$$

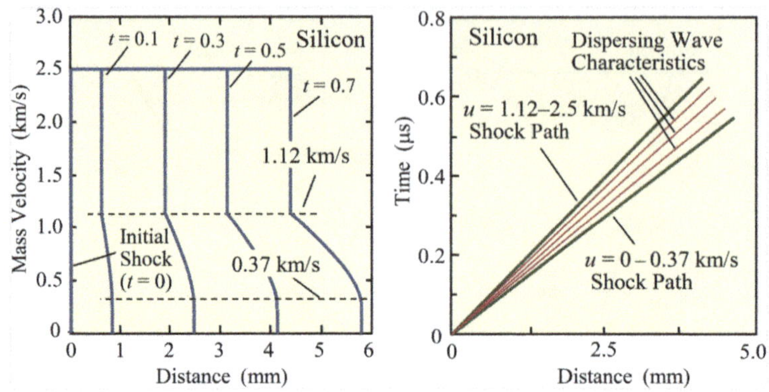

Figure 1.33. Shock input and centered simple wave propagation in crystalline silicon. Times are in microseconds. Characteristic paths from the two shock waves and the dispersive wave are illustrated in the distance–time plot on the right.

Eliminating the term in u yields

$$(c^2 - \xi^2)\frac{d\varepsilon}{d\xi} = 0 \tag{1.108}$$

and finally

$$c = \pm\xi = \pm\frac{h}{t}, \tag{1.109}$$

for either left-facing or right-facing centered simple waves.

1.5.2.1 Centered simple wave propagation in silicon

Centered wave propagation in silicon resulting from an initial shock loading to the same driving particle velocity as the previous ramp wave solution is illustrated in figure 1.33. Using the previous cubic EOS provided in figure 1.29, two steady shocks over the velocity range of 0–0.37 km s^{-1} and 1.12–2.50 km s^{-1} are calculated to emerge and propagate at their respective steady shock velocities of 8.21 and 6.35 km s^{-1}. The spreading C_+ characteristics enveloped by the two shock characteristics are shown on the right in figure 1.33 and distinguish the dispersive wave segment connecting the two shock waves.

1.5.2.2 Centered rarefaction and vaporization on shock emergence at a free surface

A further example of a centered simple wave occurs when a shock wave emerges at a free surface. A boundary condition of zero pressure at the free surface results in the propagation of a decompression (rarefaction) wave that unloads the material from the pressure state imparted by the shock wave to the zero pressure state. The decompression wave is centered at the free surface and at the time of shock arrival. Pressure, specific volume and particle velocity are constants on the straight-line characteristics emerging from the (h,t) origin of the centered wave, as illustrated

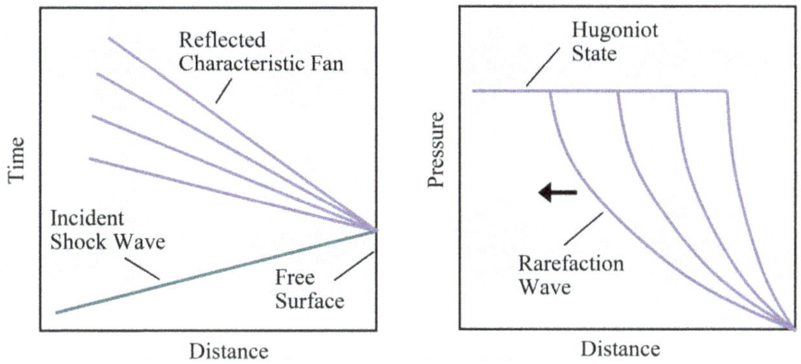

Figure 1.34. Shock compression wave incident on a free surface and the characteristic fan for the reflected simple centered decompression wave.

in figure 1.34. A spreading decompression wave subsumes the initially shock compressed matter.

Centered decompression waves are readily produced in experiments and commonly reveal details of the EOS properties of materials. The particular EOS feature of concern here occurs when mixtures of solid and water are subjected to intense shock compression. The application here is that of geological materials where a substantial fraction of water content is commonplace. When subjected to strong shock wave loading, entropy production at the shock state leads to material temperatures that are commensurate with water vaporization and subsequent steam expansion upon decompression. If the thermodynamic physics allows decompression vaporization to occur on time scales compatible with the shock compression event, then thermomechanical influences on the decompression wave propagation are expected. Both impact cratering and explosion-induced ground shock of water-saturated geologic media would exhibit the consequences of shock vaporization.

The EOS features of shock-induced vaporization can be investigated through controlled impact shock and centered simple decompression wave studies. Such studies on a water-saturated NTS tuff (Murri *et al* 1975) are appropriate for the present discussion. On the left in figure 1.35 are shown the results of a shock compression and subsequent decompression experiment on water-saturated tuff from a Hugoniot pressure of approximately $\sigma_h = 10$ GPa.

EOS models predict that upon release from principal Hugoniot states of this magnitude, the decompression isentrope will intersect the vapor dome at pressure amplitudes of the order of tens of megapascals (Butkovich 1971, Riney *et al* 1971). The vapor dome on the left is qualitative and for illustrative purposes only. The plot on the right shows a log–log graph of a decompression isentrope that is sensibly consistent with the shock release data for tuff (Murri *et al* 1975), and expansion models in the vapor phase (Riney *et al* 1971), within the constraints of an analytically convenient functional description. A measure of the volume expansion from the shock state used as the abscissa on the right in figure 1.35 is provided by $\varepsilon = (\rho_h/\rho) - 1$ where ρ_h is the density at the Hugoniot state.

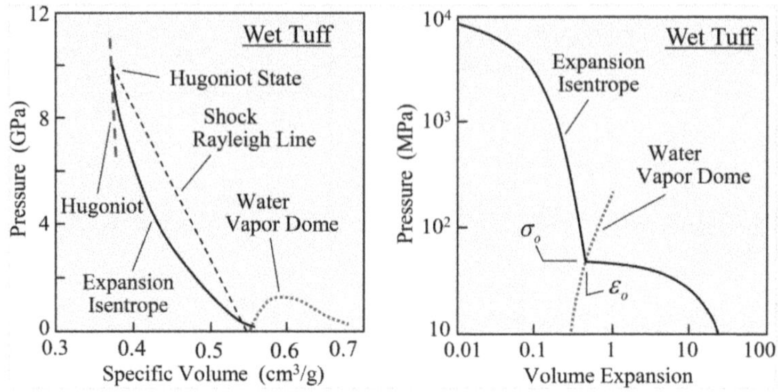

Figure 1.35. An approximately 10 GPa shock Hugoniot state in water-saturated tuff and the subsequent expansion isentrope. On the right is shown a reasonable analytic model of the expansion isentrope including vaporization.

The decompression isentrope shown on the right in figure 1.35 is represented by the exponential pressure versus expansion strain relation,

$$\sigma(\varepsilon) = \sigma_o e^{-\alpha(\varepsilon-\varepsilon_o)}, \qquad \alpha = \begin{cases} \alpha_h & \varepsilon < \varepsilon_o \\ \alpha_o & \text{otherwise}' \end{cases} \tag{1.110}$$

which applies to decompression in both the condensed phase and within the mixed-phase vapor dome. Pressure and expansion strain on entering the vapor dome, and shown in the figure, are σ_o and ε_o. Constants that reasonably approximate decompression of NTS tuff from a Hugoniot pressure of 10 GPa are $\rho_h = 2600\ \text{kg m}^{-3}$, $\sigma_o = 50$ MPa, $\varepsilon_o = 0.45$, $\alpha_h = 11.7$ and $\alpha_o = 0.07$. Note that α_h is determined from the Hugoniot pressure through the relation $\alpha_h\varepsilon_o = \ln \sigma_h/\sigma_o$, while α_o provides a measure of expansion on the isentrope through the vapor dome.

The sound speed shown in figure 1.36 follows from

$$c^2 = -\frac{\sigma'(\varepsilon)}{\rho_h} = \frac{\alpha\sigma_o}{\rho_h} e^{-\alpha(\varepsilon-\varepsilon_o)} = \left(\frac{h}{t}\right)^2. \tag{1.111}$$

The last equality is provided by the centered wave relation from equation (1.109) and yields the solution for all of the wave properties. The strain solution is,

$$\varepsilon = \varepsilon_o + \frac{1}{\alpha} \ln \frac{\alpha\sigma_o}{\rho_h}\left(\frac{h}{t}\right)^{-2}. \tag{1.112}$$

Pressure is then provided by equation (1.110). Particle velocity follows from integration of the relation

$$u(\varepsilon) = \int_0^{\varepsilon} c(\varepsilon) d\varepsilon. \tag{1.113}$$

In figure 1.36 plots of sound speed versus pressure and calculated centered rarefaction wave profiles based on the above analytic model for water-saturated

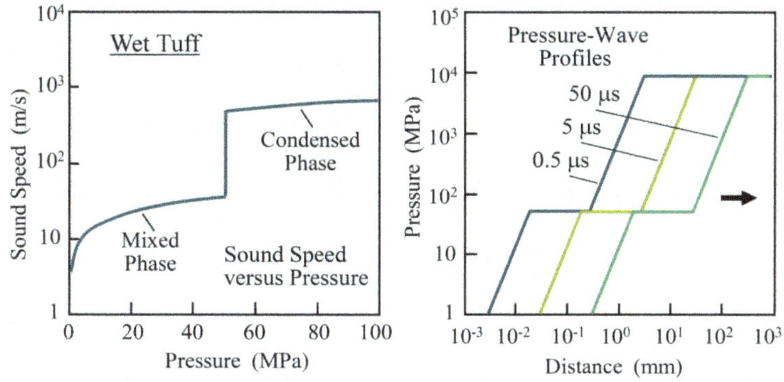

Figure 1.36. Plots of sound speed and centered rarefaction wave profiles based on analytic model for NTS tuff.

NTS tuff undergoing decompression for a Hugoniot pressure of 10 GPa are shown. An order of magnitude drop in sound speed on entering the mixed solid–water–vapor phase results in comparable separation in the decompression wave.

Integration of equation (1.113) for the particle velocity results in two solution parts. For expansion strains in excess of the onset of vaporization, $\varepsilon > \varepsilon_o$, the solution for the particle velocity provides

$$u(\varepsilon) = 2\sqrt{\frac{\sigma_o}{\rho_h \alpha_h}}\left(e^{\frac{\alpha_h}{2}\varepsilon_o} - 1\right) + 2\sqrt{\frac{\sigma_o}{\rho_h \alpha_o}}\left(1 - e^{-\frac{\alpha_o}{2}(\varepsilon - \varepsilon_o)}\right). \quad (1.114)$$

A plot of the decompression isentrope in pressure versus particle-velocity coordinates is shown in figure 1.37. This representation is of interest in that the marked effects of shock-induced vaporization on particle velocity are experimentally accessible. As observed on the left in figure 1.37, if a material such as the water-saturated tuff examined here undergoes equilibrium vaporization upon shock emergence at a free surface, then measurement of the free-surface acceleration should readily exhibit the effect through enhanced free-surface velocity. If for reasons of kinetics, the onset of boiling and vapor production does not proceed on the time scale of the shock and decompression event, then enhanced free-surface velocity on entering the vapor dome is suppressed. The alternative decompression path illustrated in the figure is traversed, and a free-surface velocity close to twice the Hugoniot particle velocity will be observed.

The issues of vaporization upon decompression from a shock state have been considered by Zeldovich and Raiser (1966). They suggest that activation energies for the onset of boiling are prohibitively high for vaporization on shock wave time scales. The arguments of Zeldovich and Raiser are applicable to homogeneous materials. Application to heterogeneous mixtures such as rock and water is debatable. Sites for activation of boiling should be markedly more prevalent in the thermally and mechanically heterogeneous shock compressed state of such mixtures.

Early shock wave experiments on a wet playa alluvium by Anderson *et al* (1965) provided measurement of free-surface velocities in excess of twice the Hugoniot particle velocity, suggesting shock vaporization on decompression. A systematic

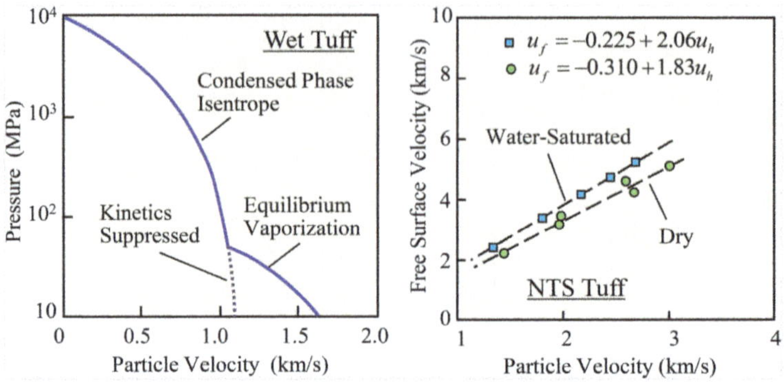

Figure 1.37. Right: centered simple wave calculation of the decompression isentrope in condensed and mixed vapor phases. Left: free-surface velocity and Hugoniot particle velocity measurements in dry and water-saturated NTS tuff spanning approximately 5–25 GPa Hugoniot pressures (Murri *et al* 1975).

shock wave study of water-saturated NTS tuff over Hugoniot pressures of 10–25 GPa using in-material and free-surface inductive particle-velocity instrumentation failed to observe enhanced free-surface velocity (Murri *et al* 1975). Free-surface velocity versus Hugoniot particle-velocity measurements for the NTS tuff are shown in figure 1.37. For comparison, the same data for dry NTS tuff are also shown in the same figure. The need for EOS models including vaporization in large-scale applications involving the intense shock compression of materials through impact or explosion remains an open question.

1.5.2.3 Centered precursor waves in uranium

Elastic precursor waves in depleted alpha-uranium have been observed by monitoring the motion of the back free surface with velocity interferometry upon emergence of an impact-induced shock wave (Grady 1986b). The unusual structure of the precursor wave in uranium is illustrated by the measurements shown in figure 1.38. In the simplest sense the HEL for this uranium metal has been identified with the distinct discontinuity in slope observed in the precursor structure. This discontinuity (the HEL) occurs at about 1.3 GPa.

In reality the deformation mechanisms responsible for the precursor structure in uranium are more complex than those captured through specification of a constant HEL amplitude. The finite ramp of the wave up to the HEL break is believed to predominantly reflect the elastic and deformation twinning properties of the material. The continuous emergence of deformation twinning with increasing stress amplitude has been described in the study of Meyers *et al* (2001). Above the HEL break, pervasive dislocation plasticity overwhelms other deformation mechanisms.

The portion of the precursor wave up to the 1.3 GPa HEL is not steady, but continues to spread in time as the propagation distance increases. This feature of the precursor wave has not been specifically verified for alpha-uranium. It has, however, been demonstrated to be the case for two other uranium alloys, uranium 3/4% titanium (Dandekar *et al* 1980) and uranium 6% niobium (Hixson *et al* 1998). Plots of this region of the precursor wave for both alloys versus the similarity parameter

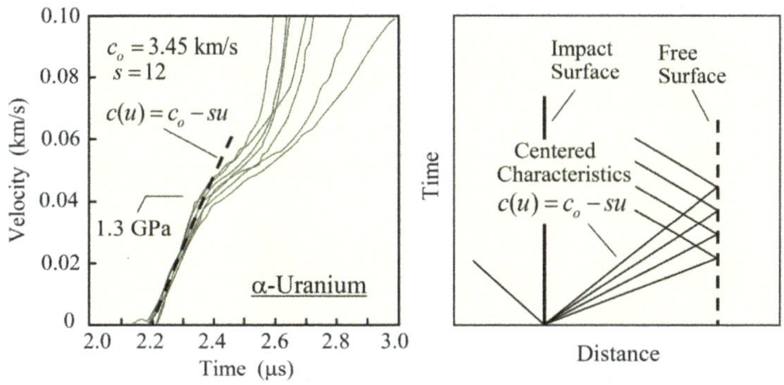

Figure 1.38. Free-surface velocity measurements of the precursor shock wave in alpha-uranium (Grady 1986b). Centered-wave-velocity versus particle-velocity amplitude relation applicable to a 1.3 GPa HEL amplitude.

$\xi = h/t$ demonstrate simple centered wave behavior (Grady 2005). It is reasonable to expect comparable behavior for alpha-uranium.

Centered wave analysis methods can be applied to the profile structure for uranium governed by elasticity and deformation twinning spanning the shock response below 1.3 GPa. (Further shock analysis methods with application to uranium metal are pursued in chapter 11.) The experimental wave profiles can be used to determine the wave velocity as a function of particle velocity within this stress range. This wave velocity is found to be adequately described by the linear function,

$$c(u) = c_o - su, \tag{1.115}$$

where $c_o = 3.45$ km s^{-1} and $s = 12$. The wave arrival predicted by this wave-velocity relation is shown as the dashed line overlying the data in figure 1.38. Centered wave propagation requires that the similarity parameter satisfies

$$h/t = \xi(u) = c_o - su. \tag{1.116}$$

The equation (1.106) relates the strain and particle velocity,

$$-\xi\frac{d\varepsilon}{d\xi} + \frac{du}{d\xi} = 0. \tag{1.117}$$

Equation (1.116) is used to eliminate particle velocity through

$$\frac{du}{d\xi} = -\frac{1}{s}, \tag{1.118}$$

and provides the differential equation

$$\xi\frac{d\varepsilon}{d\xi} + \frac{1}{s} = 0. \tag{1.119}$$

Equation (1.119) integrates to provide the strain dependence on the similarity parameter,

$$\varepsilon(\xi) = \frac{1}{s} \ln (c_o/\xi). \tag{1.120}$$

The stress–strain relation is obtained from

$$\frac{d\sigma}{d\varepsilon} = \rho_o c^2 = \rho_o \xi^2 = \rho_o c_o{}^2 e^{-2s\varepsilon}, \tag{1.121}$$

which integrates to

$$\sigma(\varepsilon) = \frac{\rho_o c_o^2}{2s} e^{-2s\varepsilon}. \tag{1.122}$$

An expansion of equation (1.122) is appropriate,

$$\rho_o c_o{}^2(\varepsilon - s\varepsilon^2 + \cdots). \tag{1.123}$$

Up to the HEL amplitude of 1.3 GPa it is readily shown by comparing the stress-compression response with the elastic response of uranium that the total strain at onset of pervasive deformation plasticity partitions into approximately 90% elastic strain and 10% deformation twinning strain.

1.5.2.4 Centered simple shear waves and impulse breach of metal plates

The analytic solutions of centered simple wave propagation can be profitably applied to the impulsive loading of ductile metal plates. Applications of interest are the deformation and possible breach of metal plates subjected to local impulsive loads, including projectile impact or explosion. Depending on a number of conditions, a metal plate subjected to such an encounter will either absorb the induced momentum, albeit probably deformed and damaged to some extent, or rupture occurs and breach of the plate results.

An illustrative example is shown in figure 1.39. A line of explosive in direct contact and covering an area of about twice the ½ inch thickness of the mild steel plate was

Figure 1.39. Breach of a mild steel plate results from detonation of a strip of explosive in direct contact with the back of the plate (Todd 2007). The momentum and energy dissipated by the plate are indicated by plastic deformation incurred in the dynamic event.

detonated and breach of the plated occurred. The sections of the breached plate are shown on the left. On the right, one section on end illustrates the extent of plastic deformation through bending incurred by the plate during the dynamic process.

The transfer of momentum in the immediate region of explosive detonation involves complex shock interactions. In the present analysis it is assumed that the initial interaction has proceeded to completion and the momentum has locally equilibrated through the thickness of the plate. A mass M (per unit length along the extent of the breach) and velocity u_M will characterize the initial momentum imparted to the plate. Subsequent deceleration is achieved through the lateral diffusion of momentum through plastic wave propagation. The conceptual model is illustrated in figure 1.40.

It is worth belaboring some of the physics that is being glossed over in the present analysis. The processes of momentum transfer are being decoupled into two independent processes. First is the local equilibration of momentum among the explosive products and the local region of the plate of comparable dimension. This equilibration should occur in several shock transits through the thickness of the plate. Second is the lateral transfer of momentum to the peripheral region of the plate. This further momentum redistribution process occurs on a substantially longer time scale associated with the plastic wave propagation—the topic of interest in this discussion.

The analysis of lateral momentum transfer is illustrated in figure 1.41. The origin of the coordinate system is a plane of traction coupling the stationary plate to the region of the initial moment imparted by the detonation as identified in figure 1.40. Traction at the contact plane, due to the relative motion, imparts a shearing force. Elasticity of the plate metal will be ignored and the material response is assumed to be rigid-plastic. The load transverse causes shear deformation and vertical displacement of the plate which propagates as a plastic wave as illustrated in figure 1.41. It should be noted that in the experiment displayed in figure 1.39 the explosive strip was specifically designed such that wave interactions led to plate breach with virtually no ejecta debris, in contrast to that depicted in figure 1.40 (Todd 2007).

A Lagrangian coordinate h specifies the distance from the contact plane. The vertical displacement as a function of the lateral position and time is $y(h,t)$, while the corresponding transverse velocity is provided by $u(h,t) = (\partial y/\partial t)_h$. A shear strain is specified in the present coordinate system as

$$\gamma = -(\partial y/\partial h)_t. \tag{1.124}$$

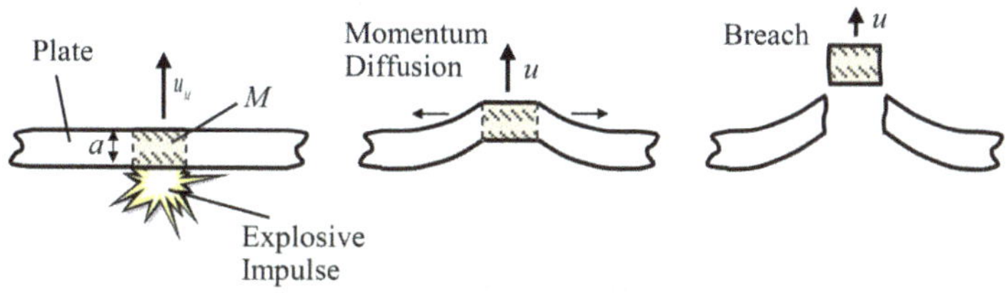

Figure 1.40. Initial impulsive loading of a metal plate, subsequent motion and momentum diffusion through plastic wave propagation, and lastly breach of the plate if shear failure is achieved.

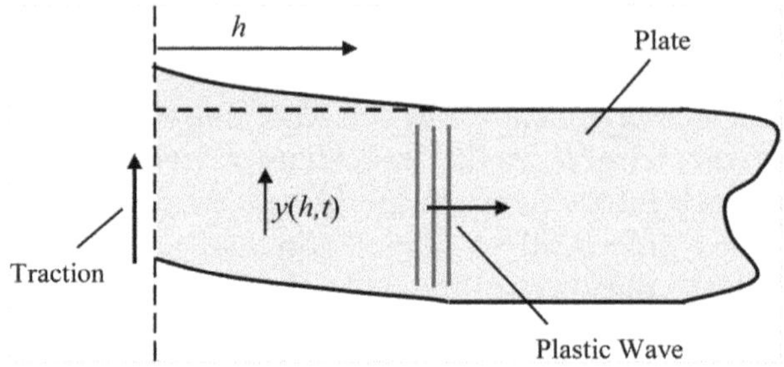

Figure 1.41. Motion and deformation imparted by the initial impulsive load and subsequent momentum diffusion through plastic shear wave propagation.

The governing equations of mass and momentum take the same form developed previously,

$$\left(\frac{\partial \gamma}{\partial t}\right)_h + \left(\frac{\partial u}{\partial h}\right)_t = 0, \tag{1.125}$$

$$\rho_o\left(\frac{\partial u}{\partial t}\right)_h + \left(\frac{\partial \tau}{\partial h}\right)_t = 0, \tag{1.126}$$

where τ is the corresponding shear stress.

One further relation is required to complete the system of governing equations and this will be some functional dependence between shear stress and shear strain. Commonly, in the high-rate deformation of metals, a dependence of shear stress on shear strain is observed to look qualitatively as shown in figure 1.42. The elastic response on initial loading is assumed to be negligible. The initial increase in shear stress with plastic shear strain is a consequence of deformation hardening and is quantified here as the local slope of the stress versus strain curve $H(\gamma)$. At some sufficiently large dynamic shear deformation, adiabatic thermal softening overcomes deformation hardening and a maximum in the shear-stress versus shear-strain curve is achieved at a strain γ_m and stress τ_m. This stress versus strain behavior is here quantified with the two-parameter hardening parameter $H(\gamma)$, shown on the right in figure 1.42. A more complex functional dependence could of course be accommodated.

The time scale over which momentum is transferred laterally through the propagation of plastic waves is expected to be sensibly greater than the time scale over which momentum is deposited during the initial impulsive loading. Thus, a centered simple wave solution for the plastic shear wave propagation is a reasonable approximation. From the earlier development for the solution for a centered simple wave,

$$c(\gamma) = h/t, \tag{1.127}$$

where the plastic wave velocity is given by,

$$c(\gamma) = \sqrt{H(\gamma)/\rho_o}, \tag{1.128}$$

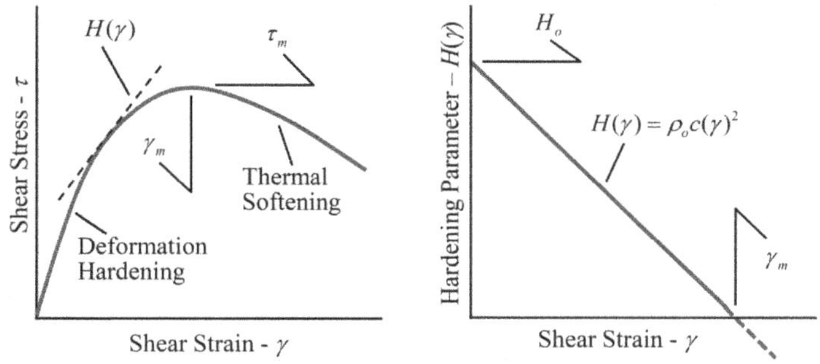

Figure 1.42. Left: dynamic shear stress versus shear strain for common metals. Right: two-parameter model for the hardening modulus.

and

$$H(\gamma) = H_o(1 - \gamma/\gamma_m). \tag{1.129}$$

The plastic shear-strain wave is solved for from equations (1.127)–(1.129),

$$\gamma = \gamma_m\left(1 - (h/c_o t)^2\right), \tag{1.130}$$

where $\rho_o c_o^2 = H_o$. The plastic shear-strain wave solution is illustrated in figure 1.43 for an initial Heaviside load to a peak strain amplitude of $\gamma_p < \gamma_m$. The strain γ_m determines the onset of shear stress reduction due to deformation-induced thermal softening.

The corresponding centered wave solutions for stress and particle velocity are obtained through the appropriate Riemann integrals provided earlier. For the shear stress,

$$\tau = \int \rho_o c^2(\gamma) d\gamma = \rho_o c_o^2 \int_0^\gamma (1 - \gamma/\gamma_m) d\gamma, \tag{1.131}$$

or

$$\tau = \rho_o c_o^2 \gamma(1 - \gamma/2\gamma_m). \tag{1.132}$$

Alternatively, since at $\gamma = \gamma_m$ the stress $\tau_m = \rho_o c_o^2 \gamma_m/2$, equation (1.132) can be written

$$\tau = \tau_m \frac{\gamma}{\gamma_m}\left(2 - \frac{\gamma}{\gamma_m}\right). \tag{1.133}$$

The solution for the particle velocity is provided through the relation

$$u = \int c(\gamma) d\gamma = c_o \int_0^\gamma (1 - \gamma/\gamma_m)^{1/2} d\gamma \tag{1.134}$$

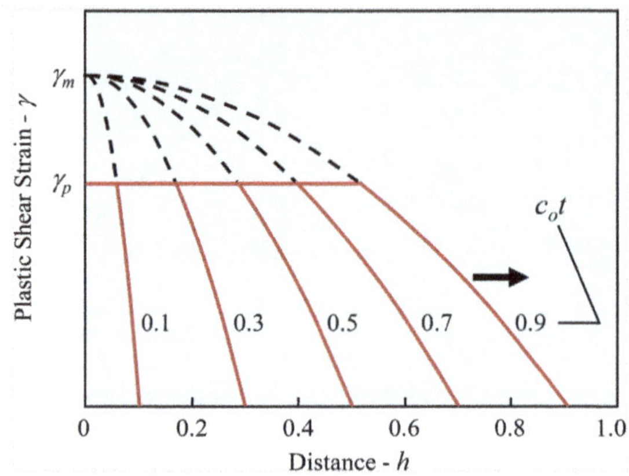

Figure 1.43. Plastic shear-strain wave at selected values of $c_o t$. The maximum strain γ_m that will propagate is determined by the plasticity model. The peak amplitude of the strain wave $\gamma_p < \gamma_m$ is determined by the traction intensity at the input plane.

and

$$u = \frac{2}{3} c_o \gamma_m [1 - (1 - \gamma / \gamma_m)^{3/2}]. \tag{1.135}$$

The maximum shear particle velocity that the traction interface can support is calculated from equation (1.135) at $\gamma = \gamma_m$,

$$u_m = \frac{2}{3} c_o \gamma_m = \sqrt{\frac{8}{9} \frac{\tau_m \gamma_m}{\rho_o}}. \tag{1.136}$$

Velocities in excess of this value cannot be transmitted though plastic shear wave propagation. When traction forces exceed the plastic shear-stress versus shear-strain load, thermal adiabatic shear deformation is localized at the traction interface and plate breach through plugging occurs. Breach through deformation localization and plugging occurred in the test illustrated in figure 1.39.

1.5.3 The steady simple waves

Steady waves represent a class of simple waves that propagate unchanged in form. All points on the wave propagate at the same wave velocity U. The paths of constant pressure and particle velocity are straight and parallel characteristics, with a slope of $1/U$ in the time–distance plane. A steady structured wave illustrating an invariant form and parallel paths is illustrated in figure 1.44. Applications in steady structured waves were described in a previous section. The focus here is on governing equations of wave propagation for the steady wave.

The governing equations of mass, momentum and energy are rewritten from equations (1.51), (1.55) and (1.58),

$$\left(\frac{\partial \varepsilon}{\partial t}\right)_h + \left(\frac{\partial u}{\partial h}\right)_t = 0, \tag{1.137}$$

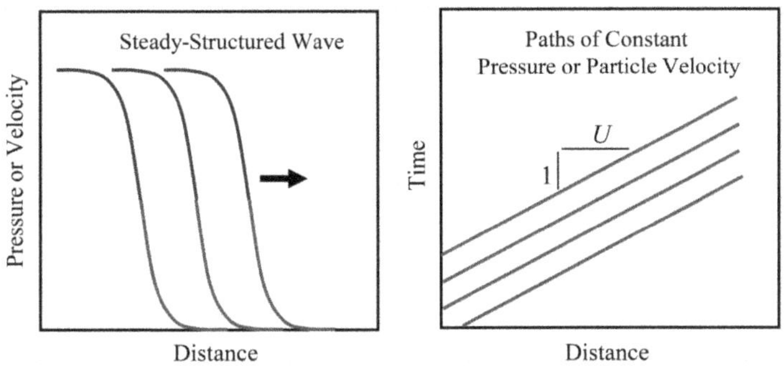

Figure 1.44. Steady structured waves propagate unchanged in form at velocity U, while wave characteristics are straight and parallel.

$$\rho_o\left(\frac{\partial u}{\partial t}\right)_h + \left(\frac{\partial \sigma}{\partial h}\right)_t = 0, \tag{1.138}$$

$$\rho_o\left(\frac{\partial E}{\partial t}\right)_h + \rho_o u\left(\frac{\partial u}{\partial t}\right)_h + \left(\frac{\partial \sigma u}{\partial h}\right)_t = 0. \tag{1.139}$$

Steady waves are functions of the single coordinate $\xi = t - h/U$. The governing relations then reduce to ordinary differential equations,

$$\frac{d}{d\xi}\left(\varepsilon - \frac{u}{U}\right) = 0, \tag{1.140}$$

$$\frac{d}{d\xi}\left(\rho_o u - \frac{\sigma}{U}\right) = 0, \tag{1.141}$$

$$\frac{d}{d\xi}\left(\rho_o E + \frac{1}{2}\rho_o u^2 - \frac{\sigma u}{U}\right) = 0. \tag{1.142}$$

Solutions of the three governing differential equations provide

$$\varepsilon - \frac{u}{U} = \text{const.}, \tag{1.143}$$

$$\rho_o u - \frac{\sigma}{U} = \text{const.}, \tag{1.144}$$

$$\rho_o E - \frac{1}{2}\sigma\varepsilon = \text{const.}. \tag{1.145}$$

Equations (1.143)–(1.145) are recognized as the same Hugoniot conservation relations that develop for a discontinuous shock wave and provided previously in equations (1.39), (1.41) and (1.44). The generalization for steady structured waves is that the above relations apply, not only from the bottom to the top of the wave, but are conserved at each and every point through the wave.

1.5.3.1 Transition to steady wave

Physically reasonable systems of equations that admit steady wave solutions are also stable. That is, if the steady wave structure is perturbed, or if a wave is introduced that is different from the steady wave structure, then the wave will evolve over time and propagation distance to the steady wave structure required by the solution. This feature is of particular interest to the experimentalist investigating structured waves. Commonly the pressure wave is introduced as a discontinuous step in pressure through the impact of a flat plate onto the test sample, as illustrated in figure 1.45. The initial Heaviside wave then propagates and evolves until recorded at some propagation distance L through the sample. The recorded structured wave is a steady structured wave only if the propagation distance L is sufficient for the wave to achieve steady wave conditions.

The evolution of an initial discontinuous pressure jump to a steady structured wave, or Taylor shock (Taylor 1910), in solids has been studied by Bland (1965). Consider the pressure versus strain response with viscous dissipation,

$$\sigma(\varepsilon) = E\varepsilon + h_1\varepsilon^2 + \eta\dot{\varepsilon}. \tag{1.146}$$

Equation (1.146) is a special case of the constitutive relation introduced in equation (1.28). The geometry shown on the right in figure 1.45 yields, through a balance of the viscous and nonlinear pressures,

$$\eta\dot{\varepsilon} = \frac{1}{4}h_1\varepsilon_h^2, \tag{1.147}$$

where ε_h is the strain at the top of the wave (intersection of the Rayleigh line and the quadratic pressure versus strain relation) and the strain rate is determined at half the wave amplitude.

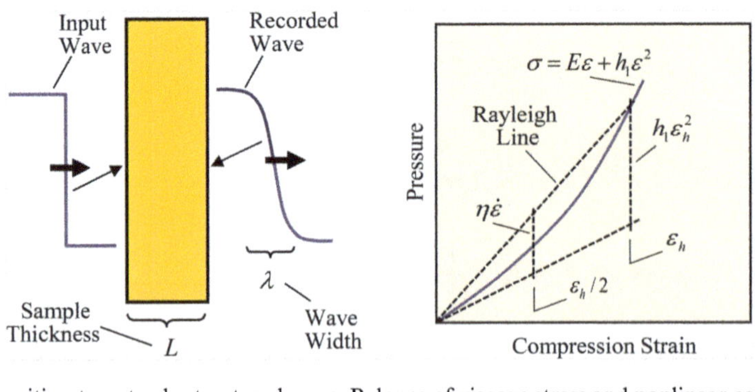

Figure 1.45. Transition to a steady structured wave. Balance of viscous stress and nonlinear compression in the steady structured wave.

The steady wave width λ is approximately related to the strain rate through $\dot{\varepsilon} = c_o \varepsilon_h / \lambda$ where $c_o = \sqrt{E/\rho_o}$ is the characteristic speed at $\varepsilon = 0$. Combining with equation (1.147) provides a measure of the steady wave thickness,

$$\lambda = 4 \frac{\eta c_o}{h_1 \varepsilon_h}. \tag{1.148}$$

The more detailed analysis of Bland (1965) arrived at a constant closer to 6, rather than 4, where λ is considered to span the 5%–95% range of the pressure change through the wave. Similar analysis can be performed if a viscosity exhibiting square-root dependence on strain rate is assumed—a dependence that has some support from experimental measurements of structured waves in solids.

The characteristic wave speed at the top of the wave is provided by

$$c_h = \sqrt{\frac{E}{\rho_o}} \left(1 + \frac{2h_1}{E} \varepsilon_h\right)^{1/2} \simeq c_o \left(1 + \frac{h_1}{E} \varepsilon_h\right). \tag{1.149}$$

Combining this expression with equation (1.48), the steady wave width takes the particularly transparent form

$$\lambda \simeq \frac{4}{\rho_o} \frac{\eta}{c_h - c_o}. \tag{1.150}$$

The wave width is directly proportional to the viscous dissipation and inversely proportional to the difference in the characteristic impedance at the top and the bottom of the wave.

Bland further shows that the propagation distance δ from an initial Heaviside input wave to that of a steady structured wave is governed by the same length scale and is given by

$$\delta = \frac{16}{3} \frac{E}{h_1 \varepsilon_h} \lambda = \frac{16}{3} \frac{c_o}{c_h - c_o} \lambda. \tag{1.151}$$

If the nonlinearity is captured through a linear shock-velocity versus particle-velocity relation, $U = C_o + Su$, or a Murnaghan equation, $K = K_o + K'\sigma$, then

$$\frac{c_h - c_o}{c_o} \simeq 2S\varepsilon_h \simeq \frac{1}{2} K' \varepsilon_h. \tag{1.152}$$

The experimental assessment of steady wave conditions is best determined through wave measurements at multiple propagation distances. The Bland analysis can be used to reasonably assess steady wave conditions from a single measurement. At a propagation distance L a dimensionless ratio, called the Bland number, $B = L/\delta$, can be formed (Swegle and Grady 1985). The propagation distance δ from equation (1.151) is calculated using the experimentally measured wave width λ. If the calculated Bland number is of order unity or greater, then steady wave conditions, according to the Bland criterion, can be presumed to sensibly hold. If B is smaller than one, steady wave conditions are questionable.

1.5.3.2 Wave velocity and the steady structured shock

All shock waves in condensed matter have a finite width. Shock waves that carry material from some uniform ambient state to a correspondingly uniform elevated state are steady structured shock waves. Such shock waves of steady finite width have been referred to as 'Taylor shocks', acknowledging Taylor's seminal study of the physical processes that structure shock waves in compressible fluid mechanics (Taylor 1910).

The Hugoniot conservation relations from equations (1.143)–(1.145) apply continuously through the steady structured shock wave. (All shock waves are structured on some length scale although experimental resolution of the shock width may not always be possible.) The shock velocity U in the conservation equations is the constant speed at each and every point on the structured wave. Kerley (2006) has noted that a perhaps more appropriate measure of the shock speed is

$$U_f = U - u. \tag{1.153}$$

Within a structured steady wave it is informative to examine this alternative measure of the wave speed. Using the relation $u = \varepsilon U$ from equation (1.143) this equation becomes

$$U_f = \frac{\rho_o}{\rho} U. \tag{1.154}$$

In this form it is apparent that U is the wave speed referenced to the coordinates in front of the wave (Lagrangian coordinates), whereas U_f is the wave speed referenced to the current state of compression in the wave (Eulerian coordinates). U and U_f are, respectively, the Lagrangian and the Eulerian measure of the speed of the wave at any point on the structured wave. U is constant throughout the wave, while U_f varies continuously from $U_f = U$ at the foot of the wave, to some value $U_f < U$ at the top.

A possibly more basic presentation of the wave speed of a structured shock is revealed by rewriting equation (1.154) as

$$\rho U_f = \rho_o U. \tag{1.155}$$

The product of density and wave speed is the wave impedance and is unaltered when expressed in either Lagrangian or Eulerian coordinates. If density is expressed in units of linear extent in the direction of wave propagation (mass per unit length—the lateral extent of the wave is of no consequence), then wave impedance has dimensions of mass per unit time. Shock impedance is in fact a velocity determining the speed at which the propagating wave subsumes mass of the body. It is independent of Lagrangian or Eulerian representation of the material. In the transition to a steady structured wave the material strives toward a constant wave impedance throughout the wave. The constant wave impedance achieved is the shock impedance.

1.5.4 Linear waves in solids

There are many applications in the analysis and evaluation of large-amplitude wave propagation where the assumption of linear wave propagation is an adequate approximation. The incidence of a wave onto a free surface or discontinuous impedance interface leading to material spall, the transit and arrival of an elastic precursor wave, and the superposition of interacting waves are all examples, among many, where the assumption of linear wave propagation is often good enough. Linearity in this context assumes that the velocity c in the governing wave equations is constant independent of the wave amplitude and frequency. It is the longitudinal elastic wave speed if the wave amplitude is below the elastic limit of the material. It could be the bulk velocity if material strength is sufficiently negligible and is ignored. Equations (1.68) and (1.69) then combine to provide the one-dimensional second-order wave equation for the particle velocity,

$$\frac{\partial^2 u}{\partial t^2} - c^2 \frac{\partial^2 u}{\partial h^2} = 0. \qquad (1.156)$$

A similar wave equation is obtained for the stress σ. The general solution of the wave equation is that due to d'Alembert (e.g. Courant and Hilbert 1962),

$$u(h,t) = f(h - ct) + g(h + ct), \qquad (1.157)$$

and for the stress

$$\frac{1}{\rho c}\sigma(h,t) = f(h - ct) - g(h + ct). \qquad (1.158)$$

An application of practical interest is that of a wave of arbitrary shape normal incident on a free surface or a discontinuous interface. The application could be that of a shock wave experiment with the wave monitored at the free surface or interface. Two elementary conditions are illustrative; namely, that of a free surface and that of a rigid wall. In that of a rigid wall a boundary condition at $h = 0$ of zero particle velocity requires, through equation (1.157),

$$g(ct) = -f(-ct), \qquad (1.159)$$

and leads to a solution for the particle velocity,

$$u(h,t) = f(h - ct) - f(-(h + ct)). \qquad (1.160)$$

For a free surface, stress is zero at the boundary and equation (1.158) yields

$$g(ct) = f(-ct) \qquad (1.161)$$

and the particle-velocity solution

$$u(h,t) = f(h - ct) + f(-(h + ct)). \qquad (1.162)$$

Corresponding solutions for the stress are provided through equation (1.158).

Analysis is modestly more complicated if the wave is normal incident on an interface separating materials of differing wave impedance. Consider a wave in a medium with impedance $z_1 = \rho_1 c_1$ coupled into a medium with wave impedance $z_2 = \rho_2 c_2$. The interaction results in a transmitted wave into medium two and a reflected wave back into medium one. Continuity of particle velocity and stress at the interface boundary at $h = 0$ requires

$$f(-c_1t) + g(c_1t) = k(-c_2t), \tag{1.163}$$

$$z_1f(-c_1t) - z_1g(c_1t) = z_2k(-c_2t), \tag{1.164}$$

where $k(h - c_2t)$ is the simple wave solution for the transmitted particle-velocity wave into medium two. These two boundary conditions provide for solution of the reflected wave and transmitted wave functional forms. Namely,

$$g(\xi) = -Rf(-\xi), \tag{1.165}$$

$$k(\xi) = (1 - R)f\left(\frac{c_1}{c_2}\xi\right). \tag{1.166}$$

In these relations $R = (z_2 - z_1)/(z_2 + z_1)$ is the elastic reflection coefficient and $(1 - R) = T$ is the transmission coefficient. The reflection coefficient $-1 < R < 1$ is defined such that no phase shift occurs for $R > 0$ (an incident compressive wave reflects as a compressive wave).

A practical example is of interest. Consider a plate of silicon carbide ceramic ($\rho_1 = 3.20\ \text{g cm}^{-3}$, $c_1 = 12.2\ \text{km s}^{-1}$) backed by lithium fluoride ($\rho_1 = 2.64\ \text{g cm}^{-3}$, $c_1 = 5.15\ \text{km s}^{-1}$). A plane elastic wave of exponential form for the particle velocity,

$$u(h,t) = f(h - c_1t) = u_oe^{(h-c_1t)/\tau}, \tag{1.167}$$

is introduced in the silicon carbide and is normal incident on the silicon carbide and lithium fluoride interface. (Such a wave can be input for example by impacting the silicon carbide with a thin plate of even higher impedance material such as tungsten metal.) The coordinate $h = 0$ at the interface and is positive into the lithium fluoride. Time $t = 0$ corresponds to the first arrival of the wave (1.167) at the interface. Applying equations (1.165) and (1.166), solutions for the particle velocity and stress in equations (1.157) and (1.158) are obtained. Namely, for the particle velocity in the interaction reflection region and in the transmitted wave,

$$u(h,t) = \begin{cases} u_o\left(e^{(h-c_1t)/\tau} - Re^{-(h+c_1t)/\tau}\right) & -c_1t < h < 0 \\ u_o(1 - R)e^{(c_1/c_2)(h-c_2t)/\tau} & 0 < h < c_2t \end{cases}, \tag{1.168}$$

and for the stress,

$$\sigma(h,t) = \begin{cases} z_1u_o\left(e^{(h-c_1t)/\tau} + Re^{-(h+c_1t)/\tau}\right) & -c_1t < h < 0 \\ z_2u_o(1 - R)e^{(c_1/c_2)(h-c_2t)/\tau} & 0 < h < c_2t \end{cases}. \tag{1.169}$$

The input wave parameters of $u_o = 100\ \text{m s}^{-1}$ and $\tau = 2\ \mu\text{s}$ yield the wave structure plotted in figure 1.46. Waveforms show the spatial-stress and particle-velocity profiles at three times subsequent to interaction at the interface. A negative reflection coefficient inverts the reflected waveforms and carries the silicon carbide into

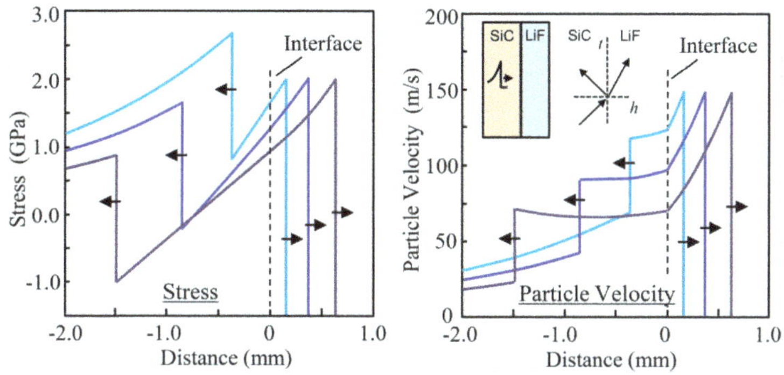

Figure 1.46. The stress (right) and particle-velocity field (right) in a silicon carbide sample backed by lithium fluoride and an exponential waveform incident on the bonding interface.

successively larger states of tensile stress. The spall strength of reasonable quality silicon carbide is approximately 1 GPa. Thus a spall plane would occur at a depth of about 1.5 mm into the sample. Formation of a spall plane would restructure the wave at this location. The consequence would subsequently transmit into the lithium fluoride with wave structure informative of the time-history of the spall failure process. Such wave interactions are central to spall wave effects in the next section.

1.6 Spall waves and the spallation of solids

Spall is the process of internal failure or rupture of condensed matter through a mechanism of cavitation due to stresses in excess of the tensile strength of the material. Usually, a dynamic spall failure is a consequence of transient states of tensile stress within the body that are brought about by the interaction of stress waves. Free surfaces are assumed to be well removed from the material point of interest and play no role in the spall event.

Spall waves emanate away from regions of spall as the process of tensile spall failure proceeds. Spall waves are key to the experimentalist. In the structure of the spall wave is carried the experimental signature of the tensile spall event, including dynamic tensile strength, energy dissipation and the time-history of the spall failure process.

A common experimental method for creating a region of spall is through the flat impact of plates of material. Such impacts lead to planar spall in which an interior planar region of the material is carried into tension, and spall failure occurs through a process of crack formation or hole cavitation. The method is illustrated in figure 1.47 where insipient spall is observed in post-test sectioned metal samples of iron and aluminum subjected to planar spall conditions (Shockey *et al* 1979, Curran *et al* 1980, 1982, 1987, Antoun *et al* 2002). Much of the experimental investigation of spall has relied on this technique. This and further aspects of planar spall are addressed in the present section.

More general wave interactions can lead to more complex domains of tensile stress and spall failure. In a Taylor impact experiment, for example, where a short cylinder of material is caused to undergo normal impact on the flat surface of a large

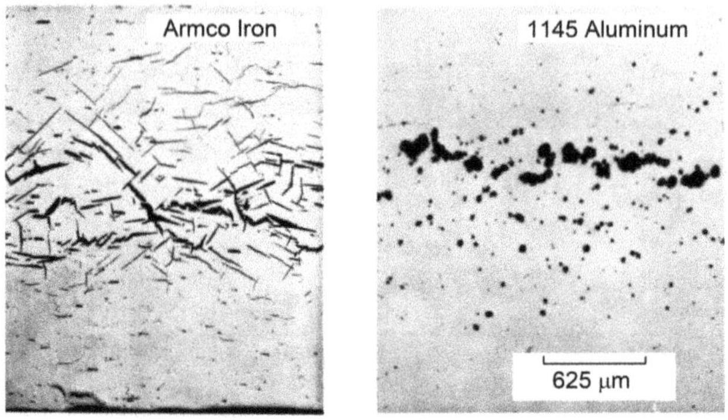

Figure 1.47. Insipient spall damage in two metals as a consequence of planar spall experiments (Shockey *et al* 1979). If subjected to marginally higher impact velocity, the upper and lower regions would separate, creating planar spall surfaces. Reprinted from Curran *et al* (1987), with permission from Elsevier.

block of material, a roughly spherical region within the cylinder is carried into dynamic tension and can undergo spall. Such a test, where two cylinders of soft aluminum experience on-axis symmetric impact, is shown in figure 1.48 (Kipp and Davison 1982). The sectioned sample reveals a region of insipient spall cavitation near the sample impact interface.

As an interior region of a condensed material is carried into tension preceding spall failure, deformation can be either elastic or plastic. The mode of deformation will depend on properties of the material and the degree of anisotropy of the strain field imposed by the interacting waves carrying the body into tension. The relative values of the spall and the yield strengths of the material, and the anisotropy of deformation strain leading to spall failure, determine the occurrence and extent of elastic or plastic deformation prior to spall. Further, inelastic deformation can occur in compression of the body before tension states are achieved that lead to spall. Inelastic deformation is of concern because of changes in metallurgical properties that can affect the spall characteristics of the material. A number of investigations have demonstrated the importance of both pre-existing and induced material microstructure in the spall process (e.g. Meyers and Aimone 1983, Zurek and Meyers 1995). In brittle materials, particularly ceramics, spall strengths have been

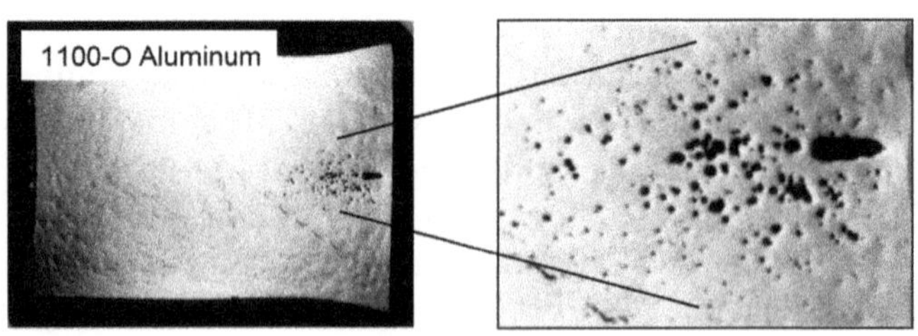

Figure 1.48. Symmetric (Taylor) impact of aluminum cylinders (12.5 mm diameter). A section through the center illustrates spall damage near the impact interface resulting from converging release waves from the radial free surface (Kipp and Davison 1982). An expanded view of the spall region is shown on the right. Reproduced with permission from AIP Conference Proceedings 78, 442 (1982), Copyright 1982, AIP Publishing LLC.

shown to depend on initial compression wave amplitude and damage sustained in the compression process (e.g. Cagnoux and Longy 1988, Yeshuran *et al* 1988).

The evolution of spall in a body subjected to transient tensile stresses is complex. A state of homogeneous tensile stress is intrinsically unstable and small perturbations in the material microstructure (microcracks, inclusions, etc) will lead to the local opening of voids and the growth of spall failure.

Evidence suggests that there is a threshold tensile stress at which void nucleation occurs and spall fracture initiates. Materials subjected to transient tensions can support tensile stresses significantly in excess of this threshold level, however. Such increased short-lived tensile stress levels are a consequence of the kinetics and inertia associated with the microscopic activation and growth of voids during the spall failure event. A substantial body of experimental and theoretical literature on the spall phenomena exists and many aspects of the effect are reasonably well understood. Representative reviews on spall include Curran *et al* (1977), Davison and Graham (1979), Novikov (1981), Curran (1982), Meyers and Aimone (1983), Grady and Kipp (1993) and Antoun *et al* (2002), and access much of the literature on the topic.

Dynamic spall failure is a process of cooperative activation, growth and coalescence of fracture damage that depends on both the pre-existing and evolving microstructure of the material. There is ample evidence, however, that frequently the continuum spall properties are less sensitive to the microstructural details of the material under study, and depend rather on a balance of the energy driving the transient spall failure process and some continuum measure of fracture resistance, such as fracture toughness, surface tension, flow stress or viscosity (Grady 1988). These continuum concepts of spall failure are expanded on in some detail in the upcoming chapters 2, 3 and 4. These ideas are touched upon in the present section, although emphasis is on wave propagation and the dynamic strength of materials in the spall event.

1.6.1 The theoretical spall strength

The spall strength of solids is motivated by first considering the theoretical tensile strength the condensed material can support in the spall event. The theoretical strength is expected to place an upper bound on measured spall strengths, or for spall strengths that are calculated from more realistic theories. In some materials, under the very rapid tensile loading rates achieved in impact applications, measured spall strengths are achieved that are not too far removed from theoretical spall strengths. Furthermore, calculations of theoretical strengths are based on energy concepts similar to those that are exploited within the fracture and fragmentation models described later.

Griffith suggested that the theoretical strength of a solid should be an appreciable fraction of the elastic modulus of the material. Griffith's arguments would advocate a theoretical tensile pressure of approximately $P_{th} = K_o/10$ achieved at an expansion volume strain of order 10%. Proportionality between the theoretical spall strength and the bulk modulus is obtained whenever a two-parameter model is chosen to represent the intermolecular potential. Other two-parameter representations of the intermolecular potential, such as the Lennard-Jones 6-12 potential, will yield a similar proportionality although the numerical coefficients will differ. The Lennard-Jones 6-12 potential yields a somewhat lower theoretical strength of $P_{th} = K_o/27$.

Although two-parameter models are rather restrictive, three-parameter models of the intermolecular potential have been developed that provide reasonable descriptions of the thermo-elastic behavior of solids. Examples include the Mie potential, the Morse potential and, more recently, a form proposed by Rose *et al* (1984), among numerous others.

With a three-parameter model of the intermolecular potential, the theoretical spall strength is not simply a constant times the bulk modulus. (Not uncommonly called a two-parameter model with the density at energy minimum implied. So be careful.) To derive an explicit expression for the theoretical spall strength, an analytic expression of the cold compressibility of a solid based on a modified Morse potential is used (Morse 1929),

$$U(v) = U_{coh}\left[\exp\left(\frac{-2(v_o - v)}{a}\right) - 2\exp\left(\frac{-(v_o - v)}{a}\right)\right]. \tag{1.170}$$

In this expression the specific cohesive energy is U_{coh} while v_o is the ambient or zero pressure specific volume. The potential curve and corresponding pressure are illustrated in figure 1.49. The scale parameter a is constrained by the relation for the ambient bulk modulus,

$$v\frac{d^2U}{dv^2}\bigg|_{v_o} = K_o, \tag{1.171}$$

yielding $a = \sqrt{2v_oU_{coh}/K_o}$.

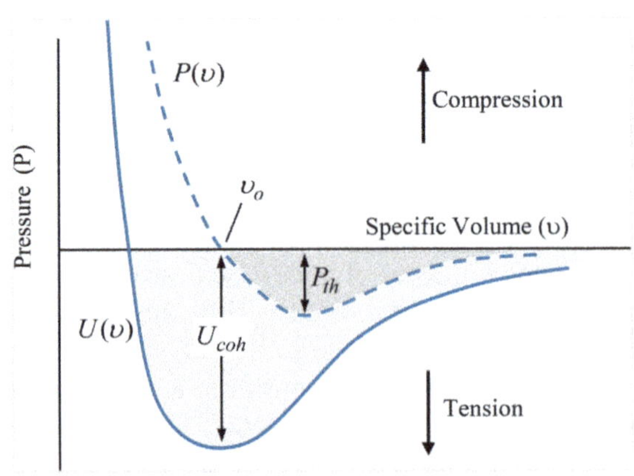

Figure 1.49. Specific volume dependence of the cold compression and tension properties of condensed matter. The cohesive energy and maximum tension (theoretical spall strength) are properties of the material.

The cold pressure is provided by

$$P(v) = -\frac{dU}{dv} = \frac{2U_{coh}}{a}\left[\exp\left(\frac{-2(v_o - v)}{a}\right) - \exp\left(\frac{-(v_o - v)}{a}\right)\right]. \tag{1.172}$$

The theoretical spall strength P_{th} results from the minimum of the cold pressure, yielding

$$P_{th} = \sqrt{\frac{U_{coh}K_o}{8v_o}}. \tag{1.173}$$

For the present Morse potential, and other three-parameter potentials, the theoretical strength depends on the geometric average of the bulk modulus and the cohesive energy density. Calculated theoretical spall strengths for selected metals based on the Morse potential are given in table 1.1 (Grady and Kipp 1993). Measured spall strengths of metals are commonly about one order of magnitude lower.

It is instructive to note that the theoretical spall strength follows from a simple balance of energies. As the material is carried into tension, a nominal elastic strain energy $P^2/2K_o$ accrues. When the elastic strain energy achieves a magnitude comparable to the cohesive energy of the material, a theoretical spall failure is predicted. Similar energy balance ideas are applied to more realistic material strength models in chapters 2, 3 and 4. The theoretical spall strength from intermolecular potential models is revisited in chapter 10.

1.6.2 The planar spall experiment

Dynamic tensile loading to spall failure can be achieved in a uniaxial strain geometry through the interaction of two opposing planar relief waves. The production and timing of such relief waves is achieved with a high-velocity impact plate launcher and proper selection of sample dimensions and ancillary target material. A method for performing a planar spall experiment is illustrated in figure 1.50.

Table 1.1. Theoretical spall strength of selected metals.

Metal	K_o(GPa)	ρ(kg m^{-3})	U_{coh}(MJ kg^{-1})	P_{th}(GPa)
Aluminum	72.2	2710	11.9	17.1
Beryllium	100	1820	35.7	28.5
Copper	137	8930	5.32	28.5
Titanium	105	4510	9.78	24.1
Iron	168	7870	7.41	35.0
Tantalum	200	16 660	4.31	42.4
Tungsten	323	19 250	4.55	59.5
Uranium	98.7	19 050	2.19	22.7
Mercury	38.7	13 530	0.33	4.60
Tin	111	7300	2.53	16.0

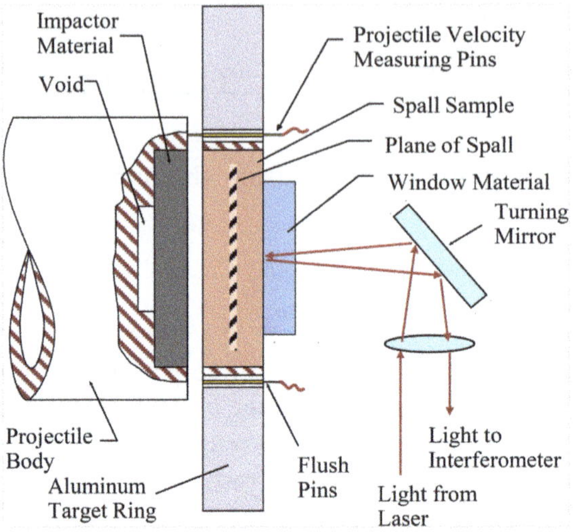

Figure 1.50. Experiment configuration for a planar impact spall experiment with velocity interferometer diagnostics and low-impedance transparent window material (fused silica, PMMA, lithium fluoride).

A sample of the test material is mounted in a stationary target holder. The test sample is subjected to impact by a plate carried within the head of a projectile that has been accelerated to a prescribed velocity by the launcher. In the experiment shown, laser-velocity interferometry is used to measure the time-resolved velocity at the back of the sample; either a free surface or an interface between the sample and a transparent window material. Fused silica, lithium fluoride or the plastic poly-methyl-methacrylate (PMMA) are common window materials that have been calibrated to the laser diagnostic method (Barker and Hollenbach 1972). It is necessary that the window material be lower wave impedance than the test material. Similar time-resolved spall measurements can be performed with piezo-resistive stress or electromagnetic velocity transducers appropriately placed within the experimental assembly.

If the tensile stress achieved at the plane of interaction of the colliding release waves is insufficient to achieve spall failure, then the velocity history returns to zero as illustrated by one consequence in the lower plot of figure 1.51. Conversely, if the sample material exhibits negligible spall strength then the velocity level identified with instantaneous spall is achieved. More commonly, the material exhibits a finite spall strength and a time history from kinetics of the spall failure process. The spall wave or pull-back signal illustrated in the figure is then recorded.

The compression wave resulting from reflection of the spall wave at the sample–window interface at time t_2 subsequently interacts and reflects from the spall plane, causing the abrupt change in particle velocity at time t_3. The stress wave trapped in the fractured sample is commonly observed to reverberate several times through the duration of the wave measurement.

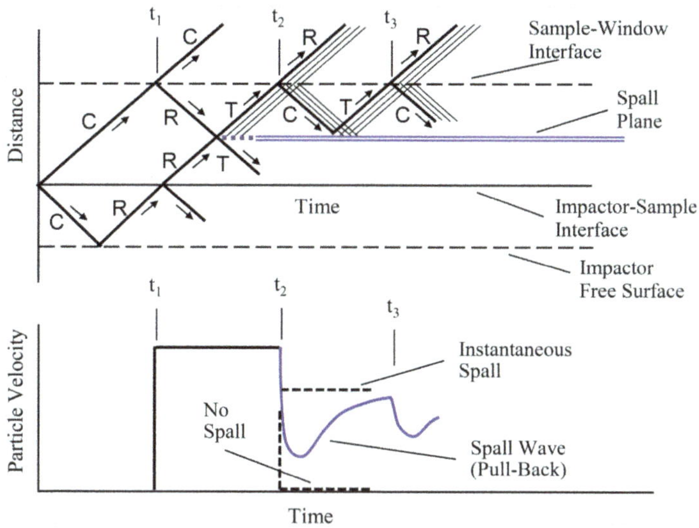

Figure 1.51. Wave interactions in a planar spall experiment. The upper plot illustrates a distance–time perspective of wave interactions where C, R and T identify compression, release and tension waves, respectively. The corresponding particle-velocity history is shown in the lower plot, where corresponding times illustrate the spall wave signal between times t_2 and t_3.

1.6.3 Experimental spall strength

One method to estimate the spall strength of a material is to conduct experiments conducive to spall failure at increasing impact intensities. Test samples are recovered, sectioned and examined for evidence of spall. For the sample that first shows evidence of spall through incipient fracture or cavitation, the tensile stress is calculated from the impact conditions and identified with the spall stress of the material. The method provides a satisfactory measure of the spall strength. It is, however, laborious and only appropriate for the more ductile materials such as metals.

More commonly, a measurement of the spall wave provides an assessment of the spall strength. Impact of a thin plate onto a test sample, which is backed by a low-impedance window material, leads to an interaction of waves that carries an interior planar region into tension. Spall failure will ensue if the tension and tension duration exceed the transient tensile strength characteristics of the material. A velocity or stress history recorded at the sample–window interface shown in figure 1.50 will have the qualitative features illustrated in figure 1.52. From such spall waves, through appropriate analysis, spall strength and other features of the spall failure process can be extracted. Some of these analysis methods are described here.

1.6.3.1 Spall-stress analysis

The stress versus particle-velocity curves for the sample material and window material through which the spall strength analysis is performed are shown in the left panel of figure 1.53. These relations are generally nonlinear and are solved

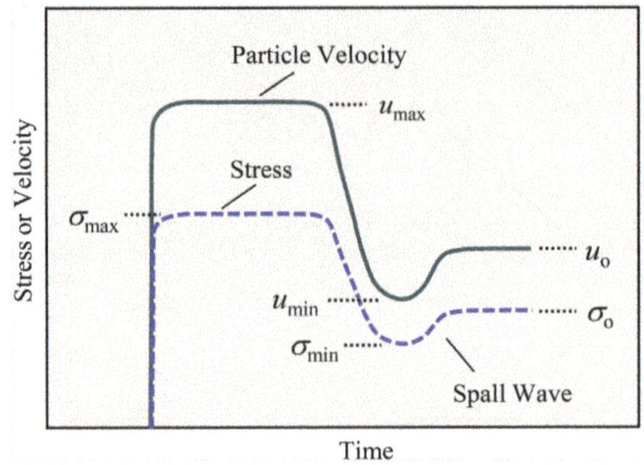

Figure 1.52. Representative time-resolved stress or particle-velocity histories illustrating features necessary to the spall strength analysis.

graphically or using a computer. If the response of the participating materials is nominally linear, and described by a characteristic impedance (z_w for the window and z_s for the test sample), then useful analytic relations for the spall strength can be derived. The appropriate shock impedance is the product of the initial density of the material and a wave speed. If, for example, the material is elastic then the longitudinal elastic wave speed is appropriate. If the material response is closer to hydrodynamic (a liquid or a metal with a low HEL), then the bulk wave speed is more suitable. (In contrast to the hydrostat, a dynamic compression curve where deviator stress is negligible or where deviator stress is chosen to be ignored will occasionally be identified as a hydrodyne through the chapters.)

From the example in the left-hand plot in figure 1.53 it follows that the spall stress σ_T, considered here to be a positive quantity, is provided by,

$$2\sigma_T = (z_s - z_w)u_{\max} - (z_s + z_w)u_{\min}. \tag{1.174}$$

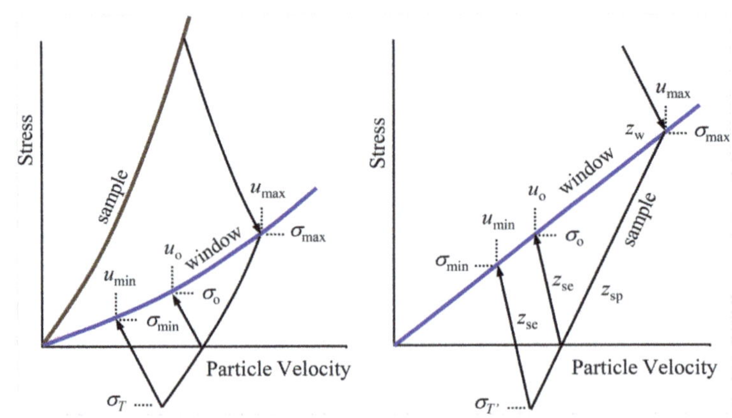

Figure 1.53. Stress versus particle-velocity plot of a typical planar spall experiment with a sample backed by a low-impedance window is illustrated on the left. The graph on the right is specific to spall in an elastic–plastic material and the Romanchenko–Stepanov analysis.

Alternatively, the following relation for the spall stress is derived,

$$2\sigma_T = (z_s + z_w)(u_o - u_{\min}). \tag{1.175}$$

The respective particle-velocity amplitudes are identified in figure 1.52. If a stress, rather than a particle-velocity, spall wave is measured, then the same equations apply with $u_{\min}$, $u_{\max}$ and u_o replaced with $\sigma_{\min}/z_w$, $\sigma_{\max}/z_w$ and σ_o/z_w.

Equation (1.175) is more appropriate when the tensile excursion, $\sigma_o - \sigma_{\min}$ is small relative to the impedance mismatch step $\sigma_{\max} - \sigma_o$. The downside is that σ_o, or u_o, is sometimes not achieved in the measured interface history before the signal reverberation from the spall plane arrives at the recording interface.

1.6.3.2 Romanchenko–Stepanov attenuation correction

Romanchenko and Stepanov (1980) recognized that spall in an elastic–plastic solid can involve plastic decompression to the tensile spall failure stress followed by elastic recovery as the tensile stress returns to zero at the spall plane. The spall wave that propagates from the spall plane to the recording interface travels with a plastic wave velocity at the wave front and an elastic wave velocity at the wave release. Illustrated in figure 1.54, the tensile stress amplitude will be reduced at the recording interface due to wave attenuation over the propagation distance. For ideal elastic–plastic behavior of the material in uniaxial-strain geometry, the plastic impedance is $z_{sp} = \rho c_o$ and elastic impedance $z_{se} = \rho c_l$, where c_o and c_l are the bulk and longitudinal wave speeds, respectively.

Relations for the spall stress comparable to equations (1.174) and (1.175) are obtained through application of the geometry to the pressure versus particle-velocity plot on the right-hand side of figure 1.53. Namely,

$$\sigma_{T'} = \frac{z_{se}(z_{sp} - z_w)u_{\max} - z_{sp}(z_{se} + z_w)u_{\min}}{z_{sp} + z_{se}}, \tag{1.176}$$

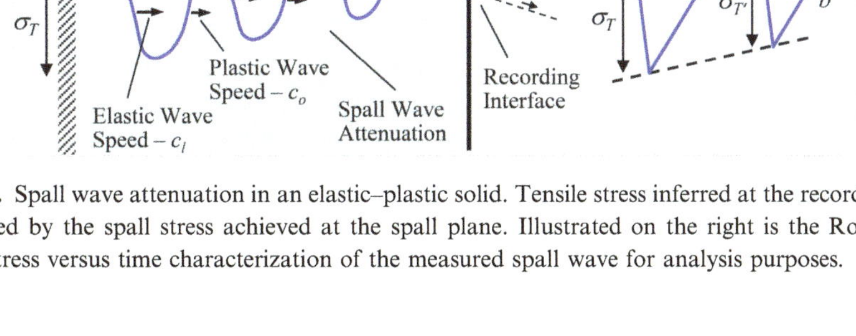

Figure 1.54. Spall wave attenuation in an elastic–plastic solid. Tensile stress inferred at the recording interface is diminished by the spall stress achieved at the spall plane. Illustrated on the right is the Romanchenko–Stepanov stress versus time characterization of the measured spall wave for analysis purposes.

or

$$\sigma_{T'} = \frac{z_{sp}}{z_{sp} + z_{se}}(z_{se} + z_w)(u_o - u_{\min}). \tag{1.177}$$

The stress $\sigma_{T'}$ provides a measure of the spall stress as inferred from the spall wave measured at the recording interface. This stress does not account for attenuation of the spall wave over the propagation distance from the spall plane to the recording interface.

The analysis of Romanchenko and Stepanov approximates the spall wave according to the illustration on the right in figure 1.54 to account for attenuation. The constants a and b provide a measure of the slope or time derivative $d\sigma/dt$ of the wave front and the wave back of the measured spall wave. Assuming that the shape of the wave imparted to the material at the spall plane maintains the shape of the recorded spall wave, straightforward geometric analysis provides for the spall stress at the spall plane

$$\sigma_T = \sigma_{T'} + \Delta\sigma, \tag{1.178}$$

where

$$\Delta\sigma = \frac{ab}{a+b}\left(\frac{1}{c_p} - \frac{1}{c_e}\right)x. \tag{1.179}$$

The propagation distance x from the spall plane to the recording interface is the thickness of the spall plate. If $a \approx b = d\sigma/dt$, then

$$\Delta\sigma \approx \frac{1}{2}\frac{d\sigma}{dt}\left(\frac{1}{c_p} - \frac{1}{c_e}\right)x \approx \frac{1}{2}\frac{\sigma_{T'}}{\tau}\left(\frac{1}{c_p} - \frac{1}{c_e}\right)x, \tag{1.180}$$

where τ is a measure of the spall wave temporal width near the midpoint.

1.6.4 Spall in metals

Metals have received by far the largest share of experimental study of the spall phenomena. The images in figure 1.47 from Shockey *et al* (1979) show micrographs of insipient spall in iron and aluminum, and are representative of the types of spall damage observed in metals. Spall experiments have been performed to explore the dependence of spall strength on the magnitude, strain rate and duration of the stress pulse imparting a tensile spallation load on the material (Antoun *et al* 2002). Microstructure features such as grain size, impurity content or initial heat treatment have been systematically varied, and the effects on spall strength subjected to experimental investigation (Zurek and Meyers 1995). From these studies, engineering and computational models of the dynamic spall strength of metals have emerged with varying degrees of sophistication (Davison *et al* 1996).

The spall strength of a metal is comparable to, and usually greater than, the HEL of the material. Table 1.2 provides experimental measures of both the HEL and the spall strength of selected metals. The micrographs shown in figure 1.47 are representative of insipient spall failure through brittle spall (Armco iron) and ductile spall (1145 aluminum). For the former, tensile stresses exceeding the strength of the material cause the activation and propagation of cracks that ultimately link up and lead to failure of the material. In the soft aluminum, tensile stresses cause the nucleation of near-spherical voids that undergo plastic cavitation until the point at which void coalescence and spall failure occur. The first five metals in table 1.2 are reported to fail through brittle spall much like the Armco iron in figure 1.47. The second five fail through ductile spall comparable to the 1145 aluminum in figure 1.47 or the 1100-O aluminum in figure 1.48.

1.6.4.1 Fracture mechanisms in spall of metals

The field of fracture mechanics identifies processes of brittle and ductile fracture. Brittle and ductile fracture is not to be equated to brittle and ductile spall, although the two processes are not unrelated. Brittle fracture is a process of cleavage on a plane of fracture of near-atomic dimensions at the crack tip. Fracture resistance is characterized by a surface energy Γ that is commonly within an order of magnitude of the theoretical surface energy of the brittle material. A tensile, or mode I, fracture toughness of the material is provided by the relation $K_c \simeq \sqrt{2\Gamma E}$ where E is the elastic modulus of the material.

Ductile fracture is characterized by a process of hole cavitation, plastic ligamentation and surface separation at the crack tip as the ductile fracture proceeds through the plane of fracture. If plasticity within the zone of failure can be described by a constant flow stress Y, then two characteristic lengths emerge from continuum properties of the ductile material. They are,

$$x_{cod} = \frac{K_c^2}{3EY}, \tag{1.181}$$

Table 1.2. Dynamic strength properties of selected metals.

Metal	K_o(GPa)	ρ(kg m^{-3})	σ_{hel}(GPa)	σ_T(GPa)
Aluminum (6061-T6)	72.2	2710	0.5–0.7	0.8–1.5
Beryllium	100	1820	0.2	0.4–1.3
Uranium	115	18 940	1.3–1.9	2.0–3.5
Titanium	105	4510	–	2.1–3.9
Steel (4340)[b]	168	7870	1.5–3.0	2.5–5.4
Aluminum (1060)[a]	72.2	2710	0.02–0.06	0.5–1.1
Copper[a]	137	8940	0.04–0.05	1.0–2.5
Tantalum	200	16 690	1.5–2.0	4.4–7.4
Tin	111	7370	0.1–0.25	0.6–0.8
Titanium (Ti–6Al–4V)	105	4420	1.6–1.8	4.1–5.0

[a]Annealed. [b]Various heat treatments.

and,

$$x_{pz} = \frac{\pi}{72}\left(\frac{K_c}{Y}\right)^2. \tag{1.182}$$

The first length scale x_{cod} is the crack opening displacement, and is the extent that the fracture surfaces must separate for tensile resistance to reduce to zero. The second length scale x_{pz} is the span of the ductile fracture process zone in the direction of crack propagation from inception to completion. The two length scales are illustrated in figure 1.55. This formalism of ductile fracture was developed by Dugdale (1960) and is succinctly described by Lawn and Wilshaw (1975). Both length scales are properties of the material and a full physical understanding requires knowledge of how they emerge from the underlying ductile metal microstructure. Here, recognition of their presence is sufficient.

The first relation in equation (1.181) reveals that the ductile fracture surface energy provided by $\Gamma \approx K_c^2/2E \sim Yx_{cod}$ is the plastic work required to separate the fracture surfaces. Equation (1.182) can be recognized as a Griffith energy balance when written as $(Y^2/2E)x_{pz}^2 l \sim (K_c^2/2E)x_{pz}l$ where an elastic strain energy within a volume $x_{pz}^2 l$ equates to a fracture surface energy over the area $x_{pz}l$ (l is an element of crack length). Features of the crack tip describing brittle and ductile fracture are illustrated on the right in figure 1.55. In equation (1.181), as the length scale x_{cod} is decreased to intermolecular dimensions, Γ approaches the lattice surface energy and Y approaches the theoretical shear strength of the metal. This concept is realized in practice when metal temperature is lowered, successively freezing out modes of dislocation mobility and achieving a ductile fracture to brittle fracture transition.

Similar energy concepts are applied to achieve an understanding of the spall failure event, and dependence of the spall strength of metals on continuum shear and toughness strength properties of the material (Grady 1988). A region of material of dimension x_{pz} is central to the processes of ductile fracture. In spall, a markedly

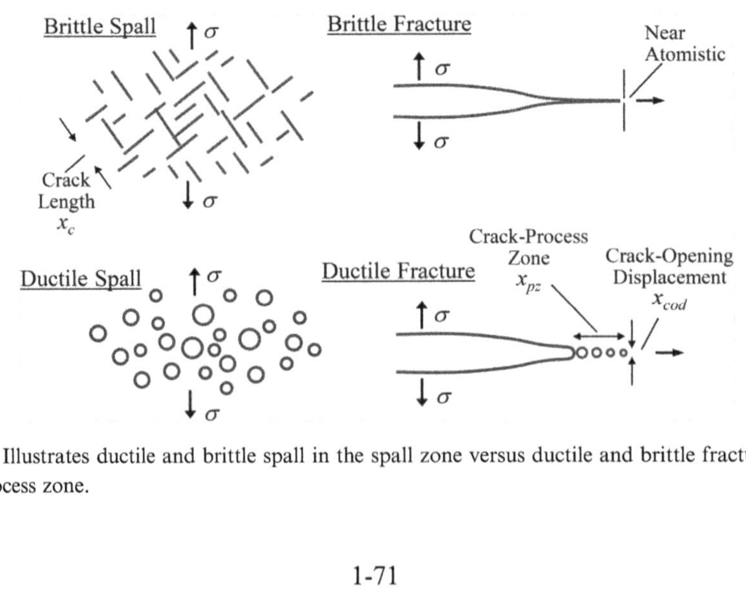

Figure 1.55. Illustrates ductile and brittle spall in the spall zone versus ductile and brittle fracture within the crack-tip process zone.

larger region of the material is carried into tension by the interacting waves responsible for the transient tensile state. For example, a volume of metal subjected to spall-inducing tension by the Taylor impact experiment illustrated in figure 1.48 is a reasonable visualization of the span of spall damage. Alternatively, contemplate a two-dimensional plane of material carried rapidly into tension by the interaction of two colliding planar release waves.

Consider ductile cavitation leading to ductile spall failure such as illustrated on the left in figure 1.55. The work of Tvergaard and Needleman (1984) and others, applying a ductile void growth model developed by Gurson (1977), have shown the work required to achieve failure from void inception to void coalescence and failure is reasonably provided by $W \approx Y\varepsilon_c$. The yield stress Y is an appropriate measure of the shear flow resistance of the ductile solid, while $\varepsilon_c \approx 0.15$ is the calculated measure of the void growth or volumetric strain to onset of failure. Since lattice volumetric strain is near zero in the spall event, ε_c similarly provides a reasonable estimate of the inelastic shear strain to spall failure. An elementary energy balance requires that the tensile loading in the spall process achieves an elastic strain energy sufficient to supply the required plastic work,

$$\frac{\sigma_T^2}{2E} \approx Y\varepsilon_c. \tag{1.183}$$

Equation (1.183) leads to a straightforward estimate of the ductile spall strength of a metal,

$$\sigma_T \approx \sqrt{2EY\varepsilon_c}. \tag{1.184}$$

The elastic modulus depends on the nature of the interacting waves leading to spall and, depending on the loading conditions, may be more appropriately described by the bulk modulus or the constrained longitudinal modulus. Equation (1.184) is shown to provide reasonable first-order estimates of the spall strength for many ductile metals (Grady 1988).

Providing an estimate of the spall strength of a brittle solid through similar energy arguments is modestly more complex. A relation comparable to equation (1.183) requires the elastic strain energy to attain a magnitude sufficient to supply the work necessary to grow and link up a system of cracks as is illustrated again on the left in figure 1.55,

$$\frac{\sigma_T^2}{2E} \approx W \approx \Gamma A. \tag{1.185}$$

The fracture surface energy is Γ, while A is the amount of fracture surface per unit volume created to accommodate the brittle spall process. An estimate of A requires the introduction of a time and distance scale into the spall fracture event. As waves interact, material in the spall zone is carried into tension at a characteristic strain rate $\dot{\varepsilon}$. The spall stress is then achieved in a time $t_c \approx \sigma_T / E\dot{\varepsilon}$. Cracks activated in the spall process will propagate a distance of order $x_c \approx ct_c$, where $c = \sqrt{E/\rho}$ is the elastic wave speed, and it is assumed that crack growth in the spall process proceeds at comparable speeds. A sensible estimate of the brittle spall crack surface area density is then $A \approx 3/x_c$. Joining these relations for A, x_c and t_c with equation (1.185) yields an estimate for the spall strength of a brittle solid,

$$\sigma_T \approx (6\rho^2 c^3 \Gamma \dot{\varepsilon})^{1/3}. \tag{1.186}$$

A measure of the fracture surface energy is provided through the material fracture toughness $\Gamma \approx K_c^2/2E$. There is evidence that the spall-stress and strain-rate dependence provided by equation (1.186) are observed in certain brittle materials (Grady 1988).

The strain rate $\dot{\varepsilon}$ provides a measure of the kinematic intensity of the deformation leading to spall failure. It is also useful to consider spall from the perspective of the pulse duration τ of the wave causing spall in the material. Replacing the strain rate in equation (1.186) with $\dot{\varepsilon} \approx \sigma_T/\rho c^2 \tau$ yields an equivalent relation for the spall-stress dependence on the pulse duration,

$$\sigma_T^2 \tau \approx 6\rho c \Gamma. \tag{1.187}$$

The definitions of brittle and ductile fracture are derived from the character of failure at the tip of the crack. The former involves a cleavage process with little or no plastic deformation, while the latter displays pervasive plasticity through local ductile hole growth and linkup.

Brittle and ductile spall, on the other hand, have been defined as dynamic failure through the activation, growth and coalescence of a network of cracks for the former, and comparable tensile failure through the plastic growth and coalescence of a collection of cavities for the latter.

Brittle spall, however, does not preclude spall failure through the activation and interaction of ductile fractures. Brittle spall is dependent on the rate at which the metal is carried into tension, as indicated by the dependence on strain rate provided by the relation for the spall strength in equation (1.186). The characteristic crack propagation distance in brittle spall, $x_c \approx ct_c \approx c\sigma_T/E\dot{\varepsilon}$, decreases as the rate of strain leading to spall increases. If the crack propagation distance is significantly greater than the dimension of the crack-tip process zone from equation (1.182), $x_c \gg x_{pz}$, then brittle spall failure through the activation, growth and coalescence of mature ductile cracks can occur. This is important because ductile fracture is a fracture toughness or K_c controlled process. Consequently, spall is also governed by the fracture toughness of the metal. As the tensile strain rate leading to spall increases, and the characteristic crack propagation distance approaches the fracture process zone dimension, $x_c \sim x_{pz}$, there is not sufficient time for ductile cracks to form and spall failure in the same metal transitions to one of ductile cavitation. Spall stress at higher rates of loading are then controlled by the flow stress Y and estimated from the relation in equation (1.184).

Thus, for a given metal with increasing spall strain rate there is a transition from a brittle spall, or fracture toughness K_c dominated spall, to ductile, or flow stress Y dominated spall. The transition occurs when $x_c \sim x_{pz}$. An estimate of the strain rate when such a transition in spall character occurs is provided by equating equations (1.184) and (1.186) yielding

$$\dot{\varepsilon}_t \approx \sqrt{\frac{2}{9}\frac{(Y\varepsilon_c)^3}{\rho\Gamma^2}}, \tag{1.188}$$

where $\Gamma \approx K_c^2/2E$ is estimated from the fracture toughness of the material. In the analysis described here, both fracture toughness and flow stress are assumed independent of strain rate over the range of strain rates appropriate to the spall phenomenon. Generally, it is known that in metals Y commonly increases with strain rate while K_c decreases. The strain-rate sensitivity of Y and K_c is relatively strong in mild steels, for example, while being somewhat weaker in aluminum alloys.

Whether or not ductile metals will reveal evidence of such a transformation in the character of the spall process over the range of strain rates achieved in applications of the spall failure event is not yet clear. The 6061-T6 aluminum alloy appears to exhibit evidence of a transition from K_c dominated spall to Y dominated spall over the strain-rate range that can be achieved in planar impact spall experiments. The study of Christman *et al* (1971, 1972) has shown microscopic evidence of incipient spall damage through isolated and intersecting cracks at modest strain rates, consistent with brittle K_c dominated spall. Tests at higher strain rates show insipient spall damage through isolated plastic hole growth, in line with ductile Y dominated spall. The results of two tests indicative of this spall transition behavior are shown in figure 1.56. SEM examination of crack-tip deformation from spall tests, as well as from slow-bend and Charpy impact tests, reveal ductile fracture dimpled surfaces with a characteristic dimple scale of 20–40 μm. Little change was observed in the crack-tip surface texture over the range of loading rates studied, suggesting little rate sensitivity of K_c.

Dimensions of the specimen and the velocity of impact provide estimates of the strain rates that correspond to the two tests illustrated in figure 1.56. The applied pulse duration is determined by two transits of the impact plate of thickness x_p. The peak particle velocity achieved from symmetric impact of an aluminum impact plate and aluminum target plate is one-half the projectile velocity u_p. Thus, the metal achieves a peak strain of $\Delta\varepsilon \approx u_p/2c$ in a time duration of $\Delta t \approx x_p/c$. An estimate of

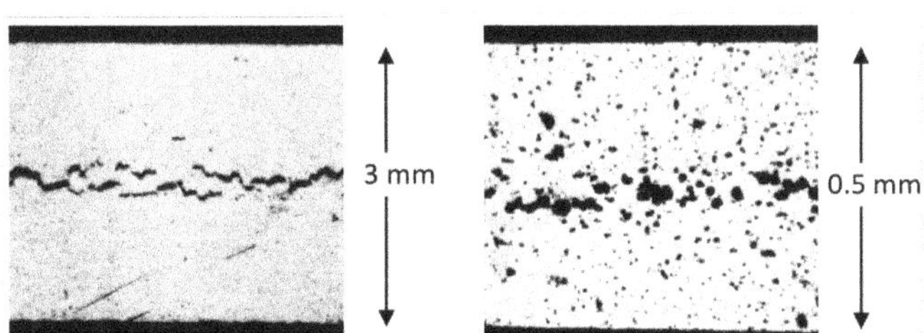

Figure 1.56. Insipient spall damage in 6061-T6 aluminum alloy (Christman *et al* 1971). Left: a 1.55 mm impact plate on a 3.02 mm target plate at 232 m s^{-1}. Right: a 0.25 mm impact plate on a 0.48 mm target plate at 345 m s^{-1}. Calculated strain rates are 7.5×10^4 s^{-1} and 6.9×10^5 s^{-1} for the left-hand experiment and right-hand experiment, respectively, nearly one order of magnitude difference.

the spallation strain rate is provided by $\dot{\varepsilon} \approx \Delta\varepsilon/\Delta t \approx u_p/2x_p$. The dimensions and impact velocities for the experiments are provided in the figure caption and give strain-rate estimates of $7.5 \times 10^4\ \mathrm{s}^{-1}$ and $6.9 \times 10^5\ \mathrm{s}^{-1}$ for the two tests, respectively.

Material properties for 6061-T6 aluminum are $\rho = 2710\ \mathrm{kg\,m^{-3}}$, $E = 7 \times 10^{10}\ \mathrm{N\,m^{-2}}$, $Y = 3 \times 10^8\ \mathrm{N\,m^{-2}}$ and $K_c = 3 \times 10^7\ \mathrm{N\,m^{-3/2}}$. Equation (1.188) yields an estimate of $\dot{\varepsilon}_t \approx 4.4 \times 10^5\ \mathrm{s}^{-1}$ for a transition from fracture-toughness dominated spall to flow-stress dominated spall in 6061-T6 aluminum. This transition strain rate is within the range of the strain rates estimated for the two tests in figure 1.56.

Spall strength data for 6061-T6 aluminum alloy are plotted in figure 1.57. Strain rates are estimated from sample dimensions and impact velocities by the same methods described earlier. Spall strength from measured spall waves are reported by Christman *et al* (1971, 1972) and Ek and Asay (1986). Spall strengths from Kreer (1971) were determined by experimental threshold spall and computational wave propagation analysis. A comparison illustrates that differences in spall strength from the different data analysis methods are not significant.

The plot also shows the spall strengths from equations (1.184) and (1.186), along with the transition strain rate from equation (1.188) at which a transition from brittle fracture-toughness dominated spall to ductile flow-stress dominated spall is predicted. It is quite remarkable that the spall strength relations provided in equations (1.183) and (1.186), based on continuum strength properties of the material and simple energy concepts, are in reasonable accord with the experimental spall measurements. One implication of this result is that the activation of ductile cavitation or brittle fracture is less important to the spall failure process than the availability of adequate energy necessary to carry spall to completion. This is shown not to be true for all solids in later chapters. Some brittle solids (glass is a good example) require spall-stress levels that are well in excess of that predicted by energy methods before internal rupture leading to spall failure is achieved.

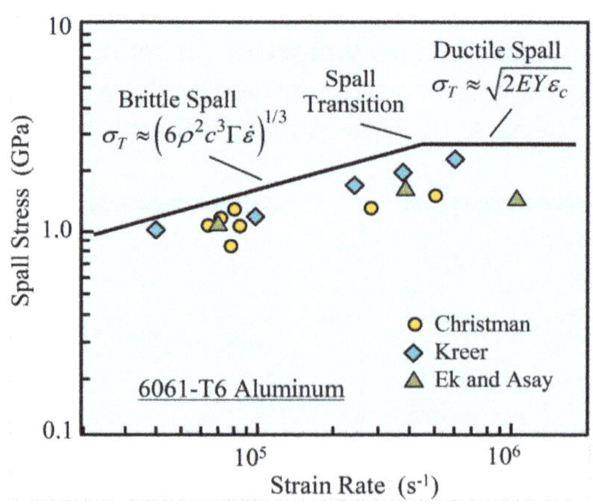

Figure 1.57. Experimental spall data for 6061-T6 aluminum alloy and comparison with spall strength predictions based on energy balance and continuum material strength properties. Spall strength data (Christman *et al* 1971, 1972, Kreer 1971, Ek and Asay 1986).

1.6.4.2 Application to spall in uranium

Impact-induced spall fracture in α-uranium is representative of the spall characteristics of a number of metals and alloys. In the study described, plates of uranium are subjected to planar impact by plates of single-crystal sapphire with free surfaces opposing the impact interface. The shock and subsequent release waves lead to a plane of tension within the uranium. If the duration and intensity of the tensile stress is sufficient, internal spall failure of the uranium follows.

Figure 1.50 describes the experimental configuration of the spall experiments, except that a window material was not used. Velocity histories monitored with velocity interferometry at the uranium free surface for the series of shock wave and spall experiments are shown on the left in figure 1.58. Details of the elastic precursor features from the same experiments were described in section 1.5.2 and are reproduced in figure 1.38. Crystalline sapphire remains elastic through the range of impact stress achieved except, perhaps, for the highest impact data point. The HEL of sapphire is about 13 GPa. A near-rectangular pressure pulse with duration equal to twice the elastic transit of sapphire is provided by the impact onto the uranium sample.

Hugoniot states and the maximum compressive stress achieved in the uranium before wave interactions leading to tension and spall are calculated from the maximum free-surface velocity achieved in the recorded profile. The Hugoniot for α-uranium is plotted on the right in figure 1.58. The Hugoniot data for two uranium alloys, U3/4Ti and U6Nb, are compared with those of α-uranium. The Hugoniot data are compared in turn with the hydrostatic EOS provided from extrapolation of low-pressure ultrasonic measurements of α-uranium. Offset between the Hugoniot and the hydrostatic EOS indicates a yield strength of approximately $Y \approx 1.0$ GPa for α-uranium and U3/4Ti, and modestly lower for U6Nb. The offset is somewhat lower in the higher Hugoniot pressure range; however, uncertainty in the offset is larger. Additional analysis of shock effects in uranium metals is undertaken in chapter 11.

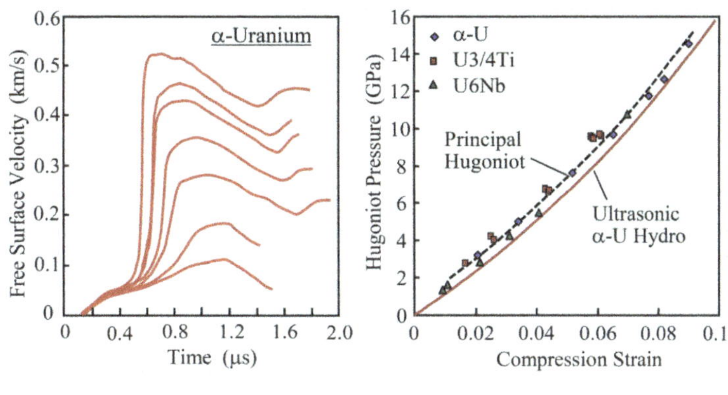

Figure 1.58. Shock-Hugoniot and spall wave measurements on α-uranium (Grady 1986b). Free-surface velocity measurements for crystalline sapphire impact plate (6.35 mm thickness) onto uranium plates (7.6 mm thickness). The impact velocities are 129, 223, 340, 426, 503, 547 and 631 m s^{-1}. Hugoniot states for α-uranium are compared on the right with Hugoniots for uranium alloys U3/4Ti (Dandekar *et al* 1980) and U6Nb (Hixson *et al* 2000), and with hydrostatic extrapolation from ultrasonic data (Abbey and Bonner 1975).

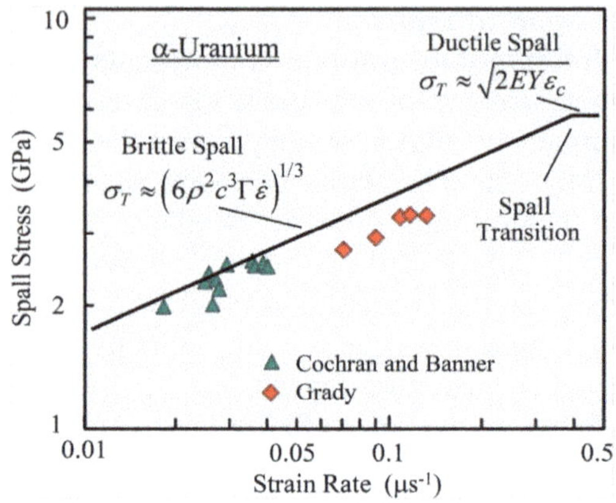

Figure 1.59. Experimental spall data for alpha uranium (Cochran and Banner 1977, Grady 1986b) and comparison with spall strength predictions based on energy balance and continuum material strength properties.

Marked spreading of the release wave following the maximum surface velocity in figure 1.58 reflects the elastic–plastic nature of the α-uranium during decompression from the Hugoniot state. A minimum particle velocity is achieved in the spall wave of five of the seven records shown in the figure. The maximum and minimum velocities are used to calculate the spall strength for uranium through the elastic–plastic spall relation provided in equation (1.176). The Romanchenko–Stepanov attenuation correction in equation (1.179) is also applied and accounts for about 20% of the reported spall strength.

The spall data for α-uranium are plotted in figure 1.59 along with comparable data from Cochran and Banner (1977). The spall data are plotted as a function of strain rate, which is calculated from the test dimensions and impact velocity using the method previously described for 6061-T6 aluminum. The spall-stress data show some evidence of increasing with increasing strain rate. It should be noted, however, that alternative dependences can be inferred. For example, shock-induced microstructure changes of the material with increasing shock intensity could be expected.

Predictions of the spall strength based on the energy balance relations provided in equations (1.184) and (1.186) are also shown with the data in figure 1.59. The brittle spall prediction of spall strength is in reasonable accord with the measured spall data. The calculated transition to ductile spall is also identified. The properties for α-uranium used in the calculation of the spall stress are ρ = 18 940 kg m^{-3}, E = 11.1 × 10^{10} N m^{-2}, Y = 1 × 10^{9} N m^{-2} and K_c = 6 × 10^{7} N m$^{-3/2}$. The longitudinal, shear and bulk elastic wave speeds for α-uranium are 3.45, 2.14 and 2.42 km s^{-1}, respectively.

1.6.5 Spall in rocks

The dynamic fracture of rock lies at the heart of a number of natural and earth-engineering applications. Natural events that are a consequence of tectonic forces such as earthquakes, landslides and volcanic eruptions are moderated by both the static and the dynamic strength of the rock medium in which the events occur. Asteroid impact causes cratering of the geology with the size and shape determined by the dynamic failure characteristics of the media. Rock blasting in applications of quarry mining, tunneling and road construction relies on a sensible understanding of the dynamic fracture properties of the rock, along with the attentive use of tools for drilling and explosive blasting.

The character of rock fracture explores a spectrum of rates and loading conditions. Concerns range from the static onset of fracture, to the terminal speed of a propagating crack, to crack branching and networking of cracks, and to the resistance and fracture failure consequences of rock subject to transient stress waves. The dynamic fracture and fragmentation of rock and other brittle solids are explored in the following several chapters. The present section concentrates on spall waves and the features of spall failure in rock when subjected to the experimental loading conditions achieved with gun impact test facilities.

The spall strength properties of rock materials that are described here were acquired by the test method pictured in figure 1.50. In all of the experimental spall measurements, the stress waves leading to spall were achieved by impact of a PMMA plate onto the test sample of rock. Velocity measurements are made with velocity interferometry at the interface between the rock sample and a PMMA window in most of the tests. A few of the rocks have wave impedance comparable to or less than PMMA, in which case velocity measurements were made at the sample free surface. Compressive stress levels on initial impact were in most cases well below the dynamic compressive strength (HEL) of the material.

Representative spall waves for several of the rocks are shown in figure 1.60 (Grady and Hollenbach 1979). Time histories of the interface velocity are recorded as the spall wave reaches the interface between the sample and window. With knowledge of the wave impedance of the sample and window materials, the velocity

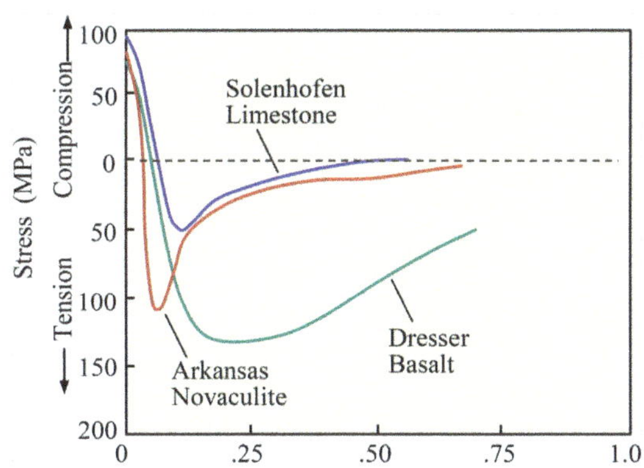

Figure 1.60. Spall wave measurements for selected rock materials (Grady and Hollenbach 1979). Using the known wave impedance properties of the rock and PMMA window material, measured velocity histories are translated into the tensile stress history achieved in the rock material through the wave interaction and subsequent spall failure event. The spall strength of the rock is assessed from the maximum tensile stress achieved in the spall wave.

history can be translated into the tensile stress within the spall zone in the sample through wave interaction and subsequent spall failure on the plane of maximum tension.

1.6.5.1 Spall strength of rock

Spall strengths of a number of rock materials are provided in table 1.3 and in figure 1.61 (Grady and Hollenbach 1979). For most of the rocks only one experiment was performed, avoiding the significant scatter in strength that is certainly present is such natural materials. A more complete study was conducted on several of the rocks. Naming of the various rocks was usually associated with the quarry or locale from which the material was acquired. Most of these rocks came into significance in the 1970s when the then Defense Nuclear Agency invested effort into the characterization of geologic media for purposes of ground-shock analysis. Other properties of these same rocks can be found in the rock mechanics literature of that era. The data on oil shale resulted from interests in the exploration of alternative energy sources during the same period. An order-of-magnitude strain rate of about 10^4 s^{-1} is associated with the rate at which the rock specimens were carried into tension and spall failure.

The spall strength for all of the rocks tested varied from about 10 MPa for Nugget sandstone and Vermont marble to 130 MPa for Dresser basalt. This is a range of about a factor of ten. For the competent low-porosity rocks, the variation is less than a factor of three, suggesting that a spall strength between about 50–150 MPa is

Table 1.3. Spall strength properties of rocks and oil shale.

Rock type	Density ($g\ cm^{-3}$)	Impact velocity ($m\ s^{-1}$)	Shock velocity ($km\ s^{-1}$)	Spall stress (MPa)
Westerly granite	2.63	99	5.28	45
Dresser basalt	2.98	83	6.41	130
Blair dolomite	2.84	85	6.70	47
Vermont marble	2.70	99	5.07	~10
Oakhall limestone	2.70	91	6.48	62
Solenhofen limestone	2.61	98	5.78	53
Indiana limestone	2.34	102	4.48	18
Nugget sandstone	2.54	98	4.05	~10
Berea sandstone	2.24	55	2.70	19
Mt. Helen tuff	1.45	57	2.22	27
Arkansas novaculite	2.63	91	5.78	91
Oil shale ($20\perp$)	2.40	98	4.41	104
Oil shale ($20\parallel$)	2.20	97	3.55	84
Oil shale ($40\perp$)	1.92	156	2.81	58
Oil shale ($40\parallel$)	1.99	157	3.11	61
Oil shale ($70\perp$)	1.70	124	2.50	78
Oil shale ($70\parallel$)	1.71	124	2.31	50

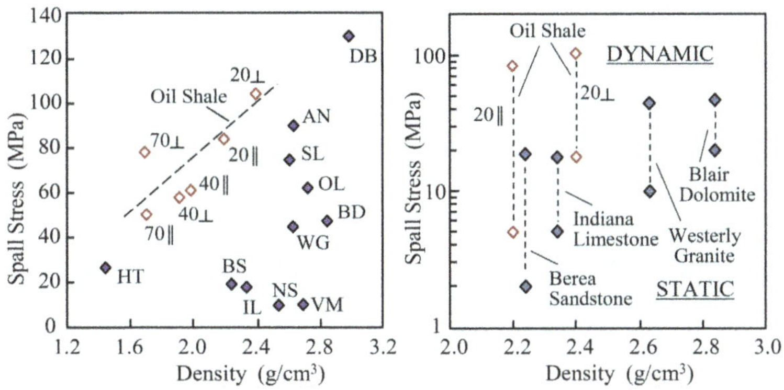

Figure 1.61. Spall strengths of selected rocks and several grades of oil shale are shown on the left (Grady and Hollenbach 1979). Double letters correlate with rock names in table 1.3. The symbols for oil shale identify spall parallel and perpendicular to the bedding planes. Static tensile (lower bound) and dynamic spall (upper bound) strengths of selected rocks are shown on the right.

appropriate for most competent rocks. The very low spall strength from the one test on Vermont marble is a significant anomaly, and further tests should have been performed. A few tens of megapascals adequately characterize the spall strength of the more porous rocks.

For a subset of the rocks tested, static tensile test experiments were performed on the same material. Comparison of static and dynamic tensile strength on the right in figure 1.61 shows that the static strength is commonly a factor of five to ten lower than the dynamic strength. Explanations for differences in fracture strength are readily found from considerations of test conditions and fracture kinetics (see chapter 9 for details of this topic). The results nonetheless provide a perspective of strength properties appropriate to diverse applications in geomechanics.

The six spall experiments on oil shale span three nominal contents of organic kerogen (20, 40 and 70 gallons/ton), and two orientations of the spall plane relative to the bedding planes of the shale. The density of the oil shale decreases as the kerogen content increases. Spall strength measurements reveal a distinct reduction in dynamic strength with increase in the shale kerogen content. There is also a slight tendency for spall strength within the bedding plane orientation to be modestly lower than spall orthogonal to the bedding plane. In contrast, static tensile strength orthogonal to the bedding plane is markedly stronger than in the bedding plane. The static strengths for oil shale in figure 1.61 are averages of a number of tensile tests in each orientation of the tensile load.

Time histories of the spall waves shown for several rocks in figure 1.60 reveal that a constant fracture strength may not adequately characterize dynamic fracture in the time-dependent failure event. The spall strength of Solenhofen limestone and Arkansas novaculite differ by about a factor of two, however the times to failure for the two rocks are about the same. The time to failure in the spall wave provides a measure of the sustained strength of the material as the displacement of the spall surfaces takes place during the failure process. Arkansas novaculite and Dresser basalt, on the other hand, exhibit comparable spall stress; however, the strength of the Dresser basalt is maintained for a markedly longer duration of time. Energy dissipation in the spall process in Dresser basalt is substantially larger. More complex time-dependent spall models are required to capture such spall kinetics, and are necessary in certain rock fracture applications to provide predictions with reasonable fidelity.

1.6.5.2 Spall dependence on impact conditions

For the three rocks for which spall waves are illustrated in figure 1.60, a more detailed assessment of the spall strength is available. The velocity of the impacting PMMA plate was systematically varied, and in turn determines the amplitude of compressive shock stress prior to spall. Further, the dimensions of the test samples were varied for one of the rock materials. In general the spall wave, and the spall stress inferred from the spall wave, is found to depend on the tensile stress intensity and duration determined by the impact conditions.

Again, the experimental spall measurements on the selected rocks were made using the experimental assembly shown in figure 1.50 (Grady and Kipp 1979, Grady 2000a). Sample plates of the rock 7 mm in thickness (6 mm for Arkansas novaculite) were subjected to planar impact by 1.5 mm thickness (1.0 mm for Arkansas novaculite) plates of PMMA. Spall waves were measured with a velocity interferometer at the opposing rock interface with a PMMA window. Representative spall waves for the three rocks are shown in figure 1.60. Solenhofen limestone is a fine-grain (50–250 μm) homogeneous calcium carbonate rock with about 3% porosity. Dresser basalt is a very hard, fine-grain (20–500 μm) serpentinized igneous silicate rock. Arkansas novaculite is a homogeneous cryptocrystalline (approximately 10 μm grain size) quartzite. An HEL of 8 GPa has been measured for Arkansas novaculite (Grady *et al* 1974). That of Dresser basalt is estimated to be about 4–5 GPa. The measured HEL for Solenhofen limestone is approximately 0.7 GPa, although evidence indicates that this is the onset of the displacive calcite I–II phase transformation rather than crystal plasticity. The densities and wave speeds for the three rocks are provided for reference in table 1.4.

Spall strength data for the three rocks are plotted in figure 1.62 as a function of the impact velocity (Grady and Kipp 1979, Grady 2000a). The linear curves on the left of the plot show the maximum tension achieved at a prescribed impact velocity if spall does not occur. Over the range tested, there is no evidence of a dependence of the spall stress on the impact intensity. Maximum compression stresses for the highest impact velocities are 0.7–0.8 GPa—well below the HEL, with the possible exception of Solenhofen limestone.

Table 1.4. Rock properties.

Rock type	Density ($g\ cm^{-3}$)	Longitudinal velocity ($km\ s^{-1}$)	Shear velocity ($km\ s^{-1}$)
Dresser basalt	2.98	6.05	3.51
Solenhofen limestone	2.61	5.48	3.02
Arkansas novaculite	2.63	5.80	4.06

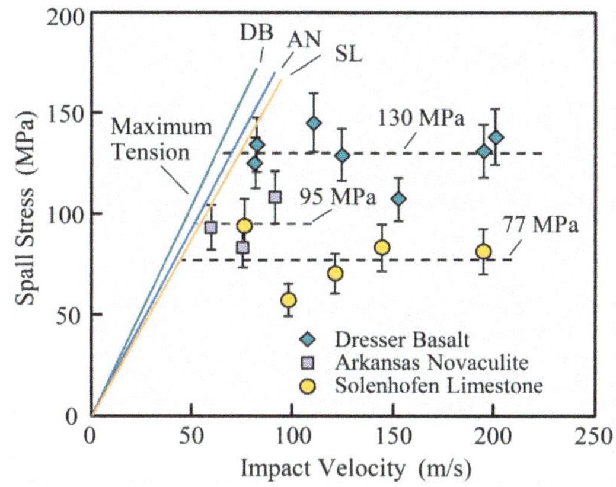

Figure 1.62. Spall-stress dependence on impact velocity for three rocks (Grady and Kipp 1979, Grady 2000a). The solid lines on the left show the maximum tension achieved at the corresponding impact velocity if the material does not spall.

Test-to-test statistical variations in the spall stress are significant. More recent studies on the spall strength of silicon carbide ceramic using line-VISAR velocity interferometer methods reveal that point-by-point variations in spall stress in a single test can be substantial. Observations point to the statistical nature of spall failure in brittle solids.

Additional experiments were performed on Arkansas novaculite that provide further insights into the spall wave and spall strength character of rock (Grady and Kipp 1979). In three spall tests, impact velocity was kept constant at 913 m s^{-1} within plus or minus several meters per second. The PMMA impactors, however, were 1.0, 2.0 and 3.0 mm in thickness. The corresponding thicknesses of the test sample plates of novaculite were 6.0, 8.4 and 10.8 mm. Spall wave profiles for the three tests are shown on the left in figure 1.63. Again, the stress–strain properties of the rock and window materials are used to translate the measured velocity history into the tensile stress history incident on the interferometer window. The error bars in figure 1.63 reflect uncertainty in consistency among the three spall wave experiments.

The regularity of the three spall waves is quite remarkable, considering the scatter in spall stress inferred from the broader set of test data. There are several interesting features in this set of three spall waves. First, there is a marked change in the rate of spreading of the decompression wave in the tension regime of the spall wave preceding spall onset. There are several reasons for this enhanced dispersion of the tension wave (Grady and Kipp 1979). The more likely is change in the elastic modulus in the tension region due to the opening of pre-existing microcracks in the novaculite (Shockey *et al* 1974).

The spall stress calculated from the three spall waves is plotted on the right in figure 1.63. The strain rate on the abscissa corresponding to each spall test is assessed from the rate of tension in the spall wave leading to spall failure. Spall strengths for

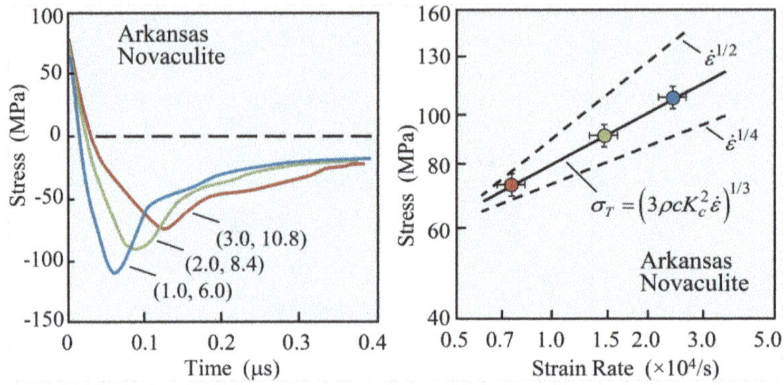

Figure 1.63. Spall waves in Arkansas novaculite at the same impact velocity of approximately 913 m s^{-1} (Grady and Kipp 1979). Thicknesses in millimeters are shown for the PMMA impact plate and sample rock plate. The spall stress for measured spall waves is plotted on the right. A comparison is shown with the energy-based spall prediction.

novaculite plotted in the figure can be compared with the energy-based relation for the spall stress of brittle solids from equation (1.186). Namely,

$$\sigma_T \approx \left(6\rho^2 c^3 \Gamma \dot{\varepsilon}\right)^{1/3} \approx \left(3\rho c K_c^2 \dot{\varepsilon}\right)^{1/3}, \tag{1.189}$$

where $\Gamma \approx K_c^2/2\rho c^2$ is introduced in the second step. The curve with a slope of one-third is calculated from equation (1.189) using a density of 2.63 g cm^{-3} and a longitudinal wave speed of 5.8 km s^{-1} from table 1.4, and by adjusting the toughness to best fit the data. A best fit of K_c = 1.06 MN m$^{-3/2}$ is remarkably close to a fracture toughness of K_c = 1.04 MN m$^{-3/2}$ for Arkansas novaculite reported by Shockey *et al* (1974). Such close agreement is surely fortuitous considering the simplicity of the analysis leading to equation (1.189). Nevertheless, the comparison certainly suggests that strain energy requirements play a paramount role in the spall failure of rock media. The agreement with the strain-rate dependence of equation (1.189) also seems fortuitous considering the marked scatter in the broader data set (see figure 1.62). A fourth experiment would certainly have been warranted. The experimental results are nonetheless noteworthy.

1.6.6 Spall in ceramics

Spall in ceramics bears resemblance to the previously described spall behavior of rock materials. The ceramics of interest are fine-grain, near-full-density polycrystalline compounds prepared through sintering or hot-isostatic pressing techniques. Measured spall strengths exceed those of rock media by about one order of magnitude. Spall strengths within the range of 0.2–2.0 GPa encompass most ceramics. As with rock, the spall strength is markedly less than the HEL strength, which is usually about one order of magnitude higher. As in the previous discussion of spall waves in solids, the spall strength features described here are measured through the experimental methods displayed in figure 1.50, or with similar techniques where strain rates to spall are a few times 10^4 s^{-1} and the applied tensile pulse duration is several microseconds.

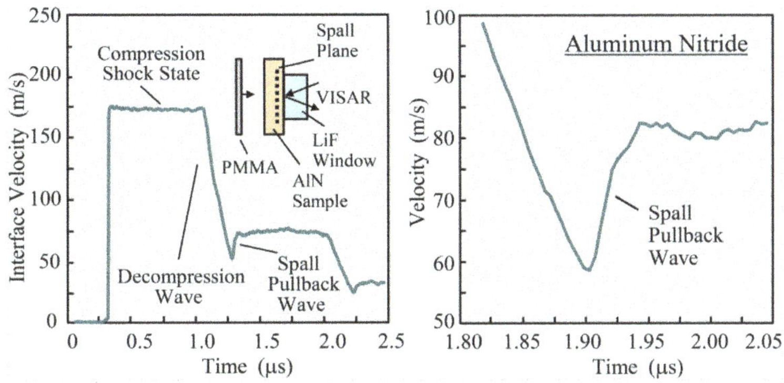

Figure 1.64. Velocity interferometry profile for a dynamic tensile fracture experiment on aluminum nitride ceramic illustrating elastic shock precompression and the subsequent spall signal (Grady 1995b). Details of the spall wave from the original VISAR record in the left-hand plot are shown on the right.

Table 1.5. Spall and HEL strength for ceramics.

Ceramic type	Density ($g\ cm^{-3}$)	Bulk modulus (GPa)	Poisson's ratio	Spall stress (GPa)	HEL stress (GPa)
Aluminum nitride (AlN)	3.25	201–204	0.24	0.5–0.6	7–9
Aluminum oxide (Al_2O_3)	3.97	251	0.23	0.5–0.8	11–12
Boron carbide (B_4C)	2.51	230–233	0.16–0.17	0.4–1.1	17–19
Silicon carbide (SiC)	3.15	213–226	0.16–0.18	0.6–1.3	15–19
Silicon dioxide (SiO_2)	2.63	36	0.08	0.07–0.11	7–9
Titanium diboride (TiB_2)	4.51	193–237	0.05–0.11	0.4–0.5	14–17
Tungsten carbide (WC)	14.9	387	0.20	1.5–3.5	4–6
Zirconium dioxide (ZrO_2)	6.03	176–194	0.30–0.31	1.2–1.4	13–16

The spall wave characteristics shown in figure 1.64 for aluminum nitride ceramic are illustrative of spall features observed in most of the ceramics investigated. The experiment configuration is illustrated in the inset. Planar impact achieves an initial compressive Hugoniot stress of about half of the 7–9 GPa HEL strength of the ceramic. Subsequent tension achieves about 0.5–0.6 GPa, as determined from the amplitude of the spall pullback wave, before spall failure ensued.

The spall strength and HEL strength for selected ceramics are provided in table 1.5 along with the density and elastic properties of the tested ceramics (Grady 1995b). Experimental measurements for the materials listed were all acquired by the methods described for the test shown in figure 1.64. Representative features of the spall phenomena are described for several carbide ceramics in the following subsections.

1.6.6.1 Boron carbide

Boron carbide is a high-dynamic-strength ceramic. The measured HEL strength of boron carbide is about 17–19 GPa for the near-full-density ceramic. The dependence of dynamic strength on ceramic density is discussed in chapter 6. Compared with the

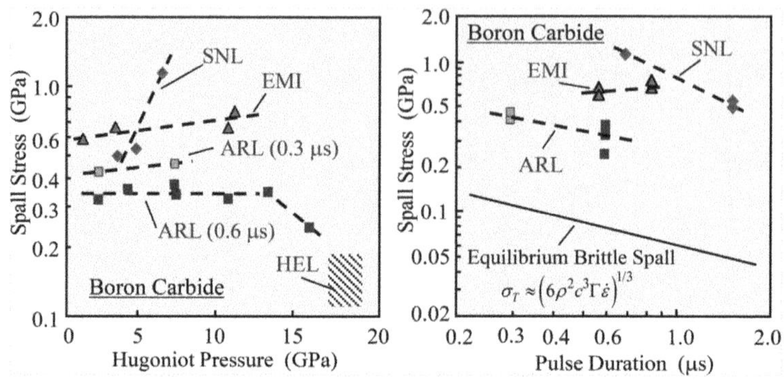

Figure 1.65. Spall strength of boron carbide ceramic (Grady 1992, Winkler and Stilp 1992, Grady and Moody 1996, Bartukowski *et al* 2002).

HEL strength of this ceramic, the spall strength of boron carbide is unremarkable. Spall strength measurements for boron carbide are provided in figure 1.65. These data address several issues concerning the spall strength character of ceramic materials.

The spall data shown are collected from several sources. Spall measurements were performed at the US Army Research Laboratory (ARL) (Bartukowski *et al* 2002), at the Ernst-Mach Institute (EMI) (Winkler and Stilp 1992) and at Sandia National Laboratories (SNL) (Grady 1992, Grady and Moody 1996). All tests were performed on high-quality and full density boron carbide ceramics from different suppliers using time-resolved VISAR diagnostics to monitor the spall wave.

On the left in figure 1.65, the spall strength measurements are plotted as a function of the Hugoniot pressure experienced in the dynamic strength test. The HEL strength of boron carbide is identified in the plot. Hugoniot pressures range from a few GPa to just below the HEL strength. The ARL data provide the broadest span and generally indicate no dependence on the amplitude of the precompression Hugoniot pressure. The exception is the one test just below the HEL strength of boron carbide, where some level of compressive shear damage is probably responsible for reducing the tensile spall strength.

A more striking observation of the spall data on the left in the figure is the marked difference in spall strength for ceramics from different suppliers. Differences of more than a factor of two suggest that differences in ceramic microstructure influence the kinetics of spall fracture activation and growth. Differences in test technique and interpretation of the data may, of course, also be a factor.

The marked difference in the several SNL tests on the left is better understood in the right-hand plot where the same spall data are plotted as a function of the time duration of the shock pressure pulse inducing spall in the material. The SNL and the ARL data both show an increase in spall strength when the pulse duration is shortened. The EMI data do not support this observation.

A prediction of the spall strength based on equation (1.189) for brittle spall and a reasonable fracture toughness of 4 MPa m$^{1/2}$ for boron carbide is shown with the data in the right-hand plot. Spall strength measurements are about one order of magnitude higher than the prediction. The data do not support an energy-based prediction of the spall properties of boron carbide ceramic.

1.6.6.2 Silicon carbide

Spall strength measurements for silicon carbide ceramic are shown in figure 1.66. Several VISAR records of the spall wave are displayed on the left. Spall data as a function of the precompression Hugoniot pressure for silicon carbide ceramic are shown in the plot on the right in figure 1.66. Measurements performed by the US ARL (Dandekar and Bartkowski 2001) were on a hot-pressed (HP) material identified as SiC–N ceramic. The three points measured at SNL (Grady and Moody 1996) were obtained on SiC–N ceramic and on a finer-grain variant of SiC–N. Spall data from the EMI (Winkler and Stilp 1992) were correspondingly obtained on a near-full-density HP silicon carbide ceramic.

The spall data from ARL and those from SNL demark an upper-bound spall strength for SiC–N silicon carbide ceramic that exhibits a dependence on the precompression Hugoniot stress that plateaus at about 1.3 GPa. The spall data from EMI exhibits a similar trend, although at about a 30% lower spall-stress level. The dependence on the precompression Hugoniot pressure probably reflects the kinetics of the spall fracture process and the modestly higher rates of tensile loading that accompany the higher Hugoniot pressure level experiments.

Spall stress measurements reduced by as much as half of the ultimate spall strength are common, however, and are believed to reveal that sites of significant microstructure weakness are inherent in the brittle ceramic materials—an effect characteristic of the statistical nature of other strength measurements in brittle solids.

An HEL compressive strength corresponding to the onset of pervasive shear failure of 18–19 GPa has been determined for SiC–N ceramic. (See chapter 6 for further discussion.) One spall experiment performed by EMI modestly exceeds this Hugoniot pressure level and exhibits sustained spall strength. The notion of plastic shear failure with sustained competence and tensile strength of the ceramic material subsequent to compressive shear failure can be considered, although the strength

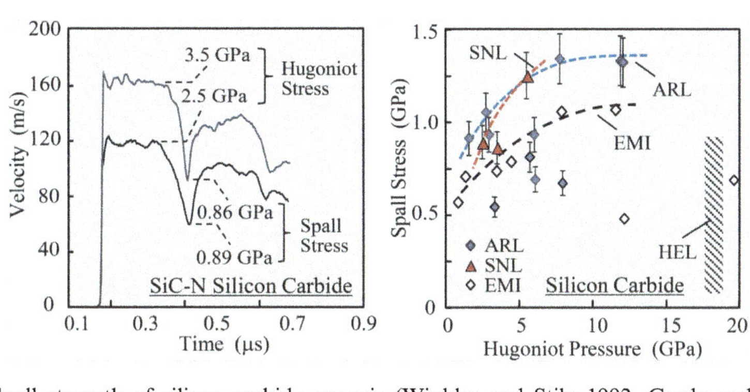

Figure 1.66. Spall strength of silicon carbide ceramic (Winkler and Stilp 1992, Grady and Moody 1996, Dandekar and Bartkowski 2001).

characteristics of the EMI ceramic are not fully known. Studies by Cagnoux (1994), for example, point toward intra-crystal plasticity mechanisms in selected ceramics dominating deformation in the shock-compression shear failure process. Similar conclusions are assessed from the divergent wave shock experiments of Paris *et al* (2011).

1.6.6.3 Tungsten carbide

Lastly, a contrast with the spall failure behavior of boron carbide (a highly brittle ceramic) and silicon carbide (a ceramic with some tendency towards ductility in the shock strength response), is the carbide ceramic tungsten carbide. Some representative measurements of the spall wave in tungsten carbide using VISAR diagnostics at the interface of a lithium fluoride window are shown in figure 1.67 (Grady 1995b, 2000b).

More commonly, tungsten carbide is prepared through liquid-phase sintering (LPS) methods resulting in a ceramic with about 5–10% by volume metal content. One profile on the right was obtained on such an LPS ceramic. Cercom Corporation prepared a unique tungsten carbide ceramic through HP techniques with nearly zero metal content. Spall properties of this ceramic provide a contrast with that of the LPS ceramic.

As a sidelight, one spall measurement on the Cercom ceramic is shown on the left in which the substantial recording time illustrates reverberation characteristics of the spall wave within the ceramic that is bounded by the spall plane on one side and the recording lithium fluoride interface on the other. The marked fidelity of this wave is maintained as the window material eventually absorbs the elastic energy.

Spall-stress measurements on tungsten carbide ceramics are provided in figure 1.68. Open symbols are for spall measurements on LPS material. The square identifies a ceramic from the Kennametal Corporation and the circles for an LPS ceramic extracted from a BS-41 armor piercing shell. Both LPS materials contained approximately 10% by volume of metal. Solid symbols are spall measurements for the Cercom HP ceramic. The data are from Dandekar (2004) with several points

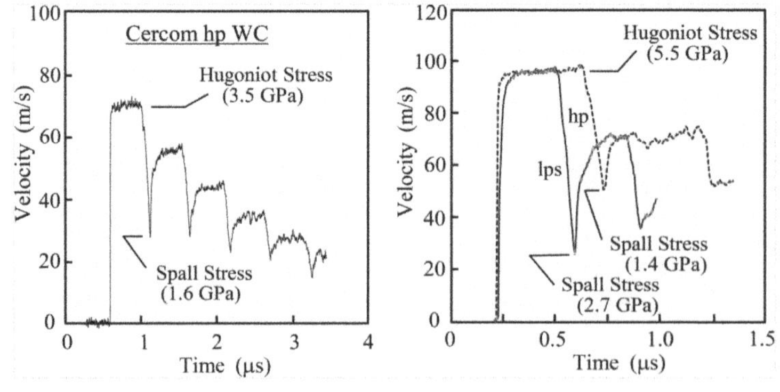

Figure 1.67. Spall pullback waves in tungsten carbide ceramic. Left: spall wave and reverberations for Cercom HP ceramic. Right: comparison of profiles for Cercom HP and Kennametal LPS ceramics (Grady 1995b, 2000b).

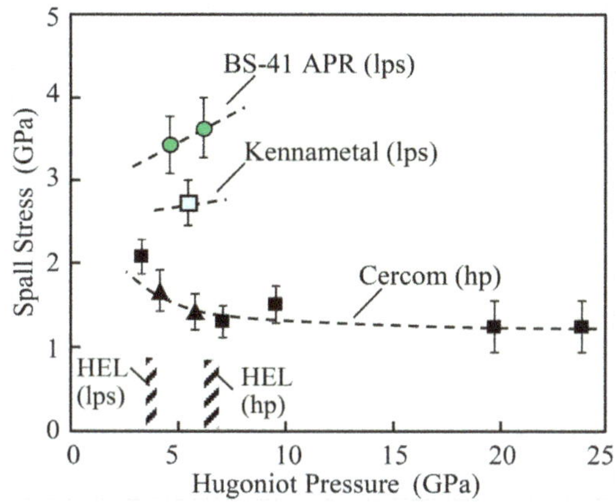

Figure 1.68. Spall strength versus impact Hugoniot pressure for selected tungsten carbide ceramics. Data from Dandekar (2004) (solid squares) and Grady (1995b, 2000b) (solid triangles and open symbols).

(triangles) obtained at SNL (Grady 1999). Spall-stress data are plotted as a function of the Hugoniot pressure achieved on impact prior to tensile loading and spall failure. HELs for both the LPS and HP ceramics are identified in the plot.

In general the LPS ceramics exhibit about twice the spall strength of the HP material. The spall strength of the LPS ceramic approaches the HEL strength of the material. With the exception of several spall measurements suggesting a modest reduction in spall strength as impact pressure approaches the HEL, the Cercom HP ceramic exhibits a nominal spall strength of about 1.2–1.5 GPa through the full range of impact pressures. This marked lack of sensitivity to compressive shear failure prior to spall is at odds with spall in other ceramics. The observation suggests a significant level of dislocation plasticity in the HEL shear failure with subsequent sustained tensile strength. Investigations of post-HEL strength through other shock physics experimental methods are pursued in chapters 6 and 9.

1.6.7 Spall energy and the spall toughness of solids

The amplitude of the spall wave and the stress inferred from this amplitude is the signature measure of the spall strength of a material. Physics-based computational models of the spall process reveal that a single spall-stress measure does not uniquely constrain the spall event. The static fracture strength of a solid is not characterized by a fracture stress. Extensive studies of the genesis and progression of static fracture have shown that measures of strain energy release and fracture toughness are more pertinent properties characteristic of the material fracture strength.

Spall strength based on energy principles has been pursued in the past several subsections with some measure of success. A balance of the strain energy driving the spall failure process and the work resisting spall fracture provides an estimate of the spall strength that is probably a reasonable lower bound. The energy analysis also provides a sensible baseline for more detailed spall models. It would be useful if appropriate analysis of the spall wave could provide an experimental measure of the spall energy as well as the spall stress.

In the present subsection effort is made to identify an experimental measure of the spall energy associated with the spall strength of a solid. The undertaking is tentative. Considerable history encompasses the determination of the spall stress associated with a spall measurement. Remarkably little work has been focused on determining the energy expended in the spall failure event.

1.6.7.1 Energy and the spall wave

Spall measurements on one further ceramic are used in investigating a method to assess energy expenditure in the spall failure process. Two velocity interferometer recordings of spall in an yttria-stabilized zirconium dioxide (ZrO_2) are shown in figure 1.69 (Grady and Mashimo 1992). Again, the experimental geometry illustrated in figure 1.50 is used in the tests, and in the recording of the spall wave at the ceramic sample and lithium fluoride window interface. The full record of the velocity–time histories is shown on the left. A magnified view of the spall pullback waves is provided on the right. The Hugoniot stresses achieved in the two experiments are approximately 11 and 12 GPa, respectively. Compression loading is elastic and well below the 16–17 GPa HEL of this zirconia.

The minimum velocity achieved in the spall pullback wave provides an experimental measure of the spall stress. The present tests yield a spall stress for the ZrO_2 of about 1.2–1.4 GPa. The time history of the spall waves on the right in figure 1.69 reveal that tensile stress in the spall zone persists for approximately 0.1 μs through the tensile loading and failure event. The impulse imparted to the ceramic through the duration of the spall event is provided by the time integral of the tensile stress history of the spall wave.

The same tensile stress history can be used to provide an experimental estimate of energy dissipation in the spall region. This analysis of the data is achieved with the model of the spall event illustrated in figure 1.70. Numerical values identified in the figure are appropriate to the test with a Hugoniot stress of approximately 11 GPa and a spall stress of 1.4 GPa. The spall wave illustrated on the left, however, has been simplified to the rectangular form displayed for the purpose of the energy analysis model. The calculation of energy dissipation in the spall plane is determined from the wave interaction plot illustrated for the zirconium dioxide sample and

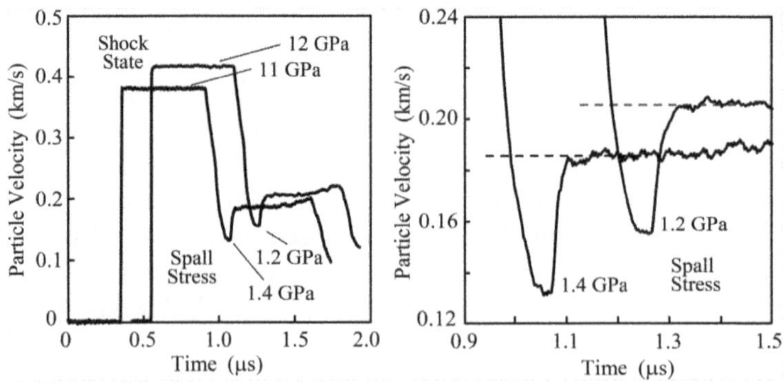

Figure 1.69. Spall waves in zirconium dioxide (Grady and Mashimo 1992).

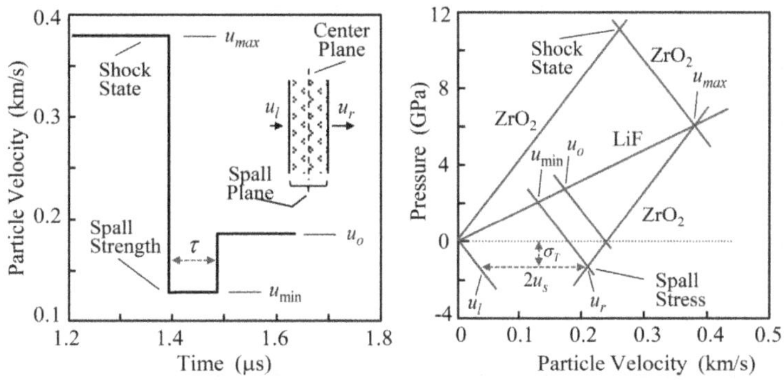

Figure 1.70. Wave interactions lead to spall failure in zirconium dioxide.

lithium fluoride window on the right in the figure. Spall is a consequence of the collision of decompression waves originating from reflection from the window interface and from the back of the impacting projectile plate (also zirconium dioxide). Two subsequent decompression waves emerge from the collision plane. A right-facing (directed toward the recording interface) decompression wave emerges from the collision plane that is limited by the tensile (spall) strength of the material. This wave provides the spall pullback signal that is recorded by the velocity interferometer. The particle velocity behind this wave is identified as u_r in the right-hand figure. A corresponding left-facing wave (towards the impacting projectile) also emerges from the collision plane and is similarly limited by the spall strength of the material. The particle velocity behind this wave is u_l also identified in the figure.

The difference between the two velocities is the speed at which the spall plane is opening. Namely,

$$2u_s = u_r - u_l. \tag{1.190}$$

The velocity u_s identifies a spall-plane-opening velocity referenced from the center of the spall plane (see the inset on the left in figure 1.70). The rate of work required in opening the spall plane, and thus the energy dissipated in the spall zone, is the product of the spall stress and opening velocity $\sigma_T u_s$, where both are assumed to be constant in the model.

The spall fracture energy is given by

$$W_s = \sigma_T u_s \tau, \tag{1.191}$$

where τ is the duration of the tensile spall wave, as illustrated on the left in figure 1.70. In keeping with common fracture lore, W_s, with dimensions of energy per unit area, refers to that of one side or the other of the spall plane. The energy density dissipated within the spall zone is not known, as the thickness of the spall region is not revealed by the experiment.

To test equation (1.191) and assess the spall energy from the zirconium dioxide spall data, the measured spall waves shown in detail on the right in figure 1.69 are used. These spall waves are not square as assumed in the model. To maintain

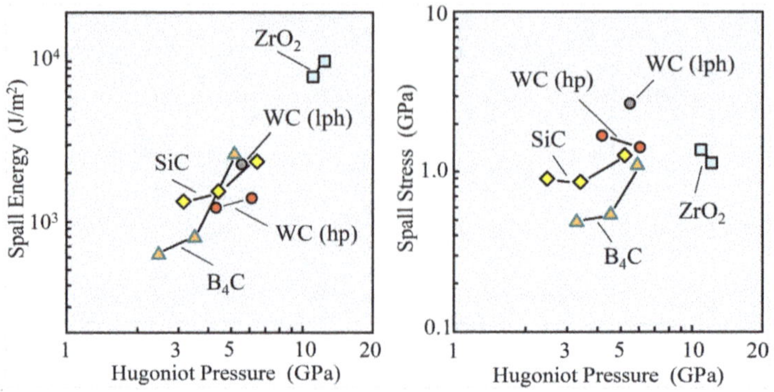

Figure 1.71. Spall energy and spall stress for selected ceramics (Grady 1995b).

consistency the stress–time integral of the area identified in the spall wave replaces the product $\sigma_T\tau$ in equation (1.191). It is not known how adequate this approximation is.

The spall energy was calculated for the zirconium dioxide and selected other ceramics discussed in the previous subsections. The spall energy calculated from equation (1.191) and the spall stress determined from the same spall pullback wave are compared in figure 1.71. The spall energy and spall stress are plotted against the Hugoniot stress state achieved in the experiment.

The results are curious. Theoretical fracture surface energies for the present compounds are of the order of 1–10 J m^{-2}. Energies from fracture toughness measurements tend to be about one order of magnitude higher. The calculated spall energies are an additional one to two orders of magnitude higher yet. There is a trend of increasing spall energy with increasing impact velocity for a specific ceramic. This trend probably reflects the increasing rate of spall failure and the tendency for refinement of fracture activation and fracture participation as the rate of fracture failure increases. The collective overall trend of the spall energy with Hugoniot pressure, if meaningful, is not understood.

1.6.7.2 The spall energy and the kinetic energy of the spall plate

Estimates of energy dissipation within the spall zone described in the previous subsection rely on measurement of the amplitude and duration of the spall wave imparted to the spall material by the spall failure process. An interesting alternative measurement to consider is that of the residual kinetic energy of the spall plate ejected in the test. The concept is illustrated by the common science exhibit referred to as Newton's cradle. The impact of a ball bearing onto a chain of penduluming ball bearings ejects the ball bearing at the end of the chain at nearly the same velocity as the initial impact. If the last two ball bearings were bonded using a suitable glue, then in principal the ejection velocity would be less, the reduction in kinetic energy related to the energy consumed in breaking the glue bond.

Accordingly, consider the experiment illustrated in figure 1.72 where a projectile plate of thickness L impacts a target plate of the same material at a velocity V.

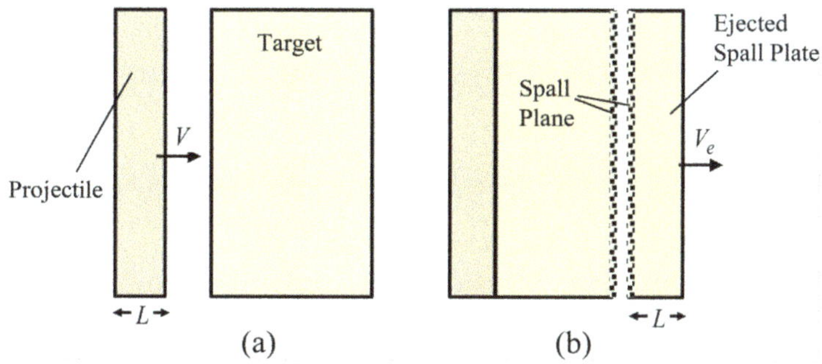

Figure 1.72. The impact conditions on the left lead to spall and the ejected spall plate on the right.

Analysis will show that the relationship between the kinetic energy lost in the spall event and the energy dissipated in the spall zone is not straightforward. There are nonetheless some positive features associated with the present experiment that may yield test results that are superior to time-resolved measurements of the spall wave. This is particularly true for brittle materials such as the rocks and ceramics where large statistical variations from local time-resolved wave measurements can be appreciable.

Assuming that the test material is elastic and reasonably linear, the impacting projectile is brought to rest and the kinetic energy is propagated as an elastic wave forward through the target sample. Reflection and interaction of the wave at the sample forward free surface leads to tensile loading at the spall plane a distance L behind the reflecting surface. If the tensile strength is exceeded a spall section of thickness L is ejected at a velocity $V_e \leqslant V$ as illustrated in figure 1.72. The reduction in kinetic energy is

$$\Delta T = \frac{1}{2}\rho V^2 L - \frac{1}{2}\rho V_e^2 L. \tag{1.192}$$

The spall wave is modeled by the two parameters spall-stress amplitude σ_T and pulse duration τ, as shown on the left in figure 1.70 (without a window material). The momentum of the ejected plate is reduced by the impulse imparted by the spall process according to

$$\rho V_e L = \rho V L - \sigma_T \tau. \tag{1.193}$$

Combining equations (1.192) and (1.193) leads to

$$\Delta T = \sigma_T \tau V - \frac{1}{2}\frac{\sigma_T^2 \tau^2}{\rho L}. \tag{1.194}$$

The spall energy (energy dissipated in the spall process) is provided by

$$2W_s = 2\sigma_T \tau u_s = \sigma_T \tau \left(V - \frac{2\sigma_T}{\rho c} \right), \tag{1.195}$$

as is illustrated on the right in figure 1.70. Calculation of the spall energy in terms of the kinetic energy decrement is then

$$2W_s = \Delta T - \left(\frac{2\sigma_T^2\tau}{\rho c} - \frac{\sigma_T^2\tau^2}{2\rho L}\right) = \Delta T - 2W_k. \tag{1.196}$$

Equation (1.196) reveals that the kinetic energy decrement of the spall plate is balanced by the fracture energy W_s and an elastic wave energy W_k that reverberates about the plate center of mass and will eventually thermalize. An equal elastic energy is also imparted to the remaining target plate.

Measurement of the kinetic energy decrement from an experiment, as depicted in figure 1.72, therefore does not provide a direct measure of the spall energy. Rather, the measurement provides the sum of the energy expended through both spall fracture dissipation and momentum transfer. As an example, a calculation is provided in figure 1.73 based on properties for silicon carbide ceramic. Spall stress and spall energy values from figure 1.71 of $\sigma_T = 1$ GPa and $W_s = 2$ kJ m^{-2} are applied in the calculation and assumed to be constant over the span of impact velocity considered. Other properties assumed in the calculation are $\rho = 3150$ kg m^{-3}, $c = 12.2$ km s^{-1} and $L = 1$ mm.

A sensible measurement of the spall energy expended requires that the kinetic energy decrement be an appreciable fraction of the initial kinetic energy imparted in the impact. A fractional decrement of $\Delta T/T$ equal to 0.4 down to 0.1 is spanned by the impact velocity range of 100–200 m s^{-1}. Over this same span the ratio of the Hugoniot stress to the spall stress σ_h/σ_T ranges from about two to four. The spall energy W_s is the principal contributor to the kinetic energy decrement towards the upper end of the velocity range. At the lower end W_s and W_k are comparable.

Through elastic wave solutions of the spall failure event more realistic models of the spall failure history can be introduced, providing further insights into

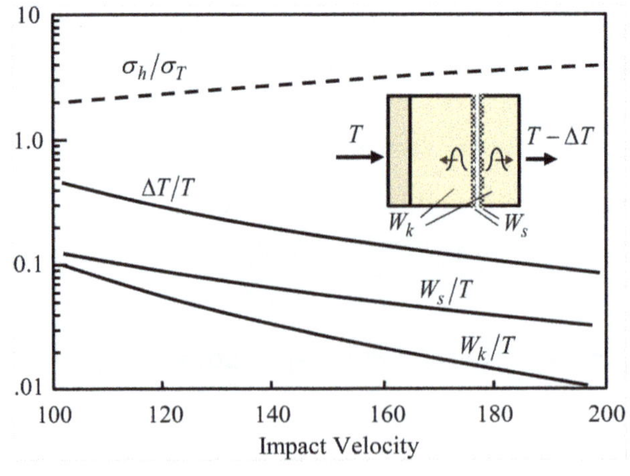

Figure 1.73. Kinetic energy decrement and partitioning between spall dissipation energy and elastic wave energy. Based on properties for silicon carbide ceramic.

partitioning of spall dissipation and kinetic energy radiation. Such analysis of spall failure is undertaken in chapters 3 and 4.

The experimental method does not provide a direct measure of the spall energy. The technique is complicated through a mixing of the measured kinetic energy decrement into spall dissipation and elastic wave energy. The experimental method is not without merit, however. Several experiments can be performed where impact velocity and/or impact plate thickness are varied. Through this expedient relative contributions can be sorted out.

Further, the more global experimental method offers attractive features for computational spall model assessment. The method avoids the difficulties of point spall wave measurements where substantial statistical point-to-point differences are common in brittle solids.

1.6.8 Spall at high temperatures

Temperature asserts a first-order influence on the spall properties of solids. Spall strength follows more common strength properties. Namely, fracture toughness K_c tends to increase with temperature while the yield stress Y decreases with temperature. Spall characteristics of metals reflect the temperature dependence of these fundamental strength properties. Spall strength lessens and the spall event becomes more ductile with increasing temperature.

Materials can be pre-heated before being subjected to spall. Alternatively, or additionally, temperature increase occurs as a consequence of shock compression and accompanying shock dissipation preceding spall. Shock heating effects are particularly significant when materials subjected to spall are initially distended or porous.

The dependence of spall strength on shock-induced temperature rise is demonstrated by the study of Lalle (1985) on metallic tin (see Grady 1988). Spall strength data for tin as a function of the pre-spall shock Hugoniot pressure are shown on the left in figure 1.74. Ambient temperature solid tin is calculated to melt on decompression from Hugoniot states exceeding 21.5 GPa. Spall strengths at a

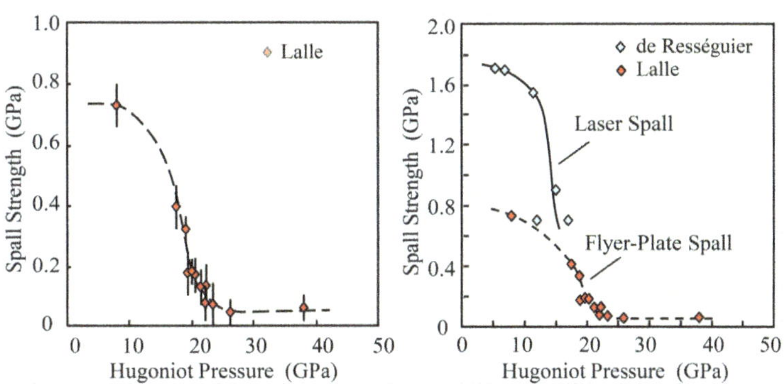

Figure 1.74. Spall strength data illustrating the effects of shock-induced heating on the spall strength of tin. Data from Lalle (1985) and de Rességuier *et al* (2007).

near-ambient shock temperature of about 0.6–0.8 GPa decrease to a spall stress of approximately 0.05–0.06 GPa near the tin melt temperature and above.

The spall properties of tin obtained in the study of Lalle (1985) were obtained with high-explosive flyer plate methods. Spall measurements on tin using laser shock loading techniques (de Rességuier *et al* 2007) achieve strain rates two to three orders of magnitude higher ($10^7\ s^{-1}$ to $10^8\ s^{-1}$). Spall strengths from laser shock are compared with the measurements of Lalle on the right in figure 1.74. The increase in spall strength with strain rate in the solid tin at lower Hugoniot pressures (and lower temperatures) is consistent with the strain-rate dependence observed in other metals. Hugoniot pressures at the spall plane in the laser spall study are provided through computational analysis of the wave interaction event. The authors suggest that Hugoniot pressures may be somewhat higher as recovery samples indicate partial melt at the highest shock amplitudes.

Kanel *et al* (1996) investigated the dependence of shock compression and spall strength properties of pre-heated aluminum and magnesium. Metal test samples pre-heated to temperatures ranging from ambient to in excess of 600 °C were subjected to planar impact by thin metal plates. Time-resolved VISAR measurements of the spall wave at the opposing free surface provide spall-stress measurements. Spall strength data for aluminum are shown in figure 1.75 as a function of pre-heat temperature. Representative spall wave measurements with the VISAR instrumentation are displayed on the left. Spall strength as a function of the initial pre-heat temperature is provided on the right.

The increased amplitude of the elastic precursor wave and corresponding HEL associated with the higher pre-heat temperature experiments shown on the left are also of interest. Such differences may relate to marked changes in Poisson ratio properties with temperature, although further studies of this effect are warranted.

In all the tests, apart from one for aluminum, an impact Hugoniot pressure near 5.8 GPa was achieved prior to spall. The reduction of about 20% in the spall strength above about 200 °C is not understood. Spall stress remains relatively insensitive to the pre-heat temperature until within about 100 degrees of melt. A subsequent

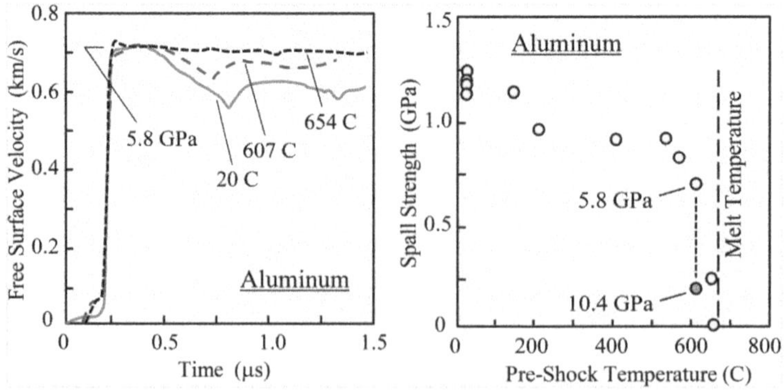

Figure 1.75. Spall strength data for aluminum and dependence on pre-heat temperature (Kanel *et al* 1996). The Hugoniot pressure preceding spall is 5.8 GPa. The Hugoniot pressure of 10.4 GPa for one aluminum test illustrates the added influence of shock-induced heating on the spall strength.

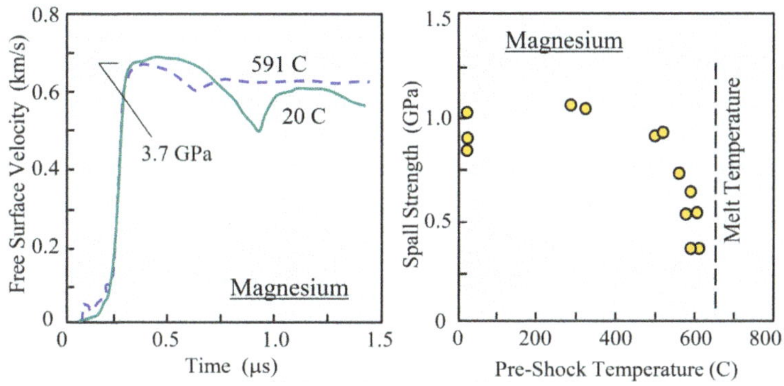

Figure 1.76. Spall strength data for magnesium (left) and dependence on the pre-heat temperature (right) (Kanel *et al* 1996). The Hugoniot pressure preceding spall is 3.7 GPa.

temperature rise results in rapid reduction in spall strength to near zero. One test performed at a higher Hugoniot pressure of 10.4 GPa results in a markedly lower spall strength, illustrating the influence of added shock-induced heating in the temperature range nearing melt.

Comparable data are shown for the pre-heat spall strength of magnesium in figure 1.76. A nominal Hugoniot pre-compression pressure before spall of 3.7 GPa was achieved in all tests on magnesium. A modest increase in spall strength is observed at intermediate temperatures. Marked reduction in spall strength occurs as the pre-heat temperature approaches the melt temperature. Increases in elastic precursor amplitude with pre-heat temperature comparable to aluminum are also observed in magnesium.

1.6.9 Spall in liquids

Spall in liquids has unique issues and a firm theoretical foundation is not yet available. Experimental data are scant and there are significant discrepancies in the available data. Consider elemental mercury. The theoretical spall strength based on a Morse potential and the cohesive energy in table 1.1 is about 4.6 GPa. A spall strength of 0.06–0.12 GPa determined from measurement of the spall wave is reported by Taylor (1982). For comparison, Carlson (1975) has measured a spall strength of about 1.9 GPa for mercury based on electron beam shock loading methods. Marked differences in the rates of loading that lead to spall in the two experimental methods probably plays a role in the reported difference in spall strength. A perspective has been provided on spall in liquid metals with assessment of available data and metal properties key to the spall event (Grady 1995a).

Spall in liquids is expected to be a process of cavity activation, growth and coalescence under transient hydrodynamic tension with thermal fluctuations or impurities providing sites for void nucleation. Fisher (1948) has put forth a theory of liquid fracture based on a process of thermally activated void nucleation and growth.

The process of spall in simple liquids appears to be that of cavitation and growth of holes or bubbles at sufficient states of transient hydrodynamic tension. This spall process was observed by photographic methods in water (Kolsky and Rader 1968). Erlich *et al* (1971) provide similar photographic evidence of spall cavitation in glycerol. Kolsky and Rader (1968) on the other hand, have shown that spall in highly viscous liquids can occur through a fracture process. A transition from spall through hole cavitation to spall through fracture that is determined by material conditions, and probably spall loading conditions, is recognized as a complicating factor of liquid spall.

More involved theories of spall failure of liquids can be supported by considerations of energy balance that is found to be fruitful in bounding the spall strength of metals, as described earlier in the section. Spall in a liquid is a consequence of transient hydrodynamic tension brought about by the interaction of decompression pressure waves. Elastic lattice strain energy in the liquid increases as the dynamic state of tension rises. As spall failure activates and goes to completion this strain energy fuels the dissipative mechanisms resisting the spall failure process. In simple liquids, two continuum sources of resistance are immediately apparent. They are the surface tension associated with the creation and growth of new surfaces in the cavitation process, and viscosity dissipation accompanying the shear deformation of cavity growth and coalescence.

A relation that balances the available strain energy to the work of surface tension and viscosity necessary to carry through the spall process is provided in Grady (1988),

$$\frac{\sigma_T^2}{2\rho c^2} = \frac{3\gamma}{x_c} + \eta\dot{\varepsilon}\varepsilon_c. \tag{1.197}$$

This corresponds, for example, to equation (1.184) for spall in a ductile metal or equation (1.186) for a brittle metal. Again, for the surface energy a length scale of order $x_c \approx ct_c \approx \sigma_T/\rho c\dot{\varepsilon}$ is required to characterize the cavitation spacing consistent with the time to spall failure. Also again, ε_c is a measure of the inelastic deformation strain accommodating the spall failure event. Joining these relations with equation (1.197) provides an expression for the spall stress of a liquid based on surface energy and viscous dissipation constraints,

$$\sigma_T^3 - 2\rho c^2\eta\varepsilon_c\dot{\varepsilon}\sigma_T - 6\rho^2c^3\gamma\dot{\varepsilon} = 0. \tag{1.198}$$

At low strain rates surface energy governs spall fracture resistance leading to the spall-stress expression,

$$\sigma_T = (6\rho^2c^3\gamma\dot{\varepsilon})^{1/3}. \tag{1.199}$$

At the highest strain rates viscous dissipation dominates and

$$\sigma_T = (2\rho c^2\eta\varepsilon_c\dot{\varepsilon})^{1/2}. \tag{1.200}$$

A measure of the strain rate at which spall fracture resistance transitions from surface energy to viscous-dissipation-governed is provided by equating the energies in equation (1.197), or the second and third term in equation (1.198), yielding

$$\dot{\varepsilon}_t \approx \frac{9}{2} \frac{\rho \gamma^2}{(\eta \varepsilon_c)^3}. \tag{1.201}$$

Spall properties for two liquids for which reasonable material properties and experimental spall data are available are shown in figure 1.77. Properties for glycerol of $\rho_o = 1200$ kg m^{-3}, $c_o = 1900$ m s^{-1}, $\gamma = 0.064$ J m^{-2} and $\eta = 1.5$ Pa s are from Erlich *et al* (1971). Properties for mercury of $\rho_o = 13\,530$ kg m^{-3}, $c_o = 1680$ m s^{-1}, $\gamma = 0.5$ J m^{-2} and $\eta = 1.5 \times 10^{-3}$ Pa s are from Carlson (1975). Predictions of spall strength from the energy-based relations of equations (1.198)–(1.200) are shown for each of the liquids. Theoretical spall strengths from table 1.1 for mercury and a value of $K_o/10$ for glycerol provide further reference in the figure.

The spall measurements for glycerol of Erlich *et al* (1971) using plate impact methods are shown in the left-hand plot. Spall predictions suggest a transition from surface-tension to viscous-dissipation-governed spall at strain rates above about 2×10^3 s^{-1}. Energy-based predictions governed by viscous dissipation are within a factor of two of the experimental measurements. The experimental point of Bull (1956) is almost an order of magnitude higher than the prediction, although the appropriate strain rate is difficult to establish.

Several experimental spall measurements for mercury are shown on the right in figure 1.77. The plate impact spall measurement of Taylor (1982) at a strain rate near 10^5 s^{-1} and the electron beam generated spall result of Carlson (1975) at a few times 10^6 s^{-1} are well above the energy prediction of spall as determined from the surface tension of liquid mercury. A tensile strength measurement obtained using a spinning capillary tube technique (Briggs 1953) is also shown in the plot.

Generally, for the two liquids examined, spall predictions based on continuum energy constraints are not very useful, although the predictions do tend to bound the experimental data from below. The implication is that under the loading rates of the

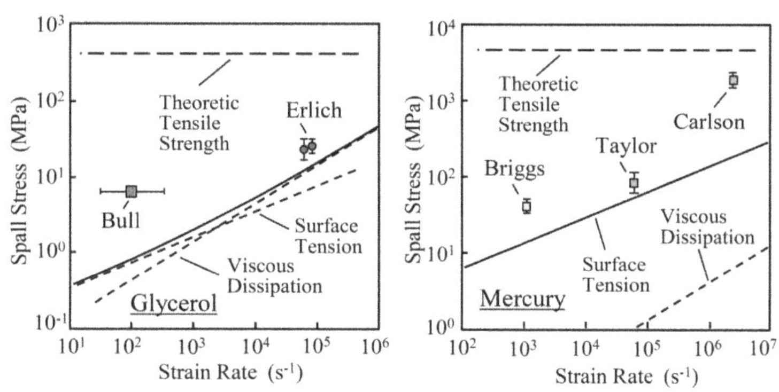

Figure 1.77. Experimental measurements and theoretical predictions of spall strength for liquid glycerol and mercury (references in text).

experiments, stress levels for fracture activation are necessary that markedly exceed the energy-based requirement. A further implication is that predictions of fragment size based on energy constraints in a liquid spall event would also be in error. The calculations of fragmentation and comparisons with experiment addressed in the next several chapters will show that the energy-based methods provide sensible predictions of dynamic fragmentation in selected liquids. Spall in liquids is a topic in need of further experimental efforts utilizing the higher-resolution of present-day diagnostics.

References

Abbey A E and Bonner B P 1975 Elastic constants of alpha uranium *J. Appl. Phys.* **46** 1427–28

Ahrens T J 1992 Equation of state *High-pressure Shock Compression of Solids* (Berlin: Springer) pp 75–113

Akobuije Chijioke D, Nellis W J and Silvera I F 2005 High pressure equation of state for Al, Cu, Ta, and W *J. Appl. Phys.* **98** 073526

Anderson G D, Fowles G R, Duvall G E, Peltzer C P and Erkman J O 1965

Antoun A, Seaman L, Curran D R, Kanel G I, Razerenov S V and Utkin A V 2002 *Spall Fracture* (Berlin: Springer)

Barker L M 1968 Fine structure of compressive and release wave shapes in aluminum measured by the velocity interferometer technique in *Behavior of Dense Media under High Dynamic Pressures* (London: Gordon and Breach) p 483

Barker L M and Hollenbach R E 1972 Laser interferometer for measuring high velocity of any reflecting surface *J. Appl. Phys.* **43** 4669–75

Barker L M 1984 High-pressure quasi-isentropic experiments *Shock Waves in Condensed Matter* ed J R Asay *et al* (Elsevier Science) pp 217–24

Bartukowski P T, Dandekar D P and Grove D J 2002 Spallation of hot-pressed boron carbide ceramic Shock *Compression of Condensed Matter* ed M Furnish, N Thadhani and Y Horie (American Institute of Physics) pp 779–82

Birch F 1978 Finite strain isotherm and velocities for single-crystal and polycrystalline NaCl at high pressures and 300 K *J. Geophys. Res.* 1257–66

Bland D R 1965 On shock structure in solids *J. Inst. Math. Appl.* **1** 56

Briggs L J 1953 The limiting negative pressure of mercury in Pyrex glass *J. Appl. Phys.* **24** 488–90

Bull T H 1956 The tensile strength of liquids under dynamic loading *Phil. Mag.* **1** 153–65

Butkovich T R 1971 The influence of water in rocks on effects of underground nuclear explosions *J. Geophys. Res.* **76** 1993–2011

Cagnoux J 1994 Analysis d'Echantillons d'Al_2O_3, TiB_2 et SiC Recuperes Apres des Chocs Allamt Jusqu'a 20 GPa, *DYMAT 94 Int. Conf. on Mechanical and Physical Behaviour of Materials under Dynamic Loading les Editions de Physique* pp 257–62

Cagnoux J and Longy F 1988 Is the dynamic strength of alumina rate dependent *Shock Waves in Condensed Matter* (Amsterdam: Elsevier) pp 293–6

Carlson G A 1975 Dynamic tensile strength of mercury *J. Appl. Phys.* **46** 4069–70

Chhabildas L C, Dunn J E, Reinhart W D and Miller J M 1993 Impact technique to accelerate flier plates to velocities over 12 km/s *Int. J. Impact Eng.* **14** 121–32

Christman D R, Isbell W M, Babcock S G, McMillan A R and Green S J 1971 *DNA Report* DASA 2501-3, AD735966.

Christman D R, Michaels T E, Isbell W M and Babcock S G 1972 *DNA Report* DASA 2510-4, AD738862.

Cochran S and Banner D 1977 Spall studies in uranium *J. Appl. Phys.* **48** 2729–39

Courant R and Hilbert D 1962 *Methods of Mathematical Physics* vol 2 (New York: Interscience)

Curran D R 1982 *Impact Dynamics* ed J A Zukas *et al* pp 333–66

Curran D R, Seaman L and Shockey D A 1977 Dynamic failure of solids *Phys. Today* **30** 1

Curran D R, Seaman L and Shockey D A 1980 Linking dynamic fracture to microstructure processes, *SRI International, Poulter Laboratory Technical Report* 003-80, June.

Curran D R, Seaman L and Shockey D A 1987 Dynamic failure of solids *Phys. Rep* **147** 253–388

Dandekar D P 2004 Spall strength of tungsten carbide *Army Research Laboratory Report* ARL-TR-3335, September.

Dandekar D P and Bartkowski P T 2001 Tensile strengths of silicon carbide under shock loading *Army Research Laboratory Report* ARL-TR-2430, March.

Dandekar D P, Martin A G and Kelly J V 1980 Deformation of depleted uranium—0.78 Ti under shock compression to 11.0 GPa at room temperature *J. Appl. Phys.* **51** 4784–9

Davison L, Grady D E and Shahinpoor M 1996 *High Pressure Shock Compression of Solids II: Dynamic Fracture and Fragmentation* (Berlin: Springer)

Davison L and Graham R A 1979 Shock compression of solids *Phys. Rep.* **55** 255–379

Dugdale D S 1960 Yielding of steel sheets containing slits *J. Mech. Phys. Solids* **8** 100

Ek D R and Asay J R 1986 The stress and strain rate dependence of spall strength in two aluminum alloys *Shock Waves in Condensed Matter* ed Y M Gupta (Plenum) pp 413–8

Erlich D C, Wooten D C and Crewdson R C 1971 Dynamic tensile strength of glycerol *J. Appl. Phys.* **42** 5495–502

Fisher J C 1948 The fracture of liquids *J. Appl. Phys.* **19** 1062–67

Fowles G R 1961 Shock wave compression of hardened and annealed 2024 aluminum *J. Appl. Phys.* **32** 1474–82

Gieske J H and Barsch G R 1968 Pressure dependence of the elastic constants of single crystal aluminum oxide *Phys. Status Solidi* **29** 121–31

Goranson R W, Bancroft D, Burton B L, Blechar T, Houston E E, Gittings E F and Landeen S A 1955 Dynamic determination of the compressibility of metals *J. Appl. Phys.* **26** 1472–79

Goto T, Sato T and Syono Y 1982 Reduction of shear strength and phase-transition in shock loaded silicon *Japan J. Appl. Phys.* **21** L369–71

Grady D E 1977a *Processes Occurring in Shock Wave Compression of Rocks and Minerals, in High Pressure Research Applications in Geophysics* (Academic) pp 389–438

Grady D E 1977b High strain rate studies in rock *Geophys. Res. Lett.* **4** 263–6

Grady D E 1979 Interrelation of flow or fracture and phase transition in the deformation of carbonate rock *J. Geophys. Res.* **84**(B13) 7549–55

Grady D E 1981 Strain-rate dependence of effective viscosity under steady-wave shock compression *Appl. Phys. Lett.* **38** 825–6

Grady D E 1988 The spall strength of condensed matter *J. Mech. Phys. Solids* **36** 353–84

Grady D E 1995a Spall and fragmentation in high-temperature metals *High-pressure Shock Compression of Solids II—Dynamic Fracture and Fragmentation* pp 219–36

Grady D E 1995b Dynamic properties of ceramic materials *Sandia National Laboratories Technical Report* SAND94-3266, February.

Grady D E 1998 Scattering as a mechanism for structured shock waves in metals *J. Mech. Phys. Solids* **46** 2017–32

Grady D E 1999 Impact failure and fragmentation properties of tungsten carbide *Int. J. Impact Eng.* **23** 307–17
Grady D E 2000b Shock and impact properties of tungsten carbide, *Applied Research Associates Technical Report* August.
Grady D E 2005 Investigation of explosively driven fragmentation of metals – metal breach, enhanced ductility, adiabatic shear, shock structure and strength, *University of California, Lawrence Livermore National Laboratory and Sandia National Laboratories, Applied Research Associates Technical Report* September.
Grady D E 2010 Structured shock waves and the fourth-power law *J. Appl. Phys.* **107** 013506
Grady D E 2011 Equation of state for solids *Shock Compression of Condensed Matter* (New York: American Institute of Physics)
Grady D E and Hollenbach R E 1979 Dynamic fracture strength of rock *Geophys. Res. Lett.* **6** 73–6
Grady D E, Hollenbach R E and Schuler K W 1978 Compression wave studies in calcite rock *J. Geophys. Res.* **83** 2839–49
Grady D E and Kipp M E 1979 The micromechanics of impact fracture of rock *Int. J. Rock Mech. Min. Sci.* **16** 293–302
Grady D E and Kipp M E 1993 Dynamic fracture and fragmentation *High-pressure Shock Compression of Solids*ed ed J R Asay and M Shahinpoor (New York: Springer) pp 265–322
Grady D E 1986b Steady-wave risetime and spall measurements on uranium (3–15 GPa) *Metallurgical Applications of Shock-Wave and High-strain-rate Phenomena* ed L E Murr, K P Staudhammer and M A Meyers (New York: Marcel Dekker) pp 763–80
Grady D E 2000a Spall properties of Solenhofen limestone and Dresser basalt *Shock Compression of Condensed Matter* ed M D Furnish, L C Chhabildas and R S Hixson (New York: American Institute of Physics) pp 1255–58
Grady D E and Mashimo T 1992 Shock and release properties of yttria-doped tetragonal and cubic zirconia *J. Appl. Phys.* **71** pp 4868–74
Grady D E and Moody R L 1985 Shock and release equation of state of calcite. *Sandia National Laboratories Technical Report*, SAND85-0947 May
Grady D E and Moody R L 1996 Shock compression profiles in ceramics, *Sandia National Laboratories Technical Report* SAND96-0551 March
Grady D E, Murri W J and Fowles G R 1974 Quartz to stishovite: wave propagation in the mixed phase region *J. Geophys. Res.* **79** 332–8
Grady D E 1992 Shock-wave properties of high strength ceramics *Shock Waves in Condensed Matter*ed ed S C Schmidt *et al* (Amsterdam: Elsevier) pp 455–8
Grady D E 1986a High-pressure release-wave measurements and phase transformation in $CaCO_3$ *Shock Waves in Condensed Matter*ed ed Y M Gupta (Plenum) pp 589–93
Gurson A L 1977 *J. Eng. Mater. Tech. Trans ASME* pp 2
Gust W H and Royce E B 1971 Axial yield strengths and two successive phase transition stresses for crystalline silicon *J. Appl. Phys.* **42** 1897–905
Hall C A, Asay J R, Knudsen M D, Stygar W A, Spielman R B, Pointon T D, Reisman D B, Toor A and Cauble R C 2001 Experimental configuration for isentropic compression of solids using pulsed magnetic loading *Rev. Sci. Instrum.* **72** 3587–95
Hixson R S, Vorthman J E, Gustavsen R L, Zurek A K, Thissel W R and Tonks D L 1998 Spall wave-profile and shock recovery experiments on depleted uranium *Shock Compression of*

Condensed Matter ed S Schmidt, D Dandekar and J Forbes (New York: American Institute of Physics) pp 479–82

Hixson R S, Vorthman J E, Gustavsen R L, Zurek A K, Anderson D L and Tonks D L 2000 Spall response of U-NB (6%) Alloy *Shock Compression of Condensed Matter* ed M D Furnish, L C Chhabildas and R S Hixson (New York: American Institute of Physics) pp 489–92

Holzapfel W B 1991 *Molecular Solids under Pressure* ed R Pucci R and Piccitto (North-Holland, Amsterdam) pp 61–8

Jeanloz R 1988 Universal equation of state *Phys. Rev.* B **38** pp 805–7

Kanel G I, Razorenov S V, Bogatch A, Utkin A V, Fortov V F and Grady D E 1996 Spall fracture properties of aluminum and magnesium at high temperatures *J. Appl. Phys.* **79** 8310–7

Kerley G I 2006 The linear U_s–u_p relation in shock-wave physics *Kerley Technical Services Research Report* KTS06-1, March.

Kipp M E and Davison L 1982 Analysis of ductile flow and fracture in two dimensions *Shock Waves in Condensed Matter* ed W J Nellis *et al* (New York: American Institute of Physics) pp 442–5

Kolsky H and Rader D (ed) 1968 *Fracture* vol 1 ed Liebowitz (Academic) pp 533–69

Kreer J R 1971 *Air Force Weapons Laboratory Technical Report* AFWL-TR-70-180.

Lalle P 1985 personal communication.

Lawn B R and Wilshaw T R 1975 *Fracture of Brittle Solids* (London: Cambridge University Press)

Li B, Chen G, Gwanmesia G and Liebermann R C 1998 *Sound* velocity measurements at mantle transition zone conditions of pressure and temperature using ultrasonic interferometry in a multi-anvil apparatus *Properties of Earth and Planetary Materials at High Pressure and Temperature* ed M H Manghnani and T Yagi (Washington, DC: American Geophysical Union) pp 41–61

Marsh S P 1980 *LASL Shock Hugoniot Data* ed S P Marsh (Berkeley: University of California Press)

McQueen R G 1968 The equation of state of mixtures, alloys and compounds, *Seismic Coupling, DARPA Proc.,* January.

Merlini M, Hanfland M and Crichton W A 2012 $CaCO_3$–III and $CaCo_3$–VI high-pressure polymorphs of calcite: possible host structures for carbon in the Earth's crust *Earth Planetary Sci. Lett.* **333-334** 265–71

Meyers M A and Aimone C T 1983 *Prog. Mater. Sci.* **28** 1

Meyers M A, Vohringer O and Lubarda V A 2001 The onset of twinning in metals: a constitutive description *Acta Mater.* **49** 4025–39

Morse P M 1929 Diatomic molecules according to the wave mechanics. II. Vibrational levels *Phys. Rev.* **34** 57–64

Murnaghan F D 1944 The compressibility of media under extreme pressures *Proc. Natl Acad. Sci* **30** 244–7

Murri W J, Grady D E and Mahrer K D 1975 Equation of state of rock, *Stanford Research Institute Report* PYU 1883, #AT(04-3)-115, July.

Novikov S A 1981 Shear stress and spall strength of materials under shock loads *Zhur. Mekh. I Tekh. Fiz.* **3** 109–20

Paris V, Frage N, Dariel M P and Zaretsky 2011 Divergent impact study of the compressive failure threshold in SiC and B_4C *Int. J. Impact Eng.* **38** 228–37 de Rességuier T, Signor L, Dragon A, Severin P and Boustie M 2007 Spallation in laser shock-loaded tin below and just above melting *J. Appl. Phys.* **102** 073535

Riney T D, Garg S K, Kirsh J W, Hastings C R and Hamilton K G 1971 Wave propagation in porous geological composites, Systems, *Science and Software, Report* 35-R-648, DASA No. 01-69-0159 (P00002), July.

Romanchenko V I and Stepanov G V 1980 *Zhur. Prik. Mekh. Tekh. Fia* **4** 141

Rose J H, Smith J R, Guinea F and Ferrante J 1984 Universal features of the equation of state of metals *Phys. Rev.*, B **29** 2963–69

Rydberg R 1931 *Z. Phys* **73** 376–85

Shockey D A, Curran D R, Seamann L, Rosenberg J T and Petersen C F 1974 Fragmentation of rock under dynamic loads *Int. J. Rock Mech. Min. Sci.* **11** 303–17

Shockey D A, Seaman L and Curran D R 1979 Microfailure models and their application to nonlinear dynamic fracture problems *Nonlinear and Dynamic Fracture Mechanics* ed N Perrone and S N Atluri (New York: ASME) pp 79–104

Sikka S K, Godwal B K and Chidambaram R 1997 Equation of state at high pressure *High-pressure Shock Compression of Solids* vol 3 (Berlin: Springer) pp 1–35

Steinberg D J 1996 Equation of state and strength properties of selected materials, *Lawrence Livermore National Laboratory Report* UCRL-MA-106439, February.

Swegle and Grady D E 1985 Shock viscosity and the prediction of shock wave rise times *J. Appl. Phys.* **58** 692–701

Taylor G I 1910 The conditions necessary for discontinuous motion of gasses *Proc. R. Soc* p 371–7

Taylor J W 1982 *Los Alamos Shock Wave Profile Data* ed C E Morris (Berkeley: University of California Press) p 38

Todd S N 2007 Dynamic fracture of a mild steel plate *Sandia National Laboratories Technical Dept.* SAND07-1403J, March

Trunin R F, Gudarenko L F, Zhernokletov M V and Simakov G V 2001 Experimental data on shock compression and adiabatic expansion of condensed matter, *Sarov* RFNC-VNIIF

Tvergaard V and Needleman A 1984 Analysis of the cup cone fracture in a round tensile bar *Acta Metall.* **32** 157–69

van Thiel M, Shaner J and Salinas E 1977 Compendium of shock wave data *Lawrence Livermore Laboratory Report* UCRL-50108, vol. 1, 2 and 3, June.

Vinet P, Ferrante J, Smith J R and Rose J H 1986 *J. Phys. C: Solid State Phys.* C **19** L467

Winkler W D and Stilp A L 1992 Pressure induced macro and micromechanical phenomena in planar impacted TiB_2 *Shock Waves in Condensed Matter* ed S C Schmidt *et al* (Amsterdam: Elsevier) pp 555–8

Wood D S 1952 *J. Appl. Mech.* **19** 521

Yeshuran Y, Brandon D G, Venkert A and Rosenberg Z 1988 The Dynamic properties of two phase aluminum glass ceramics, *DYMAT-88, Int. Conf. on Mechanical and Physical Behavior of Materials under Dynamic Loading, Les Editions de Physique* pp 11–8

Zeldovich Y B and Raiser Y P 1966 *Physics of Shock Waves and High-temperature Hydrodynamic Phenomena* vol 1 (New York: Academic)

Zurek A K and Meyers M A 1995 Microstructure aspects of dynamic failure *High-pressure Shock Compression of Solids II: Dynamic Fracture and Fragmentation* ed L Davison *et al* (Berlin: Springer) pp 25–89

Chapter 2

Dynamic fragmentation of solids and liquids

The intense shock compression of bodies and structures brought about by impact and explosion culminates in the catastrophic destruction of the partaking materials. The span of such events and applications is vast. The topic is identified here as the dynamic fragmentation of solids. The present chapter surveys a number of issues pertinent to the science and engineering aspects of dynamic fragmentation. More extended topics on dynamic fragmentation are pursued in the subsequent two chapters. Most of the issues addressed have application to the intense disruption and breakup of any condensed matter and some cases of the shock fragmentation of liquids are considered. Statistical characterization of the fragment debris resulting from a dynamic fragmentation event in terms of both fragment sizes and velocities is a key concern and this issue is explored from both a historic and a modern perspective. Intrinsic to dynamic fragmentation is the relationships between the strength of the impact or explosive event, and the intensity of the breakup in terms of the numbers and the sizes of the resulting fragment debris. Both historic and more recent efforts are examined that address this central issue of fragmentation. The catastrophic fragmentation of brittle solids has been a particularly difficult topic, and both experimental and theoretical efforts are explored that shed light on this issue. Failure and fragmentation in the spall environment are of unique interest to shock compression science, and solutions are pursued that lend further insight into the spall fragmentation event.

2.1 Introduction

The topic of dynamic fragmentation is pursued in the present chapter. Dynamic fragmentation is concerned with both the process and the outcome of an event in which a structure or body is subjected to intense and abrupt forces causing it to break up into a number of component parts (fragments). Usually the concern is with the fragmentation of an entity composed of condensed matter; i.e. a ductile or brittle solid, or a liquid. The concept of dynamic fragmentation is not alien to gaseous

doi:10.1088/978-0-7503-1254-7ch2

matter, however. One of the more thought provoking examples is that of the early Universe, theorized as initially composed of a nearly homogenous gas of elementary particles, before density perturbations and instabilities in the gravitational attraction precipitated fragmentation that eventually evolved into the distribution and fabric of galaxies and galactic clusters observed today.

The more mundane topics addressed here, however, will involve the dynamic fragmentation of condensed matter—and primarily, solid condensed matter. Some applications to the dynamic fragmentation of liquids are also pursued, although a more complete treatment is provided elsewhere (e.g. Grady 1995).

Even within this restricted consideration of dynamic fragmentation, the subject is quite broad, both in material response and in the kinematic conditions imparted by impact or explosion that lead to fragmentation. The shatter of an impacted sheet of glass differs starkly from the rupture of an explosive-driven metal shell. Impact-induced spall bears no outward similarity to the breakup of a stretching metal jet. The strain energy resident in a stone or concrete pillar subject to an excessive compressive load will cause the pillar to fail explosively and experience extensive dynamic fragmentation. A number of commonalities in these outwardly disparate fragmentation events will become apparent.

In a given dynamic fragmentation event there are a number of features that can be of concern. How many pieces does the body break into? What is the largest piece? What is the smallest piece? What is the disposition of the fragment debris in terms of momentum and kinetic energy? What is the stress, impulse or energy requirement needed to produce the fragmentation event? These and other issues are explored in the present chapter.

The next section addresses probabilistic issues of dynamic fragmentation, including both historic and more recent representations of fragment distributions, along with the arguments supporting application of the respective distributions. In section 2.3, methods used to predict fragment size in a fragmentation event are surveyed. These include empirical methods as well as techniques that are based on stress, impulse and energy driving the fragmentation process. Size prediction methods are also considered that are founded on the inherent flaw structure of engineering solids. Section 2.4 focuses on issues specific to the fragmentation of brittle solids. Of particular interest are instructive similarities between the hydrodynamic turbulence of fluids and catastrophic fragmentation of brittle solids. The comparison illuminates scale invariant features of brittle fragmentation central to the unique fragment size distributions of brittle solids. More extended topics on dynamic fragmentation are pursued in the subsequent two chapters.

2.2 Size distributions and probabilistic methods in dynamic fragmentation

The violent disruption of a formerly contiguous body into a diverging collection of discrete high-velocity fragments is a consequence of an intriguing mix of physics and statistics that is not fully understood, even with the considerable efforts directed towards this end over the past several decades. Dynamic fragmentation events range from the commonplace, such as explosive blasting of rock in the mining and quarry industry, to the more exciting issues of understanding the size distribution and clustering of the galaxies resulting from the Big Bang genesis of the Universe.

The statistical character of the sizes and velocities of fragments resulting from an explosive or impact event is self-evident. Less evident, however, is the mathematical framework most appropriate for the statistical prediction and description of a dynamic fragmentation event. The various approaches that have been pursued are considerable and no clear consensus has yet been achieved. Empiricism governed the statistical description of the size distribution of fragmentation events over most of the past century. More recently, a number of approaches have been pursued including geometric statistics, fractal physics, maximum entropy and information theory ideas, percolation theoretical methods and classical statistical mechanics.

Statistical issues in the dynamic fragmentation event are the principal concern in many engineering and scientific applications. Properties of the statistical distribution in size and shape of the fragment debris resulting from such an event are commonly the foremost features sought. In the detonation of exploding munitions what are the number, the masses and the velocities of the most lethal fragments? In the accidental or intentional detonation of a device containing hazardous material, what fraction of the disrupted material results in particulate of a size small enough to remain airborne?

Statistics of the fragmentation event will permeate all sections of the present chapter. This section reviews methods that address strictly probabilistic issues of the fragmentation event. Physical issues regarding details of the propagation and interaction of fracture leading to fragmentation are largely ignored. It is found that in many engineering applications a probabilistic approach quite adequately provides the needed answers. In other very pertinent problems, a strictly probabilistic approach is not sufficient, and more detailed physics must be pursued.

It should be noted at the outset, however, that a generally accepted theory of statistical fragmentation has not yet emerged. Consequently, a wide number of empirical and semi-theoretical approaches are currently in active use in addressing the numerous applications involving the statistical disruption and fragmentation of condensed matter.

There are a number of reasons for this. First, the physical and engineering applications of dynamic fragmentation are extremely diverse and an extensive literature on the subject going well back into the last century already exists. A number of empirical and semi-empirical theories for describing the consequences of fragmentation events are already comfortably entrenched in the various fields concerned with dynamic fragmentation. Portions of this literature have previously been reviewed (Grady and Kipp 1985). Second, it is difficult to perform experiments to select among different theories that meet the rather stringent statistical conditions assumed in the development of the theories. Methods exist for bringing the theories into accord with the more complex statistical conditions of the experiments, but uniqueness is lost and experiment no longer serves as a method for discrimination among theories.

Statistical complexities arise in experiments for several reasons. Fragmentation experiments are invariably statistically inhomogeneous. That is, the fragmentation intensity, or average fragment size, varies as a function of position within the fragmenting body because of spatial variations of the forces and energies causing fragmentation. An example is provided by the fragmentation caused when a stone strikes an untempered glass windshield of a sufficiently fast-moving automobile. Fracture intensity is severe near the point of impact where dynamic forces are high but gradually lessens in regions away from the impact point as the stresses driving fracture attenuate with distance.

Further complications arise when the size of the fragmented body is finite. If the number of fragments is few, the distribution in fragment size will depend on the total mass of the body. In addition, in certain applications of interest there is a fundamental minimum fragment size brought about by specific material conditions that skew the predicted statistical fragment distribution. This might occur, for example, when a strong aggregate is weakly bonded into a competent body. The aggregate size would then represent the minimum fragment size into which the body can be broken. The nucleus of an atom might represent such an aggregate in that the forces bonding nucleons together is weak relative to the forces necessary to disrupt an individual nucleon into its component parts.

Here we will pursue some of the probabilistic methods that have been pursued to address the statistical consequences of fragmentation events, including some of the complications noted above.

2.2.1 Early applications and empirical distributions

The reduction of a monolithic body to particulate of the same material through the application of some set of mechanical forces has been of interest in a wide range of both industrial and scientific concerns. The blasting of rock in mines and quarries; the periodic collision of planetary bodies resulting in the distribution of objects throughout the asteroid belt and other regions of the solar system; and the dynamic processes which occur where land and seas meet, producing the sand, silt and gravel beaches on Earth, are just a few examples of applications that have tested scientific thought over the past several centuries.

We will identify, in the present section, some of the more prominent scientific attempts over about the last century and a half to provide predictive order and understanding to such fragmentation processes. These efforts have generally focused on one or the other of two general issues. Namely, what are the forces and energies required to cause the fragmentation events? In addition, what are the size and distribution of particulate resulting from these breakup processes?

Over the time span considered here, most of the attention has focused on the distribution in particulate size resulting from the fragmentation. Historical efforts concerning the causes of fragmentation are addressed in later sections.

2.2.1.1 Logarithmic-normal fragment distribution

Analytic representations of the distributions of fragments resulting from various breakage events have historically been of two general types. One is of the normal distribution form (actually logarithmic-normal). The other form, which has stronger empirical roots, carries a measure of the average fragment size in the argument of an exponential function. Many forms of the exponential-based distribution have been proposed along with efforts at experimental and theoretical justification.

The logarithmic-normal fragment distribution has generally been considered inappropriate for dynamic single-event fragmentation processes, such as through impact or explosive loading. Actually, the logarithmic-normal fragment distribution has remarkable analytic flexibility, and can adequately represent most of the fragment distributions generated both in the laboratory and in nature well within the accuracy with which fragment distributions can be measured. Theoretical developments of the logarithmic-normal fragment distribution (e.g. Kolmogorov 1941, Epstein 1947) are based on the application of the asymptotic limit of the central limit theorem to continued repetitive breakage processes such as might occur in crushing and grinding, or perhaps in repeated impact events. Applicability is stated to require that the continued diminution of a body through fragmentation occurs in a self-similar manner independent of current particle size. Thus, crusher loads, geological deposits and the size frequency of meteors are representative of distributions that are expected to be appropriately described by a logarithmic-normal distribution. The functional form of the logarithmic-normal fragment distribution is

$$f(x) = \frac{1}{\sqrt{2\pi}\sigma x} e^{-\frac{1}{2\sigma^2}(\ln x - \mu)^2}, \tag{2.1}$$

where the probability density function $f(x)$ provides the number distribution of fragments of size x in terms of the scale parameter μ and shape parameter σ. The expected value of the distribution is

$$<x> = e^{\mu + \sigma^2/2}, \tag{2.2}$$

and the standard deviation, normalized by the expected value,

$$s.d./<x> = (e^{\sigma^2} - 1)^{1/2}. \tag{2.3}$$

The distribution provided in equation (2.1) is conveniently normalized by the expected value through $X = x/<x>$, yielding

$$f(X) = \frac{1}{\sqrt{2\pi}\sigma X} e^{-\frac{1}{2\sigma^2}(\ln X + \sigma^2/2)^2}. \tag{2.4}$$

Logarithmic-normal distributions from equation (2.4) for various values of the shape parameter σ are illustrated in figure 2.1. The distribution can range from the highly skewed distributions characteristic of the shatter of brittle solids, such as glass and ceramics, to a strongly centered distribution representative of the fragmentation of expanding shells or rings of ductile metals.

2.2.1.2 Rosin–Rammler and related fragment distributions

A framework for a mathematical representation of the distribution of fragments, which has found wide application, appears to have had its genesis within the field of coal and ore crushing during the 1930s. Fragment distributions are determined by

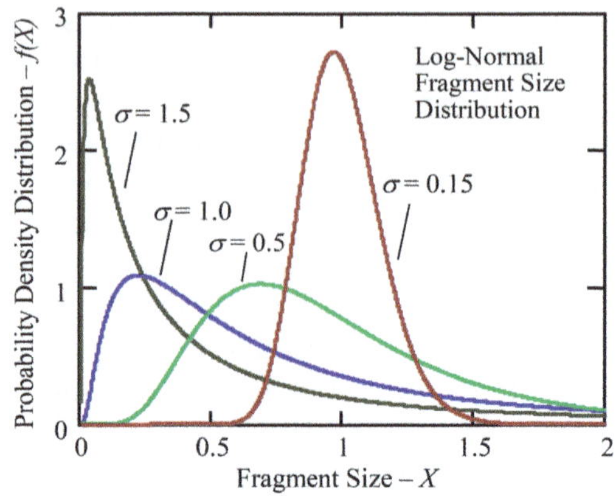

Figure 2.1. Probability density functions for logarithmic-normal fragment distribution. The expected value is unity and curves are for various values of the shape parameter.

sieving methods where fragments within a given size range are separated through a collection of variously sized sieving screens. Fragments within each size range are weighed, providing a measure of the mass fraction of fragments within each size range and consequently a measure of the fragment size distribution.

The most frequently quoted fragment distribution from this period is attributed to Rosin and Rammler (1933). The Rosin–Rammler distribution is most commonly written as a cumulative mass fraction greater than fragment size x,

$$\frac{M(x)}{M_o} = 1 - e^{-(x/\lambda)^n}, \tag{2.5}$$

where λ is the size scale parameter of the distribution and the shape parameter n typically ranges between about 0.5 and 1.5. The Rosin–Rammler fragment mass distribution is of the Weibull extreme value statistical distribution form (Hahn and Shapiro 1967). The limiting form of the Rosin–Rammler distribution, for $x \ll \lambda$,

$$\frac{M(x)}{M_o} = (x/\lambda)^n, \tag{2.6}$$

is usually attributed to Schuhmann (1940), and is quite appropriate in most applications (see also Gaudin 1926).

Attempts were made to theoretically derive the distribution of Rosin and Rammler, and most are based on the elementary fragmentation algorithm explored extensively by Lienau (1936). Lienau investigated the one-dimensional problem of the distribution in fragment lengths resulting from the random placement of breaks on a line. Provided the average fragment length λ is small compared to the total initial length L, it is readily shown that the fragment length probability density distribution is of the Poisson form,

$$f(x) = \frac{1}{\lambda}e^{-x/\lambda}. \tag{2.7}$$

If relatively few breaks occur on the line length L it can also be shown that the density distribution in fragment lengths corresponding to equation (2.7) is binomial (e.g. Grady 1990).

Bennett (1936) used the Poisson relation in equation (2.7) in an effort to justify the Rosin–Rammler distribution during an investigation of crushing and shatter of hard coals. He assumed that the Poisson process applied to the random placement of fracture planes partitioning the length, width and breadth of a fractured body, and arrived at the Rosin–Rammler relation in equation (2.5) with a value of the shape parameter (his distribution index) of $n = 1$. He offered arguments of non-ideality in the breakage process to justify values of n both less than and greater than unity.

Gilvarry (1961) noted inconsistencies in the derivation of Bennett (1936) and undertook a physics-based statistical development of fragment distributions resulting from single fracture events in brittle solids. Gilvarry postulated the existence of independent edge, surface and volume flaws underlying the breakage process in brittle bodies. By ascribing a Poisson process to each type of fracture-initiating defect he arrived at the Gilvarry distribution,

$$\frac{M(x)}{M_o} = 1 - e^{-\left(\frac{x}{\lambda_1} + \left(\frac{x}{\lambda_2}\right)^2 + \left(\frac{x}{\lambda_3}\right)^3\right)}, \tag{2.8}$$

where, again, $M(x)/M_o$ is the cumulative mass fraction of fragments of size measure x. Gilvarry noted that in most instances edge flaws (the first term in the exponential) dominated, thus providing the Rosin–Rammler equation with $n = 1$. Variations from $n = 1$ occurred, according to Gilvarry, when other flaw types contributed, or in repeated breakage processes such as crushing where the single fracture condition was not met.

The present author has also noted difficulties in the Gilvarry (1961) development, particularly in his method of mass normalization (equation (16) in his paper) of the derived statistical relation. It seems that Gilvarry replicates the inconsistencies noted in the analysis of Bennett (1936). Consequently, the Rosin–Rammler fragment distribution should probably remain as a historically interesting and experimentally convenient empirical relation for the characterization of fragment debris.

Gaudin and Meloy (1962) have provided an alternative distribution in the same vein. They worked with the binomial distribution as the basis for randomization of the fragment size, rather than with the Poisson distribution, and arrived at the cumulative mass fraction distribution,

$$\frac{M(x)}{M_o} = 1 - \left(1 - \frac{x}{\lambda}\right)^r, \tag{2.9}$$

where λ is regarded as a characteristic feed size (initial body size) and r is a measure of the reduction ratio in the breakage process. Again, the Gaudin–Meloy development appears to suffer from mass normalization inconsistencies similar to the Gilvarry efforts.

Several of the distributions described here and used in the coal and ore industry are illustrated in figure 2.2. The distributions have been normalized to a mass average fragment size of unity. A shape index of $n = 1$ is used in the Schuhmann and

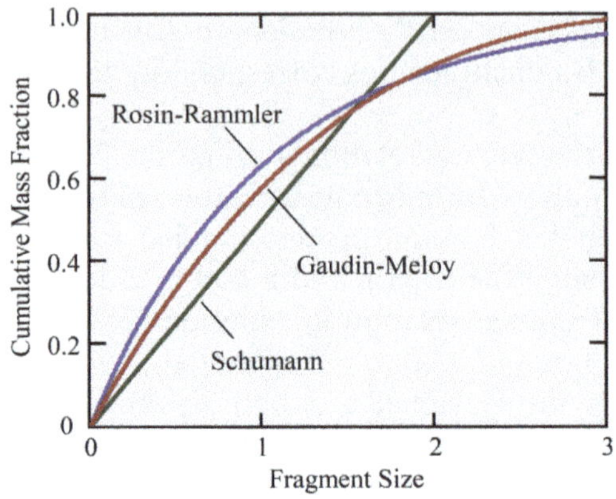

Figure 2.2. Cumulative fragment distributions used to describe the consequences of shatter and crushing in the coal and ore industry.

the Rosin–Rammler relations. A relatively modest value of $r = 1$ is used in the Gaudin–Meloy equation. As r increases, the Gaudin–Meloy and the Rosin–Rammler relations become coincident.

2.2.1.3 The Mott–Linfoot fragment distribution

To researchers working in the field of exploding munitions fragmentation, and in a number of areas closely akin to munitions fragmentation, no theoretical work has had more profound impact than that of Nevill Mott. This seminal work of Mott is found in a series of several reports published during the World War II years (Mott and Linfoot 1943, Mott 1943, 1944) and in several papers written shortly after (Mott 1947, 1948). The fragmentation work of Mott is surveyed in Grady (2006).

The Mott distribution, or the Mott–Linfoot distribution as it will be referred to here, has been used widely to characterize fragment distributions from the breakup of exploding bombs and shells (Mott and Linfoot 1943). Mott and Linfoot also used, as did earlier workers, the one-dimensional Poisson distribution derived by Lienau (1936) as the statistical basis for their development.

Their acceptance of the Lienau distribution was bolstered by the fragmenting munitions data available to them at the time, which were found to plot reasonably linearly in a log number versus cube root of the fragment mass representation. Since $m^{1/3}$ is proportional to a length measure of the fragment they reasoned that the same random variable considered in the Lienau one-dimensional development also applied in the multidimensional fragmentation event. Further, in examining fragments from the available data, they observed that substantial portions retained inner and outer surfaces of the original munitions case. This suggested that in the fragmentation of a thin plate the appropriate length scale would be proportional to $m^{1/2}$. Thus, a plot of log number versus $m^{1/2}$ should provide, by the arguments given, a better fit to the fragment distribution data. Thus, the fragment cumulative number probability distribution proposed by Mott and Linfoot (1943) was,

$$F(m) = 1 - e^{-(m/\mu)^{1/2}}, \tag{2.10}$$

where the characteristic mass μ is the distribution scale parameter. The corresponding probability density distribution is then,

$$f(m) = \frac{1}{2\mu}\left(\frac{m}{\mu}\right)^{-1/2} e^{-(m/\mu)^{1/2}}. \tag{2.11}$$

Numerous researchers have successfully used this distribution in various forms to organize and compare vast amounts of exploding munitions fragmentation data. Mott expended considerable subsequent effort in a quest to justify the functional form assumed in the equations above. In general, he did not succeed. However, his efforts resulted in some of the most thoughtful physics-based analysis of the dynamic fragmentation process that exists to the present.

2.2.2 Poisson processes and geometric methods

The idea of randomly placing points within a body (a Poisson process) and using these points to randomly partition the body through some algorithm seems intuitively reasonable. The method was pursued by a number of early workers to provide a rationale for observed fragment distributions, or to suggest a distribution when experiment data were not available. The body to be randomly fragmented could be a line (one dimension), a surface (two dimensions) or a volume (three dimensions).

As noted earlier, the seminal effort within this approach to random fragmentation is probably the study of Lienau (1936). The emphasis of the Lienau study was on the random fragmentation of a one-dimensional body or a line. Later work attempted to extend the methods of Lienau to the fragmentation of higher-dimensional bodies (surfaces and volumes). The efforts of Bennett (1936), Gilvarry (1961), and Gaudin and Meloy (1962), in attempting to justify observed fragment distributions in brittle solids, relied heavily on the one-dimensional Lienau distribution. Mott and Linfoot (1943) also explored the random geometric fragmentation of a body. Also relevant to methods of geometric fragmentation are the classic tessellation techniques of Voronoi and Dirichlet.

The random geometric fragmentation methods described here do not exhaust the various methods that have been explored. Additionally, it becomes apparent that these geometric fragmentation approaches, divorced from the fragmentation physics, do not explain the many statistical observations in the experimental fragmentation of real materials. When applied with care however, geometric fragmentation methods have useful application to practical fragmentation processes.

2.2.2.1 Lienau distribution

The theoretical study of Lienau (1936) illustrates the geometric fragmentation concept. He considered the elementary problem of an extended one-dimensional body such as a glass rod or a stretching wire subjected to forces that result in the multiple random fracturing of the body. If any point on the body is as likely as any other point to fracture, then the problem is statistically well posed. The problem is

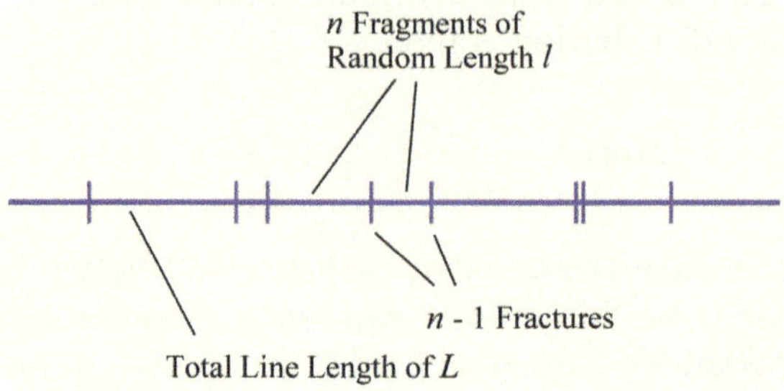

Figure 2.3. Line of total length L broken at random into fragments of variable length l by $n - 1$ fractures.

modeled as that of an infinite one-dimensional line in which breaks are introduced with equal probability at any point on the line, as illustrated in figure 2.3.

As stated, the random geometric fragmentation of a one-dimensional body is reasonably unambiguous. An analytic solution requires only a proper probabilistic description of the random breaks and the lengths of the segments delineated by these breaks. It is shown later that even this simple prescription for the random statistical fragmentation of a one-dimensional body is debatable. The following analysis leads to the one-dimensional Lienau fragment size distribution.

Consider a line of length L in which breaks on the line are introduced at random (Grady 1990). Since we are interested in partitioning the line into a large number of fragments (the average length is small compared to the total length L), the finite length of the line is not of consequence and is considered infinite. The average spacing between breaks λ, or equivalently the frequency of breaks per unit length $h_o = 1/\lambda$, characterizes the statistical distribution. The random distribution of points on a line is described by Poisson statistics. If an arbitrary length l of the line is examined, then the probability of finding n points (fractures) within the length l is given by the Poisson relation,

$$P(n,l) = \frac{(l/\lambda)^n e^{-l/\lambda}}{n!}. \tag{2.12}$$

The probability of occurrence of fragments of length l within a tolerance dl is then,

$$f(l)dl = P(0,l)P(1,dl) = (1/\lambda)e^{-l/\lambda}dl. \tag{2.13}$$

The one-dimensional probability density distribution is,

$$f(l) = (1/\lambda)e^{-l/\lambda}, \tag{2.14}$$

and the cumulative fragment distribution,

$$F(l) = 1 - e^{-l/\lambda}. \tag{2.15}$$

2.2.2.2 Binomial distribution

If, in figure 2.3, the line length L is broken into relatively few pieces, then the random placement of fractures is no longer governed by Poisson statistics. Probability

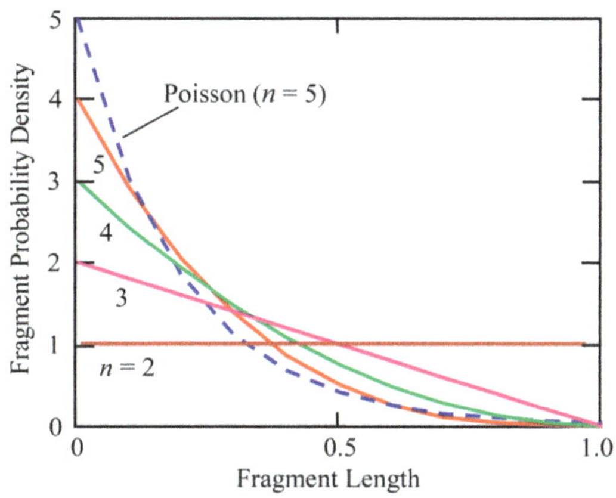

Figure 2.4. The fragment probability distributions for fragmentation of a body of unit length into either $n = 2$, 3, 4 or 5 fragments. The dashed line shows the Poisson distribution approximation to the $n = 5$ fragments case.

aspects of the fragmentation are instead governed by the binomial probability function,

$$P_{j,k}(p) = \frac{k!}{j!(k-j)!}p^j(1-p)^{k-j}. \tag{2.16}$$

Given n fragments ($n - 1$ fractures) the probable occurrence of a fragment of length l within the interval dl is again the product of the probability of no fractures in length l and one fracture within length dl where, respectively, $p = l/L$ and $dp = dl/(L - l)$,

$$f(l)dl = P_{0,n-1}(p)P_{1,n-1}(dp) = \frac{n-1}{L}\left(1 - \frac{l}{L}\right)^{n-2} dl, \tag{2.17}$$

with $f(l)$ providing the requisite probability density function. The cumulative distribution function is then,

$$F(l) = 1 - (1 - l/L)^{n-1}, \tag{2.18}$$

and the average fragment size is $\lambda = L/n$. Binomial probability densities are shown in figure 2.4 and compared with the corresponding Poisson distribution. The Poisson and binomial distributions converge as n becomes large.

2.2.2.3 Discrete fracture sites distribution

Further modification of the fragment distribution occurs when the body has a limiting grain size, which constrains the minimum spacing between fractures. Setting this minimum size to δ, the number of possible fracture sites per unit length is $N = 1/\delta$. If fractures occur at $N_f < N$ sites then the probability of fracture at any specific site is $p = N_f/N = N_f\delta$. The probability of finding a fragment of length $l = n\delta$ is then

$$P(n\delta) = (1 - p)^{n-1}p. \tag{2.19}$$

The average fragment length is $\lambda = 1/N_f = \delta/p$ and the discrete fragment probability density relation becomes

$$f(l) = \frac{\delta}{\lambda}\left(1 - \frac{\delta}{\lambda}\right)^{(l/\delta)-1}. \tag{2.20}$$

The cumulative fraction of fragments of size $l > n\delta$ is provided by

$$\sum_{i=n}^{\infty}\frac{\delta}{\lambda}\left(1 - \frac{\delta}{\lambda}\right)^{n}. \tag{2.21}$$

The relation $\sum_{n=r}^{\infty} x^n = x^r/(1 - x)$ then provides for the cumulative fragment distribution

$$F(l) = 1 - \left(1 - \frac{\delta}{\lambda}\right)^{l/\delta}. \tag{2.22}$$

The above relations converge to the exponential Lienau distribution as δ/λ becomes small.

Probability fragment-size relations can also be developed for a one-dimensional body of finite length and with discrete fracture sites (Grady 1990). The relations are more complex but tractable.

2.2.3 Geometric fragmentation of a surface

Later workers, to address the statistical fragmentation of more complex bodies, pursued the geometric fragmentation of a one-dimensional body studied by Lienau (1936). Bennett (1936) considered fragmentation of a brittle solid through the random distribution of mutually orthogonal planes distributed according to the Lienau method. Gilvarry (1961) pursued the similar, but more complex, admixture of one-, two- and three-dimensional Poisson fragmentation processes. The approach of Gaudin and Meloy (1962) was similar to that of Bennett except they chose the one-dimensional binomial distribution as the basis for a more general statistical development.

Mott and Linfoot (1943) also considered the Lienau distribution in exploring fragmentation statistics in the explosive breakup of cylindrical metal shells. Clearly, the fragmentation of bodies through random geometric methods had appealed to these earlier workers. Pursuing some of these efforts is worthwhile. The fragmentation of thin-walled cylindrical munitions led Mott to consider the geometric breakup of a surface, or a two-dimensional body. Mott and Linfoot showed that certain two-dimensional geometric statistical fragmentation methods are amenable to rigorous analytic solutions.

2.2.3.1 Mott random line fragmentation

In figure 2.5 selected geometric fragmentation methods are illustrated which have been pursued to explain experimental fragment distributions. The first illustrates the partitioning of a plane with vertical and horizontal parallel lines, while the second

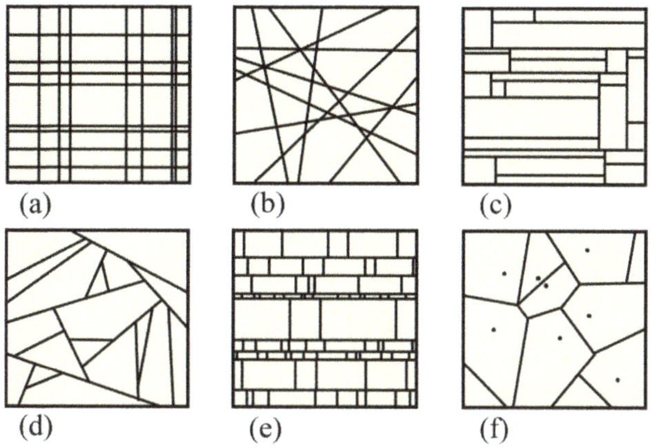

Figure 2.5. Geometric random fragmentation algorithms explored by Mott and others.

with randomly oriented straight lines. Both geometric fragmentation algorithms were explored in the report of Mott and Linfoot (1943). The first is seen to be the two-dimensional version of the process pursued by Bennett (1936) and others.

The solution of the random horizontal and vertical line fragmentation geometry put forth by Mott and Linfoot proceeds as follows (e.g. Grady 2006). The probability density distribution over fragment lengths and widths is provided by the juxtaposition of the one-dimensional Lienau distribution over both the x (horizontal) and y (vertical) directions,

$$f(x,y) = \frac{1}{x_o y_o} e^{-x/x_o - y/y_o}. \tag{2.23}$$

The cumulative distribution $F(z)$ over all fragments of size greater than $z = \sqrt{xy}$ is then provided by the integral

$$1 - F(z) = \iint\limits_{xy>z^2} \frac{1}{x_o y_o} e^{-x/x_o - y/y_o} dxdy. \tag{2.24}$$

The relation can be integrated over the y coordinate yielding

$$\frac{1}{x_o} \int_0^{\infty} e^{-\frac{1}{x_o}\left(x + \frac{x_o}{y_o}\frac{z^2}{x}\right)} dx. \tag{2.25}$$

Not immediately obvious is the transformation

$$x = z\sqrt{\frac{x_o}{y_o}} e^{\theta}, \tag{2.26}$$

which leads in turn to the integral

$$2\frac{z}{z_o} \int_0^{\infty} e^{-2\frac{z}{z_o}\cosh\theta} \cosh\theta d\theta, \tag{2.27}$$

where $z_o = \sqrt{x_o y_o}$. Mott and Linfoot recognized this as one of a series of integral solutions for the modified Bessel functions of integer order (Abramowitz and Stegun 1954) and arrived at the cumulative fragment size probability distribution

$$F(z) = 1 - 2\frac{z}{z_o}K_1(2z/z_o), \tag{2.28}$$

and probability density distribution

$$f(z) = 4\frac{z}{z_o^2}K_o(2z/z_o). \tag{2.29}$$

The intent of the exercise by Mott and Linfoot was to rationalize experimentally observed distributions of fragments resulting from exploding munitions, and a more intuitive analytic distribution (the well-known Mott distribution) developed by them in which they predicted the fragment size dependence $\ln f(z) \sim z$. They were not fully satisfied when this expectation was compared to equation (2.29).

Mott and Linfoot (1943) also briefly explored the second algorithm in figure 2.5 in which partitioning lines are oriented at random. They were unable to find an analytic solution except at the small fragment limit, where they suggested that $\ln f(z) \sim z$ was satisfied. Grady and Kipp (1985) explored this algorithm through computer methods and found that the distribution did not differ markedly from the Bessel solution obtained for the vertical and horizontal line fragmentation algorithm.

A useful and interesting feature of the vertical and horizontal line fragmentation method is the ability to vary the density of lines in the two orientations, and assess the distribution in fragment aspect ratio as well as size. This distribution has application, for example, in the breakup of biaxial stretching metal shells where stretching rates in orthogonal directions differ. Pursuing the previous solution, the distribution

$$k(r) = \frac{1}{r_o}\frac{1}{(1 + r/r_o)^2} \tag{2.30}$$

over the fragment aspect ratio $r = x/y$ is readily obtained.

2.2.3.2 Voronoi–Dirichlet fragmentation

The present discussions of random geometric fragmentation would be remiss without consideration of the Voronoi–Dirichlet construction (e.g. Boots and Murdoch 1983). This method for the random partitioning of space has received by far the lion's share of attention in a broader spectrum of literature. The resulting distributions have been proposed for such diverse applications as the distribution of galactic matter throughout the Universe (Kiang 1966) and the formation of geologic columnar structures such as the Giant's Causeway in Northern Ireland (Weaire and Rivier 1984). The Voronoi–Dirichlet construction in two dimensions is illustrated in panel (f) of figure 2.5.

Analytic relations for the fragment size distributions resulting from the Voronoi–Dirichlet construction have not been directly determined. A computational determination of the resulting fragment size distributions has been widely pursued, however (e.g. Crain 1978), and analytic expressions which successfully reproduce the computational distributions have been developed by inductive methods (Kiang 1966).

Both the analytic distributions and the inductive development of them are of interest to the present outline of statistical fracture through geometric methods. Kiang (1966) first considered the one-dimensional Voronoi–Dirichlet construction where points are distributed at random on a line (a Poisson process) and then the degenerate (one-dimensional) Voronoi cell from the perpendicular bisector (the midpoint) of each point pair is determined. Thus, the Voronoi–Dirichlet distribution on a line is the dual of the Lienau distribution considered earlier (or the degenerate Delauney distribution). Whereas in the Lienau distribution random points on the line were considered as breaks or fractures, in the present one-dimensional Voronoi–Dirichlet distribution the same random points constitute in some sense the centroid of fragments with fractures occurring at the bisector points.

The fragment size distribution for the one-dimensional Voronoi–Dirichlet distribution can be readily determined (Grady 2006). The probability of finding a length l_i between a Poisson point pair is given by the Lienau expression

$$f(l_i)dl_i = \frac{1}{\lambda}e^{-l_i/\lambda}. \tag{2.31}$$

The probability of finding a point pair of length l_1 adjacent to a point pair of length l_2 is then the product

$$f(l_1)f(l_2)dl_1dl_2 = \frac{1}{\lambda^2}e^{-(l_1+l_2)}dl_1dl_2. \tag{2.32}$$

Implementing the transformation

$$l = (l_1 + l_2)/2, \tag{2.33}$$

$$\xi = (l_1 - l_2)/2, \tag{2.34}$$

and integrating over the variable ξ leads to the one-dimensional Voronoi–Dirichlet distribution,

$$f(l) = \frac{2}{\lambda}\left(\frac{2l}{\lambda}\right)e^{-2l/\lambda}. \tag{2.35}$$

The distribution in equation (2.35) is a gamma function of order $n = 2$. Kiang (1966) offered without proof that symmetrically higher-order gamma functions would provide analytic fragment distributions for Voronoi–Dirichlet partitioning of an area or a volume. Following Kiang, write the general expression for fragment distributions over fragment mass,

$$f(m) = \frac{1}{\mu}\frac{n}{\Gamma(n)}\left(\frac{nm}{\mu}\right)^{n-1}e^{-nm/\mu}, \tag{2.36}$$

where $n = 2$, 4 or 6 is for fragmentation of a line, surface or volume, respectively.

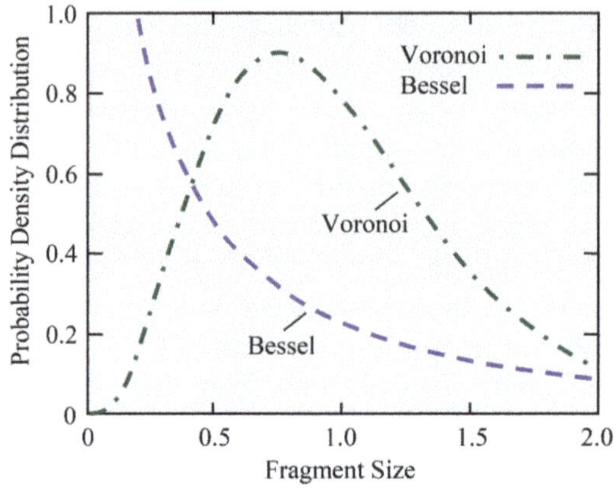

Figure 2.6. Comparison of fragment size distributions resulting from the Mott–Linfoot random horizontal and vertical line algorithm and Voronoi–Dirichlet tessellation in the geometric fragmentation of a surface.

Computational distributions from Voronoi–Dirichlet constructions on an area performed by Kiang (1966) were in acceptable agreement with equation (2.36) for $n = 4$. A degree of controversy was generated by Kiang's proposal among subsequent authors as to the adequacy of equation (2.36); both for and against. Apparently, the construction of computer algorithms to generate Voronoi–Dirichlet fragment distributions is not a trivial exercise. In any case, for the present geometric fragmentation investigations, equation (2.36) is an adequate analytic representation of Voronoi–Dirichlet distributions in line and area fragmentation.

Fragment size (area) distributions resulting from both the Voronoi–Dirichlet algorithm and the horizontal and vertical lines algorithm (Bessel distribution) are compared in figure 2.6. The two density distributions are normalized to unit expected value. The comparison reveals the stark differences resulting from differing randomization algorithms and certainly raises questions as to the applicability of statistical geometric methods to fragmentation.

2.2.3.3 Mott cylinder segmentation

Mott was clearly not satisfied with the results from the fragment distributions resulting from the random placement of infinite lines on a surface. It was becoming apparent to him that such methods did not lead to unique distributions, and were very much dependent on the algorithm selected. Nevertheless, he made a final attempt at the use of a random geometric method before abandoning the approach. Perhaps he reasoned that if the method was algorithm-dependent, then if a method was selected which more reasonably approximated the fracture event of interest, that more credence could be attributed to the resulting distribution.

From the elongated and sliver-shaped fragments recovered from exploding munitions tests Mott surmised that fracture in an end-detonated metal shell with cylindrical symmetry would occur through longitudinal running cracks with occasional crack branching and crack intersection resulting in the observed fragments.

Thus he proposed the random partitioning algorithm illustrated in panel (e) of figure 2.5. For the random distribution of longitudinal and transverse lines on the plane, Mott made the interesting selection of the one-dimensional Voronoi–Dirichlet distribution discussed in the previous subsection rather than the Lienau distribution used in his earlier geometric pursuits. He may have made this choice for one of several reasons. It is possible that he carried through the analysis using the Lienau distribution and found, as can be shown, that the distribution does not converge in the small fragment limit. Alternatively, ideas to emerge in his later work, to be discussed in a later section, may have influenced this selection; namely, that the physics of fracture interaction precludes the close proximity of parallel fractures and thus limits the number of smaller fragments. The Voronoi–Dirichlet distribution is observed to better provide a statistical constraint limiting the number of the close parallel fractures and hence the number of smaller fragments.

The method, shown in panel (e) of figure 2.5, consists of first inscribing randomly positioned horizontal lines, according to the one-dimensional Voronoi–Dirichlet distribution from equation (2.35), and then segmenting each horizontal strip with randomly positioned vertical lines where the average spacing within any strip is proportional to the width of that strip, $x_o = py$.

Following methods similar to the solution resulting from the horizontal and vertical random line algorithm, one arrives at the fragment density distribution as a function of the fragment area a,

$$f(a) = \frac{2}{a_o}\frac{1}{\sqrt{4a/a_o}}\int_0^\infty \left(\xi\sqrt{4a/a_o} - 1\right)\left(1 + 1/\xi^2\right)e^{-\xi\sqrt{4a/a_o} - 1/\xi^2}d\xi. \qquad (2.37)$$

This distribution is provided in Mott's third internal report (Mott 1943), and differs only in the distribution variable $\lambda = \sqrt{a/a_o}$ used by him. The scale parameter is $a_o = py_o^2$, where y_o is the average spacing of the axial strips. The constant p is the aspect ratio of the average length of fragments in a strip to the width of the strip. Mott suggested p of approximately five.

Mott found that the present geometric distribution was again close to the Mott–Linfoot fragment distribution ($\ln N \sim \sqrt{a}$) but not markedly better than the earlier attempt with random infinite lines. The two geometric distributions are compared with the Mott–Linfoot distribution in figure 2.7. It is interesting that the distributions from the two algorithms tend to straddle the Mott–Linfoot proposed distribution, each with respectively larger and smaller variance. It is likely that little else can be said.

2.2.3.4 Other geometric algorithms

Other algorithms for randomly partitioning a surface are readily constructed (e.g. Grady and Kipp 1985). Several alternative algorithms are illustrated in figure 2.5. Mott and Linfoot (1943) also considered the algorithm shown in panel (b). They were not able to generate the resulting fragment distribution except in the small fragment limit where the fragment probability distribution proportional to the

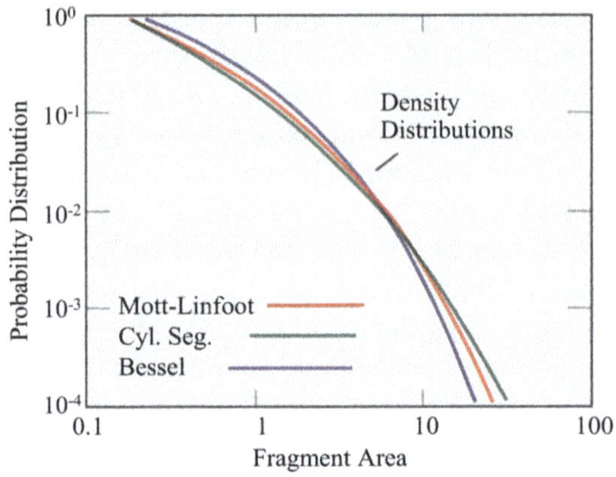

Figure 2.7. A comparison of the fragment size distributions initially proposed by Mott and Linfoot with the two random geometric fragmentation constructions attempted by them. Namely, the cylinder segmentation distribution and the random horizontal and vertical line (Bessel) distribution.

square root of the fragment area was demonstrated. The fragmentation algorithms shown in panels (c) and (d) were explored in Grady and Kipp (1985). Both algorithms result in an exponential fragment size distribution. Other random construction algorithms have been explored (Grady and Kipp 1985).

At this point, a degree of healthy suspicion should be growing as to the applicability of random geometric fragmentation algorithms to actual physical fragmentation phenomena. Later it will be shown that some utility of these methods can be made use of in modeling the statistical fragmentation phenomena, but they should be employed only with a sensible understanding of the underlying physics.

2.2.4 Alternative fragment-size statistics methods

The geometric fragmentation methods that were explored in the previous subsections pursue techniques for randomly distributing the fragment boundaries through some partitioning algorithm in the breakup of a competent body. Although the results are useful in some applications, one finds that there are many ways to randomly partition a body, and that the resulting fragment size distributions are not unique. Approaches other than geometric methods have been pursued in attempting to develop fragment size distributions appropriate to experimental observation. Some of these methods are described here.

2.2.4.1 The Mott–Linfoot distribution

The original distribution developed by Mott and Linfoot (1943) was not based on geometric arguments. Existing fragmenting munitions data and some impetus from the work of Lienau (1936) led them to propose that a measure of the fragment size $x \sim \sqrt{a}$ was a random variable distributed according to a Poisson process. The fragment area is a and is proportional to the mass m of the fragment since Mott and

Linfoot were concerned with the breakup of thin shells. This statistical assumption led to the Mott–Linfoot distribution in fragment mass,

$$f(m) = \frac{1}{2\mu}\left(\frac{m}{\mu}\right)^{-1/2} e^{-(m/\mu)^{1/2}}, \tag{2.38}$$

where μ is the distribution scale parameter and uniquely defines the distribution. The corresponding cumulative fragment distribution is then

$$F(m) = 1 - e^{-(m/\mu)^{1/2}}. \tag{2.39}$$

The Mott–Linfoot distribution has been remarkably successful in both correlating munitions fragmentation data and in developing predictive fragmentation relations. Mott and Linfoot also suggested that if the preponderance of fragments were of a size smaller than the thickness of the exploding shell, then the fractional power in the exponent should be 1/3 rather than 1/2. This argument is consistent with their assumption above that a length measure of the fragment was a random Poisson variable.

2.2.4.2 The exponential distribution

Grady and Kipp (1985) proposed an alternative assumption for the statistical distribution in fragment size observed in munitions fragmentation. They suggested that if such fragmentation can be represented by a mechanism-independent statistical description, then the fragment mass m, as opposed to fragment size (either $m^{1/2}$ or $m^{1/3}$ in the Mott–Linfoot development), is the more appropriate random variable. They then distributed the mass of the fragment over fragment number according to a Poisson process (or binomial process if the fragment number is small).

Thus, if the fragment mass is viewed as a random scalar variable, then the random fragmentation of the mass is analogous to the one-dimensional Lienau problem. Fragmentation is determined by breaks distributed randomly over the scalar measure of mass. The breaks determine a Poisson variate and lead to a cumulative fragment probability distribution

$$F(m) = 1 - e^{-m/\mu}, \tag{2.40}$$

and density distribution

$$f(m) = \frac{1}{\mu} e^{-m/\mu}. \tag{2.41}$$

In contrast to the Mott distribution, the present distribution keeps the same linear exponential functional form for both area and volume fragmentation. Some support for application of an exponential distribution to munitions fragmentation is provided by the experimental study of Mock and Holt (1983) and in observations made by Mott in a later report (Mott 1944). In the words of Mott '... it is doubtful whether it [equation (2.39)] can be said to have any theoretical justification for the fragmentation of shell casings ... Actually an alternative formula due to Payman [equation (2.40)] gives a rather better fit...' (see Mott 1944, page 6).

2.2.4.3 Statistical inhomogeneity and Poisson mixtures

This is an appropriate place to discuss the topic of statistical inhomogeneity as it relates to the fragment size distribution. When a body of finite size is broken into a number of fragments that constitute the fragment distribution, statistical homogeneity implies that every point within that body is described by the same mean fragment size as well as the other moments of the distribution. In practice, this is almost never the case. Fragmentation of a uniformly expanding ring or spherical shell would perhaps be the closest to approaching statistical homogeneity. The introduction to this section referred to the example of a stone impact onto the windshield of a car where the average fracture spacing increases with distance from the point of impact as a consequence of the decreasing intensity of forces causing fracture. The fragment size distribution would in this case be statistically inhomogeneous. One approach to addressing statistical inhomogeneity in fragmentation is the proposition that the heterogeneous distribution is an additive mixture of component homogeneous distributions (Grady and Kipp 1985). In principle, any basis distribution can be used for the homogeneous fragment distribution. If the exponential distribution resulting from the Poisson process described in the preceding subsection is used then the inhomogeneous cumulative distribution,

$$N(m) = \sum_{i=1}^{n} g_i N_i e^{-m/\mu_i}, \tag{2.42}$$

results, with g_i and N_i, respectively, the mass fraction and fragment number per unit mass of each homogeneous component and $\mu_i = 1/N_i$. Such a superposition approach to statistical heterogeneity is referred to as the method of Poisson mixtures (Puri and Goldie 1979). The simplest distribution of this form is the bilinear (in semi-log space) distribution,

$$N(m) = g_1 N_1 e^{-m/\mu_1} + g_2 N_2 e^{-m/\mu_2}. \tag{2.43}$$

Application of Poisson mixtures to fragment distributions is illustrated with the exponential and the bi-linear distribution in figure 2.8.

The bilinear distribution has found application in describing arena test data of warhead fragmentation. See for example Elek and Jaramaz (2009). Improved representation of fragmentation data is certainly aided through the added parameters provided in the distribution. The underlying physics of fragmentation, however, can also lead to bilinearity of the measured distribution. Explosive failure of metal cases can for example lead to fragmentation through both spall and radial fracture (adiabatic shear or tension). The two failure modes occur at markedly different loading rates with accompanying different average fragment size.

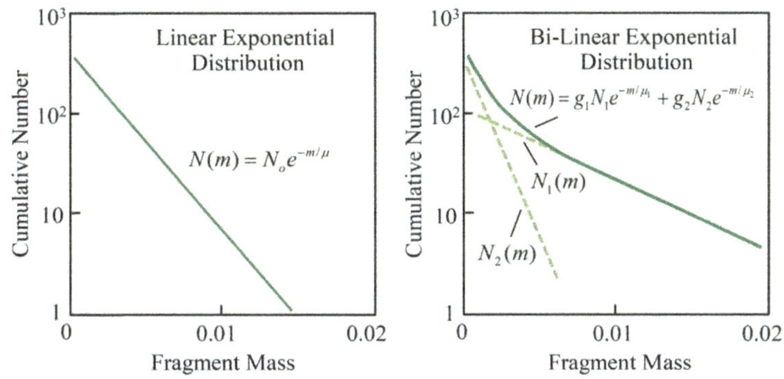

Figure 2.8. Representative application of the method of Poisson mixtures with a statistically homogeneous distribution shown on the left with $N_o = 400$ and a statistically inhomogeneous distribution on the right with $N_1 = 160$, $N_2 = 800$, $g_1 = 5/8$ and $g_2 = 3/8$, also totaling to 400 fragments (Grady and Kipp 1985).

2.2.4.4 The hazard function distribution

The fragment distribution of Mott and Linfoot, and the exponential distribution suggested by Grady and Kipp, can be generalized through standard methods of hazard (or survival) statistics (e.g. Hahn and Shapiro 1967). Consider a collection of fragments with an occurrence frequency, or hazard function, $h(m)$ where m is the fragment mass. The hazard function $h(m)$ is also referred to as the likelihood function. Then according to the tenets of survival statistics the probability of finding a fragment of size m within interval dm is just the product of finding no fragments smaller than m and that of finding one fragment in the interval between m and $m + dm$, or

$$f(m)dm = e^{-\int_0^m h(m)dm} h(m)dm. \tag{2.44}$$

The fragment probability density distribution is then

$$f(m) = h(m)e^{-\int_0^m h(m)dm}, \tag{2.45}$$

and the cumulative fragment distribution is

$$F(m) = 1 - e^{-\int_0^m h(m)dm}. \tag{2.46}$$

The number density $n(m) = N_o f(m)$ and cumulative number $N(m) = N_o F(m)$ follow directly, where the total number of fragments in the distribution is N_o. This derivation of the fragment distribution is fully rigorous. The unknown of course is the functional form of the hazard function $h(m)$. This function is determined either by exploring physics specific to the fragmentation process of interest, or else by hypothesizing a functional form and validating through experiment.

For example, given no additional insight into the fragmentation process there is little reason to assume a bias of $h(m)$ toward either the large or the small masses and the simplest assumption is $h(m) = h_o$, a constant (a Poisson process). Thus $f(m)$ in equation (2.45) is

$$f(m) = h_o e^{-h_o m}, \tag{2.47}$$

and the cumulative fragment number distribution is

$$F(m) = 1 - e^{-m/\mu}, \tag{2.48}$$

where $\mu = 1/h_o$ is the distribution scale parameter and the average fragment mass. Grady and Kipp (1985) suggested the applicability of this distribution to certain dynamic fragmentation processes.

Mott and Linfoot (1943), following the theoretical work of Lienau (1936) and inspection of some exploding munitions fragmentation data, were led to assume that a linear size measure x of the fragment dimension is a random variable and hence $h_o dx$ provides the chance of fracture determining a fragment of mass m. Mott was interested in the breakup of thin-walled cylinders where fragment dimensions were large compared to wall thickness. Hence $x \sim m^{1/2}$ and $dx \sim (1/2)m^{-1/2}dm$. Thus, an appropriate hazard function,

$$h(m) = \frac{1}{2\mu}\left(\frac{m}{\mu}\right)^{-1/2}, \tag{2.49}$$

follows from Mott's assumption, and equation (2.45) becomes

$$f(m) = \frac{1}{2\mu}\left(\frac{m}{\mu}\right)^{-1/2} e^{-(m/\mu)^{1/2}}, \tag{2.50}$$

and the corresponding cumulative fragment distribution is

$$F(m) = 1 - e^{-(m/\mu)^{1/2}}. \tag{2.51}$$

It is reasonable to consider limiting the functional dependence of $h(m)$ to the power-law relation,

$$h(m) = \frac{\beta}{\mu}\left(\frac{m}{\mu}\right)^{\beta-1}, \tag{2.52}$$

which leads to the cumulative fragment size distribution of the Weibull form,

$$F(m) = 1 - e^{-(m/\mu)^{\beta}}, \tag{2.53}$$

where μ and β are the scale and shape parameters of the distribution, respectively. This relation clearly encompasses the previous examples. Namely, $\beta = 1/2$ corresponds to the distribution arrived at by Mott and Linfoot (1943) while $\beta = 1$ corresponds to that suggested by Grady and Kipp (1985).

Generality of the parameter β for a munitions-specific scaling equation is warranted for several reasons. Mott and Linfoot argued that when fragments of a size smaller than the case thickness dominate the fragment distribution, then $\beta = 1/3$

is probably more appropriate. Although apparently observed under certain testing conditions, such statistical behavior was not found in the extensive cylinder fragmentation tests of Mock and Holt (1983) for example. For specific munitions, a range of expansion strain rates will lead to statistical heterogeneity (a different size parameter at different positions along the munitions case). This breakup feature will broaden the distribution leading to smaller effective values of β. On the other hand, a degree of case scores, or other processing, with the intention of biasing the distribution toward a unique size has the effect of increasing the distribution shape parameter β. (Note that as β approaches infinity equation (2.53) approaches a Heaviside function with all fragments the same size.)

Some enlightenment is offered through examination of several of the previous random geometry fragmentation algorithms within the framework of the present hazard statistics description of the fragment size distribution relation. Three geometry fragmentation algorithms from the previous figure 2.5 are replotted in figure 2.9. First in the figure is the random line algorithm first examined by Mott and Linfoot (1943). The distribution of fragment sizes from this algorithm was shown by Mott and Linfoot to be described by a Bessel function (equation (2.29)). Actually, the Mott–Linfoot solution applied only to the random placement of vertical and horizontal lines (panel (a) in figure 2.5). Computer generated distributions (Grady and Kipp 1985) have shown that the random orientation algorithm is also well described by the Bessel function. The second fragmentation algorithm in figure 2.9, identified as sequential segmentation (Grady and Kipp 1985), leads to a fragment size distribution rigorously described by a simple exponential (equation (2.41)). The third Voronoi–Dirichlet partitioning of the surface leads to a gamma function distribution (equation (2.36) with $n = 4$). The three fragmentation algorithms originate from the same initial random placement of Poisson points.

Fragment size distribution functions for the three fragmentation algorithms in figure 2.9 are shown in figure 2.10 along with the corresponding fragmentation likelihood function (hazard function) $h(m)$. The likelihood functions are particularly revealing. As pointed out previously, the simple exponential distribution results from a constant likelihood function $h(m) = h_o$. The gamma function distribution results from a likelihood function for which probability of occurrence diminishes at

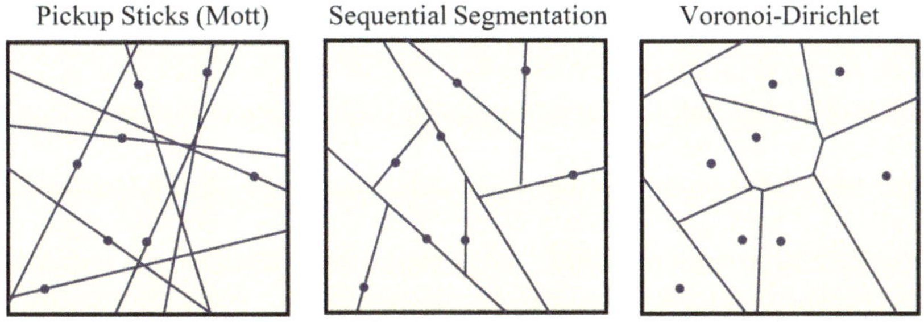

Figure 2.9. Three geometric algorithms for randomly fragmenting a surface generated from the same set of Poisson points. That on the left was studied by Mott and Linfoot (1943) and that in the center by Grady and Kipp (1985). The Voronoi–Dirichlet tessellation is illustrated on the right.

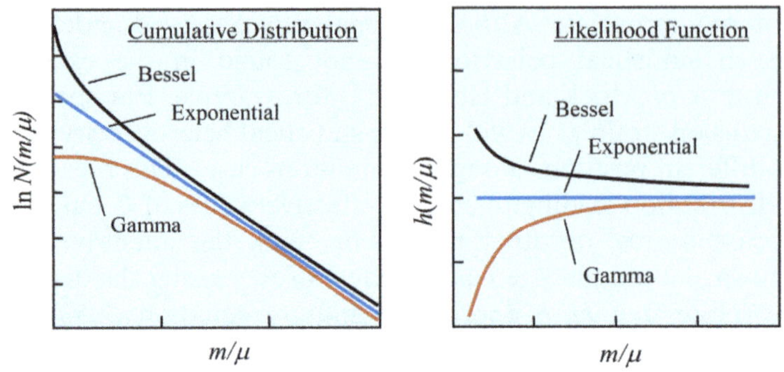

Figure 2.10. Fragment size distribution and fragment likelihood function for the Mott algorithm, the sequential segmentation algorithm, and the Voronoi–Dirichlet algorithm, random fragmentation from the previous figure.

the smaller fragment sizes, leading to a corresponding reduction of fragments in the smaller end of the distribution. In contrast, the Mott–Linfoot fragmentation algorithm follows from a likelihood function that increases the probability of smaller fragments. One or the other of the several geometric fragment distributions may be a more appropriate representation of the experimental distribution under certain material type and fragmentation conditions. None, however, provides a deeper description of the underlying fragmentation physics.

2.2.4.5 Mott statistical fracture distribution

Several technical reports published within the year of 1943 reveal the maturing of Mott's understanding of the dynamic fragmentation process and the outline of a statistical theory of fragmentation that is still one of the leading theories available. The Mott fragmentation theory was published several years later in the open literature (Mott 1947). The development of Mott's theory is summarized here and leads to statistical fragment distributions that differ starkly from the earlier distribution proposed by Mott and Linfoot (1943) and used by others since.

The Mott theory of fragmentation is conceptualized by considering the Mott cylinder (or ring) as illustrated in figure 2.11 (Grady 2006). The Mott cylinder is an idealization of an expanding cylindrical shell whose outward motion is imparted by some symmetric radial impulse. Mott, in particular, focused on the natural fragmentation of exploding pipe bombs. The model is applicable to other test conditions such as the radial acceleration and fragmentation of metal rings through explosive or magnetic methods.

An explosively driven expanding metal cylinder is a decidedly multidimensional fragmentation event. Fragmentation of the Mott cylinder is only an approximation to this event. The theory attempts to capture the characteristic circumferential spacing of fractures and the statistical distribution in that spacing. The theory is not intended to account for the axial propagation and interaction of cracks within a finite length cylinder. The Mott cylinder is an expanding metal body with radial velocity u and radius r at the time when multiple fracture and breakup of the cylinder

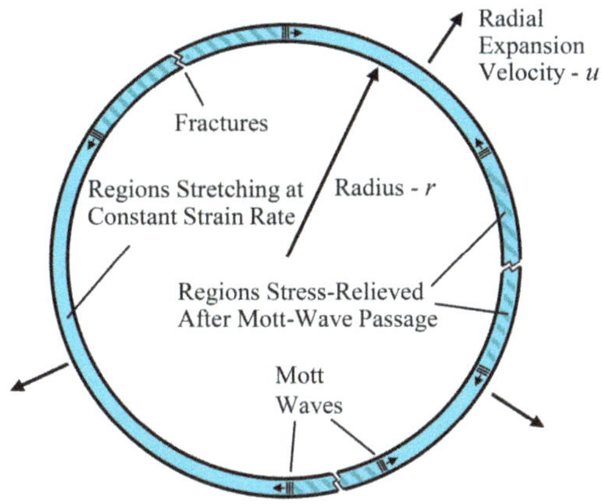

Figure 2.11. Mott cylinder illustrating the one-dimensional activation and interaction of fractures leading to the statistical fragmentation of the body (Grady 2006).

proceeds. Just before break up, the cylinder body is in circumferential tension and undergoing uniform circumferential stretching at a rate given by the ratio $\dot{\varepsilon} = u/r$.

Mott proposed that fragmentation proceeds through the random spatial and temporal occurrence of fractures resulting in a distribution in fracture spacing. Release waves propagate away from sites of fracture relieving the tension and precluding the need for further fracture within the regions encompassed by tension release waves. Fragmentation is complete when fracture-induced release waves circumferentially subsume the entire cylinder.

Thus, within the model for dynamic fragmentation proposed by Mott, two physical issues are addressed. First, is the issue of when and where fractures activate on the Mott cylinder. Second, is the nature of propagation of tensile release waves (Mott waves) from the sites of fracture.

Mott fracture activation. Mott put forth arguments that energy dissipation was not of consequence in the fracture process and proposed instead a statistical strain-to-fracture criterion. Mott assumed that fractures occurred with equal probability at any point around the circumference of the cylinder at a frequency governed by a strain-dependent hazard function $\lambda(\varepsilon)$ (e.g. Hahn and Shapiro 1967) such that $\lambda(\varepsilon)d\varepsilon$ provides the statistical number of fractures occurring within a unit length of the cylinder circumference within the strain interval $d\varepsilon$.

Mott expected $\lambda(\varepsilon)$ to be a strongly increasing function of strain and suggested both an exponential hazard function,

$$\lambda(\varepsilon) = Ae^{\gamma\varepsilon}, \tag{2.54}$$

and a power-law hazard function,

$$\lambda(\varepsilon) = \frac{n}{\sigma}\left(\frac{\varepsilon}{\sigma}\right)^{n-1}. \tag{2.55}$$

The former leads to a Gumbel form of extreme value statistics, while the latter to the Weibull form. Mott pursued in detail the exponential hazard function. Here, for purposes of contrast, the two-parameter power-law hazard function is explored. For sensibly large n the parameter σ is the expected value of the strain to fracture in a unit length while σ/n is proportional to the standard deviation in the strain to fracture.

Mott tension release. The region of the material that has experienced tensile release at strain ε is given by

$$D_x(\varepsilon) = \int_0^{\varepsilon} 2g(\varepsilon - \eta)\lambda(\eta)d\eta, \tag{2.56}$$

where $\lambda(\eta)d\eta$ is the statistical number of fractures activated on the Mott cylinder at a strain η within interval $d\eta$. The function $g(\varepsilon - \eta)$ is the distance traveled by a tensile stress release wave over the strain interval $\varepsilon - \eta$ from a fracture. The factor of two accounts for right- and left-facing release waves. (Since strain rate is assumed constant over the duration of the fracture process, strain and time are synonymous through $\varepsilon = \dot{\varepsilon}t$.)

In equation (2.56) $D_x(\varepsilon)$ provides the circumferential fraction of the Mott cylinder that has been encompassed by stress release waves emanating from sites of fracture at a current strain ε. The equation also determines the fraction of the cylinder in which further fracture is precluded. A form of equation (2.56) was derived by Mott in the original 1943 reports.

Correspondingly, the number of fractures activated at strain ε is,

$$N(\varepsilon) = \int_0^{\varepsilon} (1 - D_x(\eta))\lambda(\eta)d\eta, \tag{2.57}$$

where the factor $D_x(\eta)$ accounts for the fraction of the Mott cylinder no longer deforming in tension.

The functional form of the stress release relation $g(\varepsilon)$ must be specified. There are several possibilities. If the expanding Mott cylinder is elastic at the time of fracture then a constant elastic release wave velocity governed by the elastic modulus is sensible. Mott, however, considered an expanding ductile metal cylinder and assumed a material on the tensile yield surface governed by a constant flow stress Y. Instantaneous fracture and a rigid ideally plastic constitutive response leads to the stress release function, derived later in this chapter,

$$g(\varepsilon) = \sqrt{\frac{2Y}{\rho\dot{\varepsilon}^2}\varepsilon}\,. \tag{2.58}$$

Stress wave diffusion, rather than wave propagation, governs stress release under the assumed plastic deformation conditions.

The physical and statistical principles outlined above were then used by Mott to determine a distribution in fragment length (or fracture spacing) resulting from the plastic fracture of the expanding Mott ring. It should be emphasized that this one-dimensional distribution bears no relationship to the earlier two-dimensional Mott distribution arrived at intuitively by Mott and Linfoot.

The Mott graphical solution. Mott was not able to develop an analytic solution and proceeded with a graphical method. It is likely that with a modest amount of additional time Mott would have developed an analytic solution, which was his nature. It is interesting that nearly concurrently, other workers (Johnson and Mehl 1939) were pursuing transformation reaction kinetics in metals and developed analytic tools ideally suited to Mott's statistical fragmentation theory.

The graphical solution to the statistical fragment size distribution derived by Mott is illustrated in figure 2.12 and proceeds as follows. A parameter D is defined which is a function of time t (or strain). At any time, $D(t)$ is the fraction of the stretching body that has been subsumed by the Mott release waves propagating from points of fracture. Since further fractures are assumed to occur only in the fraction of the body not yet encompassed by these release waves, $1 - D$, the number of fractures that will appear in the time increment t to $t + dt$ is,

$$dN = (1 - D)\lambda(\dot{\varepsilon}t)\dot{\varepsilon}dt, \tag{2.59}$$

where, as noted previously, Mott used the exponential fracture frequency relation $\lambda(\dot{\varepsilon}t) = A\exp(\gamma\dot{\varepsilon}t)$.

The tension release fraction $D(t)$ of the plastic stretching body is determined by the collective Mott waves emanating from fractures initiating prior to time t. Each Mott wave propagates according to (see equation (2.58)),

$$x_i = \sqrt{2Y/\rho\dot{\varepsilon}}(t - t_i)^{1/2}, \tag{2.60}$$

where t_i is the initiation time of the ith fracture. Introducing a dimensionless time through the expression $\xi = \gamma\varepsilon = \gamma\dot{\varepsilon}t$, equation (2.60) becomes

$$x_i = \sqrt{2Y/\rho\dot{\varepsilon}^2\gamma}(\xi - \xi_i)^{1/2}, \tag{2.61}$$

where Mott recognized the leading expression on the right-hand side as the normalizing length scale x_o (proportional to the average fragment length) for the distribution in fragment lengths. The constant γ is the Mott fragmentation parameter commonly used by those who work with cylinder arena fragmentation data through the Mott method.

Figure 2.12. The graphical statistical fragmentation undertaken by Mott is illustrated from top to bottom as fractures randomly activate and tensile release waves subsume the body.

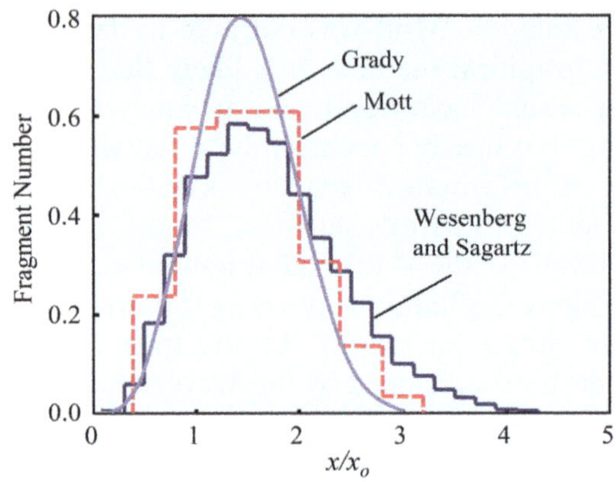

Figure 2.13. Mott distribution for the spacing of fractures on a fragmenting expanding cylinder. Comparisons are shown to fragment distributions calculated by Wesenberg and Sagartz (1977), and by the analytic methods of Grady (1981a, 1981b).

Mott worked with equations (2.59) and (2.61) using a deck of cards and graph paper, to determine the fragment distribution, and arrived at the solution shown in figure 2.13. The statistical sequence shown in figure 2.12 is illustrative of the method performed numerous times by Mott to acquire the statistical distribution. Mott noted that the average fragment size was about $1.5x_o$. Again, this distribution bears no relation to the earlier distribution inferred by Mott and Linfoot for fragmenting munitions data.

Wesenberg and Sagartz (1977) performed fragmentation experiments through impulsive magnetic inductive expansion of aluminum cylindrical shells. Using computer methods and an appropriate random number generator, they produced fragment size distributions by solving the same pair of equations as Mott. Wesenberg and Sagartz displayed their calculated distributions as the average of the individual results of 10 rings, 100 rings and 1000 rings, respectively, and concluded that a reasonably large number of calculations are required to achieve sensible convergence. Their distribution resulting from the average of 1000 rings is compared with the Mott distribution in figure 2.13. Elek and Jaramaz (2004) have undertaken a more extensive computational assessment of the Mott statistical analysis and demonstrated sensible agreement with the original Mott analysis. They also make comparisons to the expanding ring data of Wesenberg and Sagartz (1977) and Grady and Benson (1983).

Analytic solution of the Mott distribution. Analytic solutions of the Mott model have been obtained (Grady 1981a, 1981b, 2006). An inspection of equation (2.56) reveals that the function $D_x(\varepsilon)$ will exceed unity at sufficiently large strain. This non-physical result is a consequence of not accounting for two factors in the fracture activation and stress wave propagation process. First, the fracture activation function $\lambda(\varepsilon)$ does not exclude the activation of further fractures within regions previously stress relieved. Second, the stress release function $g(\varepsilon)$ does not account for the impingement and the overlap of opposing release waves from separate neighboring fractures. Thus, equation (2.56) is only applicable for a dilute number of fractures early in the fracture and release process. To account for fracture exclusion and wave impingement in the statistically random Mott model, a statistical method introduced by Johnson and Mehl (1939) in their pursuit of reaction kinetics in metals can be used. Exclusion and impingement is accounted for through the relation,

$$D(\varepsilon) = 1 - e^{-D_x(\varepsilon)}, \tag{2.62}$$

which provides the fraction of the Mott cylinder encompassed by fracture stress release waves at any strain ε. $D(\varepsilon)$ and $D_x(\varepsilon)$ are equivalent at early times as they should be. The function $D(\varepsilon)$, however, approaches unity as ε becomes large. The relation corresponding to equation (2.57) becomes

$$N(\varepsilon) = \int_0^{\varepsilon} (1 - D(\eta))\lambda(\eta)d\eta = \int_0^{\varepsilon} e^{-D_x(\eta)}\lambda(\eta)d\eta. \tag{2.63}$$

With this contribution to the statistical fragmentation theory, it is possible to develop analytic solutions to the distribution in fracture spacing for fragmentation of the Mott cylinder. Details of this analytic solution are provided elsewhere (Grady 1981a, 1981b, 2006). The calculated distribution in fracture spacing by this method is

$$f(x) = \frac{\beta^2}{4}\frac{1}{x_o}\left(\frac{x}{x_o}\right)^3 e^{-\frac{1}{4}(x/x_o)^3} \int_0^1 (1 - y^2)e^{-\frac{3}{4}(x/x_o)^3 y^2} dy, \tag{2.64}$$

where $\beta = 3/\Gamma(2/3)$ and x_o is the corresponding distribution scale parameter. This analytic distribution is also shown with the Mott distribution in figure 2.13.

2.2.4.6 Fragmentation of metal cylinders

Several sets of fragmentation data bring to light the underlying physics of the fragmentation theory of Mott. The first is a series of tests in which thin cylindrical shells of aluminum are subjected to magnetic forces that lead to rapid expansion and fragmentation of the shells. The second are arena tests where cylindrical shells of steel are subjected to explosive loading and fragmentation.

Aluminum cylinder fragmentation test. Wesenberg and Sagartz (1977) performed fragmentation experiments that come close to replicating the fragmentation conditions posed for the theoretical analysis undertaken by Mott. In the study of Wesenberg and Sagartz, magnetic inductive methods powered by a large capacitor discharge system were used to rapidly expand and fracture thin cylindrical shells of 6061-T6 aluminum. An internal copper conductor carries current. The copper conductor applies a radial driving pressure approaching two GPa for several microseconds accelerating the aluminum cylinder to a terminal radial velocity. The cylinders were 127 mm in diameter and 102 mm in length with a wall thickness of 1.27 mm. Dynamic fracture was observed photographically to occur at about 30% expansion strain and at a strain rate of approximately 10^4 s^{-1}. A photograph illustrating the dynamic fragmentation process is shown in figure 2.14.

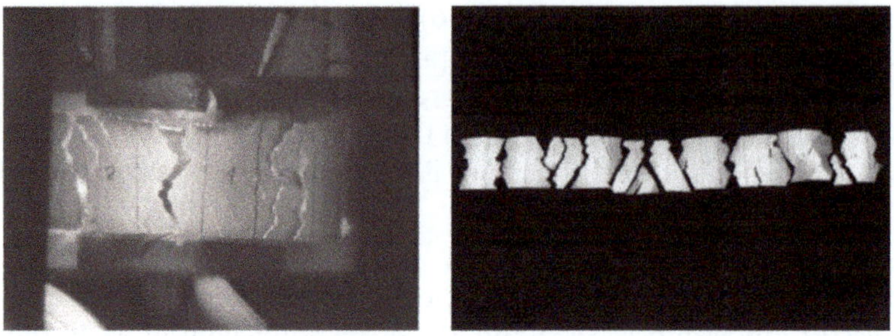

Figure 2.14. High-speed photograph illustrating deformation and fracture features in the expanding aluminum cylinder tests of Wesenberg and Sagartz (1977). Fragments from one magnetic expanding cylinder test on 6061-T6 aluminum.

A total of 125 fragments were collected from 11 equivalent tests. All fragments were assigned an effective length by weight. A histogram of the statistical size distribution for the aluminum cylinder data is shown in figure 2.15. In this figure, the ordinate identifies the number of fragments normalized to a single ring that occurs within the corresponding fragment size (length) interval. The abscissa provides the fragment size as a fraction of the original circumference of the ring.

The experimental fragment size data of Wesenberg and Sagartz are compared with Mott's graphical distribution in figure 2.15, and with the analytic statistical distribution for plastic fracture provided by equation (2.64). Both theoretical distributions provide a satisfactory representation of the relatively narrow spectrum of fragment lengths observed in the experiments.

Explosive fragmentation of steel cylinders. The physics-based fragmentation theory of Mott was developed to address the breakup of exploding munitions. The theory has not, however, found wide use in this application. In fact, the earlier fragment distribution relation of Mott and Linfoot that was arrived at through more

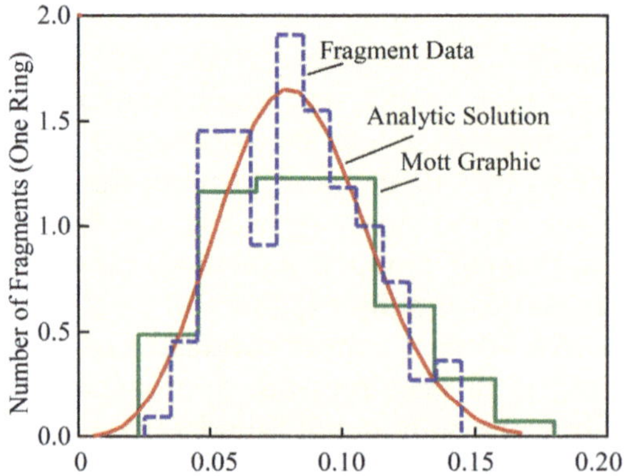

Figure 2.15. Experimental fragment size data for expanding aluminum ring tests of Wesenberg and Sagartz (1977) and comparisons with graphical and analytic solutions of the Mott fragmentation theory (Grady 1981a, 1981b).

intuitive arguments has been markedly more popular. The Mott–Linfoot distribution, and subsequent variations, has been extensively used to describe the fragment size distributions measured in exploding munitions testing.

There are several possible reasons for this, The Mott theory suggests a relatively narrow spread in fragment sizes as indicated by the fragment size distributions shown in figures 2.13 and 2.15. A markedly broader range of fragment sizes is measured in arena testing of munitions. It has been suggested that secondary fragmentation during fragment collection in arena testing provides fragment distributions that are not representative of the initial breakup. The processes of fracture and breakup leading to fragmentation may not be well represented by the plastic fracture model posed by Mott. Munitions metals may be too brittle. Mechanisms of adiabatic shear and extension fracture commonly observed in munitions metal fragments may not be described by the Mott fragmentation model.

A further reason, however, for the apparent disparity between arena test data and the Mott fragmentation theory lies in the nature of munitions test data and the statistical predictions provided by the Mott fragmentation model. Arena test data are typically a plot of cumulative fragment number versus fragment mass—a Mott plot. The physics-based Mott theory predicts a measure of the statistical distribution of circumferential spacing of fractures of a fragmenting cylindrical shell. In the following discussion, arena fragmentation test data, and an analysis of those data, are described in which efforts are focused on relating the test data to specifics of the Mott fragmentation theory. Further details of this study can be found in Grady (2006).

Smooth-walled explosively loaded metal cylinders fragment without the aid of fragmentation enhancing techniques. In the natural fragmentation of exploding cylinders the statistical distribution in the size and velocity of fragments is determined through a complex interplay among the explosive characteristics, the geometry of the explosive-case system and mechanical properties of the case metal. Discussion is focused on one study in which the explosive and case metal were particularly well characterized, and quite thorough diagnostics were used to interrogate the explosive natural fragmentation event (Grady and Hightower 1992, Grady 2006). These results are considered in light of Mott's statistical theory of fragmentation.

An explosive fragmentation experiment was performed on a 15.2 cm diameter smooth wall metal cylinder. The cylinder was 38.1 cm in length with a wall thickness of 5.7 mm. The cylinder was machined of 4140 steel and heat treated to a Rockwell hardness of 40 ($Y = 1.1$ GPa). The cylinder was filled with RX-35-AN explosive and the cylinder ends were confined. The explosive was center-detonated at one end. High-speed front-lit photography using a CORDIN framing camera with 5 μs frame intervals was used to observe acceleration and breakup of the expanding cylinder. The opening of fractures and the emergence of explosive gases are consistent with the 1.20–1.25 strain to fracture measured on recovered fragments. An expansion velocity of 1760–1830 m s^{-1} determined from the photographs compares well with the limiting Gurney velocity of about 1800 m s^{-1} calculated for this cylinder.

Multiple flash radiography provides the fragment velocity, trajectory and pattern for a 40° sector of the cylinder. Fragments from this sector were captured in fiberboard bundles. The 161 recovered fragments represent 90% of the weight of the 40° cylinder sector. From the recovered fragments, it was observed that fracture was predominantly along elongated strips with the fracture parallel to the axis of the cylinder. A number of the fragments were 4–5 times longer than wide. Both tensile and shear fracture was observed from examination of fracture surfaces. Shear fracture appeared to be the dominant breakup mechanism. Fragment size statistics from the recovered fragments are now compared with the fragmentation theory of Mott.

For comparison with the Mott theory, it is necessary to establish the radial expansion velocity of the exploding cylinder at the radius of fracture to provide a measure of the strain rate at breakup. An early theory used to calculate the fracture strain of explosively expanding cylinders is due to Taylor (1963). The Taylor theory leads to a relation for the circumferential stress in the shell subjected to an internal pressure P given by $\sigma(y) = Y - P(1 - y/h)$ where Y is the yield stress in simple tension, h is the shell thickness and y is a coordinate through the thickness, $0 \leqslant y \leqslant h$. Thus $\sigma(0) = Y - P$ (compression) at the inner surface and $\sigma(h) = Y$ (tension) at the outer surface. The crossover point occurs at an interior point of the shell. The Taylor theory assumes that failure occurs when the internal pressure within the expanding cylinder decreases to a value such that tension is just achieved at the inner surface. For the RX-35-AN explosive the measured velocity history, combined with appropriate Gurney expressions, provided the calculate pressure versus radius. An internal pressure of $P = Y = 1.1$ GPa, corresponding to the yield stress of 4140 steel, is calculated at an expansion radius of $R/R_o = 1.24$. This fracture strain calculated through the Taylor criterion is compared with through-the-thickness measurements of fragments recovered from the test on the left in figure 2.16. Results are sensibly consistent with the Taylor prediction.

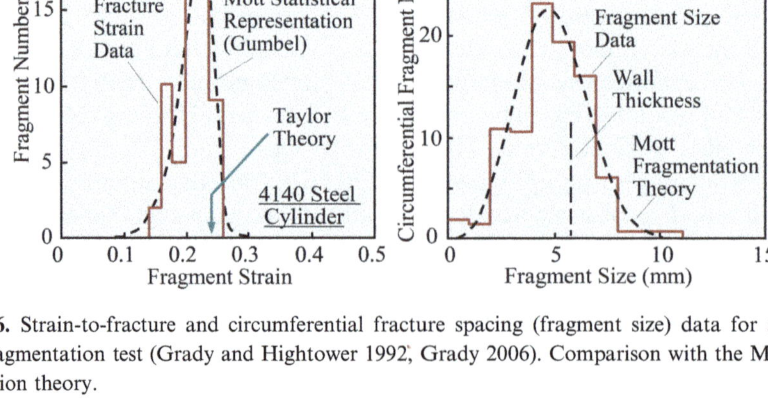

Figure 2.16. Strain-to-fracture and circumferential fracture spacing (fragment size) data for an exploding cylinder fragmentation test (Grady and Hightower 1992, Grady 2006). Comparison with the Mott statistical fragmentation theory.

The Mott strain-to-fracture function from equation (2.54) is written as

$$\lambda(\varepsilon) = Ae^{\gamma\varepsilon} = \frac{1}{\sigma}e^{(\varepsilon-\mu)/\sigma}, \tag{2.65}$$

where the second equality explicitly displays the Gumbel distribution location parameter μ and scale parameter σ. A fit to the strain-to-fracture data shown in figure 2.16 yields a distribution location parameter (and mode) of $\mu = 0.22$, and scale parameter $\sigma = 0.025$. (The standard deviation is 1.283σ and the Mott gamma parameter is $\gamma = 1/\sigma = 40$). The Gurney analysis provides an expansion velocity of 1530 m s^{-1} at the radius of fragmentation. The expansion velocity in turn establishes a strain rate at the point of fragmentation of $\dot{\varepsilon} \approx 1.6 \times 10^4$ s^{-1}.

To make a statistical comparison of the fragment size data with the theory of Mott, the following analysis of the data is performed. Every fragment is weighed and the length of every fragment is measured. An effective rectangle is assumed for a fragment such that the mass is given by $m = \rho wtl$ where ρ is the density and w, t and l are the width, thickness and length, respectively. A circumferential fragment width is then calculated for each fragment from $w = m/\rho tl$, where m and l are the measured values and t is the initial wall thickness of the cylinder. The width w provides an effective average measure of fracture spacing for that fragment in terms of the initial cylinder dimensions. A fragment of length l is then considered to be a fraction of a strip of length L, given by $n = l/L$, where L is the length of the cylinder. Through this method, a number (of fracture strips) versus circumferential width distribution is determined for the fracture spacing data.

The circumferential fragment number data are plotted as a number histogram on the right in figure 2.16. The number data on the ordinate are scaled from the 40 degree sector to a full cylinder. The initial wall thickness of the cylinder is identified in the figure. The statistical fracture spacing data are compared with the statistical Mott distribution on the right in figure 2.16. The data are in quite reasonable agreement with the Mott theory. The data are also in good agreement with fragment size calculated from the Mott theory—a topic covered in detail in a later section. Disconcerting, however, is the observation that fracture spacing in the present test spans the thickness of the cylinder case, bringing into question the issues of a thin case assumed in the development of the Mott theory.

Mott recognized that the theory predicting the statistical spacing of axial fractures was not readily testable. Analysis of the data just described is, with some effort, specifically performed to test the fragmentation theory of Mott. The data, excepting some lingering concerns, sensibly support Mott's theory. To relate this to the available test data, Mott proposed that an axial strip of width x would, on average, fracture into fragments with length ax. The parameter a is an aspect ratio with a value of perhaps 3–5, indicative of the elongated fragments observed in arena test data. Mott provided some simple estimates of the expected aspect ratio.

Figure 2.17 depicts an expanded cylinder of circumference C and length L, illustrating a representative fractured strip of width x and a representative fragment of length ax. Using the analytic expression for the Mott distribution of

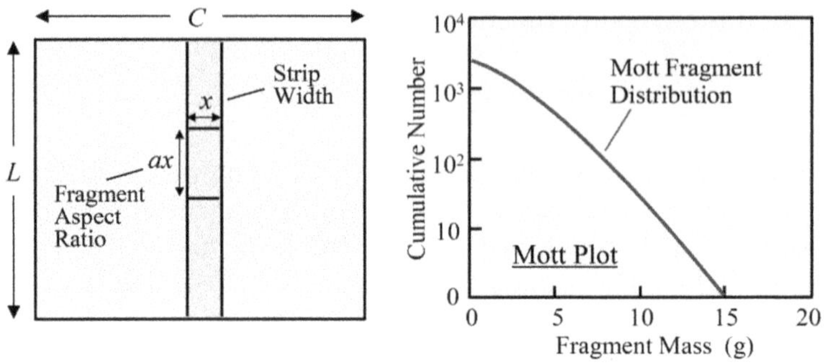

Figure 2.17. Left: a cylinder shell of circumference C and length L is fragmented into strips of nominal width x and the strips into fragments of nominal length ax. Right: cumulative fragment distribution (Mott plot) from the Mott theory.

fracture spacing from equation (2.64), the number of strips of width x within interval dx is,

$$dn = \frac{C}{x_o} f(x) dx, \tag{2.66}$$

where the distribution scale parameter x_o provides an approximate average of the fracture spacing. The number of fragments associated with the same interval is then,

$$dN = \frac{L}{ax} dn = \frac{LC}{ax_o} \frac{f(x)}{x} dx, \tag{2.67}$$

providing a cumulative number greater than

$$N(x) = \frac{LC}{ax_o} \int_x^\infty \frac{f(x)}{x} dx. \tag{2.68}$$

The mass corresponding to a fragment of width x is

$$m(x) = \rho a x^2 t, \tag{2.69}$$

where t is the thickness of the cylinder wall.

A plot of the cumulative fragment number versus fragment mass (a Mott plot) is provided by equations (2.68) and (2.69), offering a ready comparison of theory with fragmenting cylinder test data. An aspect ratio of $a = 3$ is assumed and properties for the 4140 steel cylinder test are used to provide the Mott fragment distribution plot shown in figure 2.17.

The theoretical distribution is not out of line with arena test data, although the data commonly exhibit more small fragments than is predicted by the Mott theory (see Grady 2006, for example). (Also compare with figure 2.10 and the discussion of the fragment-size likelihood function.) The assumption of a constant fragment aspect ratio in the theory of Mott is probably too simplistic. Alternatively, it is not unreasonable to question test data, as fragment recovery without causing secondary breakup is challenging.

2.2.4.7 Mass normalized fragment size distributions

Statistical fragmentation is appropriately posed through the specification of the hazard function $h(m)$ that expresses the likelihood of occurrence of a fragment of size m within a fragment distribution. As developed in section 2.2.4.4, the statistical framework leads to analytic expressions for the number density $n(m)$ and cumulative number $N(m)$ descriptions of the fragment distribution.

Frequently, the total number of fragments in the distribution is not readily available. In experimental studies such as arena tests of fragmenting shells, the total fragment number N_o is difficult to assess. Smaller fragments are lost. Quantifying the fine end of the distribution is tedious and is often of less concern.

The total fragment mass M_o is commonly better known. Either the entire device participates in fragmentation and is weighed beforehand, or enough fragments are collected to adequately bound the total mass of fragments. Recognizing this difficulty in the statistical characterization of dynamic fragmentation of warhead systems, Held (1990) proposed the empirical distribution relating cumulative fragment mass to cumulative number of largest fragments,

$$M(N) = M_o(1 - e^{-bN^{\lambda}}). \tag{2.70}$$

The Held distribution has found application in adequately describing fragment size distributions from exploding warheads.

Alternatively, the issue is addressed through application of the previous statistical development. Again, working with a power-law hazard function,

$$h(m) = \frac{\beta}{\mu}\left(\frac{m}{\mu}\right)^{\beta-1}, \tag{2.71}$$

which leads to distribution functions of the general Mott type. The number density distribution is

$$\frac{dN}{dm} = n(m) = N_o\frac{\beta}{\mu}\left(\frac{m}{\mu}\right)^{\beta-1} e^{-(m/\mu)^{\beta}}, \tag{2.72}$$

with N_o the total number of fragments in the distribution. The mass differential is

$$dM = mdN = N_o\beta\left(\frac{m}{\mu}\right)^{\beta} e^{-(m/\mu)^{\beta}}dm, \tag{2.73}$$

and integration provides the cumulative mass of all fragments greater than size m,

$$M(m) = N_o\beta\mu\int_{m/\mu}^{\infty} \xi^{\beta}e^{-\xi^{\beta}}d\xi. \tag{2.74}$$

When expressed in the form of the Euler integral, equation (2.74) has the solution (Grady 2001)

$$M(m) = M_o \frac{\Gamma\big((1+\beta)/\beta(m/\mu)^\beta\big)}{\Gamma((1+\beta)/\beta)}, \tag{2.75}$$

in terms of the Euler gamma function. Total fragment mass and fragment number are related through

$$M_o = N_o \mu \Gamma((1+\beta)/\beta) = N_o \bar{\mu}, \tag{2.76}$$

where $\bar{\mu}$ is the expected value of the fragment mass.

With an adequate number of large fragments collected from a fragmentation event, the cumulative mass distribution from equation (2.75) is fit to the data. Total fragment number N_o is calculated from equation (2.76) and the cumulative fragment number calculated from the Weibull distribution,

$$N(m) = N_o e^{-(m/\mu)^\beta}. \tag{2.77}$$

Fragment mass can be eliminated between equations (2.75) and (2.77) yielding the same functional dependence $M(N)$ as the Held distribution, and although complex, provides an equally adequate description.

2.2.4.8 Application to metal case fragmentation

It is appropriate at this point to consider statistical fragment distributions developed in the past several subsections with experimental data. Much of the historical development has provided focus through efforts to better characterize the fragmentation of warhead systems. Accordingly, test fragmentation data on well-characterized explosively driven metal cylinders is suitable for this purpose. The photograph in figure 2.18 illustrates the fragmentation in progress of a metal cylinder approximately 50 μs after explosive detonation. This test is one of a series of

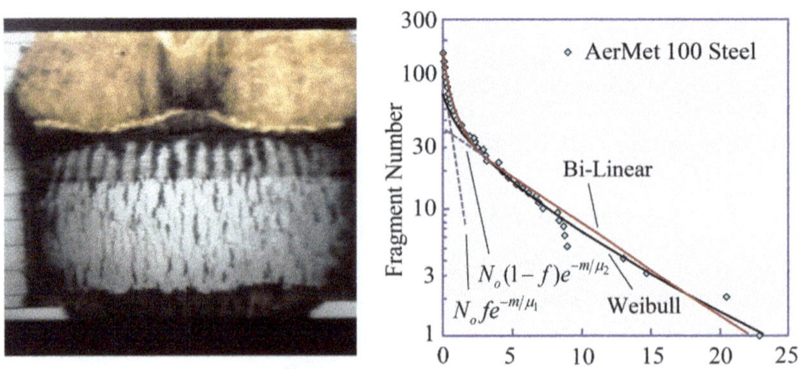

Figure 2.18. Fragmentation of Aermet 100 steel cylinders and comparison with statistical models (Grady 2001). Courtesy of Naval Surface Warfare Center, USA.

experiments performed on AerMet 100 steel to investigate the mechanics of dynamic fracture, and the size and statistics of the resulting fragment debris (e.g. Grady 2001). The test shown is on a steel cylinder 8 inches in length and 8 inches in outer diameter, with a 5/16 inch cylinder wall thickness. The upper two inches of the outer cylinder surface was scribed leading to the observed regular fragmentation in this region. Although of interest in its own right it is not the concern here. The lower six inches of the cylinder reveals the processes of natural fragmentation.

The experimental fragment distribution plotted on the right in the figure was obtained on the same geometry cylinder without the scribing. Careful fragment soft recover of an angular segment of the cylinder provided 160 fragments totaling approximately 300 g. Recovered fragments range from over 20 g to fractions of a gram. The test is representative of results obtained from the broader experimental series. A best fit of a Weibull distribution provided by equation (2.77), with statistical weighting towards the larger fragments, is shown with the data in the figure. The statistical fit requires a mean fragment size $\mu = 2.4$ g, a Weibull shape parameter $\beta = 0.65$, and a total fragment number $N_o = 80$. The distribution provides a total mass of fragments $M_o = 260$ g, about 40 g less than recovered, and under-predicts the fragment number by a factor of two. Clearly, the Weibull distribution provides a poor description of the small fragment end of the experimental distribution. Similar results are achieved with a Held fragment distribution. The Weibull or the Held distributions remain appropriate when concern is strictly with the larger fragments constituting most of the mass of the distribution. When statistical representation of the full span of the fragment debris is required, such distributions are not adequate.

A bi-linear Poisson (BLP) distribution from equation (2.43) is also shown with the data in the figure. The distribution is written here in the form

$$N(m) = N_o f e^{-m/\mu_1} + N_o(1 - f)e^{-m/\mu_2}, \tag{2.78}$$

with $N_o = 160$ the total fragment number, $f = 0.75$ the number fraction of the small fragment component, while $\mu_1 = 0.6$ g and $\mu_2 = 6.0$ g. The total mass of $M_o = 310$ g is calculated from the distribution, only modestly higher than the measured experimental mass. The two linear components of the composite distribution are also shown in the plot. When an analytic description of the full size range of the distribution is required, clearly the BLP distribution is better suited to the task.

The BLP distribution emerges from recognition of statistical heterogeneity of the fragment distribution, which in turn results from heterogeneity in the forces, leading to fracture and fragmentation. Possible sources leading to statistical heterogeneity in the test pictured in figure 2.18 are not apparent. Axial cracking resulting from circumferential rupture is clearly visible in the photograph. Possible spall failure resulting from the sharp shock coupled by the explosive into the cylinder would lead to more intense fracture and smaller fragments. Failure through spall is not yet evident in the photograph.

2.2.5 Entropy and energy approaches to fragment distributions

Concepts from classical statistical mechanics have found limited application to fragment size statistics and have not been extensively explored. Such statistical methods are nonetheless intriguing and may lend insight into a topic that is not yet fully understood. Several examples are described that pursue entropy maximum methods and classical energy-based statistical mechanical methods.

2.2.5.1 Entropy methods

An alternative method of developing a fragment size distribution relation is based on the proposition of maximizing the information entropy associated with a fragmentation event. Efforts to constrain fragment distributions through entropy methods have been pursued in several studies (e.g. Grady and Kipp 1985, Englman *et al* 1987, Cohen 1993). In the earliest approach, a fragmentation event is considered in which a body of mass M is broken into N fragments. In the development, energy is ignored and the proposition is made that any distinct fragment distribution is equally likely subjected to the constraint of N fragments of total mass M.

The problem is formulated by partitioning the mass M into equal intervals $\delta m = M/\eta$. If n_s fragments have mass within the range m_s to $m_s + \delta m$, then the number of ways that the N fragments can be distributed is

$$P(n_1, n_2, \ldots n_\eta) = N! \prod_{s=1}^{\eta} 1/n_s!. \tag{2.79}$$

The entropy is written as

$$\ln P = \ln N! - \sum_{s=1}^{\eta} \ln n_s!, \tag{2.80}$$

or in the limit $\delta m \ll M$,

$$\ln P = \ln N! - \int_0^\infty \ln n(m)! dm, \tag{2.81}$$

where $n(m)$ provides the fragment number distribution over mass. The entropy $\ln P$ is maximized subject to constraints,

$$N = \int_0^\infty n(m) dm, \tag{2.82}$$

and

$$M = \int_0^\infty m n(m) dm. \tag{2.83}$$

Variations provide

$$\delta \ln P = -\int_0^\infty \delta \ln n! dm = 0,$$

$$\delta N = \int_0^\infty \delta n dm = 0, \tag{2.84}$$

$$\delta M = \int_0^\infty m \delta n dm = 0.$$

Introduction of the undetermined multipliers α and β, and making use of the approximation $\delta \ln n! \simeq \ln \delta n$, provides

$$\ln n(m) + \alpha + \beta m = 0, \tag{2.85}$$

or

$$n(m) = e^{-\alpha - \beta m}. \tag{2.86}$$

Constraints provide $N = e^{-\alpha}/\beta$ and $M = N/\beta$ yielding

$$n(m) = \frac{N^2}{M} e^{-Nm/M}. \tag{2.87}$$

Finally, identifying the average fragment mass $\mu = M/N$ provides the probability density distribution over fragment mass,

$$p(m) = \frac{1}{\mu} e^{-m/\mu}. \tag{2.88}$$

It is observed that the present maximum entropy distribution provides the same exponential distribution as the earlier Poisson process with a uniform hazard function.

Englman *et al* (1987) have also pursued a maximum entropy formulation in which energy dependence in the fragmentation event is retained and included as a further constraint. This development leads to a more complex distribution that is power-law over a limited range of fragment size. Cohen (1993) develops a fragmentation entropy expression and provides a method, through the second law of thermodynamics, for examining the applicability of commonly used empirical fragment distribution relations.

2.2.5.2 Energy and statistical fragment distributions

A number of the statistical formulations that provide relations for the distributions in fragment size pay little or no attention to energetics in the breakup event. These include the various geometric fragmentation algorithms, the hazard function methods, fractal considerations, as well as the previous entropy maximum approach.

Even the Mott strain-to-failure model, while accounting for key physical details of the fracture interaction process, does not address the energy consumed in the fragmentation event.

There is evidence that fracture energy governs the characteristic fragment size in the breakup event as attested to, for example, by the repeated consistency with experimental data of the von Rittinger (1867) fragment size relation, or the more recent energy-based fragmentation models (applied to spall failure in chapter 1 and explored further in later sections). It is also reasonable to consider energy principles in the processes governing the statistical distribution of fragments.

Griffith (1943) put forth an interesting energy-based statistical formulation leading to functional expressions for fragment size distributions. (Note that this is not the same Griffith whose seminal works were instrumental in the development of classical fracture mechanics.) The approach proposed by Griffith was investigated briefly by Grady and Kipp (1987), and pursued further in Grady *et al* (2000) and Grady and Winfree (2001). The resulting distribution is found to differ starkly from the exponential distribution over fragment mass provided by the previous entropy maximum method as well as several of the earlier developments.

In the Griffith analysis, consider a body of total mass M. Conceptualize the partitioning of that body into arbitrary but equal-sized mass elements (referred to as molecules by Griffith). Now consider the random partitioning of a total energy E among the mass elements. If the energy content of an element is viewed as discrete and indexed by an integer j, then classical statistical mechanics provides the statistical number of mass elements n_j having an energy e_j as

$$n_j = Ae^{-\beta e_j}, \tag{2.89}$$

where A and β are normalizing parameters constrained by the total mass M and energy E. The number n_j is proportional to the mass fraction of the body with energy e_j. In the limit of a continuous function,

$$m(e) = Ae^{-\beta e}, \tag{2.90}$$

where $m(e)$ is the mass distribution over energy. Constraining A and β yields,

$$\lambda(e) = \frac{1}{\bar{e}}e^{-e/\bar{e}}, \tag{2.91}$$

where $\lambda(e) = m(e)/M$ is the spectral mass fraction and $\bar{e} = E/M$ is the average specific energy.

It is argued that the specific energy e associated with a mass element can be related to the size x of the fragment within which it resides. A fragment of diameter x has surface energy Γx^2 and mass ρx^3 (numerical coefficients are ignored). Thus, the specific energy and size are related through

$$e = \frac{\Gamma}{\rho x}, \tag{2.92}$$

where the surface energy Γ is assumed to be independent of the size of the fragment.

Identifying a size scale through $\sigma = \Gamma/\rho\bar{e}$, equation (2.91) transforms to a mass fraction distribution over fragment size through the relation $\lambda(e(x))dx = \lambda(e)de$, yielding

$$\lambda(x) = \frac{\sigma}{x^2} e^{-\sigma/x}. \tag{2.93}$$

Equation (2.93) is a well-behaved distribution that nicely captures the statistical size cut-off in the small particle limit. Equally important, equation (2.93) leads to a finite total surface area (and energy) of the distribution of fragments, a feature not found in several more prominent empirical size distribution relations. The transformation from equation (2.91) to equation (2.93) appears to be the appropriate analytic method. The functional transformation differs importantly from the same step presented by Griffith. The significance is that the resulting relations of Griffith are more in line with the early empirical fragment distributions for brittle solids. Those interested in pursuing energetic approaches to fragment sizes and distributions are encouraged to inspect in detail the original analysis of Griffith (1943).

Equation (2.93) can be related to the empirical Rosin–Rammler (1933) mass versus fragment size density distribution,

$$m(x) = \frac{M}{\sigma}\left(\frac{x}{\sigma}\right)^{n-1} e^{-(x/\sigma)^n}. \tag{2.94}$$

The energy-based distribution from equation (2.93) is equivalent to equation (2.94) for the special case of $n = -1$ and provides, for this special case, a physical statistical mechanical basis for the above empirical distribution. An exponent of $n = -1$ falls outside the range of about $0.5 \leqslant n \leqslant 1.5$ commonly associated with the Rosin and Rammler distribution. Brown and Wohletz (1995) also defended a Weibull (Rosin–Rammler) form for fragment size distributions on physical grounds.

For $n = -1$ (energy-based statistics) equation (2.94) yields

$$m(x) = \frac{M}{\sigma}\left(\frac{\sigma}{x}\right)^2 e^{-(\sigma/x)}. \tag{2.95}$$

In the limit of large fragments $x \gg \sigma$ this distribution becomes

$$m(x) = M\frac{\sigma}{x^2}, \tag{2.96}$$

and the number density

$$n(x) = \frac{m(x)}{\rho x^3} = \frac{M}{\rho}\frac{\sigma}{x^5}. \tag{2.97}$$

Equation (2.97) is an energy-based statistical relation intended for providing probabilistic predictions of the large fragments produced in an impulsive fragmentation event. Applications, for example, would be hypervelocity impacts on space debris shielding where concerns are with most lethal (probably largest) fragments generated in such events (Grady *et al* 2000, Grady and Winfree 2001). It is also of interest that the energy-based relation of equation (2.97) provides a power-law dependence between number and size (hence fractal) in contrast to the Rosin–Rammler relation which is exponential over the same size range. Fractal processes in fragmentation are discussed in a later section.

Justification for the applicability of an energy-based statistical mechanical approach to the size distribution in a mechanical breakage is problematic. A number of theoretical efforts have been pursued based on various randomization processes not associated with energetics and have made sensible correlations with experimental data lending some credence to these methods. None is capable of satisfactorily describing the broad range of experimental fragment size distributions that have been observed. A number of approaches including geometric statistics and percolation theories lead in the limit to exponential fragment distributions. Such distributions are shown to sensibly describe the results of very simple fragmentation experiments such as the breakup of expanding rings and shells. They do not adequately represent fragment size results from more complex breakup events. It is argued (Grady and Kipp 1985) that the added complexity is a consequence of a spectrum of strain rates participating in the fragmentation event and that a superposition of exponentials (Poisson mixtures) adequately characterizes the distribution. This proposition, however, does not lead to a predictive theory.

Empirical distributions continue use in application, with the Rosin–Rammler (Weibull) distribution one of the most common. A comparison of the energy-based statistical mechanical distribution with the Rosin–Rammler distribution is shown in figure 2.19. Several curves for the latter cover the observed values of the distribution index n. Although quite adequate within the range of most of the data, the empirical distribution is problematic in the tails of the distribution. In the small particle limit, the Rosin–Rammler distribution integrates to infinite fragment number and, depending on the value of n, infinite surface area. In addition, in the large particle limit the expected length-scale-invariant power-law dependence is not observed.

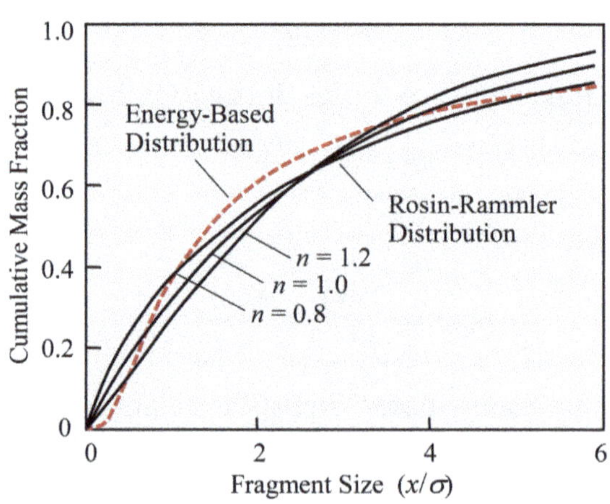

Figure 2.19. Cumulative fragment mass versus size comparing the energy-based statistical mechanical distribution with the empirical Rosin–Rammler distribution with n = 0.8, 1.0, 1.2 (Grady 2009).

Discrepancies between the statistical mechanical and the empirical distributions in the large fragment size limit are commonly not of concern. This tail of the distribution is important to applications where statistical prediction of the large fragments produced in the breakup process is a central objective. As observed in the figure, the statistical mechanical distribution predicts more frequent occurrence of large fragments than does the empirical Rosin–Rammler distribution.

2.2.5.3 Distribution extremes

Issues of the statistical distribution of fragment size in a dynamic fragmentation event become critical when the distribution extremes are of concern, either at the small end or the large end of the fragment distribution. For example, as briefly discussed in the previous subsection, concern may be with the few largest fragments produced in the breakup event. Large fragments are a problem in quarry and mine blasting where oversize material requires secondary fragmentation efforts. Successful application of shielding with spaced armor can depend on the size of the largest fragments produced (e.g. Grady *et al* 2000).

If a collection of fragments is reasonably described by a probability density distribution $f(m)$, with m the fragment mass, then a common and expedient method for estimating the statistical largest fragment follows from the complementary cumulative number distribution function,

$$N(m) = \int_{m}^{\infty} N_o f(m) dm, \tag{2.98}$$

with N_o the total number of fragments in the distribution. An estimate of the mass of the largest fragment $m_{\max}$ follows from solution of the relation

$$N(m_{\max}) = \int_{m_{\max}}^{\infty} N_o f(m) dm = 1. \tag{2.99}$$

For example let $f(m) = (1/\mu)\exp(-m/\mu)$ be the Poisson distribution and

$$N(m) = N_o e^{-m/\mu}. \tag{2.100}$$

Set $N(m_{\max}) = 1$ and the solution,

$$m_{\max} = \mu \ln N_o, \tag{2.101}$$

provides an estimate of the largest fragment in the distribution. For $N_o = 10$ the largest fragment is about 2.3 times the mean; for $N_o = 100$ the largest fragment is approximately 4.6 μ.

A more careful statistical assessment of the largest fragments is possible, however. The field of extreme value statistics is concerned with the largest and smallest (or extreme) values in a distribution (e.g. Ang and Tang 1984). Let the cumulative probability function be

$$F(m) = \int_{0}^{m} f(m) dm, \tag{2.102}$$

then in the statistical treatment the mass $\bar{m}$ of the largest fragment in a sample of N_o fragments will have a cumulative probability distribution function given by

$$F_{N_o}(\bar{m}) = (F(\bar{m}))^{N_o}, \tag{2.103}$$

where $\bar{m}$ spans the same range as m. The corresponding probability density function for the largest fragment is then provided by

$$f_{N_o}(\bar{m}) = \frac{d}{d\bar{m}}\left(F_{N_o}(\bar{m})\right). \tag{2.104}$$

Again, consider the Poisson distribution $f(m) = (1/\mu)\exp(-m/\mu)$, for which the largest fragment cumulative distribution for N_o fragments is

$$F_{N_o}(\bar{m}) = (1 - e^{-\bar{m}/\mu})^{N_o}. \tag{2.105}$$

The Poisson largest fragment cumulative distribution from equation (2.105) and the corresponding probability density distribution derived from equation (2.104) are shown in figure 2.20 for a collection of $N_o = 10$ and $N_o = 100$ fragments. It is interesting that the estimate of largest fragment size provided by equation (2.101) corresponds to the distribution mode (identified by the arrows in the figure) and approaches a cumulative probability $F_{N_o}(m_{\max}) = 1/e \simeq 0.368$ for sensibly large N_o.

As the number of fragments in the distribution becomes large equation (2.103) approaches,

$$F_{N_o}(\bar{m}) = e^{-N_o(1-F(\bar{m}))}. \tag{2.106}$$

Again, for the Poisson distribution, equation (2.106) yields,

$$F_{N_o}(\bar{m}) = e^{-N_o e^{-\bar{m}/\mu}}, \tag{2.107}$$

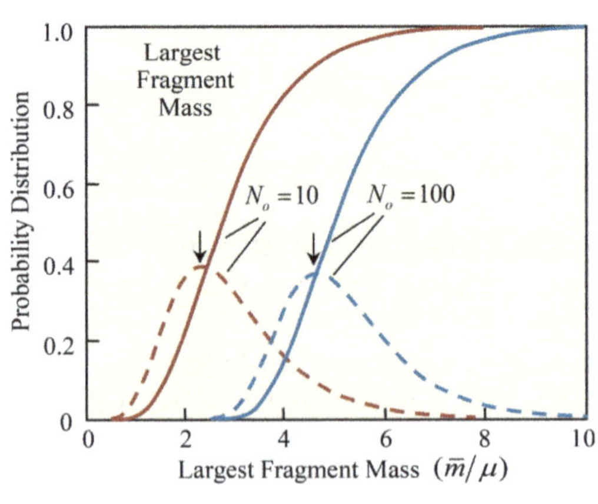

Figure 2.20. Cumulative probability and probability density functions for largest fragment mass are illustrated for Poisson fragment distribution of 10 and 100 fragments (Grady 2009). The arrows identify the distribution mode and corresponds to estimates based on equation (2.101).

the Gumbel extreme value distribution. For $N_o = 100$ equation (2.107) reproduces the curve in figure 2.20, and even for as low as $N_o = 10$ the equation is a close approximation to the corresponding curve in the figure.

2.2.5.4 Fragment velocity statistics

The topics addressed in the present section up to this point have focused strictly on the statistics of the size (or the mass) of fragments produced in a dynamic fragmentation event. Such events—the explosive fragmentation of a metal shell or the spall debris resulting from a high-velocity impact—eject fragments at substantial velocities. There is observed to be a statistical spread in the ejecta velocity resulting from these events. Very little theoretical effort has focused on the statistical aspect of ejecta velocity in dynamic fragmentation.

Two examples are illustrated in figure 2.21. The first shows measured experimental fragment velocities resulting from the natural fragmentation of an exploding tungsten cylindrical shell (Reedal *et al* 1999). The ordinate is the angle of the ejection trajectory relative to the equator of the cylinder. Of interest is the significant spread in velocities at any specific ejection angle. The second example is from a laboratory study attempting to replicate planetesimal collision events through the high-speed impact of steel projectiles onto spheres of basaltic rock (Waza *et al* 1983). The graph provides a cumulative distribution by mass of the velocity of the ejected fragments in a center-of-momentum coordinate system.

After a dynamic fragmentation event, expansion kinetic energy resides in the outward directed motion relative to some center of momentum associated with the fragmentation event. For an expanding cylindrical shell, for example, it would be relative to the average terminal case velocity.

The present development will consider an easily visualized example. Consider a spherical solid at the center of which an explosion leads to the fragmentation of the solid and outward expulsion of the resulting fragments. Depending on the energy of the explosion, the resulting collection may consist of two, three, four, five or any larger number of fragments. The event is illustrated for three fragments in figure 2.22.

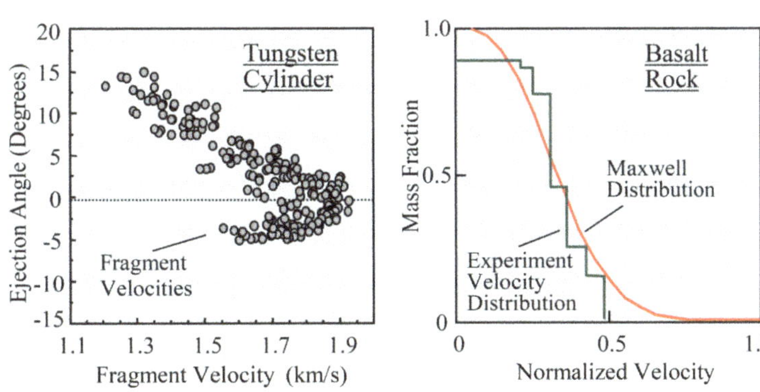

Figure 2.21. Experimental observations of statistical variations in fragmentation ejecta velocity. Explosive fragmentation of a tungsten cylinder (Reedal *et al* 1999). Impact fragmentation of a basalt rock sphere (Waza *et al* 1983).

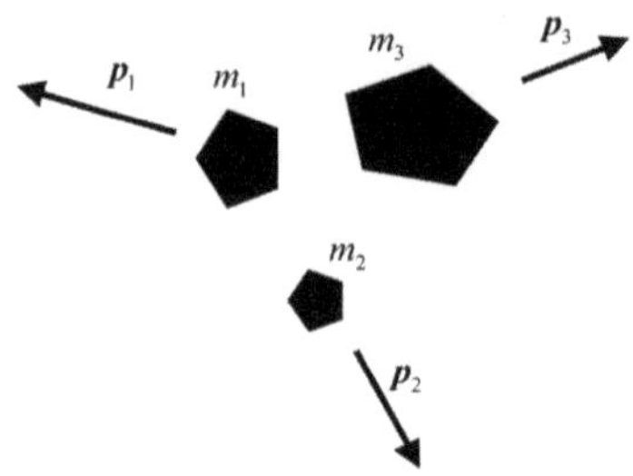

Figure 2.22. Fragmentation event results in three fragments with statistically determined masses and momenta.

An impulsive fragmentation event has led to the expulsion of $N = 3$ fragments resulting from the break up of a body of mass M into individual fragment masses m_i. Each fragment emerges from a stationary center-of-momentum point with velocity u_i. The total kinetic energy of the fragment collection is E. The issue of interest is the probabilistic distribution of velocities (or momenta) over the collection of fragments subject to the following constraints on mass, momentum and energy (bold type indicates a vector quantity whereas normal type provides the scalar value of the same property),

$$\sum_{i=1}^{N} m_i = M, \tag{2.108}$$

$$\sum_{i=1}^{N} \boldsymbol{p}_i = \sum_{i=1}^{N} m_i \boldsymbol{u}_i = 0, \tag{2.109}$$

$$\sum_{i=1}^{N} e_i = E. \tag{2.110}$$

The quantities e_i and $\boldsymbol{p}_i$ are the individual fragment kinetic energies and vector momenta related through

$$e_i = \frac{\boldsymbol{p}_i^2}{2m_i} = \frac{m_i \boldsymbol{u}_i^2}{2}. \tag{2.111}$$

In certain applications, physical conditions resulting from details of the fragmentation event will constrain the distribution of velocities over the resulting fragments. Here the premise is that no knowledge is available as to how the energy E is distributed over the N fragments, and simply seeks the most probable distribution given the specified constraints. The problem as posed is amenable to classical statistical mechanics. The methods are described in such classical papers as those of Maxwell (1879) and Jeans (1921). The present development follows closely the derivation of Levesey (1980) on the statistical treatment of disintegration in nuclear reactions.

The analysis is most easily followed by considering a limited number of fragments. The solution is pursued for $N = 3$ and then generalized to arbitrary fragment number. Momentum conservation from equation (2.109) yields the vector condition

$$\boldsymbol{p}_1 + \boldsymbol{p}_2 + \boldsymbol{p}_3 = 0. \tag{2.112}$$

The rather obscure substitutions are made,

$$\boldsymbol{q}_1 = \sqrt{\frac{M}{M - m_1}}\boldsymbol{p}_1, \tag{2.113}$$

$$\boldsymbol{Q} = \sqrt{\frac{m_1(m_2 + m_3)}{m_2 m_3}}\left(\frac{m_2}{m_2 + m_3}\boldsymbol{p}_1 + \boldsymbol{p}_2\right), \tag{2.114}$$

that, when combined with energy conservation from equation (2.110) yield the expression

$$2m_1 E = q_{1x}^2 + q_{1y}^2 + q_{1z}^2 + Q_x^2 + Q_y^2 + Q_z^2. \tag{2.115}$$

From this point the subscript on m_1 will be dropped as the intention was to isolate attention to one fragment as constrained by the remaining fragments through the relations provided in equations (2.108)–(2.110).

Equation (2.115) is a six-dimensional hyper-sphere in the space of the six independent momentum components with radius R, given by

$$R = \sqrt{2mE}. \tag{2.116}$$

It is the classic functional form aspired to before ascribing equal statistical weights to equal volumes in momentum phase space. The statistical distribution of states for a fragment of mass m is obtained by integrating the remaining momentum variables over the hyper-sphere. The process results in a probability distribution for kinetic energy of a fragment of mass m of

$$f(e) \propto e^{1/2}\left(E - \frac{eM}{M - m}\right)^{1/2}. \tag{2.117}$$

Equation (2.117) provides the distribution for a mass M fragmented into $N = 3$ fragments with total kinetic energy E. From the equation, the maximum kinetic energy (hence velocity) available to a fragment of mass m is

$$e_{\max} = \frac{M - m}{M}E. \tag{2.118}$$

The analysis carried through for an arbitrary number N of fragments leads to a kinetic energy probability distribution for fragments (Levesey 1980),

$$f(e) \propto e^{1/2}\left(E - \frac{eM}{M - m}\right)^{\frac{3N-8}{2}}. \tag{2.119}$$

The transition to a Maxwell energy distribution follows when N becomes large and $M \gg m$. Equation (2.119) becomes

$$f(e) \propto e^{1/2}(E - e)^{\frac{3N}{2}}, \tag{2.120}$$

and in turn for large N is the exponential Maxwell kinetic energy distribution

$$f(e) \propto e^{1/2} \exp\left(-3e/2\bar{e}\right), \tag{2.121}$$

where $\bar{e} = E/N$ is the average kinetic energy per fragment. For large N the energy distribution becomes independent of the fragment mass m. The corresponding velocity distribution resulting from the appropriate substitution in equation (2.121) does depend on the fragment mass, weighting the distribution for heavier fragments toward correspondingly lower velocities.

The distribution from equation (2.119) is illustrated in figure 2.23 for selected values of the fragment number. In the left-hand plot, the fragment size remains the same at $m = M/4$ while in the right-hand plot all fragments are equal size $m = M/N$ in each distribution. Also illustrated in the right-hand plot is the limiting Maxwell distribution from equation (2.121) for an average energy of $\bar{e} = E/5$, showing that for five fragments the distributions are already close.

There are many impulsive fragmentation events for which the statistical velocity distributions developed in this section would not apply. Detailed geometry and impact conditions would lead to a spectrum of velocities which may differ markedly from the Maxwell representation of equation (2.121). In events where initial conditions are not precisely determined, or in the average response of many similar events, the present approach can provide useful statistical predictions. An application of the statistical theory has been applied to the data of Waza *et al* (1983) and is shown with the data in figure 2.21 (Grady *et al* 2000, Grady and Winfree 2001).

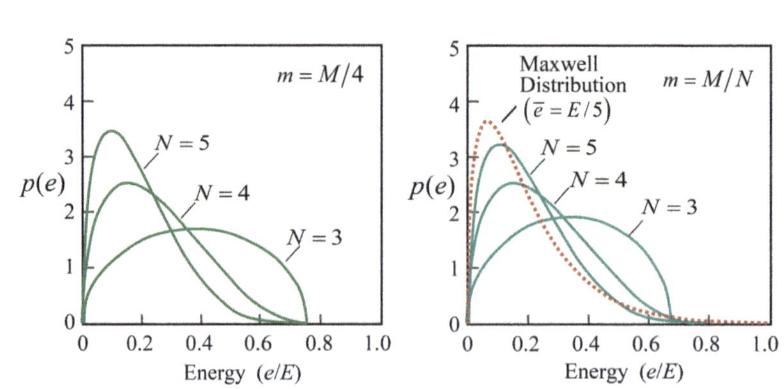

Figure 2.23. Kinetic energy probability for three, four and five fragments. Fragment mass is $M/4$ in the left panel and all fragments are equal size, M/N, on the right. The Maxwell distribution for an average mass of $E/5$ is shown for comparison (Grady 2009).

2.3 Fragment size in dynamic fragmentation

In a dynamic fragmentation event, a principal issue is characterization of the intensity of fragmentation. That is, in the simplest sense, the number of fragments the disrupted object breaks into or, conversely, the average fragment size. This question can take different forms with differing directions of concern. In some applications, prediction of the amount of new surface area created in the event may be of interest. In others, how this surface area is distributed over fragment size may be more pressing. The concern may be with the few largest fragments created, or it may be with finest airborne particulate. Issues relating to fragment size distributions were pursued in the previous section. The present section will focus on the prediction of fragment size in a dynamic fragmentation event.

The conditions driving the fragmentation intensity are many and are characterized in various ways by analysts that have explored the problem. Energy imparted to the body through the loading process leading to disruption, and attempts to correlate energy with the fragmentation intensity, was common in earlier approaches. Stress amplitude and impulse criteria have also been pursued. A kinematic criterion based on the rate of strain carrying the body to disruption has had some success. The geometry of the fragmenting body also plays an important role.

The material properties governing the resistance (or lack thereof) of the structure or body to the fracture processes leading to fragmentation are also quite varied. In some materials, continuum strength properties, such as fracture toughness, yield strength and surface tension, appear to correlate with the intensity of fragmentation. In others, characterization of microscopic defect structure appears necessary to prediction of the fragmentation consequences. Mott's seminal theory of fragmentation is based on the determination of statistical strain to failure of the sites responsible for fracture activation.

In the present section, some of the theoretical efforts are described that have been undertaken to predict the size of fragments, or the newly created fracture surface area, in the dynamic fragmentation event. This survey is not exhaustive, but provides a perspective of some of the avenues that have been pursued.

2.3.1 Historical theories of fragment size prediction

Much of the early work in attempting to quantify fragment size in dynamic fragmentation was stimulated by needs in the mine and quarry blasting, and rock crushing industries. In the arena of military application, the studies of Mott during the World War II years are seminal.

2.3.1.1 Fragment size laws of von Rittinger, Kick and Bond

Early efforts in predicting the reduction in size of a solid material subjected to fragmenting forces focused on the energy input to the system in the breakage process. The workers in this area were concerned with brittle and semi-brittle materials, and processes such as mine and quarry blasting, rock and mineral crushing, and grinding were the applications of interest.

Apparently the earliest, and most widely recognized effort, is the size reduction law of von Rittinger (1867) relating the proportionality between the specific energy input E to the increase in new fragment surface area,

$$dA \sim dE. \tag{2.122}$$

A characteristic fragment size x is proportional to the length scale of the fragment distribution and it follows that $x \sim 1/A$, where A is the fragment surface area per unit volume. The Rittinger law leads to a reduction in fragment size with a dependence on energy of the form,

$$x = \frac{x_o}{1 + \alpha E}, \tag{2.123}$$

where x_o is the initial size of the body and α is an efficiency index of the fragmentation process.

An alternative theory of fragmentation was proposed by Kick (1885) whereby equivalent amounts of energy resulted in equivalent geometric reductions in the sizes of fragments. The Kick hypothesis leads to

$$E \sim \ln(x_o/x), \tag{2.124}$$

or

$$x = x_o e^{-\alpha E}. \tag{2.125}$$

Some years later, Bond (1952) recognized that neither Rittinger's law nor Kick's law fully worked in the engineering applications of interest and proposed a relation that is a compromise of the two earlier relations,

$$x = \frac{x_o}{(1 + \alpha E)^2}. \tag{2.126}$$

2.3.1.2 The summarization of Charles

The fragment size relations of von Rittinger, Kick, and Bond are summarized in a study of Charles (1957) where he generalized the earlier relations into the differential expression

$$dE = -C dx/x^{\eta}, \tag{2.127}$$

where dE is the increment of specific energy required to achieve size reduction dx. Values of $\eta = 1$, 3/2 and 2 retrieve the relations derived by Kick, Bond and von Rittinger, respectively.

There is ample evidence over the diverse applications of interest to support any of the three fragment size laws under certain conditions. In testing any of the relations, however, the difficulty is, first, quantifying the energy input responsible for fragmentation, and second, characterizing the appropriate fragment size. Size was commonly characterized as the modulus or length scale k of a measured size distribution, such as the commonly used Schuhmann (1940) cumulative weight versus fragment size relation,

$$M(x) = (x/k)^n. \tag{2.128}$$

Energy characterization usually reduces to quantifying the proportional work in the breakage event, such as the relative time in a particular crusher, or the relative height in a drop-tower fragmentation apparatus. Efforts that more carefully quantify the strain energy fueling fracture in the brittle fragmentation process appear few. One extensive effort by Bergstrom *et al* (1961) is discussed in a later section on brittle fragmentation.

Charles (1957) examined diverse fragmentation data and found that the power η in his equation (2.127) depended on the breakage method applied, but ranged within the bounds provided by the Rittinger and the Kick fragment-size-reduction laws. Combining the Schuhmann distribution in equation (2.128) with his equation (2.127), he provided a size reduction relation of the form

$$k = \frac{x_o}{(1 + \alpha E)^{1/(\eta - 1)}}. \tag{2.129}$$

A relation for α in terms of the energy reduction constants η and C, and the Schuhmann index n, is attempted, but Charles identifies α as an empirical comminution constant. Although largely empirical, equations (2.128) and (2.129) provided a means of systemizing a broad body of fragmentation data on brittle materials and were eminently useful in application.

2.3.1.3 Mott statistical strain-to-fracture theory

Mott in 1943 addressed an entirely different application in dynamic fragmentation and his theoretical approach to the problem differs starkly from the previous authors. Mott addressed the issue of fragmenting munitions and undertook an effort to understand the influences of case metal, explosive properties and system geometry, on the processes of natural explosive-driven fragmentation. His efforts within the months of 1943 resulted in a remarkably sophisticated theory governing the size and statistics of the fragmentation process in exploding warheads. His seminal efforts have continued to influence this area of dynamic fragmentation to the present day. In this section, these early theoretical ideas of Mott relating to prediction of fragment size are briefly described. Mott's work is discussed further in both previous and later sections.

Mott was interested in the explosion-driven fragmentation of tubular munitions and his theoretical efforts primarily focus on the statistical frequency of axial fractures around the circumference of the cylindrical shell. The steel case metals considered by Mott underwent extensive plastic expansion prior to fracture onset and subsequent fragmentation. Mott concluded that an ultimate tensile strength Y of the steel is a governing fragmentation property and showed that when a fracture occurred, as illustrated in figure 2.11, regions away from the fracture were relieved of the driving stress by a diffusion-like wave which propagated according to the relation

$$x \sim \sqrt{\frac{Yt}{\rho\dot{\varepsilon}}} \sim \sqrt{\frac{Y}{\rho\dot{\varepsilon}^2}}\varepsilon^{1/2}. \tag{2.130}$$

In this relation x is the propagation distance of the diffusive release wave after a time t from fracture initiation. The metal density is ρ, and $\dot{\varepsilon} = u/r$, where u is the radial velocity of the case and $\dot{\varepsilon}$ is the circumferential stretching rate. Derivation of the Mott wave motion is developed in section 2.3.5.

Equation (2.130) is a reasonable representation of the physics considered by Mott for the fracture of a plastically deforming steel shell. It is also instructive to consider an alternative assumption of elastic release propagation from similar points of fracture. This would be more appropriate for brittle fracture, for example, where the elastic limit is not achieved before onset of fracture. The relation for the distance of stress release corresponding to equation (2.130) would then be

$$x \sim ct \sim \frac{c}{\dot{\varepsilon}}\varepsilon, \tag{2.131}$$

where c is a measure of the elastic wave speed in the material. In equation (2.130) or (2.131) the distance x is the nominal distance that the release waves emanating from a fracture will propagate before interacting with an adjacent fracture. Consequently, the distance x is proportional to the average circumferential spacing of the fractures. The time t and the strain $\varepsilon = \dot{\varepsilon}t$ are the time and strain accumulated over the dynamic fracture process.

Mott proposed that fracture in the expanding shell was dependent on statistical variations in the strain to fracture and that this fracture property could be characterized by an analytic function $\lambda(\varepsilon)$ that provides the statistical rate of fracture activation. As discussed in the previous section, Mott supposed that $\lambda(\varepsilon)$ should be a rapidly increasing function of strain, and proposed both a power law,

$$\lambda(\varepsilon) = \frac{n}{\sigma}\left(\frac{\varepsilon}{\sigma}\right)^{n-1}, \tag{2.132}$$

and an exponential law,

$$\lambda(\varepsilon) = \frac{1}{\sigma}e^{(\varepsilon-\mu)/\sigma}. \tag{2.133}$$

The power-law activation function in equation (2.132), or the exponential function in equation (2.133), lead to extreme value statistics of the Weibull and the Gumbel form, respectively.

The strain interval in equation (2.130) or (2.131) is determined by the statistical spread in strain to fracture allowed by either the Weibull or the Gumbel distribution and is proportional to the standard deviation in activation strain provided by either distribution. For an exponential activation function, and the Gumbel distribution, the standard deviation is of order σ and

$$\varepsilon \sim \sigma. \tag{2.134}$$

Thus, equation (2.130) provides for the characteristic fragment size

$$x \sim \sqrt{\frac{Y\sigma}{\rho\dot{\varepsilon}^2}}, \tag{2.135}$$

which is, within a constant, the same expression developed by Mott (1947), where $\gamma = 1/\sigma$ is the gamma parameter introduced by Mott. Alternatively, similar arguments using equation (2.131) for elastic release leads to

$$x \sim \frac{c\sigma}{\dot{\varepsilon}}. \tag{2.136}$$

Equation (2.135) or (2.136) predicts that the characteristic fragment size decreases as the first power of the expansion strain rate $\dot{\varepsilon}$.

When the power-law activation function in equation (2.132) is used to describe the strain-to-fracture statistics, the same solution for fragment size is more subtle as was pointed out by Mott. The distribution moments for the Weibull distribution are dependent on the length scale of interest. Consequently, the strain interval corresponding to equation (2.134) for the Gumbel distribution is

$$\varepsilon \sim \frac{\sigma}{x^{1/n}}. \tag{2.137}$$

Combining equations (2.130) and (2.137) leads to

$$x \sim \left(\frac{Y\sigma}{\rho\dot{\varepsilon}^2}\right)^{\frac{n}{2n+1}}, \tag{2.138}$$

which provides a dependence of fragment size on $\dot{\varepsilon}$ ranging between an inverse first and an inverse two-thirds power, as was also noted by Mott. Equation (2.131) for elastic fracture yields

$$x \sim \left(\frac{c\sigma}{\dot{\varepsilon}}\right)^{\frac{n}{n+1}}, \tag{2.139}$$

yielding for fragment size an inverse first to inverse one-half power dependence on strain rate.

2.3.2 Weibull critical flaw fragmentation

A model of dynamic fracture and fragmentation for brittle solids that provides a method for predicting average and distributed fragment size in impact or explosive loading applications is based on Weibull critical flaw concepts. The model has been implemented into computational codes and used to predict the consequences of blasting in hard rock, explosive fragmentation of oil shale and impact fracture of ceramics (Grady and Kipp 1980a). The model includes features of the Weibull critical state theory for the stress-load-induced failure of solid bodies. The relationship between the Weibull and the fragmentation theory offers alternative methods for assessing the material fracture parameters essential to the model. Similarities of the present model to the statistical activation and interaction fragmentation theory of Mott (1943) is also apparent.

Central to the model is a statistical population of fracture-producing flaws characterized by the tensile elastic strain ε (or stress) at which fracture activates. In computational applications this parameter is usually the mean tensile strain, or alternatively that collinear with the maximum principal tensile stress. Fracture-producing flaws are characterized through a cumulative number density expression of the power-law form

$$N(\varepsilon) = (\varepsilon/\eta)^m. \tag{2.140}$$

The parameter η has dimensions of length to the power $3/m$. At a site of fracture activation, the subsequent release of tensile stress is assumed to propagate spherically outward with a velocity c_g. The velocity c_g is the limiting fracture growth velocity in the brittle material. The volume of time-dependent stress release about a site after fracture activation is

$$v(t) = (c_g t)^3. \tag{2.141}$$

The dynamics of the tensile fracture event is characterized by an expansion strain rate $\dot{\varepsilon}$ such that

$$\varepsilon = \dot{\varepsilon}t. \tag{2.142}$$

In computational implementation, the expansion rate is either an average of the principal strain rates, or that corresponding to the maximum principal tension.

Material properties for the fracture and fragmentation model require the flaw structure parameters, η and m, the fracture stress release velocity, c_g, and the elastic modulus and density, E and ρ. The model relations expressed in equations (2.140)–(2.142) combine to provide the fraction of stress relieved material as dynamic fracture proceeds,

$$D(t) = \int_0^t v(t - \tau)dN = \int_0^t v(t - \tau)N'(\varepsilon)\dot{\varepsilon}d\tau. \tag{2.143}$$

In the model $D(t)$ is identified as the fracture damage, and is the volume fraction of the fractured and stress-released material. The damage satisfies $0 \leqslant D(t) \leqslant 1$. The damage parameter D is used to determine both the frequency of fracture and the continuum stress state as a function of time through the fracture process. A straightforward linear energy function, with the damage D identified as an isotropic internal state variable,

$$U = \frac{1}{2}E\varepsilon^2(1 - D), \tag{2.144}$$

provides both the fracture strain energy release,

$$\Gamma = -\frac{\partial U}{\partial D}\bigg|_{\varepsilon} = \frac{1}{2}E\varepsilon^2, \tag{2.145}$$

and the continuum stress,

$$\sigma = \frac{\partial U}{\partial \varepsilon}\bigg|_{D} = E(1 - D)\varepsilon, \tag{2.146}$$

with $E(1 - D)$ the fracture-damage reduced elastic modulus of the material.

The number density of fractures activated in the dynamic event are provided through the relations,

$$N_g(t) = \int_0^t (1 - D)N'(\varepsilon)\dot{\varepsilon}d\tau \simeq \int_0^t e^{-D}N'(\varepsilon)\dot{\varepsilon}d\tau. \tag{2.147}$$

The second integral relation with the exponential expression approximates the first through application of the statistical scheme of Johnson and Mehl (1939) outlined in an earlier section, and accounts for fracture release overlap and thus softens the rate that damage approaches one. The spectrum of fracture number activation over time is provided by the derivative of equation (2.147),

$$n_g(t) = (1 - D)N'(\varepsilon)\dot{\varepsilon}. \tag{2.148}$$

A measure of the spectrum of fragment sizes is provided by relating fragment size to the nominal spacing between fractures during the damage growth time,

$$x \sim c_g(t_D - t), \tag{2.149}$$

and eliminating time between equations (2.148) and (2.149). The time t_D is provided by $D(t_D) = 1$ in equation (2.143). The model assumes that fractures activate statistically over time, but arrest approximately together during the final catastrophic growth phase occurring at time t_D. Although the model does not boast the rigor of the one-dimensional statistical theory of Mott, application of the model in practice has been fruitful (Grady and Kipp 1980a). Like the Mott solution, the theory leads to a rather narrow spread in fragment size. The theory does not predict the Schuhmann-like power-law distribution observed in experimental brittle fracture, or the fractal behavior discussed in later sections. Selection of the mean, or the mode, of the distribution provides a reasonable measure of the average fragment size.

Under a condition of uniform strain-rate loading a maximum stress (the fracture stress) is achieved,

$$\sigma_M = \alpha\dot{\varepsilon}^{3/(m+3)}, \tag{2.150}$$

with a characteristic fragment size

$$x_m = \beta\dot{\varepsilon}^{-m/(m+3)}. \tag{2.151}$$

The fracture model constants α, β and m are provided by experimental data.

The physics underlying the model and leading to the strain-rate fracture stress in equation (2.150) provide a sensible explanation for the increased dynamic strength in the explosive or percussive breakage of rock. Early studies of Rinehart (1965) provide evidence that the dynamic fracture strength of rock can exceed the static strength by as much as an order of magnitude. Chapter 1 provides further characterization of static and dynamic strengths of rock and rock-like materials. In rock-blasting calculations, the known static fracture strength is often arbitrarily increased by a factor of seven or eight to account for dynamic conditions.

Perhaps the first material for which quantitative properties within the context of a strain-rate-dependent strength and fragmentation model were acquired is oil shale (Grady and Kipp 1979). Experimental fracture stress and fragment size measurements as a function of strain rate for this material are shown in figure 2.24. The data are for an approximately 20 gallon per ton oil shale from the Anvil Points mine of the Green River formation in the state of Wyoming. The matrix material interspersed with laminar plains of kerogen (the oil content) is primarily dolomitic in a form referred to as marlstone.

Correspondence of the present dynamic fracture and fragmentation model with Weibull strength of materials theory is revealed by noting that the derivative of the cumulative number density in equation (2.140),

$$N'(\varepsilon) = \frac{m}{\eta}\left(\frac{\varepsilon}{\eta}\right)^{m-1} = h(\varepsilon), \tag{2.152}$$

provides the power-law statistical failure hazard function $h(\varepsilon)$ integral to the Weibull theory. The cumulative probability of failure $P(\varepsilon)$ of a body of volume V, subjected to a stress load $\sigma = E\varepsilon$, follows from

$$dP = (1 - P)Vh(\varepsilon)d\varepsilon, \tag{2.153}$$

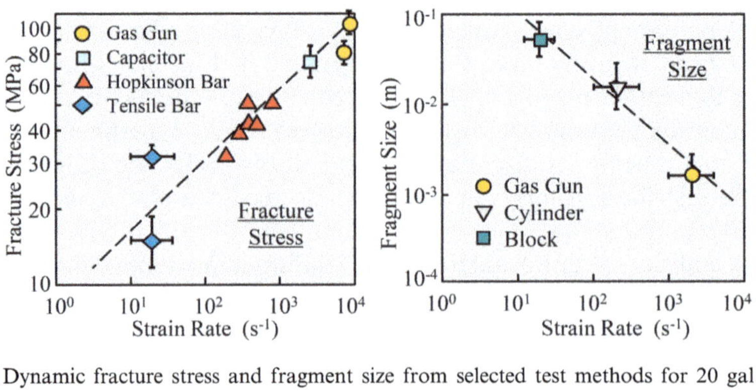

Figure 2.24. Dynamic fracture stress and fragment size from selected test methods for 20 gal ton^{-1} Green River oil shale (Grady and Kipp 1979, 1980a).

and

$$P(\varepsilon) = 1 - e^{-V(\varepsilon/\eta)^m}. \tag{2.154}$$

The scale dependence of the Weibull strength is observed since equivalent probabilities of failure require the exponent in equation (2.154) be constant and hence,

$$V(\varepsilon/\eta)^m = \text{constant}. \tag{2.155}$$

2.3.3 An impulse and energy criterion for fragment size

A nominal fragment size resulting from a dynamic fragmentation event can be estimated from one of several basic physical arguments. Here two methods are described in which elementary momentum and energy concepts local to a forming fragment are considered.

The one-dimensional fracture event pictured in figure 2.25 identifies a fragment of length x_o that is in the process of forming. A coordinate x delineates the distance from the center of the fragment. The body before fracture is stretching at a uniform strain rate $\dot{\varepsilon}$. The region to the right of center contains a momentum $\rho x_o{}^2\dot{\varepsilon}/8$. Through the fracture event, momentum in the fragment is reduced to zero when

$$\int_0^{t_f} \left(\sigma_f - \sigma_{pz}(t)\right)dt = \rho x_o{}^2\dot{\varepsilon}/8. \tag{2.156}$$

where σ_f is the constant stress at the fragment center and $\sigma_{pz}(t)$ is the time-dependent stress in the fracture zone. Assuming stress $\sigma_{pz}(t)$ in the fracture zone reduces effectively instantaneously to zero, equation (2.156) yields

$$\sigma_f t_f = \sigma_f x_o/2c = \rho x_o{}^2\dot{\varepsilon}/8, \tag{2.157}$$

where c is the elastic wave speed that determines the time over which the momentum is reduced to zero and σ_f is the tensile fracture stress. Solving for x_o gives

$$x_o = 4\sigma_f/\rho c\dot{\varepsilon}, \tag{2.158}$$

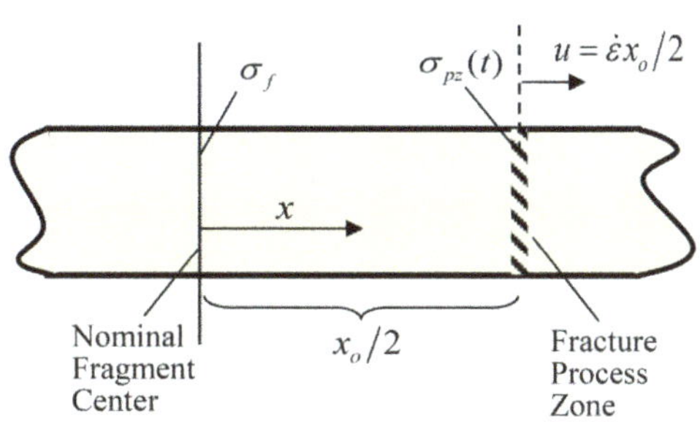

Figure 2.25. A fragment in the formation process and the bases for analysis of momentum and energy balance estimates of fragment size.

for a measure of the nominal fragment size based on the premise that residual individual fragment momentum is brought to rest upon completion of fragment formation.

An alternative fragment size criterion is based on the assumption that the kinetic energy in the same region identified in the figure is fully dissipated in the fracture zone. Identifying this dissipation with fracture energy Γ, energy balance yields the relation

$$\frac{1}{24}\rho\dot{\varepsilon}^2(x_o/2)^3 = \Gamma, \tag{2.159}$$

and provides a nominal fragment length of

$$x_o = 2(24\Gamma/\rho\dot{\varepsilon}^2)^{1/3}. \tag{2.160}$$

Relating Γ to a more familiar fracture toughness through $\Gamma = K_c^2/2E$ yields

$$x_o = 2\left(\sqrt{12}\,K_c/\rho c\dot{\varepsilon}\right)^{2/3}. \tag{2.161}$$

Considering the relatively elementary momentum and energy principles used to estimate a nominal fragment size in equations (2.158) and (2.161), it is instructive to apply the relations to a reasonably well characterized material under representative fragmentation conditions. A 4140 steel heat treated to a Rockwell hardness of 40 was used to investigate exploding case fragmentation (Grady and Hightower 1992). The same study provides illustrations of fragmentation statistics in the previous section. Properties for the material include a density of 7870 kg m^{-3}, an elastic modulus of 200 GPa, a yield stress of 1.1 GPa and a fracture toughness of 80 MPa $m^{1/2}$. Using the yield stress as a measure of the fracture stress (probably somewhat low due to the hardening properties of the steel), the average fragment dimension (fracture spacing) in the strain-rate range in the neighborhood of 10^4 s^{-1} from the equation (2.158) impulse-based (σ_f = 1.1 GPa) prediction and the equation (2.161) energy-based prediction are shown in figure 2.26. The plots span

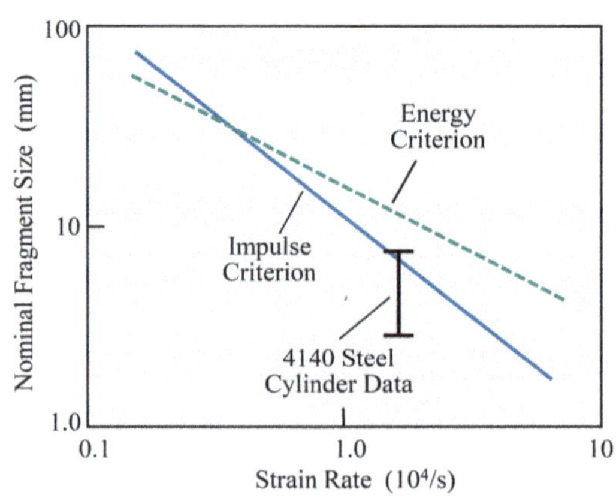

Figure 2.26. Comparison of impulse and energy-based fragment size prediction and data from 4140 steel cylinder fragmentation (Grady and Hightower 1992).

fragment widths covering about 80% of the mass of the fragments in one expanding cylinder fragmentation test on the 4140 steel. Quantitative agreement between the two predictive curves is remarkable considering the use of markedly different static properties (yield strength versus fracture toughness) in the functionally different relations for fragment size. The different strain-rate dependences for the two simple theories (a minus first power for impulse and minus two-thirds power for energy) are found to span the range of strain-rate dependence calculated from more detailed fragment size analysis.

2.3.4 Mott fracture with dissipation

As summarized earlier, Mott (1943, 1947) developed a fragmentation model for the breakup of an expanding ductile metal shell. The statistical strain-to-fracture model yields an average fracture spacing of

$$x \sim \sqrt{\frac{Y\sigma}{\rho\dot{\varepsilon}^2}}, \qquad (2.162)$$

when an exponential strain-to-fracture function (Gumbel statistics) is assumed, or

$$x \sim \left(\frac{Y\sigma}{\rho\dot{\varepsilon}^2}\right)^{\frac{n}{2n+1}}, \qquad (2.163)$$

for a power-law function (Weibull statistics). In both expressions, Y is the tensile flow stress at fracture, ρ is the metal density and $\dot{\varepsilon}$ is the cylinder rate of expansion. Also, in both cases, σ is a measure of the standard deviation in strain to fracture, and is the key material parameter governing fragmentation in the Mott statistical theory. As noted earlier $\sigma \sim 1/\gamma$, where γ is the fragmentation parameter explicitly used in Mott's reports.

A different application of the Mott fragmentation model has led to markedly different physics governing the characteristic fracture spacing (Grady *et al* 1984, Kipp and Grady 1985). Mott assumed upon achieving the strain-to-failure criterion at a circumferential point that fracture was instantaneous and, based on momentum balance, that the release of tensile stress propagates away from the point of fracture according to (solutions to propagation of Mott waves will be developed shortly)

$$\rho\dot{\varepsilon}\xi\frac{d\xi}{dt} = Y, \qquad (2.164)$$

where ξ is the propagation distance of the Mott stress release wave from the point of fracture. The combination of equation (2.164) with the statistical strain-to-failure relations leads to the Mott predictions of characteristic fragment size provided in either equation (2.162) or (2.163).

An alternative approach assumes that fracture is not instantaneous, but that stress reduces continuously with some measure of the opening of the fracture. A straightforward approach is to assume a crack opening model in which stress is reduced linearly from Y to zero with some fracture opening displacement parameter $0 \leqslant \eta \leqslant \eta_c$. In this case, the work of fracture $W = Y\eta_c/2$ is dissipated in completing the fracture process. An alternative relation corresponding to equation (2.164), again considering momentum balance, that governs the motion of the stress release wave is obtained,

$$\rho\dot{\varepsilon}\xi\frac{d\xi}{dt} = \frac{Y^2}{2W}\eta, \tag{2.165}$$

where

$$\frac{d\eta}{dt} = \dot{\varepsilon}\xi \tag{2.166}$$

provides the crack opening displacement η. Solutions to this coupled system of ordinary differential equations are

$$\xi(t) = \frac{1}{12}\frac{Y^2}{\rho W}t^2 \tag{2.167}$$

for the propagation distance of the tensile release wave and

$$\eta(t) = \frac{1}{36}\frac{\dot{\varepsilon}Y^2}{\rho W}t^3 \tag{2.168}$$

for the crack-opening displacement over the interval $0 \leqslant \eta \leqslant \eta_c$. Rather than instantaneous fracture, a time

$$t_f = \left(\frac{72\rho W^2}{Y^3\dot{\varepsilon}}\right)^{1/3} \tag{2.169}$$

is required for fracture completion.

Substitution of this fracture time into equation (2.167), a distance of propagation of tensile release $\xi(t_f)$ is calculated. It is proposed that twice this distance provides a reasonable measure of the nominal fracture spacing (Kipp and Grady 1985),

$$x = \left(\frac{24W}{\rho\dot{\varepsilon}^2}\right)^{1/3}. \tag{2.170}$$

Relating W to the Γ in the previous energy criterion in equation (2.16) finds that the present fragment size prediction is functionally the same (dimensional arguments would yield this conclusion) and a factor of two smaller.

The present relation for fragment size and that put forth by Mott in equation (2.163) differ starkly in the underlying physics, the functional dependence and the governing material parameters. In the present equation (2.170), fracture energy controls the fragment size. In Mott's relation, a statistical strain-to-fracture parameter is the governing property.

Reconciliation of the two outwardly different theories for characteristic fragment size in the present one-dimensional ductile fracture model has been pursued

(Grady and Olsen 2003). The difference in the predicted strain-rate dependence is understood, at least in part. Mott chose to use a Gumbel distribution to represent statistical strain to failure. The Gumbel distribution lacks a distribution shape parameter and leads to a unique inverse first power strain-rate dependence. Use of the more flexible Weibull distribution, which has a distribution shape parameter, leads instead to a non-unique strain-rate dependence ranging between an inverse two-thirds and first-power dependence, encompassing the energy-based prediction in equation (2.170). Noteworthy, however, is that identical strain-rate dependence between equations (2.163) and (2.170) requires $n = 1$ in the Weibull distribution, or a constant (uniform) fracture activation function.

Experimental data appear to support the predicted strain-rate dependence over the range of nominal fragment sizes achieved in the fragmentation of ductile metals. Experimental examples supporting either bound (two-thirds or first power) can be found, however. For example, fragmentation of ductile uranium rings exhibits very nearly an inverse two-thirds power dependence on strain rate (Grady and Olsen 2003) (see section 2.3.7.3), whereas selected exploding steel cylinders show an inverse first power behavior (Grady *et al* 2001) (see section 2.2.4.6).

Concerning the characteristic fragment size, Mott's statistical fragmentation theory and the energy-based theory can be sensibly reconciled if Mott's statistical fracture activation function is not equated to a fracture completion function. If, at some loading strain rate $\dot{\varepsilon}_o$, Mott's theory predicts an average fragment size smaller than that predicted by the energy-based theory, then it is shown that all fractures activated by the Mott function cannot reach completion because of energy limitations. It is also readily shown that, for an exponential or strong power law ($n \gg 1$) fracture activation function, a characteristic strain rate is determined from

$$\dot{\varepsilon}_o \sim \sqrt{\frac{Y^3\sigma^3}{\rho W^2}}. \tag{2.171}$$

At strain rates greater than $\dot{\varepsilon}_o$ the fracture hazard function, provided by equation (2.132) for fracture activation, is constant, or nearly constant, and of the form $\lambda(\varepsilon) = \lambda_o = 1/\sigma$. The statistical scale parameter σ is a continuum material length scale of order, $\sigma \sim W/Y$ with Y the flow stress at breakup and W the fracture energy.

A one-dimensional computational simulation of the Mott fragmentation process performed by Kipp and Grady (1986) provides support for this interpretation of the two theories. In that work it was recognized that interplay between the dynamics of the fragmentation event and the population of flaws seeding the multiple fracture process can lead to conditions in which flaw structure controlled the extent of fragmentation on one hand while energy limitations controlled fragmentation on the other.

Computer simulations of dynamic fragmentation were performed to support experimental fragmenting ring studies (Grady *et al* 1984). A one-dimensional finite difference wave code was used to calculate the response of an aluminum rod or wire 0.1 m in length stretching plastically at a flow stress $Y = 100$ MPa and at a uniform

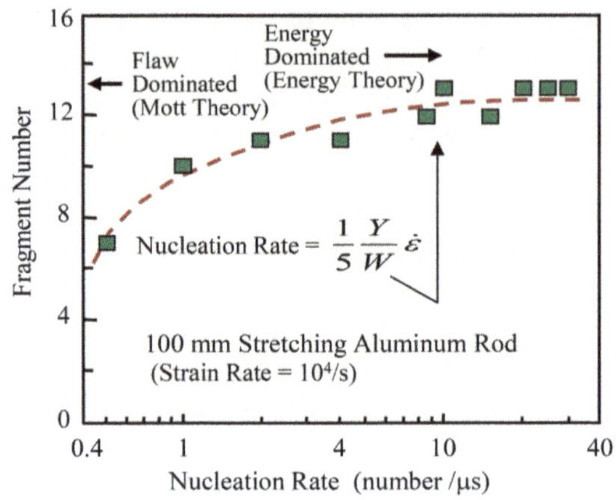

Figure 2.27. Fragment number versus fracture activation rate from computational simulations illustrating transition from flaw-site dominated fragmentation to energy dominated fragmentation (Grady 2006).

stretching rate of $\dot{\varepsilon} = 10^4\ s^{-1}$. Fracture sites were introduced randomly in time according to a constant nucleation rate parameter $\lambda(\varepsilon) = \lambda_o$, and randomly positioned within the length of the rod. The nucleation rate λ_o was the only parameter varied over the series of calculations. When fracture was nucleated at a computational cell, stress in that cell was relaxed from the flow stress Y to zero as the cell distended, such that the plastic fracture energy within that cell of $W = 2 \times 10^4\ J\ m^{-2}$ was dissipated. The number of fragments produced as the nucleation rate λ_o was varied over approximately one order of magnitude was determined from the simulations and is shown in figure 2.27.

At the lower nucleation rates, every fracture nucleation site is computed to grow to full fracture. The number of fractures and the corresponding characteristic fragment size is then governed fully by the flaw structure and the fracture activation (seeding) function. As the nucleation rate is increased, the number of activated fractures that fail to grow to completion increases correspondingly. At the highest nucleation rates, the number of fragments becomes independent of the number of fracture sites nucleated and is determined strictly by the fracture energy W resisting fracture growth. The calculated energy-governed constant fracture survival rate, $\lambda_o \cong Y\dot{\varepsilon}/5W$, identified in the figure, is sensibly consistent with the computed transition from flaw-limited to energy-limited fragmentation. Distributions in fragment lengths from the computer simulations were also found to be consistent with predictions of the Mott statistical theory and with experiment.

2.3.5 Mott wave solutions

In Mott's seminal study of the explosive-driven fragmentation of metal shells, an analysis of the propagation of stress release waves away from sites of fracture is integral to the theory. The analysis was undertaken by Mott to assess the interacting of multiple fractures and to establish the characteristic spacing between fractures. Other features of the Mott theory were introduced earlier in the chapter. The focus

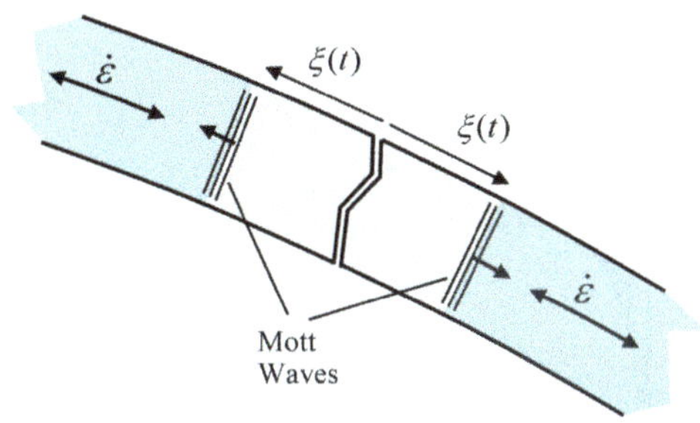

Figure 2.28. Diffusion-like Mott waves propagating from point of fracture into regions stretching at a rate governed by the expansion velocity and at a flow stress Y.

here is on the solutions of such waves (Mott waves), including the original plastic wave solution derived by Mott. The initial calculation of Mott assumed a rigid constant-flow-stress plasticity model. Lee (1967), in the same time frame, but published much later, extended this calculation to an elastic–plastic body, showing more clearly the range of validity of the Mott solution. Comparison of the Mott solution with heat-like diffusion equations demonstrates the decidedly diffusive character of the Mott wave. Hardening and softening plasticity can also be introduced into the Mott wave calculation. Fracture dissipation in the Mott wave analysis is described in the previous subsection.

The dynamic fracture analysis undertaken by Mott is one-dimensional. It is best visualized as that of a uniformly stretching rod or expanding ring such as previously shown in figure 2.11 and as also illustrated in figure 2.28. The model is applicable to problems such as fragmentation of a rapidly expanding cylinder in which the circumferential stretching rate substantially exceeds the axial. Similar wave propagation occurs in expanding spherical shells where principal strains are closer to equiaxial, although the Mott-like analysis becomes more complex. Here, for clarity, the analysis focuses on a stretching one-dimensional filament or wire. Prior to fracture the body is uniformly stretched to an axial instability strain ε and continues to increase at a constant stretching rate $\dot{\varepsilon}$.

2.3.5.1 Mott rigid–plastic solution

Mott considered the body to be rigid perfectly plastic and straining in tension under a constant flow stress Y. A more realistic model of an expanding and thinning body (rigid plastic) might be as illustrated in figure 2.29, where the equivalent tension T (equivalent tensile stress times shell thickness) initially increases due to deformation hardening, and subsequently decreases when thinning overcomes hardening. Adiabatic heating or cavitation damage may further enhance the onset of tension instability. Mott assumed that fracture occurred at or near point 2 in the figure (load maximum).

At the onset of breakup, fractures are considered to occur at random in both time (or equivalently strain) and in spatial location on the stretching body. Mott assumed that fractures occur instantaneously relieving the tensile stress at the point of fracture abruptly to zero. Fracture resistance at the point of failure and corresponding fracture dissipation are ignored.

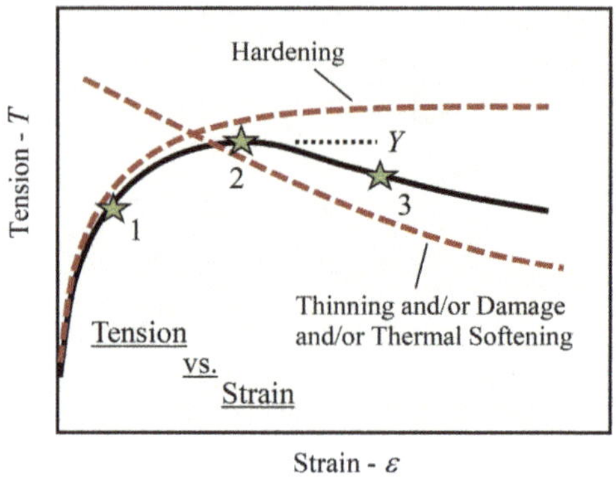

Figure 2.29. The in-plane tension as a function of thinning strain for expansion of a deformation-hardening metal filament or wire.

Accordingly, Mott considered a one-dimensional body stretching plastically under a tensile stress Y and uniformly at a constant stretching rate $\dot{\varepsilon}$. Fracture is initiated by setting the tensile stress to zero at time $t = 0$ at some Lagrangian position $h = 0$. Regions of the body experiencing tensile stress less than Y are considered rigid. A moving boundary (the Mott wave) separates the material stretching uniformly at stress Y in front of the boundary from rigid material moving at a uniform velocity behind the boundary. This Mott wave propagates outward from the point of fracture $h = 0$. The stress and velocity field associated with the Mott wave at a time $t > 0$ are illustrated in figure 2.30. The position of the Mott wave is $\xi(t)$ while crack opening displacement is $\eta(t)$. The velocity field is then

$$u(h,t) = \begin{cases} \dot{\varepsilon}\xi(t) & 0 \leqslant h < \xi(t) \\ \dot{\varepsilon}h & \xi(t) \leqslant h \leqslant h_o \end{cases}, \tag{2.172}$$

where h_o is some arbitrary distance greater than the propagation distance of interest.

The corresponding stress field is apparent. The total momentum of the rod within the region $0 \leqslant h \leqslant h_o$ is given by

$$\rho\dot{\varepsilon}\xi^2 + \int_{\xi}^{h_o} \rho\dot{\varepsilon}h dh = \frac{1}{2}\rho\dot{\varepsilon}\left(\xi^2 + h_o{}^2\right). \tag{2.173}$$

Equating the time rate of change of momentum to the imbalance in tensile stress yields the differential equation

$$\rho\dot{\varepsilon}\xi\frac{d\xi}{dt} = Y, \tag{2.174}$$

for the position $\xi(t)$ of the Mott wave at time t. Integration yields

$$\xi(t) = \sqrt{\frac{2Yt}{\rho\dot{\varepsilon}}}. \tag{2.175}$$

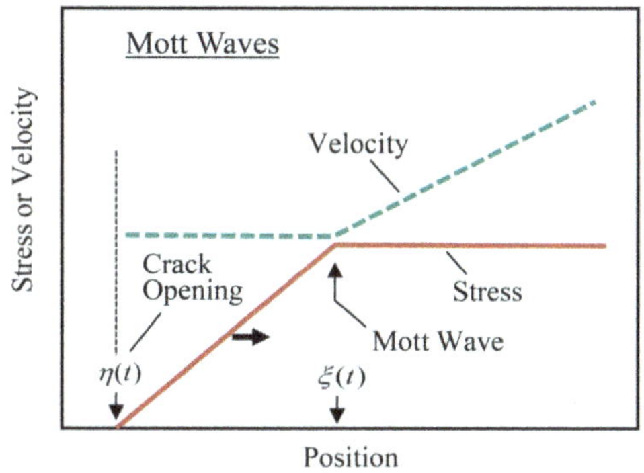

Figure 2.30. The Mott wave stress and velocity field at time t after fracture at position $h = 0$ reduces tensile stress from Y to zero.

Fractures occurring in the stretching body lead to the propagation of waves away from points of fracture, which relieve the tensile stress. The time-dependent motions of these Mott waves are governed by both material properties and kinematic conditions through the relations above. Within regions subsumed by Mott waves, further fracture will not occur. Subsequent fracture will only occur in regions not yet reached by the Mott waves, and which continue to stretch unimpeded at the rate $\dot{\varepsilon}$ and flow stress Y. This statistical process was depicted in figure 2.12.

2.3.5.2 Lee elastic–plastic solution

Mott recognized that excessively high velocities of the interface $\xi(t)$ at early times was a consequence of the rigid–plastic assumption and inconsistent with a more rigorous elastic–plastic treatment of the problem. Mott acknowledged in his war years a report of an analysis due to E H Lee, which was published some years later (Lee 1967). Lee considered the same initial and boundary conditions posed by Mott, but treated the material response behind the Mott wave interface as elastic. His analysis showed that the initial drop in stress from $\sigma = Y$ to $\sigma = 0$ at the origin $h = 0$ propagates as a decaying shock discontinuity in stress and particle velocity at an elastic wave speed c. This shock discontinuity decays to zero at a distance

$$h = \frac{2Y}{\rho c \dot{\varepsilon}} = \lambda, \tag{2.176}$$

at a time

$$\tau = \lambda / c. \tag{2.177}$$

Subsequent reflected elastic waves and the interface $\xi(t)$ are acceleration discontinuities (discontinuities in the slopes of stress and particle velocities).

Continuation of the solution reveals that the interface $\xi(t)$ is a polygon in the h versus t plane with vertices

$$h = n\lambda, \quad t = n^2\lambda / c, \quad n = 1, 2, 3, \ldots, \tag{2.178}$$

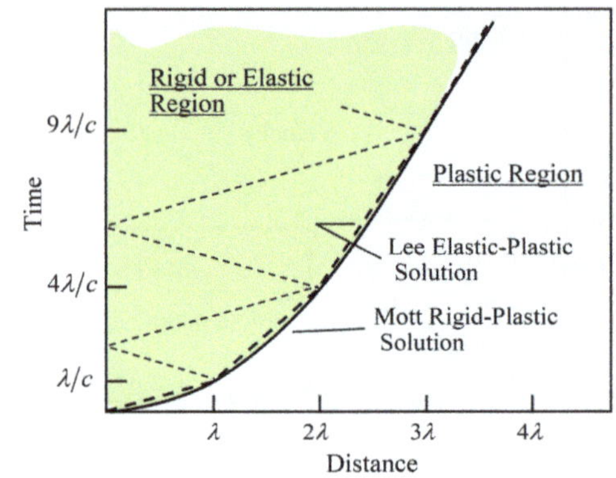

Figure 2.31. Comparisons of the interface separating the plastic region and rigid–plastic region according to the solutions of Mott (1947) and Lee (1967).

where each segment propagates at a velocity of

$$c_n = \frac{c}{2n - 1}. \tag{2.179}$$

The rigid–plastic solution of Mott (1947) and the elastic–plastic solution of Lee (1967) are compared in figure 2.31. The rigid–plastic solution of Mott envelopes the elastic–plastic solution, touching at the vertices. Within several characteristic distances λ, the rigid–plastic solution is a good approximation to both the position of the interface, and to the stress and velocity field behind the interface.

2.3.5.3 Stress diffusion solution

The simplicity of Mott's analysis aptly reveals the underlying physics. The Mott solution applies to the point where hardening in the stretching rod saturates and the tension versus strain loses hyperbolicity. At this point of stationary tension the governing equations transition to parabolic and the diffusive nature implicit in the motion of the Mott wave emerges. The diffusive character of the stress release is demonstrated by writing the linear diffusion relation

$$\frac{\partial^2 \sigma}{\partial h^2} - \frac{1}{\kappa}\frac{\partial \sigma}{\partial t} = 0, \tag{2.180}$$

with the diffusion constant

$$\kappa = Y/2\rho\dot{\varepsilon}. \tag{2.181}$$

Consider the same problem treated by Mott in which fracture at $t = 0$ and $h = 0$ instantly decreases the tensile stress from $\sigma = Y$ to $\sigma = 0$. This classic solution (e.g. Matthews and Walker 1964) is for the stress

$$\sigma/Y = erf(\zeta). \tag{2.182}$$

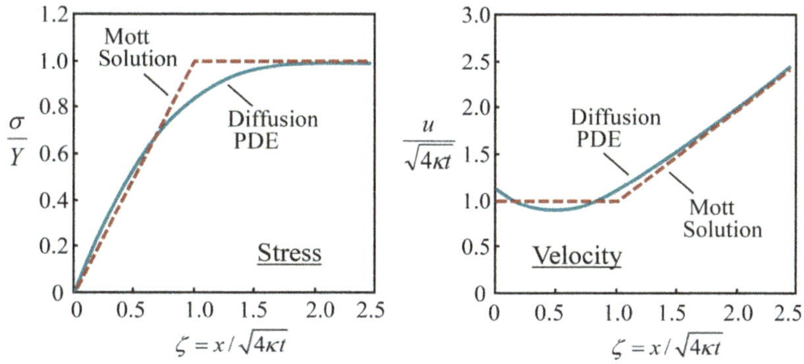

Figure 2.32. Comparison of solutions for Mott's rigid–plastic model and a linear stress-diffusion model of fracture in a stretching plastic filament or wire (Grady 2006).

The velocity is determined through the momentum relation

$$u/\sqrt{4\kappa t} = \sqrt{\frac{4}{\pi}}\exp\left(-\zeta^2\right) + 2\zeta erf(\zeta) - \zeta. \tag{2.183}$$

The functional dependence in equations (2.182) and (2.183) is on the similarity parameter,

$$\zeta = h/\sqrt{4\kappa t}. \tag{2.184}$$

The diffusion equation solution and the rigid–plastic solution of Mott are compared in figure 2.32.

The rigid–plastic solution of Mott with or without fracture dissipation, the elastic–plastic solution of Lee and the solution to the diffusion equation are, of course, only models of the actual processes of fracture and the stress unloading occurring in the rupture of a rapidly stretching ductile ring or shell. Which model most accurately depicts reality is debatable. All, however, reveal physics of the fracture process and point to the decidedly diffusive nature of stress wave propagation in the ductile solid.

2.3.5.4 Hardening and softening in the Mott wave solution

Mott introduced a model of, and a solution for, the speed of propagation of tensile stress unloading in expanding thin sheets (or wires) of metal undergoing uniform plastic deformation at a constant stretching rate. Mott chose to examine the case where the tension was approximately constant $T = Y$ over the problem time of interest (point 2 on the stress versus expansion strain plot in figure 2.29). It is not difficult to imagine fracture-inducing defects such that fracture and subsequent stress release occurs during the hardening phase of the deformation (point 1 in figure 2.29). Similarly, fracture delayed until well into the softening phase (point 3) is also possible. In such cases the equation corresponding to equation (2.174) governing propagation of the Mott wave generalizes to

$$\rho\dot{\varepsilon}\xi\frac{d\xi}{dt} = T(\varepsilon) = Y + H\dot{\varepsilon}t. \tag{2.185}$$

Here, the solution is for the special case where Y is the initial tension and H is a linear strain hardening (or softening) constant. Equation (2.185) integrates to the solution

$$\xi(t) = \sqrt{\frac{2Yt}{\rho\dot{\varepsilon}}}\left(1 + \frac{1}{2}\frac{H\dot{\varepsilon}}{Y}t\right)^{1/2} \tag{2.186}$$

for the position of the Mott wave at any time t.

For positive H (hardening), the time-dependent motion for constant Y occurs at early time. At late time the Mott wave asymptotically approaches a constant velocity of $\sqrt{H/\rho}$. For negative H (softening) the Mott wave arrests after a time $t = -Y/H\dot{\varepsilon}$ and travels a distance $\xi = (-Y^2/\rho\dot{\varepsilon}^2 H)^{1/2}$.

Models for Mott wave propagation in other than an axial one dimension are also possible. For example, equal stretching rates in orthogonal in-plane directions (an expanding spherical shell), where tension at a point is reduced from Y to zero (initial rupture is a hole in the metal membrane), can also be solved. Continued dissipation in material behind the circular Mott wave due to further hoop expansion must be accounted for, but the solution provides further demonstration of the diffusive nature of the stress release propagation process.

2.3.6 Energy-based criteria for fragment size

Fracture of solids occurs at sites of weakness inherent in all engineering materials. It follows that many theories of dynamic fracture and fragmentation attempt to capture this inherent flaw structure as an intimate part of the theory. Characterization of the flaw structure for specific materials is difficult in practice, however, and usually parameter fitting to indirect data is the only available approach. Further, fracture and fragmentation models based on inherent flaw activation have commonly ignored energy consumed in fracture activation and growth through the creation of new surface area, plastic work or viscous dissipation. Dissipation comes from the dynamic load driving fracture and fragmentation, and it is reasonable to expect that this dissipation energy plays a key role in the intensity of fragmentation.

Consequently, efforts have been pursued to develop theories of fragment size based on energy principles, with less attention attributed to the underlying microstructure involved in the fracture process. Toward this end Grady (1982a, 1988), and Glenn and Chudnovsky (1986), pursued methods in which the work from the dynamic load driving fracture is balanced against energy consumed in the fracture process to provide first-order estimates of fragment size in the dynamic event. More recently, Drugan (2001) and Zhou *et al* (2006) have undertaken approaches where attention to details of the fracture dissipation and interaction processes leads to increased structure in the predicted dependence of fragment size on the dissipation dynamics.

2.3.6.1 Kinetic and strain energy criteria

An energy-based theory found to provide sensible predictions of characteristic fracture spacing and fragment size in a fragmentation event focuses on the kinetic energy associated with the velocity divergence within the fragmenting zone (Grady 1982a). By considering the fragment size as a variable, a local kinetic energy relative to the center of mass of a fragment is calculated. Assuming a model for fracture energy dissipation in the fragmentation process, and further assuming that the local kinetic energy fuels the dissipation, a balance is struck between the driving energy and the fracture energy. This energy balance leads to an optimum fragment surface area or, equivalently, a characteristic fragment size as illustrated in figure 2.33. Fragment surface area and fragment size are related through the fragment distribution and fragment geometry. Different assumptions of fragment geometry (spheres, cubes, etc) lead to differences in the magnitude of the local kinetic energy. Different implementations of the energy balance in the theory can be explored. These uncertainties lead to variations in predicted average fragment size of as much as a factor of two.

Consider a uniform elastic dilational expansion within the interior of a solid (or a liquid) body. A spherical region of radius a will have a kinetic energy referenced to its mass center of

$$T = \frac{2\pi}{45}\frac{\dot{\rho}^2}{\rho}a^5. \tag{2.187}$$

This kinetic energy, in terms of a volume measure of the expansion strain rate $\dot{\varepsilon} = \dot{\rho}/\rho$, and a diameter of the insipient fragment $s = 2a$, is

$$T = \frac{\pi}{720}\rho\dot{\varepsilon}^2 s^5. \tag{2.188}$$

Without specifying a particular fracture mechanism, if Γ is identified as the energy per unit area required to produce new fragment surface, then the work required to create a spherical fragment of diameter s is

$$U = \pi s^2 \Gamma. \tag{2.189}$$

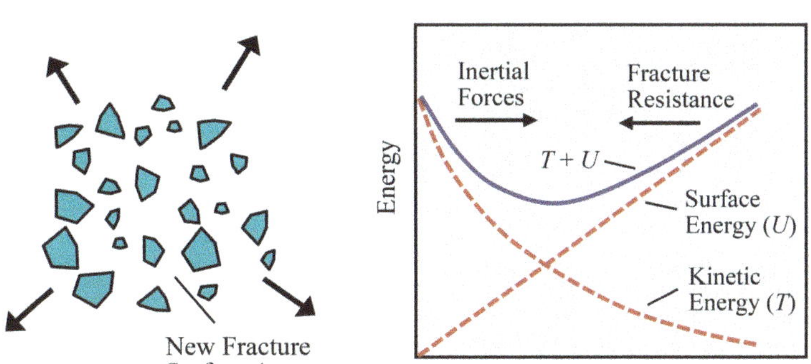

Figure 2.33. Thermodynamic balance of local kinetic energy and fragment surface energy provides a measure of the new fracture surface area, or equivalently the average fragment size, in a dynamic fragmentation event (Grady 1982a, 2009).

Assuming a complete transfer of kinetic energy T into work of fragmentation U, an expression for the characteristic fragment size is obtained,

$$s = 2\left(\frac{90\Gamma}{\rho\dot{\varepsilon}^2}\right)^{1/3}. \tag{2.190}$$

An early application of the kinetic-energy-based model of dynamic fragmentation was on the explosion-induced fragmentation of liquid uranium metal (Grady 1982b). Hollow cylindrical metal shells were imploded through detonation of explosive on the outside of the shell. Inward directed motion of the shell leads to intense shock compression when the cylinder collapses on the centerline. Shock dissipation is sufficiently intense that most of the metal either melts or achieves a very hot solid state. Material motion is reversed on rebound at the centerline. Fragmentation ensues with hot liquid or solid metal particulating into the tens to hundreds of micrometers particle size range. Ejected high-velocity fragments undergo subsequent aero-thermal burning (discussed further in chapter 3). A photograph of one experiment is shown on the left in figure 2.34. Particle size distributions in such experiments are determined by the placement of witness plates that intercept the flux of fragments. An average fragment size and corresponding expansion strain rate for several experiments is provided on the right in the figure.

Equation (2.190) provides a prediction of the nominal fragment size that is compared with the data in figure 2.34. The fragmentation energy Γ required for the calculation is provided by the surface tension $\gamma = 1.75$ J m^{-2} for liquid uranium. A reasonable comparison of theory with experiment is shown in the right-hand panel of the figure.

Glenn and Chudnovsky (1986) extended the present energy balance theory by including an elastic energy of dilation,

$$E = \frac{\pi P_{so}{}^2}{12\rho c^2} s^3, \tag{2.191}$$

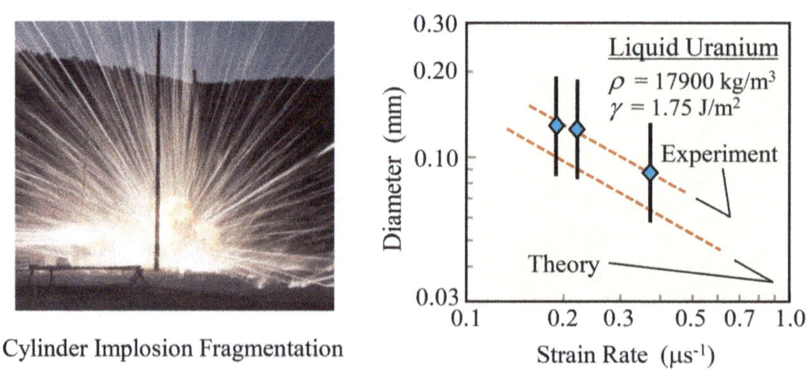

Figure 2.34. Liquid metal fragmentation from imploding uranium metal cylinder fragmentation tests is shown on the left. The fragment size data are from witness plate measurements. Average fragment size versus expansion strain rate for three tests is displayed on the right and compared with the fragmentation theory (Grady 1982b, 2010). Courtesy of Sandia National Laboratories.

in addition to the kinetic energy in fueling the fragmentation process. This extension of the theory requires an additional material property, namely a dilatational tensile spall strength P_{so}.

An energy balance equivalent to the previous analysis yields

$$\frac{\pi}{720}\rho\dot{\varepsilon}^2 s^5 + \frac{\pi P_{so}{}^2}{12\rho c^2} s^3 = \pi s^2 \Gamma. \tag{2.192}$$

The resulting cubic can be solved for the nominal fragment size s. At high loading rates the solution converges to that of the kinetic energy balance of equation (2.190), while at lower loading rates the predicted fragment size approaches a constant governed by the tensile spall strength P_{so}. At the lower limit the relation of Glenn and Chudnovsky approaches the energy-based expression of von Rittinger (1867) discussed in earlier sections.

In the present development of the fragmentation model, where an interior region undergoing dilatational expansion until fracture occurs through internal cavitation, the model applies equally well to a brittle or a ductile solid, or to a liquid, although the nature of the fracture resistance and dissipation may differ. The present energy balance has also been applied to expanding rings (or stretching jets) and to expanding shells (or stretching sheets) to predict characteristic fragment size. Here ductile metals are more commonly of interest, and deformation is well into the plastic region when fragmentation occurs. In these fragmentation geometries, elastic energy is less likely to contribute significantly to the breakup process. Predicted fragment sizes based on the kinetic energy balance theory agree well with the earlier Mott wave fragmentation analysis when fracture energy is included.

2.3.6.2 Energy-horizon theory of fragmentation

An alternative energy-based theory sheds additional light on the dynamic fragmentation processes (Grady 1988). Like the previous several theories, it addresses only the end states of the fragmentation process, and does not speak to the evolution of transition states through the fracture process.

Two physically reasonable propositions are assumed in the energy-horizon theory of fragmentation. The first is a horizon condition (a correlation distance) that, through time and communication constraints, establishes a characteristic region within which the fracture process must be independent of the environment surrounding that region. The horizon length scale at the time of failure places an upper bound on the size of fragments. The second proposition is an energy condition that requires that the elastic potential and kinetic energies within the region bounded by the temporal horizon exceed the fracture energy dissipated in the failure event. These two conditions, when combined with a kinematic description of constant dilatant strain-rate loading, lead to a set of inequalities that constrain the desired spall fragmentation properties. Similar relations can be posed for a body subjected to a Heaviside loading to a constant and uniform tensile pressure amplitude.

Assuming a spherical region within the body of radius ct, these relations are

$$s \leqslant 2ct, \tag{2.193}$$

$$\frac{1}{2}\frac{P^2}{\rho c^2} + \frac{1}{120}\rho\dot{\varepsilon}^2 s^2 \geqslant \frac{6\Gamma}{s}, \tag{2.194}$$

$$P = \rho c^2 \dot{\varepsilon} t. \tag{2.195}$$

The tensile pressure is P, while t and c are the time and elastic wave speed, respectively. Again, s is the fragment size, $\dot{\varepsilon}$ the dilatational rate of expansion, and Γ the energy required to create new fragment surface area. Equation (2.193) states that, if spall fragmentation occurs within a time t from the onset of fracture instability, then regions of diameter $2ct$ must fail independent of the adjacent environment. Fragments produced within this region must be smaller than the region itself. Equation (2.194) requires that, for fragmentation to proceed, the elastic plus kinetic energy at failure exceed the required fragmentation energy in this same region. Equation (2.195) is the kinematic relation expressing the time history of tensile pressure leading to spall failure.

The functional character of the governing inequalities is illustrated in figure 2.35. From the horizon criterion in equation (2.193), failure at time t requires that the fragmentation energy expended in creating new fracture surface area exceeds $6\Gamma/2ct$. This fragmentation energy curve is shown in figure 2.35. Combining equation (2.193) with the left-hand side of equation (2.194) yields for the elastic plus kinetic energy available to fuel fragmentation at time t,

$$\left(\frac{1}{2} + \frac{1}{30}\right)\rho c^2 \dot{\varepsilon}^2 t^2. \tag{2.196}$$

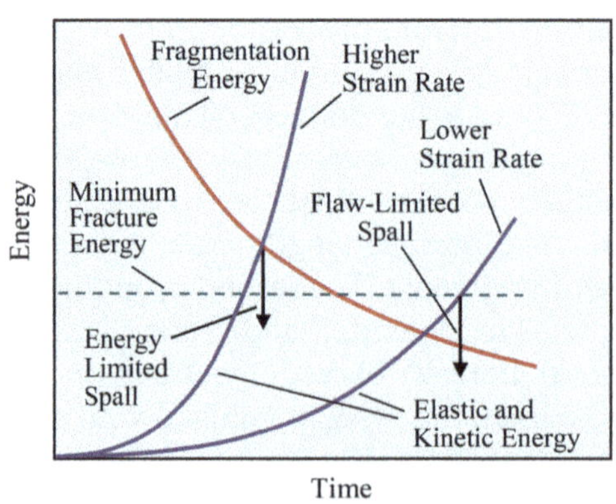

Figure 2.35. Time dependence of the fracture energy and the driving elastic plus kinetic energy within the horizon-limited region. Energy-limited spall and flaw-limited spall are illustrated. The limiting fracture energy corresponds to a constant spall stress P_s that must be exceeded before fracture failure can occur (Grady 1988, 2010).

The larger portion is provided by elastic strain energy, the smaller by the kinetic energy. The increase in energy in equation (2.196) depends on the strain rate. Two curves plotting this energy as functions of time are illustrated in figure 2.35; one at a higher strain rate, the other at a lower strain rate. Also, a minimum fracture energy value is identified (the dashed line). This corresponds, for example, to the constant fracture tensile pressure P_{so} identified in the model of Glenn and Chudnovsky (1986).

The two energy curves leading to spall failure and fragmentation are selected to illustrate two distinctly different cases of fragmentation. At the higher strain rate, the minimum fracture energy is exceeded. However, further time is required to satisfy the fragmentation energy requirement. Spall failure does not occur until the elastic and kinetic energy curve intersects the fragmentation energy curve as illustrated in the figure. This case is identified as energy-limited spall fragmentation. The lower-strain-rate energy curve, in contrast, achieves the fragmentation energy criteria below the constant limiting fracture energy level (or constant spall strength P_{so}). Further tensile loading is needed to achieve this minimum fracture energy (spall stress) requirement. This case is identified as flaw-limited fragmentation.

For energy-limited fragmentation, equality in the governing relations applies and specific relations for the spall strength and fragment size are derived. For flaw-limited fragmentation, the full inequalities apply and additional features must be added to the model before spall and fragmentation properties can be determined.

2.3.6.3 Energy-limited fragmentation

If the material carried into tension is predisposed to spall through a sufficient inherent flaw structure when the energy criterion is satisfied, then the fragmentation equations (2.193)–(2.195) as equalities can be solved for the spall failure and fragmentation properties (Grady 1988). These properties are, for the fragment size

$$s = \left(\frac{48\Gamma}{\rho\dot{\varepsilon}^2}\right)^{1/3}, \tag{2.197}$$

for the tensile spall pressure

$$P_s = (6\Gamma\rho^2 c^3 \dot{\varepsilon})^{1/3}, \tag{2.198}$$

and the time to spall

$$t_s = \left(\frac{6\Gamma}{\rho c^3 \dot{\varepsilon}^2}\right)^{1/3}. \tag{2.199}$$

In solving for the fragmentation properties in equations (2.197)–(2.199), the kinetic energy term (the factor of 1/30 in equation (2.196)) is ignored in favor of the elastic energy term to simplify the expressions. The kinetic energy is less than 10% of the total driving energy. (In other fragmentation geometries, the expanding ductile ring for example, kinetic energy is paramount in fueling the fragmentation event.)

Several further relations are derived. Combining equations (2.198) and (2.199) yields the expression

$$P_s^2 t_s = 6\rho c\Gamma. \tag{2.200}$$

This product of the spall stress and time is a constant when Γ is itself a constant independent of strain rate. The relation is useful in estimating fragmentation energies from spall data. Also, from equations (2.197) and (2.198), an alternative relation for the fragment size is derived,

$$s = \frac{2P_s}{\rho c\dot{\varepsilon}}, \tag{2.201}$$

dependent on the spall stress and the strain rate.

2.3.6.4 A comparison of energy fragmentation theories

The present energy-horizon theory of fragment size contrasts with the earlier kinetic-energy-based theory and the later modification by Glen and Chudnovsky. In the development of the kinetic energy model (Grady 1982a), the elastic strain energy accumulated during the tensile dilation process was ignored. In contrast, the model of Glenn and Chudnovsky (1986), and the later energy-horizon model, include elastic strain energy but in uniquely different ways. Glenn and Chudnovsky introduce a material-dependent constant tensile spall pressure P_{so}, which limits the magnitude of the dilatant elastic tensile strain energy that can contribute to the fracture process. Tensile pressure P in the energy-horizon theory is not so constrained. Tension continues to increase under dilatant loading until the energy-governed fragmentation failure criterion is satisfied. The spall strength in the energy-limited model is not a material constant, but is a derived property (see equation (2.198)) dependent on the tensile loading rate. The fragment size predictions based on the several energy models as a function of tensile loading rate are illustrated in figure 2.36. Both the kinetic energy model and the energy-horizon model exhibit the same $s \sim \dot{\varepsilon}^{-2/3}$ dependence of the fragment size. Fragment size at a specified strain rate for the energy-horizon model is about a factor of 2.5 smaller due to the added contribution of elastic strain energy.

A material dependent constant spall pressure P_{so} in the Glenn–Chudnovsky model introduces both a length scale $s_o = 12\rho c^2\Gamma/P_{so}^2$ and strain-rate scale $\dot{\varepsilon}_o = P_{so}^3/\rho^2 c^3\Gamma$. These scales provide the respective normalization parameters in figure 2.36. The Glenn–Chudnovsky model and the kinetic model converge, as expected, when the elastic energy becomes insignificant at the higher strain rates.

Differences between the Glen–Chudnovsky and the energy-horizon models at all strain rates is worthy of note as both include elastic strain energy but in different ways. Neither is necessarily wrong. Each implies a different structure to the fracture-producing flaws leading to failure in the fragmentation process. The Glen–Chudnovsky model suggests an avalanche of flaws, which activate when tensile pressure approaches P_{so} and limits the magnitude of tensile strain energy that the body can support regardless of the loading rate. The energy-horizon model, in

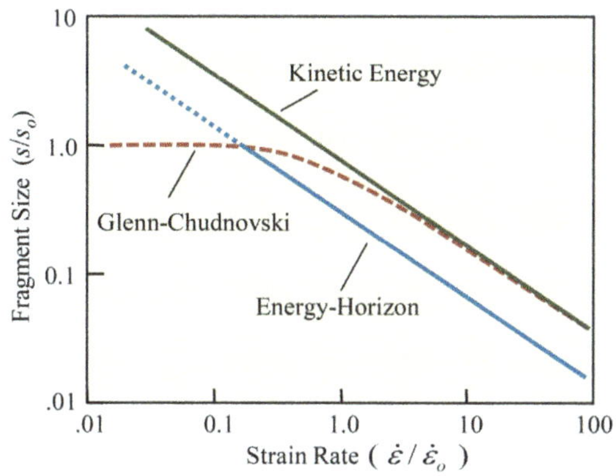

Figure 2.36. Comparison of fragment size prediction from several energy-based theories of dynamic fragmentation (Grady 1982a, Glenn and Chudnovsky 1986, Grady 1988, 2009). Ordinate and abscissa are normalized by characteristic size and strain-rate scales, respectively.

contrast, requires a flaw structure that allows for increasingly higher levels of tensile strain energy (and spall strength) as the loading rate is increased.

2.3.6.5 A general fragmentation strain rate

The rate of material straining is a convenient kinematic parameter in the previous several theories for characterizing the dynamics leading to fragmentation. It determines the local rate of application of energy in the functional form $\rho s^2 \dot{\varepsilon}^2$, where s is a length scale characteristic of the fragmentation event. In the theories, a uniform rate of volumetric dilation carries the material to a tensile pressure at which dynamic failure and fragmentation ensues. This restriction is not necessary. The same energy-based theories have found successful application where the local driving energy is strictly distortional, such as the breakup of a rapidly expanding shell or stretching jet. Consequently, a more consistent expression of the rate of deformation appropriate to the post-process computational simulations of fragmentation has proven useful. The following development has provided a consistent measure of the strain rate under more general fragmentation conditions. The analysis is specific to the earlier kinetic energy theories (Grady 1982a, Glenn and Chudnovsky 1986). It is readily extended to the energy-horizon theory.

Consider a body in motion characterized by a velocity field $u(x)$. Consider further, a small spherical region of volume υ at an arbitrary point within the body. The kinetic energy of this region is given by

$$T = \frac{1}{2}\rho \int_{\upsilon} u^2(x) d\upsilon. \tag{2.202}$$

The density ρ is considered constant over the region. The velocity can be related to the center of mass of the spherical region through a series expansion,

$$u(x_{cm} + r) = u(x_{cm}) + G \cdot r + \cdots, \tag{2.203}$$

where G is the velocity gradient tensor. Ignoring higher-order terms and introducing the symmetric stretching tensor,

$$D = \frac{1}{2}(G + G^T), \tag{2.204}$$

and the antisymmetric spin tensor,

$$W + \frac{1}{2}(G - G^T), \tag{2.205}$$

yields

$$u(x_{cm} + r) = u(x_{cm}) + W \cdot r + D \cdot r + \cdots. \tag{2.206}$$

The kinetic energy of the spherical region, neglecting higher-order terms, is then

$$T = \frac{1}{2}\rho \int_v (u(x_{cm}) + W \cdot r + D \cdot r)^2 dv. \tag{2.207}$$

Lastly, the kinetic energy can be decomposed into the sum

$$T = \frac{1}{2}\rho \int_v u^2(x_{cm}) dv + \frac{1}{2}\rho \int_v (W \cdot r)^2 dv + \frac{1}{2}\rho \int_v (D \cdot r)^2 dv, \tag{2.208}$$

for the center-of-mass, rotational and distortional components, respectively. The first decomposition follows from the theorem in mechanics that allows separation into the center-of-mass kinetic energy plus the kinetic energy of motion about the center of mass. The decomposition into rotational and distortional kinetic energy is likewise demonstrated.

Following the earlier kinetic energy fragmentation development, assume that center-of-mass and rotational kinetic energy at a point remain constant through the fragmentation process, and only the distortional kinetic energy fuels the failure and fragmentation process.

The distortional kinetic energy can be written

$$T^d = \frac{1}{2}\rho D_{ij} D_{ik} \int_v r_j r_k dv, \tag{2.209}$$

where D is evaluated at the center of mass and removed from the integral. Evaluating the integral,

$$\int_v r_j r_k dv = \frac{1}{3}\delta_{jk} \int_v r^2 dv = \frac{4\pi}{15} a^5 \delta_{jk}, \tag{2.210}$$

over a sphere of radius a, the distortional kinetic energy becomes

$$T^d = \frac{2\pi}{15}\rho a^5 \mathrm{Tr} D^2. \tag{2.211}$$

Identify the local effective strain rate with

$$\dot{\varepsilon} = \sqrt{\mathrm{Tr}D^2} = \sqrt{\dot{\varepsilon}_x{}^2 + \dot{\varepsilon}_y{}^2 + \dot{\varepsilon}_z{}^2}. \tag{2.212}$$

For uniform shear-free dilation with $\dot{\varepsilon}_x = \dot{\varepsilon}_y = \dot{\varepsilon}_z$, the effective strain rate is $\dot{\varepsilon} = \sqrt{1/3}\,\dot{v}/v$, whereas for uniform volume-conserving extension with $\dot{\varepsilon}_y = \dot{\varepsilon}_z = -\dot{\varepsilon}_x/2$ the effective strain rate is $\dot{\varepsilon} = \sqrt{3/2}\,\dot{\varepsilon}_{eq}$, where $\dot{\varepsilon}_{eq}$ is the equivalent plastic strain.

Assuming as earlier that the distortional kinetic energy is consumed in fragment surface energy,

$$W = 4\pi a^2 \Gamma, \tag{2.213}$$

a nominal fragments size $s = 2a$ results,

$$s = 2\left(\frac{30\Gamma}{\rho\dot{\varepsilon}^2}\right)^{1/3}. \tag{2.214}$$

2.3.7 Fragmentation application

This section examines an experimental fragmentation study within the context of two different theories of fragmentation described earlier in the chapter. The theories are appropriate to the dynamic fragmentation of rapidly expanding metal shells. Both theories have attractive features and demonstrate predictive abilities with available experimental data. The two theories are based on strikingly different premises. Here it is shown that, with a clearer statement of the initial premises, the theories merge and provide a richer and more predictive fragmentation theory. The statistics-based theory of Mott (1947) as outlined in section 2.2.4.5 provides one theoretical assessment the experimental data. The energy-horizon theory described in section 2.3.6.2 provides and alternative comparison with the data. Overlap of the Mott and the energy theory is reconciled in section 2.3.4.

A series of dynamic fragmentation experiments was performed by Olsen (see Grady and Olsen 2003) on rings of uranium-6%-niobium (U6Nb) approximately 30 mm in diameter and 0.75 mm in square cross section. The technique is illustrated in figure 2.37. An electric current pulsed solenoid and an intermediate metal driver ring magnetically accelerate the U6Nb rings to maximum radial velocities ranging over 50–300 m s^{-1}. Laser velocity interferometry provides measurements of radial displacement and velocity histories. Strain to fracture and dynamic flow stress are obtained from the data, along with the number of fractures and distributions in fracture spacing. A determination is made of the dependence of these measured observables on the velocity at fracture (or, equivalently, the strain rate at fracture).

2.3.7.1 Mott statistical fragmentation

Mott put forth arguments that energy dissipation at the site of fracture was not of consequence in the fracture process of expanding cylinders and proposed instead a statistical strain-to-fracture criterion. Mott assumed that fractures occurred at

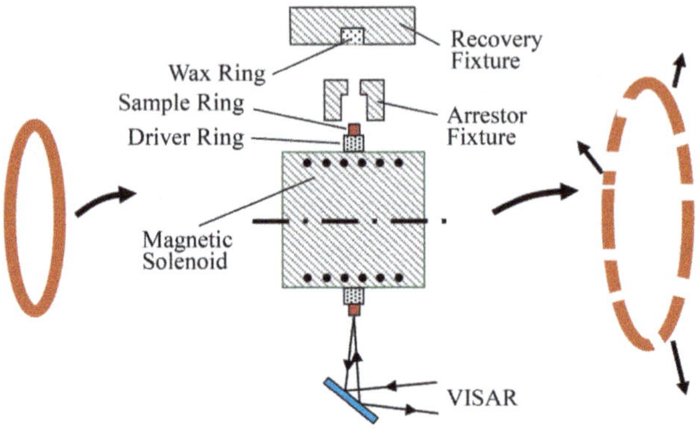

Figure 2.37. Schematic of experimental assembly and consequences of impulsive expanding ring fragmentation of uranium 6% niobium metal (Grady and Olsen 2003).

random around the circumference of the cylinder (the ring in the present analysis) at a frequency governed by a strain-dependent hazard function $h(\varepsilon)$, such that $h(\varepsilon)d\varepsilon$ provided the statistical number of fractures occurring within a unit length of the circumference of the ring over the strain interval $d\varepsilon$. Second, Mott proposed that stress relief ensues from a fracture site precluding further fracture in the regions of stress release. Waves of stress release (Mott waves) are illustrated in figure 2.11. Stress release is captured by a function $d(\varepsilon - \eta)$, which is the distance traveled from the fracture site by a tensile stress release wave over the strain interval $\varepsilon - \eta$. (Since strain rate is assumed constant over the duration of the fracture process, strain and time are synonymous through $\varepsilon = \dot{\varepsilon}t$.) Accordingly, the strain- (or time)-dependent relation provides the fraction of the stress release region of the fragmenting ring,

$$F(\varepsilon) = \int_0^{\varepsilon} 2d(\varepsilon - \eta)h(\eta)d\eta, \tag{2.215}$$

incorporating both the statistical fracture activation hazard function $h(\eta)$ and the stress release function $d(\varepsilon - \eta)$.

An inspection of equation (2.215) reveals that the function $F(\varepsilon)$ will exceed unity at sufficiently large strain. This is a consequence of not accounting for two factors in the fracture activation and stress wave propagation process. First, the fracture activation function $h(\varepsilon)$ does not exclude the activation of further fractures within regions previously stress-relieved. Second, the stress release function $d(\varepsilon)$ does not account for the impingement and the overlap of opposing release waves from separate neighboring fractures. Thus, equation (2.215) is only applicable for a dilute number of fractures early in the fracture and release process.

To account for fracture exclusion and wave impingement in the Mott model, a statistical method introduced by Johnson and Mehl (1939) is applied. Exclusion and impingement is accounted for through the relation

$$D(\varepsilon) = 1 - e^{-F(\varepsilon)}. \tag{2.216}$$

The relation provides the fraction of the Mott cylinder $D(\varepsilon)$ encompassed by fracture stress release waves at any strain ε. $D(\varepsilon)$ and $F(\varepsilon)$ are equivalent at early times. The function $D(\varepsilon)$ approach unity as ε becomes large.

A strain-to-fracture hazard function of the power-law form,

$$h(\varepsilon) = \frac{n}{\sigma}\left(\frac{\varepsilon}{n}\right)^{n-1}, \tag{2.217}$$

is appropriate. Mott used an exponential Gumbel hazard function. There is no evidence that the form of the hazard function significantly affects the predictions of the model.

Second, the Mott analysis of plastic release in ductile metals from sites of fracture results in the stress release function

$$d(\varepsilon) = \sqrt{\frac{2Y}{\rho\dot{\varepsilon}^2}}\varepsilon. \tag{2.218}$$

With the model specified by equations (2.217) and (2.218), the number of fracture sites as the accelerated ring expands is given by the strain-dependent function,

$$N(\varepsilon) = \int_0^{\varepsilon} (1 - D(\eta))h(\eta)d\eta = \int_0^{\varepsilon} e^{-F(\eta)}h(\eta)d\eta. \tag{2.219}$$

Integration of equation (2.219) to large strain yields the model prediction of the final number of fractures per unit circumferential length of the ring,

$$N = \beta_n\left(\frac{\rho\dot{\varepsilon}^2}{2\pi Y}\frac{n}{\sigma}\right)^{n/(2n+1)}, \tag{2.220}$$

where

$$\beta_n = \left(\frac{2n}{2n+1}\right)^{1/(2n+1)}\left(\frac{1}{\sqrt{n}}\frac{\Gamma(n+1/2)}{\Gamma(n)}\right)^{2n/(2n+1)}\Gamma\left(\frac{2n}{2n+1}\right). \tag{2.221}$$

The constant β_n is a number close to unity for all values of $n \geqslant 1$. Over this range of n the strain-rate dependence of the fragment number in equation (2.220) spans $2/3 \leqslant \dot{\varepsilon} \leqslant 1$.

The random activation of fractures on the ring both in space and in time, as assumed in the Mott model, allows for calculation of the statistical distribution of fracture spacing (fragment lengths). This calculation was performed graphically by Mott, and has been completed by analytic methods for the special case of $n = 1$ in the power-law hazard function. The calculated analytic distribution in fracture spacing developed previously in section 2.2.4.5 by this analytic method is,

$$f(x) = \frac{\gamma^2}{4}\frac{1}{x_o}\left(\frac{x}{x_o}\right)^3 e^{-\frac{1}{4}(x/x_o)^3}\int_0^1 (1 - y^2)e^{-\frac{3}{4}(x/x_o)^3 y^2}dy, \tag{2.222}$$

where $\gamma = 3\Gamma(2/3)$ and $x_o = (3\sigma Y/2\rho\dot{\varepsilon}^2)^{1/3}$. The sensitivity of the distribution to the functional form of the fracture hazard function $h(\varepsilon)$ is not known, but comparison of the analytic distribution from equation (2.222) and the graphical distribution of Mott suggests that this sensitivity is probably small. Both the analytic and the graphical distribution are compared in figure 2.13.

In summary, the statistics-based theory of dynamic fragmentation developed in the seminal study of Mott provides a physically plausible and intellectually satisfying description of the fragmentation process. Within the one-dimensional model of Mott's expanding ring, the theory is fully predictive, providing the average fragment size and the distribution about the average, as well as the statistical and temporal history of fracture, and the strain to fracture.

2.3.7.2 Energy-based fragmentation

A theory of dynamic fragmentation based on markedly different initial assumptions is pursued through the energy-horizon model developed in section 2.3.6.2. Again, one-dimensional fracture and fragmentation of an expanding ring is assumed. The fundamental difference in the two theories is that Mott assumes that energy dissipated in the fracture process is not of concern, and fracture at a site on the ring would be instantaneous on achieving critical strain levels. In contrast, energy dissipation and an associated fracture delay lie at the heart of the energy-based fragmentation theory.

Formulation of the energy-based fragmentation theory specialized to the ductile expanding ring proceeds by extending the stress release analysis developed by Mott. The analysis is used to calculate the time history of plastic release waves (Mott waves) emanating from sites of fracture accounting for fracture resistance and dissipation. It is noteworthy that a diffusion wave, rather than an elastic wave, determines the extent of the correlation horizon accommodating dynamic failure. The extension of the analysis proceeds by considering, rather than instantaneous fracture at the fracture site, a fracture resistance which reduces from the flow stress Y to zero as a crack-opening-displacement parameter y goes from zero to some critical crack opening displacement y_c. An assumption of a linear reduction of the fracture resistance then leads to a fracture energy dissipation $\Gamma = Yy_c/2$. The assumption simplifies the analysis somewhat, although other functional forms for the reduction of fracture resistance do not markedly alter the value of Γ. This analysis is carried out in section 2.3.4 and results in an average fracture spacing (the symbol W replaces Γ in the earlier analysis)

$$x_o = \left(\frac{24\Gamma}{\rho\dot{\varepsilon}^2}\right)^{1/3}. \tag{2.223}$$

The fracture resistance Γ is considered a property of the material. It is sensible in the present analysis to identify a further property with the dimensions of fracture toughness through

$$K_c = \sqrt{2E\Gamma}, \tag{2.224}$$

where E is the appropriate elastic modulus. This exercise is useful if for no other purpose than for comparison of the model fracture resistance inferred from experimental data to that determined from more common fracture failure modes. The expression for the characteristic fracture spacing then becomes

$$x_o = \left(\frac{\sqrt{12}\,K_c}{\rho c\dot{\varepsilon}}\right)^{2/3}. \tag{2.225}$$

The energy-based theory does not address the issue of the statistical distribution of fragment sizes. It is assumed that equation (2.225) provides a measure of the average fragment size and that the fragment number per unit length is provided by the inverse,

$$N = \left(\frac{\rho c\dot{\varepsilon}}{\sqrt{12}\,K_c}\right)^{2/3}. \tag{2.226}$$

It is instructive at this point to compare equation (2.220) resulting from the Mott statistical strain-to-failure fragmentation theory and equation (2.226) from the energy-horizon theory with the experimental fragmenting ring data on U6Nb. The comparison leads to a deeper understanding of the underlying physics and points to a more inclusive theory of the fragmentation process.

2.3.7.3 Experiment and theory

In the experimental study, U6Nb metal rings are accelerated as illustrated in figure 2.37 by a pulsed magnetic field to radial velocities spanning the range of 50–300 m s^{-1}. Measured deceleration of the freely expanding ring prior to fragmentation, provides calculation of a tensile flow stress of nominally 1 GPa. Fragmentation occurs at a load-maximum expansion strain of approximately 30%.

The number of fragments produced (equivalently, the number of fractures) is determined for each test. Fragment number versus the expansion velocity at fragmentation are shown for the series of U6Nb expanding ring experiments on the left in figure 2.38. The anomalous point high on the graph is the consequence of one test on a markedly differently heat-treated U6Nb sample (discussed later in the section). A least squares fit, excluding the one anomalous point, provides the power-law representation of the data shown in the figure. A second curve representing the data assumes a power of two-thirds. In one representative test, each fragment was separately weighed and the resulting cumulative fragment size distribution is shown on the right in figure 2.38.

Weibull parameters σ and n necessary for predicting of the fragment number in the Mott statistical theory are not known for the U6Nb material tested. Hence, only sensibility of the experimental results can be examined. The observed power-law dependence of fragment number on expansion velocity close to two-thirds indicates that the Weibull parameter n in equation (2.217) is quite close to unity. Assuming $n = 1$, the second Weibull parameter is calculated to be $\sigma = 7.7 \times 10^{-5}$ m. The standard deviation in strain to fracture in a sample of length L is σ/L. Considering

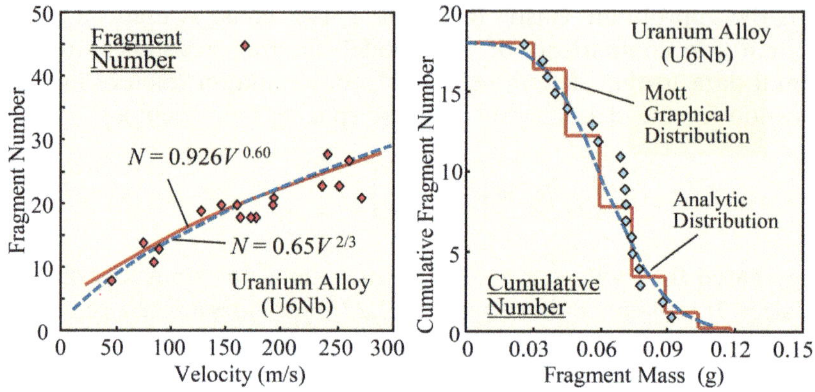

Figure 2.38. Left: fragment number versus expansion velocity at fracture for U6Nb expanding ring fragmentation tests (Grady and Olsen 2003). Right: comparison of cumulative fragment distribution for one representative U6Nb expanding ring fragmentation test with the theoretical Mott fragment size distribution.

specimens of length of order one-half to one centimeter, the nominal length of fragments in the present ring tests, a spread in strain to fracture of approximately 0.008 to 0.016 is calculated, a few percent of the observed 0.30 load-maximum strain to failure. It is of interest that the Gumbel strain-to-failure distribution assumed by Mott, which does not have a statistical shape parameter, does not predict the two-thirds power dependence on expansion velocity observed in the data. The Weibull and Gumbel distribution and relationship to fragment size is discussed in section 2.3.1.3.

Prediction of the distribution in fracture spacing is also a facet of Mott's statistical theory. Comparison of both the graphic distribution generated by Mott and the analytic distribution from equation (2.222) are compared with the experimental distribution in figure 2.38. The observed distribution and the theoretical distributions based on the Mott statistical fracture theory are in reasonable accord, as has also been demonstrated on several other sets of data earlier in the chapter.

The fragment number data are replotted in figure 2.39 as a function of strain rate for direct comparison with equation (2.226) resulting from the energy-based theory of fragmentation. A dynamic fracture toughness of K_c = 60 MPa m$^{1/2}$ provides the best fit shown in the figure. The toughness inferred from the data is modestly less than static fracture toughness measurements of K_c = 90–110 MPa m$^{1/2}$ on similar U6Nb alloys.

Other features observed in the U6Nb ring fragmentation experiments also attest to the importance of energy dissipation and finite fracture time in the dynamic fracture process. Inspection of recovered ring fragments reveal arrested necking regions; a signature of fractures enveloped by tensile release (Mott) waves and fracture growth arrested before full fracture and separation.

The present study provides added insight through examination of the recovered ring fragments. A metallographic image of a representative fracture is shown in figure 2.39. The outer deformed surface of the stretched ring exhibits an undulating,

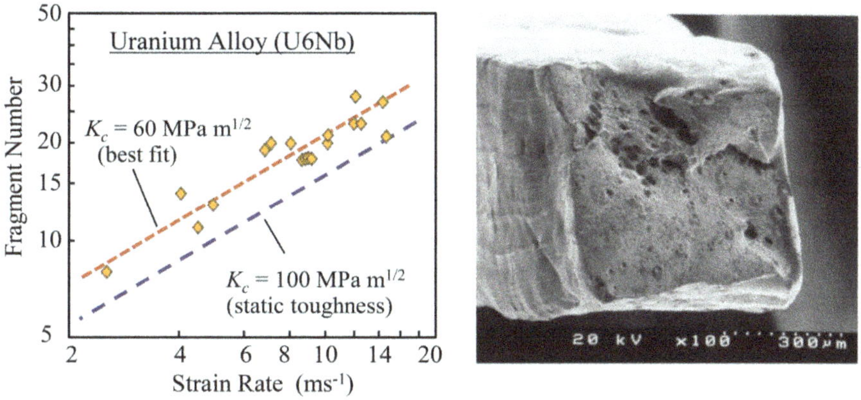

Figure 2.39. Fragment number data plotted against expansion strain rate and predictions based on energy fragmentation theory (Grady and Olsen 2003). Metallographic image of a representative U6Nb ring fracture.

uneven surface characteristic of heterogeneous slip-line plasticity. The reduced area of the fracture surface points to a pre-fracture deformation localization through plastic necking. Failure is completed through pervasive ductile extension fracture through the narrowest portion of the necking region. Arrested necking regions are observed in most of the recovered fragments.

Ductile deformation through dimple plasticity is evident in the fracture surface. The dimple morphology appears tri-model. Dimple sizes range from 2 to 3 μm for the smallest, up to 15–30 μm for the largest. Inclusions observed within larger dimples are probably fracture initiation sites. Dimple size may reflect local fracture speed in the failure process. The present fragmentation experiments were conducted on U6Nb metal quenched from elevated temperature into the more ductile gamma phase.

The fracture character in the U6Nb ring fragmentation tests shows similarity to observations of spall fracture in U6Nb (Tonks *et al* 1999, Zurek *et al* 2000). Similarities extend to both the character of dimple plasticity on fracture surfaces and identification of carbide inclusion fracture initiation sites. The similarity is not surprising as strain rates at fracture in necking regions of the expanding ring tests exceed 10^4 s^{-1}, approaching that of the spall experiments.

The anomalous data point observed in figure 2.38 corresponds to a sample annealed from the elevated temperature state leaving most of the metal in the more brittle alpha phase. The fracture surface from a ring experiment fragment from this sample revealed predominantly cleavage fracture, differing starkly from the more ductile fracture of the gamma-phase material.

A comment is warranted on the comparison of the dynamic toughness inferred through the energy-based theory and the static fracture toughness in figure 2.39. The reasonably close agreement between the dynamic and static values is noteworthy and speaks well for energy-controlled mechanisms governing the characteristic fracture spacing and fragment number. On the other hand, one questions the factor of two difference. First, there are simplifying theoretical assumptions in the energy theory that could account for the difference. Alternatively, a material issue may contribute to the disparity. The present metal rings with substantially less than a square millimeter cross section, combined with the statistical selectivity of weaker fracture sites, could lead to statistically lower dynamic fracture energy.

2.3.7.4 Reconciling two theories

The statistical fragmentation theory of Mott is based on two functional properties governing response of the material. The first is a strain-dependent fracture activation function $h(\varepsilon)$. Here a power-law form is used,

$$h(\varepsilon) = \frac{n}{\sigma}\left(\frac{\varepsilon}{\sigma}\right)^{n-1}. \tag{2.227}$$

The second is the function determining diffusion-like tensile release from sites of fracture,

$$d(\varepsilon) = \sqrt{\frac{2Y}{\rho\dot{\varepsilon}^2}}\varepsilon. \tag{2.218}$$

The two functions form the basis for the Mott theory and yield the fracture frequency,

$$N = \beta_n\left(\frac{\rho\dot{\varepsilon}^2}{2\pi Y}\frac{n}{\sigma}\right)^{n/(2n+1)}. \tag{2.229}$$

The energy-based theory, in contrast, yields for the average fracture frequency,

$$N = \left(\frac{\rho\dot{\varepsilon}^2}{24\Gamma}\right)^{1/3}. \tag{2.230}$$

The theories are equivalent if the Weibull constants have the values,

$$n = 1, \qquad \sigma = \beta_1^3\frac{12}{\pi}\frac{\Gamma}{Y} \cong 5\frac{\Gamma}{Y}. \tag{2.231}$$

Using $K_c = 60$ MPa m$^{1/2}$ to calculate Γ, and a value of $Y = 1$ GPa, yields a value of $\sigma = 8.2 \times 10^{-5}$ m, remarkably close to the previous calculation of σ.

The two theories are brought into accord through sensible interpretation of the underlying physics. In the theory of Mott, a fracture activated (through equation (2.227)) is a fracture completed. The Mott analysis is viable if the fracture energy is sufficiently low and there is a sufficient paucity of activation sites. If fracture energy is not negligible and fracture activation sites are numerous, the theory of Mott fails. The energy-based theory applies.

Joining the theories is discussed further in section 2.3.4. The merging is illustrated in figure 2.40. Dynamic strain to failure increases with the intensity of the fragmentation event (the strain rate). From the Mott statistical development the cumulative strain to fracture is calculated from

$$\varepsilon_f = \int_0^\infty (1 - D)d\eta. \tag{2.232}$$

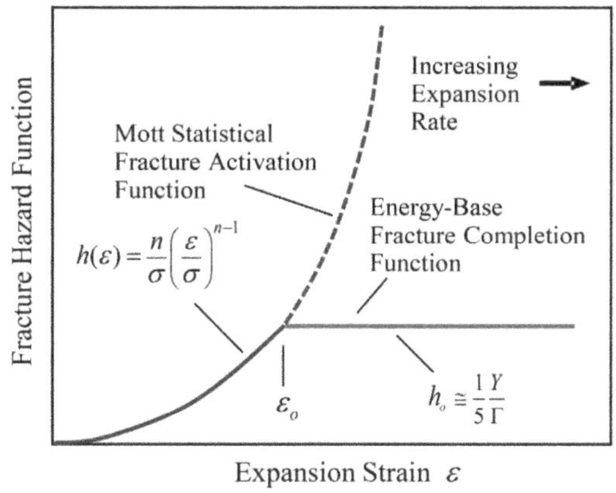

Figure 2.40. Merging of the Mott and the energy-based fragmentation theories. The transition from flaw-limited fragmentation to energy-limited fragmentation is identified (Grady and Kipp 1993).

The Mott strain to fracture increases with expansion strain rate according to $\varepsilon_f = \alpha_n \dot{\varepsilon}^{2/(2n+1)}$ where α_n is a constant. If the fracture strain ε_f remains less than some transition strain ε_o identified in the figure, then the Mott fragmentation theory applies. Fracture energy is sufficiently small such that fracture spacing is governed by the statistical activation function.

Energy of fracture enters the analysis when fracture strain achieves levels in excess of the transition strain,

$$\varepsilon_o \simeq \left(\frac{1}{5}\frac{\sigma^n}{n}\frac{Y}{\Gamma}\right)^{\frac{1}{n-1}}. \tag{2.233}$$

In excess of the transition strain the strain to fracture increases according to the energy-based theory as $\varepsilon_f \sim \dot{\varepsilon}^{2/3}$. Fracture spacing and statistics when the strain to fracture exceeds the transition strain is governed by fracture energy properties of the material. In figure 2.40 the difference between the Mott activation function and the fracture completion function above the transition strain is identified with arrested necking in the ring fragmentation experiments. Namely, fractures are activated but energy transport to the fracture sites is insufficient for fracture completion.

Experimental results of the ring fragmentation tests on U6Nb indicate that fragmentation occurs in the range in which the strain to fracture exceeds the transition strain over the span of expansion velocities investigated. Fragment numbers are consistent with energy-based calculations. Numerous arrested necking regions attest to fracture activation levels in excess of the number of competed fractures. A computational study of ring fragmentation that spans the transition strain is illustrated in figure 2.27 and is described in the accompanying text.

2.4 Dynamic fragmentation in brittle materials

The failure and fragmentation of brittle solids, particularly highly brittle materials such as glass, ceramic and high-strength rock, differ starkly from that of the ductile metals. The features of brittle fracture and fragmentation are unique and warrant independent discussion. Much of the historical work in brittle solids centers on fragmentation concerns in the mining, blasting and crushing industries. Some of this history is surveyed in earlier sections. In the present section, several studies are explored that exemplify failure and fragmentation of brittle solids. Particularly insightful is the seminal experimental study of Gilvarry and Bergstrom (Gilvarry and Bergstrom 1961, Bergstrom *et al* 1961, Bergstrom 1962) on the catastrophic fragmentation of compressed spheres of glass and other brittle solids. Hopkinson bar techniques offer a further experimental method for assessing the character of dynamic fragmentation of brittle solids. Examination of the fractal nature of brittle fragmentation (Turcotte 1986) offers a distinctive perspective on the dynamic failure process. Insightful comparisons are made on similarities between hydrodynamic turbulence and brittle fragmentation.

2.4.1 Gilvarry–Bergstrom brittle spheres

An instructive study is provided into the nature of brittle dynamic fragmentation by the investigation of Gilvarry (Gilvarry 1961, Gilvarry and Bergstrom 1961) and Bergstrom (Bergstrom *et al* 1961, Bergstrom 1962). The authors study the elastic strain-induced shatter of solid spheres of brittle material. Experimental fragmentation results are generated that offer unique insight into both the energy issues governing the catastrophic breakup of a brittle solid, and the character of the resulting fragment size distributions. Supporting theoretical studies of Gilvarry (1961) on the statistics of fragmentation are discussed in an earlier section.

2.4.1.1 Distribution characteristics of brittle fracture

The fragmentation tests consist of spheres of brittle solids (glass, sapphire, quartz, flint) of sizes ranging from 1/8 to 1 inches in diameter that are compressed between platens of tungsten carbide. Failure occurred through the explosive release of the strain energy, accompanied by intense fracture and fragmentation of the samples. Stored elastic strain energies prior to fracture are determined through measurement of the compressive load and elastic strain of the test samples. High-speed photography along with collection and analysis of the fragment debris provide the principal test diagnostics.

After slow compression of the spherical samples, failure is an abrupt and explosive event. Catastrophic fragmentation and release of strain energy ensues. Fragments are accelerated outward at high speeds. Large fragments originate from the outer regions of the sphere and tend toward lunar, or orange segment, in shape. The larger fragments constitute most of the mass (~90%). High-speed photography of the ejecta fragments yields estimates of fragment kinetic energy of about one-half the stored elastic strain energy prior to fracture.

Fragments from the tests provide size distributions through sieving techniques. The authors noted that for tests performed in a steel collection chamber, substantial secondary fragmentation occurs due to impact of the high-velocity fragments. Subsequent tests used a gelatin recovery medium to precluded secondary fragmentation. Fragment distribution data from tests on one-inch soda-lime glass spheres are plotted in figure 2.41. The data are sensibly described by a distribution of the Schuhmann type (Schuhmann 1940),

$$M(x)/M_o = (x/x_o)^n, \tag{2.234}$$

with a value of n close to unity. It is noteworthy that both strain-energy-induced (initial burst) and kinetic-energy-induced (secondary impact) fragmentation lead to the same functional dependence.

Sieving methods are reasonably accurate down to particle sizes of about 100 μm. Below this size, continued quantitative assessment of the fragment distribution becomes difficult. Gilvarry and Bergstrom (1961) used an alternative Coulter counter method to assess the small particle characteristics of the distribution. A power-law distribution of the form of equation (2.234), with a Schuhmann distribution index of $n = 1$, was found to provide a good description of the glass sphere particle size distributions down to sizes smaller than about one micrometer. The key observation is that a power-law distribution of the Schuhmann type with $n = 1$, characterizes fragment distributions in the brittle fracture of glass over at least four decades in fragment size (twelve decades in fragment mass). The authors note that similar observations were made for the other brittle materials investigated.

2.4.1.2 Strain energy in the brittle failure process

Key to the study was assessment of the elastic strain energy stored in the compressed sphere at failure. Relations for the strain energy and the maximum stress in the test sphere are from the paper of Bergstrom (1962) and are attributed to the elasticity text of Timoshenko and Goodier (1951). The equations are, for the specific strain energy E (total strain energy divided by the mass of the sphere),

$$E = \frac{4}{5}\left(\frac{3}{4}\right)^{5/9}(\rho\pi)^{1/9}\left(\frac{1-\nu_1^2}{E_1}+\frac{1-\nu_2^2}{E_2}\right)^{2/3}\left(\frac{P}{M^{\frac{2}{3}}}\right)^{5/3}, \tag{2.235}$$

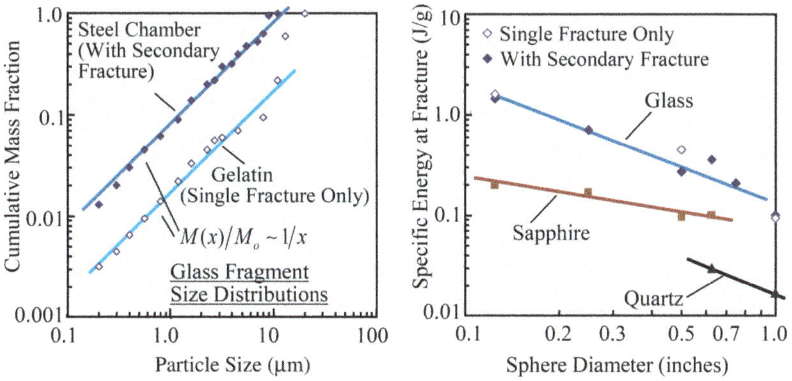

Figure 2.41. Experimental fragment size distribution curves for compression failure and fragmentation of one-inch diameter spheres of soda-lime glass (left plot). Specific elastic strain energy at fracture versus the test sphere diameter (right plot) (Bergstrom *et al* 1961).

and the maximum compressive stress at the center of the sphere–plate contact zone,

$$\sigma_m = \left(\frac{40}{\pi^4}\right)^{1/5}\left(\frac{1-\nu_1^2}{E_1}+\frac{1-\nu_2^2}{E_2}\right)^{-4/5}(\rho E)^{1/5}. \tag{2.236}$$

In these relations ρ is the density, P is the load, M is the mass of the sphere, and ν_1, ν_2, E_1, E_2 are Poisson's ratios and elastic moduli for the respective sphere and platen materials. Compression-to-fracture tests were performed over a range of sphere sizes. The size dependence of the strain energy to fracture was assessed. For the glass sphere tests a maximum stress from equation (2.236) of approximately 5 GPa and 10 GPa was achieved in the one inch and the one-eighth inch glass spheres, respectively.

Specific strain energy at failure is plotted as a function of sphere diameter for the brittle materials in figure 2.41. Most of the measurements plotted represent an average over a number of tests. The open symbols for glass are tests performed in gelatin that precluded secondary fragmentation. All others were performed in a steel collection chamber. All materials exhibited an increase in specific strain energy with decrease in size at failure. Sapphire shows somewhat less sensitivity than glass or quartz.

The Schuhmann distribution size parameter (x_o in equation (2.234)) from a fit of equation (2.234) to the cumulative mass versus size data is shown in figure 2.42. Failure strain energy versus distribution size parameter is cross-plotted on the right in figure 2.42. The important point is that all of the data do not differ with statistical significance from an inverse first-power dependence,

$$E \sim x_o^{-1}. \tag{2.237}$$

This result is significance. The distribution size scale x_o is inversely proportional to the new fracture surface area created in the fragmentation process. The results of this

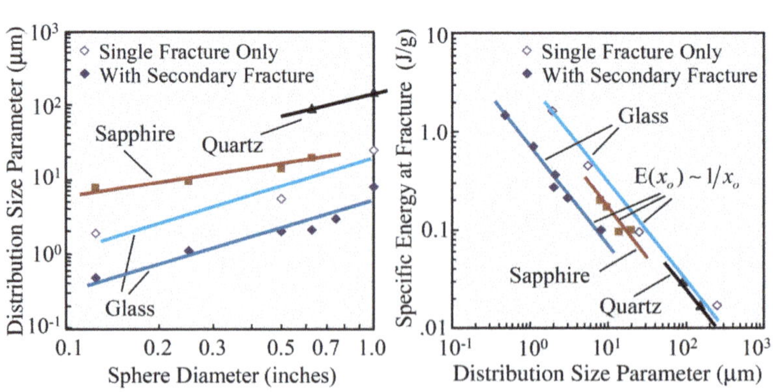

Figure 2.42. Left: Schuhmann distribution size parameter for compression failure and fragmentation of test spheres of soda-lime glass, single crystal sapphire and polycrystalline quartz. Right: specific elastic strain energy at fracture versus the Schuhmann distribution size parameter for the same test data (Bergstrom *et al* 1961).

work support the historical von Rittinger energy-based fragmentation relation (von Rittinger 1867) discussed in an earlier section.

2.4.2 Hopkinson bar fragmentation

Quasi-static compression of the Gilvarry–Bergstrom spheres of brittle solids leads to catastrophic fragmentation. Fragmentation of brittle solids can also be explored with dynamic test methods. Such experimental methods include explosive loading, projectile impact and Hopkinson bar loading among others. The Hopkinson bar test methods, originally developed to characterize the dynamic strength properties of metals, have been used to investigate the strength, failure and fragmentation of brittle solids.

Early application of Hopkinson bar techniques on brittle materials focused on natural rocks and minerals (e.g. Green and Perkins 1968, Lundberg 1976). Features of dynamic failure from these and similar studies are summarized in Grady and Lipkin (1980b). Subsequently, Hopkinson bar methods were extended to examine the failure and fragmentation of ultra-high strength ceramics. The studies of Lankford (1979), Bourne *et al* (1997), Chen and Ravichandran (1997), Wang and Ramash (1997) and Shih *et al* (2000) are representative.

Generally observed is a high elastic-failure stress (and elastic strain energy), followed by catastrophic failure through fracture and fragmentation, not unlike the Gilvarry–Bergstrom spheres. Unique to the Hopkinson bar experiment is a transition to loading rates at which the kinetics of fracture activation and growth are not initially able to keep pace with the rate of loading. Increasing levels of elastic stress, and strain energy, are achieved before catastrophic fracture and failure overwhelms the process. An increase of the effective strength of brittle solids with loading rate initiates at strain rates in the neighborhood of $10^3\ s^{-1}$ to $10^4\ s^{-1}$. A power-law dependence of failure stress on strain rate up to a maximum power of about one-third is observed. Various theories including crack activation and growth kinetics, and energy arguments have been offered to explain the observed strain-rate dependence.

Here, the focus is on one Hopkinson bar study that investigates the failure and fragmentation of brittle solids (Costin and Grady 1983, 1984). The study is appropriate to the discussions of fragment size statistics in the present chapter. Fragmentation experiments performed on ceramic and ceramic-like materials used a Kolsky torsion bar. There was, at the time, no specific reason for using a torsion bar apparatus other than availability. Emphasis in the testing focused on the size distribution of fragments produced in the failure event. Methods were used to assess particle sizes from a few millimeters down to tens of nanometers.

Materials tested included a ferroelectric ceramic lead zirconate titanate (PZT), depleted uranium dioxide (UO_2), quartz glass (SiO_2), isotropic graphite, and four grades of oil shale with nominal kerogen contents of 10, 20, 30 and 40 gallons per ton (GPT). Test samples were cartwheel shaped with the active section 10 mm in diameter and 7.5 mm in length, as shown in the photograph in figure 2.43. Flange sections were bonded to the input and transmission bar. An input torque pulse of 120 N·m was maintained constant over the test series providing a nominal

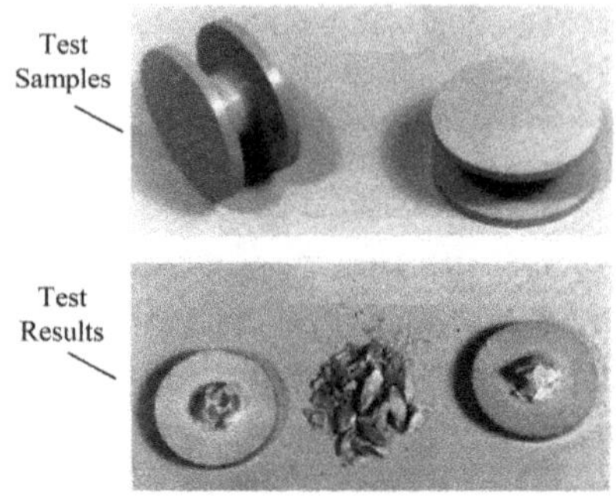

Figure 2.43. Photograph of a test sample (PZT) illustrating sample geometry and consequences of the torsion Kolsky bar failure event (Costin and Grady 1983, 1984). Reproduced with permission from Costin and Grady (1984).

loading strain rate of 500 s^{-1}. Strain gage measurements performed on the input and transmission bar provided the stress and strain history through the test sample loading and failure event.

The test section of the bar was enclosed in a sealed metal chamber to capture all of the fragment debris. For one material, the uranium dioxide, concern extended to airborne particulate generated in the failure event. In these tests, filtered air was circulated through the chamber with a vacuum pump and into a series of instruments designed to measure the quantity and size of the airborne particulate. These instruments included a cascade impactor, a condensation nuclei counter and an electrostatic precipitator, allowing characterization of the mass and size distribution down to a few tens of nanometers.

Larger particles were sorted with mechanical sieves ranging from 38 μm to 2000 μm. Particles larger than 2000 μm were individually weighed and their effective size calculated. Fragment size data are plotted in the Schuhmann cumulative mass-versus-size representation. One set of data for PZT ceramic, shown on the left in figure 2.44, is representative of all of the data collected excepting the graphite. Only graphite did not plot linearly in this display, exhibiting instead a bi-linear curve and suggesting more complex dynamic fracture behavior.

Properties for the materials tested are provided in table 2.1 along with the principal experimental measurement; namely, the distribution shape parameter or Schuhmann index n. Several features are observed. The lowest Schuhmann index of $n = 1.3$ is for glass and, although slightly higher, is not out of line with the measurements of Gilvarry and Bergstrom (1961) on the compression failure of glass spheres discussed earlier in the section. A trend is observed for oil shale. As the kerogen content of the oil shale increases (presumably with increasingly less brittle behavior) the Schuhmann index increases. An increase in the value of n indicates proportionally less mass in fine fragments.

A trend is also observed in the PZT ceramic (Costin and Grady 1983). Fracture toughness differs for the four formulations. The mass of fragments in the finer sieve sizes increases systematically with decreasing fracture toughness. A Schuhmann size scale x_o in equation (2.234) was determined for the four ceramics. This size scale is

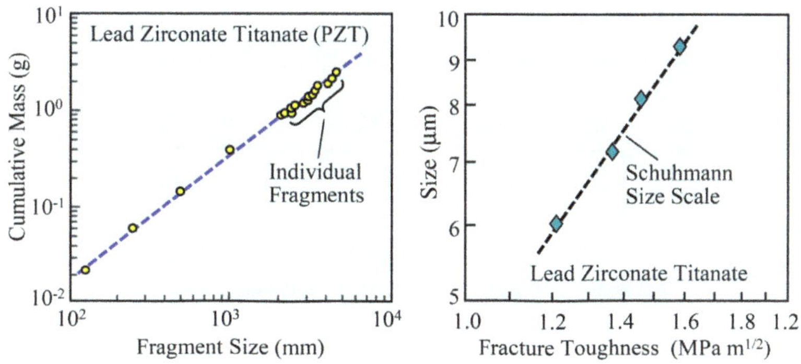

Figure 2.44. Torsion Kolsky bar fragmentation experimental particle size distribution for PZT ceramic. Schuhmann size scale versus fracture toughness for four preparations of the PZT ceramic (Costin and Grady 1983, 1984).

Table 2.1. Fragmentation properties for selected brittle solids.

Material	Density ρ (kg m^{-3})	Young's modulus E(GPa)	Fracture K_{IC} (MPa m$^{-1/2}$)	Schuhmann index n
UO_2	10 860	192	1.6	1.5
Glass	2200	37	0.54	1.2–1.3
Oil shale				
10 GPT	2590	20	1.2	2.1
20 GPT	2360	15	1.0	2.2
30 GPT	2140	10	0.8	2.9
40 GPT	1850	6	0.65	3.2
Graphite	1850	12	1.4	2.6
PZT-1	7250	110	1.22	1.3–1.4
PZT-2	7370	114	1.38	1.4–1.5
PZT-3	7300	112	1.47	1.25–1.35
PZT-4	7430	110	1.59	1.4–1.5

plotted against fracture toughness on the right in figure 2.44. A decrease in the Schuhmann size scale with decreasing toughness is observed.

Sieving results for uranium dioxide (UO_2) are shown in figure 2.45. Notable was the self-similar or fractal character of fragments. Fragment shape, angularity and surface texture were visually the same over the sieve size range. Also undertaken was the collection of aerosol particulate down to the tens of nanometer size with the associated collection devices. Uranium dioxide particulate collected with the electrostatic precipitation device is illustrated in figure 2.45.

The graphite exhibited bi-linear behavior in a Schuhmann plot of the fragment size data. Fragments of size greater than 1000 μm, which constituted about 95% of the mass of the sample, exhibited the larger $n = 2.6$ reported in the table. The finer fragment particulate provided a value of n close to unity.

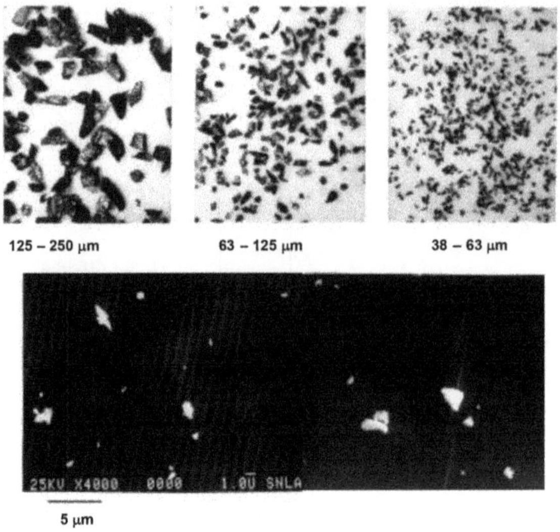

Figure 2.45. Torsion Kolsky bar uranium dioxide fragmentation tests. Selected sieve sizes are shown in the upper photographs, and electrostatic precipitator aerosol particulate in the lower (Costin and Grady 1983, 1984). Reproduced with permission from Costin and Grady (1984).

2.4.3 Fragmentation as a fractal process

Turcotte (1986) noted that the process of fragmentation of a solid body results, in some cases, in a fragment distribution that is fractal. A distribution of objects is fractal if their number N greater than some linear dimension x is represented by a power-law relation,

$$N(x) \sim x^{-D}, \tag{2.238}$$

with $D > 0$. D is the fractal dimension. An example is provided by the number of oceanic islands with linear dimension greater than x equal to the square root of the island area. The example satisfies equation (2.238) with a fractal dimension of about 1.3 (Mandelbrot 1982).

In applications dealing with the breakup of brittle solids the Rosin–Rammler distribution (Rosin and Rammler 1933) is commonly found to be appropriate. The Rosin–Rammler distribution is written in terms of the mass of fragments M with linear size smaller than x,

$$M(x) = M_o\left(1 - e^{-(x/x_o)^n}\right), \tag{2.239}$$

and is more often applied in the form attributed to Schuhmann (1940),

$$M(x) = M_o(x/x_o)^n. \tag{2.240}$$

Relating the fragment number to the mass increment, $dN \simeq dM/\rho x^3$ yields,

$$N(x) \simeq N_o(x/x_o)^{n-3}, \tag{2.241}$$

with

$$N_o = \frac{n}{3 - n}\frac{M_o}{\rho x_o^3}. \tag{2.242}$$

Comparing equation (2.241) with equation (2.238), relates the fractal dimension to the Schuhmann index through

$$D = 3 - n. \tag{2.243}$$

The distribution is fractal provided n is less than three. Typically n ranges over 0.5–1.5 for most rock and mineral breakage processes (e.g. Bennett 1936), and n is found to be very close to unity for the compression shatter experiments of Bergstrom (1962) on spheres of glass, quartz and sapphire.

Many breakage processes of brittle solids lead to fractal-like fragment distributions with a fractal dimension D ranging between about 1.5 and 2.5. A number of examples are provided in the paper of Turcotte (1986) including various rock breakage processes along with distributions of natural earth and asteroid materials. Kolsky bar experiments on brittle materials including glass, uranium dioxide, PZT and oil shale in the previous section also showed values of n in this range.

Implications of a power-law distribution and fractal behavior resulting from fragmentation in brittle solids is that the mechanisms of fracture leading to breakup remain invariant over the span of size scales achieved. This in turn implies that fragment sizes within the range of the distribution remain large relative to the intrinsic length scales of the fracture process. The relation

$$x_{pz} = \frac{\pi}{72}\left(\frac{K_c}{Y}\right)^2, \tag{2.244}$$

provides an estimate of the crack-tip process-zone length scale (Lawn and Wilshaw 1975). Based on values of K_c and Y for glass, a measure of $x_{pz} \sim 5$ nm is obtained, well below the length scales measured in most assessments of brittle fracture distributions. This length scale will range upward for less brittle materials.

If fragmentation of brittle solids is a fractal process leading to $N \sim x^{-D}$ or, correspondingly, to $M \sim x^{3-D}$, then the underlying physics of interest is in explaining the mechanisms responsible for the specific value of the fractal dimension D. Turcotte (1986) has explored renormalization group methods used by others to characterize various scale invariant critical state phenomena. A parameter p_m is identified that determines the probability of fracturing of a cell of the solid body. The fractal dimension D is calculated from p_m, which in turn relates to the fragility of the brittle material. The fractal nature of fragmentation is addressed again later in the chapter.

2.4.4 Unlikelihood that brittle fragmentation is a Poisson process

A fragment size distribution of the Schuhmann form

$$M(x) = M_o(x/x_o)^n, \tag{2.245}$$

where $M(x) \leqslant M_o$ is the cumulative mass of fragments less than size x, is known to describe the consequences of the fragmentation of brittle solids reasonably well in most instances. The distribution index n is often close to unity and almost certainly within the range of $0.5 < n < 2.5$. The distribution applies to both single-event fragmentation such as high-velocity impact or explosive breakage, as well as to comminution processes such as crushing or grinding.

A number of the earlier authors, including Bennett (1936), Lienau (1936), Gilvarry (1961), and Gaudin and Meloy (1962), among others, attempted to rationalize equation (2.245) through a model of brittle fracture as a random placement of fractures through a Poisson process. As described earlier in the chapter, Lienau (1936) showed that random fragmentation of a line (a one-dimensional body) through a Poisson process leads to the cumulative number distribution,

$$N(x) = N_o(1 - e^{-x/\lambda}). \tag{2.246}$$

Bennett (1936) used a relation of the form of equation (2.246) to distribute fracture planes in three mutually orthogonal directions and attempted to rationalize the Schuhmann distribution in equation (2.245). Gilvarry (1961) explored similar arguments. Gaudin and Meloy (1962) pursued the same approach over a finite dimension body in which case equation (2.246) is replaced with a binomial distribution.

Difficulties in these earlier analytic studies are illustrated in the following. The mass density distribution over size provided by the derivative of the Schuhmann equation (2.245),

$$\frac{dM}{dx} \sim x^{n-1}, \tag{2.247}$$

is observed to have a weak dependence on fragment size. It would be constant if $n = 1$ (equal masses in equal size intervals).

Theoretical distributions based on a Poisson statistical process differ markedly from this observation. In one dimension the Lienau relation from equation (2.246) yields

$$dN \sim e^{-x/\lambda}dx, \tag{2.248}$$

and the fragment mass distribution

$$\frac{dM}{dx} \sim \frac{xdN}{dx} \sim xe^{-x/\lambda} \sim x. \tag{2.249}$$

The last step examines the small fragment limit of the distribution. In one dimension the mass density increases as the first power of fragment size.

Mott and Linfoot (1943) were apparently the first to provide a rigorous analytic solution of the two-dimensional Poisson fragmentation process (the random partitioning of an area by the random distribution of orthogonal horizontal and vertical lines according to equation (2.246)). They obtained the number density distribution

$$\frac{dN}{dx} = \frac{4x}{\lambda^2} K_o(2x/\lambda), \tag{2.250}$$

where K_o is a modified Bessel function. The mass density in two dimensions,

$$\frac{dM}{dx} \sim \frac{x^2 dN}{dx} \sim x^3 K_o(x), \tag{2.251}$$

is plotted in figure 2.46. A fit to the lower five decades in the plot yields approximately

$$\frac{dM}{dx} \simeq x^{2.8}. \tag{2.252}$$

If fragment area is partitioned by a Poisson process as suggested by Grady and Kipp (1985), then

$$\frac{dM}{dx} \sim \frac{x^2 dN}{dx} \sim x^3 e^{-(x/\lambda)^2} \sim x^3, \tag{2.253}$$

with the power-law dependence applying to the small fragment portion of the distribution. Equation (2.253) is in reasonable agreement with the analytic Bessel solution in this range as illustrated in figure 2.46 and compared with equation (2.252). Note that the Mott–Linfoot assumption of the form

$$N(m) \sim 1 - e^{-\sqrt{m/\mu}}, \tag{2.254}$$

where the mass $m \sim x^2$ leads to

$$\frac{dM}{dx} \sim \frac{x^2 dN}{dx} \sim x^2, \tag{2.255}$$

exhibits poorer agreement with the analytic solution.

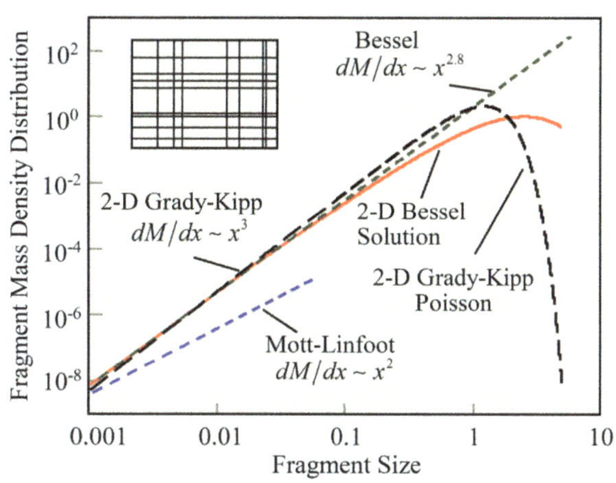

Figure 2.46. Solutions to Poisson two-dimensional fragmentation through partitioning of an area by random orthogonal horizontal and vertical lines, as illustrated in the inset (Grady 2009).

The analytic solution to random planes in three orthogonal directions has not apparently been attempted but is unlikely to support the use of the model in justifying the mass density dependence on size required by the Schuhmann distribution in equation (2.247). A linear dependence on size for the one-dimensional fragmentation algorithm and the cubic dependence of the two-dimensional fragmentation would suggest

$$\frac{dM}{dx} \sim x^5, \tag{2.256}$$

for the three-dimensional orthogonal planes random fragmentation of a body. In fact, the Poisson process assumed by Grady and Kipp (1985) leads to

$$\frac{dM}{dx} \sim \frac{x^3 dN}{dx} \sim x^5 e^{-(x/\lambda)^3}. \tag{2.257}$$

In any case the preponderance of data on fragment size distributions for brittle solids suggest that brittle fragmentation is not well described by a Poisson process.

2.4.5 Scale invariance in the catastrophic fracture of brittle solids

Statistical features of brittle fracture are not captured through the probabilistic approaches discussed earlier in the chapter. The efforts of earlier workers (Bennett 1936, Gilvarry 1961, Gaudin and Meloy 1962) to characterize brittle fracture through a Poisson statistical process are probably in error. Some understanding of dynamic fragmentation of brittle solids is achieved through parallels with hydrodynamic turbulence in fluids. The latter is regarded as one of the remaining unsolved problems of classical physics. It is perhaps not surprising that a satisfactory statistical theory of brittle fracture and fragmentation continues to elude a sound theoretical basis.

2.4.5.1 Hydrodynamic turbulence

Turbulence can occur in fluids described by the Navier–Stokes equations where governing material properties are relatively few in number. A macroscopic length scale L, a characteristic velocity V, and a viscosity η are adequate to reveal the essential features. The length L is, for example, the size of the structure imparting motion to the fluid, while V is the nominal velocity impressed on the fluid. When the dimensionless Reynolds number $R_e = LV/\eta$ is sufficiently large, turbulence arises in the fluid motion.

Turbulence in the flow emerges when large-scale laminar flow is not sufficient to dissipate the energy through viscous friction. Turbulence is then the activation, the growth and the motion of irregular fluid disturbances on successively smaller length scales necessitated by requirements of dynamic fluid dissipation. The essence of turbulence is the transient and steady state motion of a hierarchy of sub-motions over a wide range of length scales. This cascade to successively smaller length scales in the turbulence process proceeds until velocity gradients of order V/λ are achieved. The dimension λ is the limiting substructure length scale that is adequate to support the necessary viscous dissipation. The range of length scales between L and λ is determined by the Reynolds number and increases with increasing Reynolds

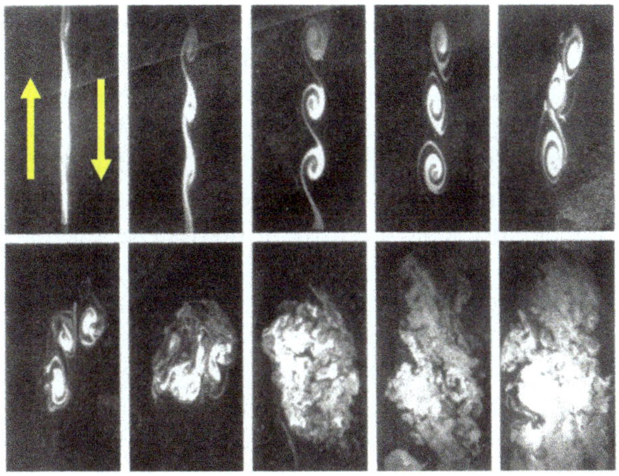

Figure 2.47. This image of turbulence in a fluid is reproduced from the review article of Falkovich and Sreenivasan (2006). Increasing time is from left to right and top to bottom. The orientation of shear flow is shown in the first image. The history of turbulence is observed through evolution of the light color die introduced on a plane in the fluid. Reproduced with permission from Falkovich and Sreenivasan (2006). Copyright 2006, AIP Publishing LLC.

number. The processes of turbulence in fluids are strikingly illustrated by the example shown in figure 2.47, taken from the *Physics Today* review article of Falkovich and Sreenivasan (2006). The authors provide a succinct discussion of the physics underlying hydrodynamic turbulence. Similar intermediate scale invariance has been ascribed to the topic of fatigue fracture (Barenblatt 2006). Scale invariance is there identified with a law of intermediate asymptotics.

The range of sub-motions of length scales bounded by L and λ, and reasonably removed from either, is identified as the inertial range. Physics of turbulence is expected to be invariant to length scale through the inertial range. Kolmogorov (1941) argues for scale invariance on dimensional grounds, and arrives at a power-law dependence of features of the turbulent motions on length scale over the inertial range. Recent theoretical work on turbulence suggests that the scale invariance of Kolmogorov is not fully realized (Chen *et al* 2005). For present purposes, however, such scale invariance, or near-scale invariance, is a crucial observation of hydrodynamic turbulence, and offers a perspective for understanding the nature of dynamic brittle fragmentation.

2.4.5.2 Catastrophic fragmentation

Parallels to hydrodynamic turbulence are seen in the catastrophic fracture and fragmentation of brittle solids (Grady 2008, 2010). Consider a solid body of characteristic size L composed of a brittle material such as glass or a high-strength ceramic. A compressive load is impressed on the body inducing nominal elastic strain energy per unit mass U. Fracture failure initiates when critical stress conditions are achieved at some site in the body. Once initiated, fracture proceeds rapidly and explosively, converting the elastic strain energy into surface fracture energy Γ and kinetic energy of the ejected fragments.

Fracture in brittle solids is weakly dissipative, however, and failure through one or several through-going cracks is far from adequate to account for dissipation of the initial stored elastic strain energy. Consequently, during failure, fracture on successively finer length scales proceeds though a cascade of crack branching until length scales adequate to the dissipation of the initial elastic strain energy are achieved. This limiting length scale is expected to scale as $\lambda \sim \Gamma/\rho U$, with ρ the material density. The length scale λ can be a number of decades smaller than the characteristic size L of the body.

Within the inertial range of $\lambda < x < L$ the physical processes become scale invariant as the catastrophic fracture cascades to successively smaller length scales. (Terminology is freely borrowed from the theoretical description of hydrodynamic turbulence.) Consequently, the fragment count within this range is expected to exhibit a power-law dependence on fragment size, reflecting the requirement of self-similar scaling. As fragment size approaches the limiting dissipation length scale of λ, the functional dependence will diverge from a power-law dependence exhibiting an emerging dependence on the limiting length scale λ.

Although the specific functional form is not known for the fragment number distribution, an appropriate relation for the number distribution is readily guessed. The functional form,

$$N(x) = \frac{N_o}{1 + (x/\lambda)^{\delta}}, \tag{2.258}$$

exhibits the necessary power-law dependence for $x \gg \lambda$ and the appropriate limiting behavior within the range of x in the neighborhood of λ. $N(x)$ is the complementary cumulative number distribution (number greater than) of fragments while N_o is the total number of fragments. The exponent δ is the fractal dimension in the power-law limit of the cumulative fragment relation. The relation readily transforms to the Schuhmann equation in a mass distribution representation of the fragments.

Similarities between hydrodynamic turbulence and catastrophic brittle fracture also are evident in the propagation of a single crack. The laminar flow of a rapidly shearing fluid becomes unsettled by the emergence of insipient eddies and vortices as the intensity of shearing is increased. Dissipation through laminar viscosity is increasingly inadequate to balance the power input. Similarly, the energy dissipated at the tip of a fast running crack in a brittle solid is a strong function of the velocity of the crack. Increased dissipation is a consequence of the onset of microcrack branching instability at sufficiently high velocity (Sharon *et al* 1996). Microcrack instability is the physical equivalent of hydrodynamic turbulence exhibited by the crack in exploring crack-tip deformations capable of the required dissipation.

2.4.5.3 Modeling the brittle fragmentation phenomenon

In this subsection the scale invariance concepts of catastrophic fracture are enfolded into earlier models of dynamic fragmentation. The theory is based on the energy-horizon concept introduced in the previous section.

The first requirement of the theory is establishing the size of the body considered in the energy balance. For the two experimental examples discussed earlier, it is the size L of the glass sphere subjected to compressive failure, or that of the ceramic cylinder fractured through loading by the Hopkinson bar. For a body subjected to rapid loading this size scale is captured through characterization of an expansion strain rate $\dot{\epsilon}$ leading to tensile failure and fragmentation. In the expansion process, a characteristic time τ must pass from the onset of tension until failure through fracture occurs. If the material has an elastic wave speed c, then a region of size

$$s = 2c\tau \tag{2.259}$$

determines the correlation length scale. Such a region is, for example, a sphere centered at any arbitrary point in the deforming media with radius $c\tau$. Every region of size s within the domain of the body undergoing tensile expansion fails independently.

The second requirement of the theory is that the elastic strain energy in the region be adequate to fuel the fracture process. The dissipation energy Γ required to create new fracture surface is assumed to be constant. As a criterion, consider the energy required to fracture a cube of size s into equal cubes of size $s/2$. The new fracture surface area requires that the available strain energy be no less than,

$$\rho U = \frac{1}{2}\frac{P^2}{\rho c^2} = \frac{6\Gamma}{s}. \tag{2.260}$$

The term on the left is the elastic strain energy U at a hydrotension P. The term on the right provides the new fracture surface energy requirement. The expression yields the minimum tension capable of breaking the region into fragments of size s.

This energy-based model is illustrated with the plot shown in figure 2.48. The energy criterion provided by equation (2.260) is identified as the equilibrium energy curve. Several tensile pressure versus correlation horizon loading paths are shown. One is a quasi-static loading path with the horizon the fixed finite size L of the body. Several dynamic loading paths are shown that correspond to successively higher loading strain rates $\dot{\epsilon}$. If the material is disposed to fail when the equilibrium energy requirement is met (intersection of the loading path with the energy-equilibrium curve from equation (2.260)), then energy-equilibrated failure and fragmentation occurs. This intersection of the dynamic loading path with the equilibrium energy relation yields the energy-based fragment length scale $\lambda \sim (\Gamma/\rho\dot{\epsilon}^2)^{1/3}$.

Competent brittle solids seldom fail when an equilibrium energy requirement is satisfied. Some alternative failure criterion must be achieved as indicated in the figure by a non-equilibrium failure curve. Failure within the correlation region occurs when tensile stress exceeds some material strength property determined by the intrinsic defect structure of the material. In glasses, certain crystalline solids and high-quality ceramics, spall experiments reveal tensile stresses approaching a significant fraction of theoretical strength before onset of spall failure.

One dynamic stress load path in the figure illustrates bifurcation of the energy-equilibrium length scale λ into the two length scales λ_1 and λ_2 upon brittle failure at

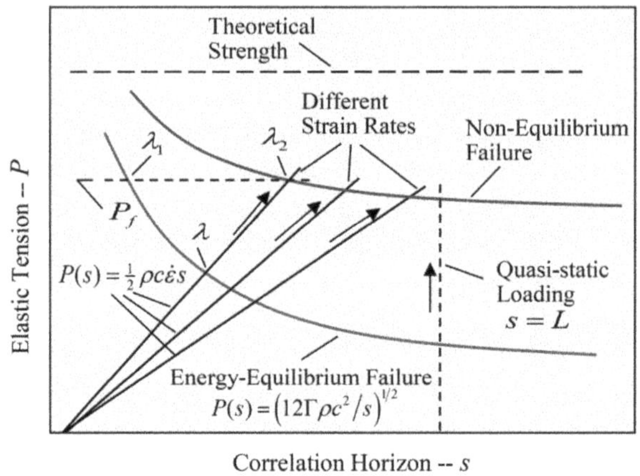

Figure 2.48. Graphical representation of energy-based theory of dynamic fragmentation as applied to brittle solids (Grady 2008, 2010).

some elevated failure tension P_f. For quasi-static loading λ_2 corresponds to the specimen size L. For dynamic loading it is the correlation length at onset of failure. The length scales λ_1 and λ_2 are bounds for the ensuing fragment distribution. For competent ceramics that can support significant elastic strain energy before onset of failure, the bounding length scales can be decades apart. The length scales λ_1 and λ_2 determine the inertial range of the dynamic fracture event within which the processes of fracture must be independent of scale. Within this size range the fragment distribution exhibits the power-law behavior required by the fractal nature of the breakup event.

2.4.5.4 Fragmentation of brittle solids

Some comparisons of the model of fragmentation in the previous subsection can be made with experimental data. Elastic strain energy at failure was measured for both the glass-sphere and the ferroelectric-ceramic fragmentation experiments for which the size distribution data are shown in figures 2.41 and 2.44. The compression fragmentation test on glass was performed on a one-inch diameter sphere. The upper length scale is of order $\lambda_2 \simeq 25$ mm. The torsion Hopkinson bar tests on ferroelectric ceramic were conducted on cylinders 10 mm in diameter and 7.5 mm in length. An upper length scale of order $\lambda_2 \simeq 7.5$ mm is appropriate for the ceramic.

For the glass sphere a fracture energy of $\Gamma \simeq 4.5\ \mathrm{J\ m^{-2}}$ is calculated from a fracture toughness of 0.75 MPa $\mathrm{m^{1/2}}$, a glass density of 2500 kg $\mathrm{m^{-3}}$ and an elastic modulus of 63 GPa. In the compression failure test elastic strain energy of $\rho U \simeq 0.5\ \mathrm{J\ cm^{-3}}$ can be calculated at the onset of catastrophic fracture. A lower length scale of $\lambda_1 \simeq 50\ \mu\mathrm{m}$ is estimated from equation (2.260).

For the ceramic cylinder, a fracture toughness of 1.5 MPa $\mathrm{m^{1/2}}$, a density of 7300 kg $\mathrm{m^{-3}}$ and an elastic modulus of 110 GPa provide a fracture energy of $\Gamma \simeq 10.2\ \mathrm{J\ m^{-2}}$. A measured elastic strain energy at fracture of $\rho U \simeq 0.75\ \mathrm{J\ cm^{-3}}$ provides a lower length scale of $\lambda_1 \simeq 80\ \mu\mathrm{m}$.

For both tests the length scales, $\lambda_1$and λ_2 span nearly three decades in fragment size, and brackets the power-law range of the experimental data. This observation is consistent with the theory. Fragment size data near the lower length scale λ_1, which would be of considerable interest, were not collected in either of the experiments.

Consider now an experiment in which brittle fragmentation data were collected under very high strain-rate (shock) loading (Grady and Kipp 1987). In this experiment a test sample of fused silica was subjected to shock loading to a stress level below, but an appreciable fraction of, the Hugoniot elastic limit (approximately 8.0 GPa) in a confining container. A representative sample of the resulting fragment debris was extracted from the container and a fragment size distribution determined by scanning electron microscopy and image analysis methods. A scanning electron microscope image of the shocked fused-silica fragment debris from which the distribution was determined is shown in figure 2.49.

The fragment size distribution for this fused silica data is shown in figure 2.50 as both a cumulative mass fraction distribution, as was used to describe the previous glass and ceramic data, and also as a differential distribution showing the mass fraction in each incremental size range. The overlying curve on the latter spectral distribution is from a theory due to Griffith (1943) that was discussed in an earlier section (Grady and Kipp 1987). A distribution of the form suggested in equation (2.258) fits the data equally well. A Schuhmann distribution curve overlays the cumulative distribution on the left.

The energy-based theoretical fragmentation plot shown in figure 2.48 is shown again in figure 2.51, but specific to the present experiment and fused silica material. The properties of fused silica necessary to construct the plot include a density of 2200 kg m^{-3}, an elastic wave speed of 5000 m s^{-1} and a fracture toughness of 1.0 MPa m$^{1/2}$ providing a fracture energy of $\Gamma = 9.0$ J m^{-2}. A strain rate of $\dot{\varepsilon} \simeq 10^6$ s^{-1} is representative of the impact velocity and sample size used in the experiment.

At this strain rate, equilibrium fragmentation is inferred from the point of intersection of the two curves in the plot to occur at a tensile stress of approximately

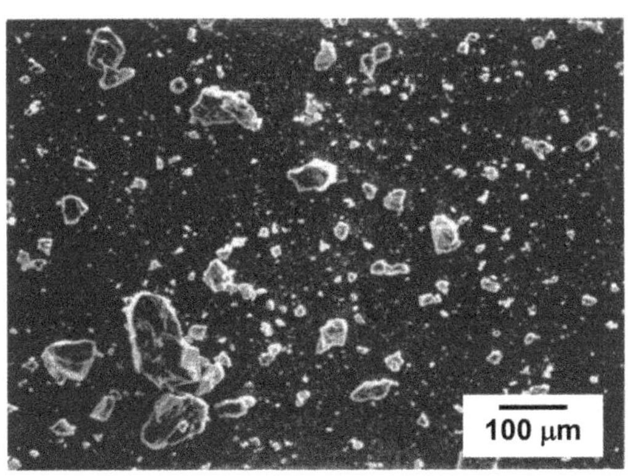

Figure 2.49. Scanning electron microscope image of fused silica fragment debris resulting from shock wave loading. Reprinted from Grady and Kipp (1987). Copyright 1987, with permission from Elsevier.

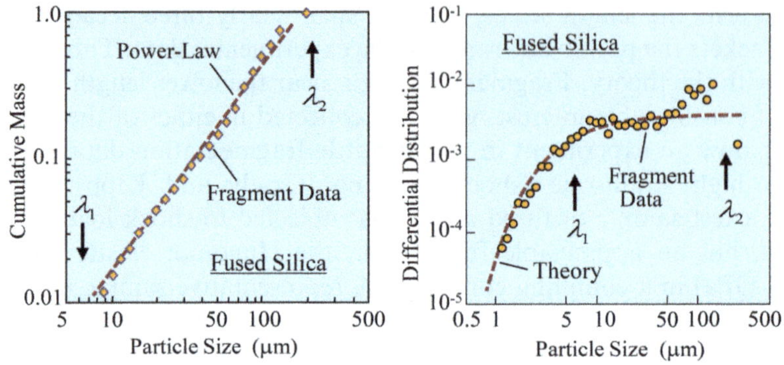

Figure 2.50. Experimental fragment distribution data for shock impact fragmentation of fused silica (Grady and Kipp 1987). A cumulative mass distribution is shown on the left and the same data in a differential representation on the right. The theoretical curve on the right is based on the work of Griffith (1943). A straight line Schuhmann power-law curve fits the data on the left.

0.3 GPa and with a characteristic fragment size of about $\lambda \simeq 60$ μm. The fused silica, however, supports substantially higher dynamic tension before spall failure. Wave interactions in this test were not measured, but if a sensible spall strength for fused silica of 1 GPa is assumed (and a corresponding elastic strain energy), then the inertial range of length scales spanning self-similar fracture and fragmentation, as shown in the plot, is bracketed by the values of $\lambda_1 \simeq 6$ μm and $\lambda_2 \simeq 200$ μm.

The experimental data are in sensible agreement with the theory. The experimental size distribution data shown in the right-hand plot of figure 2.50 are reasonably flat over the size range of about 10–100 μm, suggesting near power-law behavior with a power of n close to one. The experimental differential distribution drops off rapidly below 10 μm, and is consistent with the estimate of a $\lambda_1 \simeq 6$ μm lower bound for the fragmentation inertial range and the limiting behavior of the distribution in the neighborhood of this size scale.

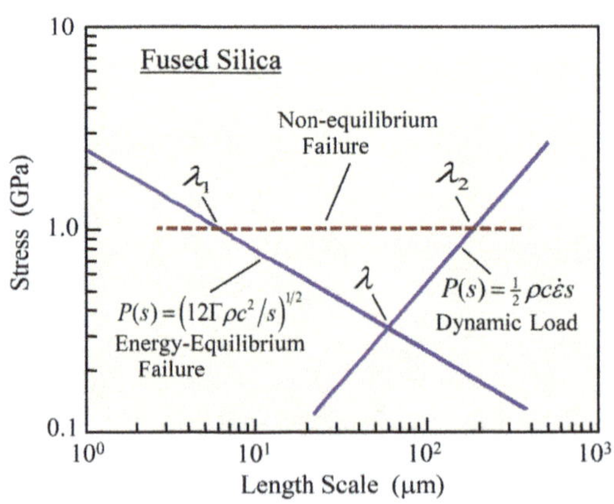

Figure 2.51. Non-equilibrium failure and fragmentation in shock wave compression of fused silica (Grady 2008, 2010).

The power n is not provided by the theory. Considerable experimental observation suggests a range of perhaps $0.5 < n < 1.5$, and perhaps higher in some applications. One attempt to provide a physical explanation for the magnitude of the power n based on statistical mechanical principles is provided in the study of Griffith (1943).

The catastrophic failure and fragmentation of competent brittle solids is not yet adequately understood. The present discussions have focused on the statistic size distribution of fragment debris resulting from the catastrophic fracture of a brittle solid. It has been known for some time that fragment distributions resulting from the breakup of brittle solids are power-law in form. This fractal character of the distribution argues for physical processes of dynamic fracture that are length-scale invariant over much of the size range of the experimental distribution. A close analogy of dynamic fragmentation of brittle solids with hydrodynamic turbulence in fluids is noted, and it is sensible to transfer some of the theoretical concepts from this field to the problem of catastrophic fragmentation of brittle solids. The extension of earlier energy-based theories of dynamic fragmentation also show application. There continues to be a need for well-characterized fracture and fragmentation data on brittle solids to further the theory and modeling efforts.

2.5 Shock fragmentation: a summary

2.5.1 Discussion

The shock compression of solids encompasses a broad span of physical phenomena and has become a mature topical area in the field of scientific research. Not fully appreciated is the fact that the shock compression event is ultimately destructive. In some cases, much effort is expended in the minimization and containment of such destruction. In other cases, such destruction can be put to constructive use. An example is that of spacecraft shielding against meteor and orbital debris. In yet other cases, such destruction is the end objective. Military application is a clear example, but also explosive blasting in quarry and mining engineering.

In many of the applications where shock compression leads to destruction of the materials and components of a system, there is a need for understanding of the processes and consequences of the breakup event. These processes and consequences involve the failure, fragmentation and dispersion of solid material and, in of themselves, constitute a range of scientific issues. This topical subject is identified here as the dynamic fragmentation of solids. The present chapter is devoted to this topical area, and surveys many of the key issues.

The statistics of dynamic fragmentation—the statistical distributions of fragments and ejecta velocities—are pursued in some depth. Although not exhaustive, a number of fragmentation statistics issues are covered from both a historical and a modern perspective. The extent of fragmentation, the average fragment size or the number of fragments produced, and how this extent relates to the intensity of the interaction and the properties of material governing the resistance to fracture, is central to dynamic fragmentation. This topic is surveyed in some detail; covering efforts of earlier workers in the fields, as well as more recent efforts. Brittle material is singled out as a class of materials with dynamic fragmentation properties that are not fully understood. Some succinct observations from the present chapter are summarized in the following.

Early efforts to understand statistical fragmentation through application of geometric partitioning algorithms and Poisson point processes appears, when examined over time, not to have been overly fruitful. Attempts to extrapolate simple algorithms to the breakup of three-dimensional bodies and predict the Schuhmann distribution (the power law between cumulative fragment mass and fragment size) for fragmentation of brittle solids, in fact, appear to be wrong.

The classic Mott distribution (log fragment number versus square-root fragment mass) emerged from similar thinking on random geometric partitioning. Subsequent efforts by Mott through physics-based analysis of the statistical fragmentation process failed to justify this original distribution. Mott concluded in a later report (Mott 1944) that there was no theoretical justification for this distribution that has since become the hallmark for the analysis of bomb and warhead fragmentation testing. The Mott distribution, however, does provide a convenient exponential form for systematic representation of test data—particularly when extended to the more general Weibull distribution (e.g. Grady 2006).

An exponential distribution of the fragment number versus fragment mass, or some form of an exponential distribution (such as the log-normal, the Voronoi, or the Weibull distribution), does sensibly describe experimental distributions in a large number of cases. An exponential distribution emerges from a random, or Poisson, partitioning of the body mass into a distribution of fragment masses. Exponential distributions appear to be more representative of the fragmentation of ductile materials; however, the full extent of applicability is not understood.

Fragment distributions resulting from the breakup of brittle solids appear to be largely power-law. This fractal character of the distribution suggests that over some range of length scales the fracture processes leading to fragmentation are scale invariant. There are some clear, and instructive, similarities between catastrophic fragmentation of brittle solids and turbulence in hydrodynamic systems.

The statistics of ejecta velocities resulting from a catastrophic fragmentation event has been largely unexplored. The approaches pursued in this chapter apply the premises of classical statistical mechanics to dynamic fragmentation and results in relations that parallel classic Maxwell velocity distributions.

Prediction of the characteristic fragment size (the intensity of fragmentation) in a dynamic fragmentation event also has an interesting historical perspective. Both von Rittinger (1867) and Kick (1985) in the latter decades of the 19th century proposed expressions relating fragment size to some measure of the energy causing breakup. There was apparently some contention among the proponents of the several theories. That of von Rittinger, relating the input energy to the new fragment surface energy, has better survived the test of time.

Again, the physics-based theory of Mott (1947) is unique in providing a prediction of both the fragment size scale and the statistical distribution of fragments sizes about this scale. The material property governing fragment size is a parameter (the Mott γ parameter) characterizing the statistical spread in strain to failure of sites of weakness in rapidly deforming metals. Although physically sensible, this parameter is difficult to determine through methods other than calibration to a set of fragmentation test data.

More recently, predictions of fragment size based on energy principals have demonstrated reasonable success and practical applicability in both scientific and engineering level calculations. The approach extends in some sense the original fragment size relation of von Rittinger through a more thorough description of both the input energy responsible for the dynamic fragmentation, and the energy expenditure resisting production of new fracture surface area in the fragmentation event.

The shock compression resulting from impact or explosion leads to transient kinetic energies and strain energies that cause failure of the material and fuel the fragmentation event. Resistance to fragmentation results from the dissipation required to propagate fracture and create new fracture surfaces. Commonly this resistance is adequately captured through a material surface energy or fracture toughness property.

For many materials, a balance of these energies provides a single fragment size scale that adequately centers the fragment distribution descriptive of the fragment debris. For others (notably the brittle materials) there is a schism leading to two length scales that bound a fragment distribution that is power-law within the range constrained by the two length scales. Both failure stress and fracture energy are required properties for such materials.

The early statistics-based theory of Mott and the more recent energy-base theory can be reconciled through inclusion of both the microscopic fracture activation physics and the dissipative energetics of dynamic fracture growth and coalescence.

2.5.2 Closure

The destruction of bodies brought about by intense shock wave loading leads to a number of interesting scientific issues. The scientific or engineering concerns will differ depending on whether the application seeks to mitigate, enhance, utilize or just understand the destruction event and the accompanying breakup consequences. The breakup and fragmentation brought about by shock-induced destruction involves both physical and statistical issues. The present chapter surveys a substantial fraction of the scientific efforts that have been put forth to provide a scientific basis for the topic.

Nonetheless, the chapter is far from complete. Shock-induced fragmentation of liquids is not addressed. Fragmentation of liquid metals is surveyed to some extent by Grady (1995). Other liquids, particularly polymeric liquids, also exhibit complex dynamic fragmentation phenomena. Borg *et al* (2001) provide some insight into the fragmentation of polymeric liquids. Some issues on the dynamic fragmentation of such liquids are touched on in the next chapter.

The shock-induced crush and particulation of brittle granular materials is a topic of interest. Dynamic crush is addressed in chapter 7. Another important topic is the aeroballistics-induced dynamic fragmentation of high-velocity bodies in atmosphere. This phenomenon can accompany a primary impact- or explosion-caused fragmentation event by further reducing the size of fragments ejected at high velocity. Thermochemical processes participating in the reduction event can further complicate aeroballistics breakup. Dynamic fragmentation initiated by, and primarily governed by, instability issues is an area deserving of attention. Instability fragmentation can include the breakup of rapidly stretching jets or thin sheets of solid metals or liquids. Papers by Romero (1989, 1991) provide some access to the literature on this subject. The breakup of the accelerating interface of two materials of dissimilar densities induced by Rayleigh–Taylor instabilities is also of interest. Lastly, in the shock event, fragmentation can be broadened to phenomena outside the mechanical regime. Any change of state (electrical, chemical, polymorphic phase, crystal failure, energetic reaction) involves fragmentation into cellular structures dictated by both the material properties, in addition to the time and spatial constraints of the shock event. Dynamic fragmentation is, and will continue to be, a fundamental aspect of the shock compression process. A number of these issues are addressed in later chapters. In particular, the following two chapters explore a number of wider ranging topics of dynamic fragmentation in physics and engineering that make use of the theoretical tools presented in this chapter.

References

Abramowitz M and Stegun I A 1954 *Handbook of Mathematical Function* No. 55 (*National Bureau of Standards Applied Mathematics Series*) US Government Printing Office

Ang A S and Tang W H 1984 *Probability Concepts in Engineering Planning and Design Volume II: Decision, Risk, and Reliability* (New York: Wiley)

Barenblatt G I 2006 Scaling phenomea in fatique and fracture *Int. J. Fracture* **138** 19–35

Bennett J G 1936 Broken coal *J. Inst. Fuel* **10** 22–39

Bergstrom H C, Sollenberger C L and Mitchel W 1961 Energy aspects of single particle crushing *Trans. AIME* **220** 367–72

Bergstrom B H 1962 Energy and size distribution aspects of single particle crushing, *Proc. 5th Symp. on Rock Mechanics*, ed C Fairhurst (Pergamon) pp 155–172.

Bond F C 1952 Third theory of comminution *AIME Trans.* **193** 484–94

Boots B N and Murdoch D J 1983 The spatial arrangement of random voronoi polygons *Comput. Geosci.* **9** 351–65

Borg J B, Grady D E and Cogar J R 2001 Instability and fragmentation of expanding liquid systems *Int. J. Impact Eng.* **26** 65–76

Boslough M B and Asay J R 1992 Basic principles of shock compression *High-pressure Shock Compression of Solids* ed J R Asay and M Shahinpoor (Berlin: Springer) pp 7–42

Bourne N, Millet J and Pickup I 1997 *J. Appl. Phys.* **81** 6019

Brown W K and Wohletz K H 1995 Derivation of the Weibull distribution based on physical principles and its connection to the Rosin–Rammler and lognormal distributions *J. Appl. Phys.* **78** 2758–63

Charles R J 1957 Energy-size reduction relationships in comminution *AIME Trans.* **208** 80–8

Chen S Y, Dhruva B, Kurien S, Sreenivasan K R and Taylor M A 2005 Anomalous scaling of low-order structure functions of turbulent velocity *J. Fluid Mech.* **533** 183–92

Chen W and Ravichandran G 1997 Dynamic compressive failure of a glass ceramic under lateral confinement *J. Mech. Phys. Solids* **45** 1303–28

Cohen R D 1993 Second law limitations on particle size distribution functions *Proc. R. Soc. Lond.* **440** 611–20

Costin L S and Grady D E 1983 Dynamic fragmentation of ferroelectric ceramics using the torsional Kolsky bar *Sandia National Laboratories Technical Report* SAND83-2004

Costin L S and Grady D E 1984 Dynamic fragmentation of brittle materials using the torsional Kolsky bar, *Proc. Conf. Mechanical Properties at High Rates of Strain (Oxford) Inst. Phys. Conf. Ser.* 70 21-328.

Courant R and Hilber D 1962 *Methods of Mathematical Physics* (New York: Interscience)

Crain I K 1978 The Monte-Carlo generation of random polygons *Comput. Geosci.* **4** 131–41

Curran D R, Seaman L and Shockey D A 1987 Dynamic failure of solids *Phys. Rep.* **147** 253–88

Drugan W J 2001 Dynamic fragmentation of brittle materials: analytical mechanics-based models *J. Mech. Phys. Solids* **49** 1181–208

Elek P and Jaramaz S 2004 Modeling of fragmentation of rapidly expanding cylinders *Theor. Appl. Mech.* **32** 113–30

Elek P and Jaramaz S 2009 Fragment mass distribution of naturally fragmenting warheads *FME Trans.* **37** 129–35

Englman R, Rivier N and Jaeger Z 1987 Fragment size distribution in disintegration by maximum entropy formalism *Phil. Mag.* B **56** 751–9

Epstein B 1947 The mathematical description of certain breakage mechanisms leading to the logarithmico-normal distribution *J. Franklin Inst.* **244** 471–7

Falkovich G and Sreenivasan K R 2006 Lessons from hydrodynamic turbulence *Phys. Today* **59** (4) 43–9

Gaudin A M 1926 An investigation of crushing phenomena *AIME Trans.* **73** 253–316

Gaudin A M and Meloy T P 1962 Model and a comminution distribution equation for single fracture *AIME Trans.* **223** 40–3

Glenn L A and Chudnovsky A 1986 Strain-energy effects on dynamic fragmentation *J. Appl. Phys.* **59** 1879–80

Gilvarry J J 1961 Fracture of brittle solids. I. Distribution function for fragment size in single fracture (theoretical) *J. Appl. Phys.* **32** 391–99

Gilvarry J J and Bergstrom B H 1961 Fracture of brittle solids. II. Distribution function for fragment size in single fracture (experimental) *J. Appl. Phys.* **32** 400–10

Grady D E 1981a Fragmentation of solids under impulsive stress loading *J. Geophys. Res.* **86** 1047–54

Grady D E 1981b Application of survival statistics to the impulsive fragmentation of ductile rings *Shock Waves and High-strain-rate Phenomena* ed M A Meyers and L E Murr (New York: Plenum) pp 181–92

Grady D E 1982a Local inertial effects in dynamic fragmentation *J. Appl. Phys.* **53** 322–5

Grady D E 1982b Analysis of prompt fragmentation in explosively-loaded uranium cylindrical shells *Sandia National Laboratories Technical Report*, SAND82-0140, February

Grady D E 1988 The spall strength of condensed matter *J. Mech. Phys. Solids* **36**(3) 353–84

Grady D E 1990 Particle size statistics in dynamic fragmentation *J. Appl. Phys.* **68**(12) 6099–105

Grady D E 1995 Spall and Fragmentation in high-temperature metals *High-pressure Shock Compression of Solids II: Dynamic Fracture and Fragmentation* pp 219–36

Grady D E 1996 Shock equation of state properties of concrete *Structures under Shock and Impact IV* ed N Jones, C A Brebbia and A J Watson (Southampton: Computational Mechanics) pp 405–14

Grady D E 1998 Scattering as a mechanism for structured shock waves in metals *J. Mech. Phys. Solids* **46** 2017–32

Grady D E 2001 Modeling and analysis tools for fragmenting munitions, *Naval Surface Warfare Center, Dahlgren Division, Applied Research Associates Technical Report* May.

Grady D E 2005 Comparison of hypervelocity fragmentation and spall experiments with Tuler–Butcher spall and fragment size criteria *Int. J. Impact Eng.* **33** 305–15

Grady D E 2006 *Fragmentation of Rings and Shells* (Berlin: Springer)

Grady D E 2008 Fragment size distributions from the dynamic fragmentation of brittle solids *Int. J. Impact Eng.* **35** 1557–62

Grady D E 2009 Dynamic fragmentation of solids *Shock Wave Science Technology Reference Library: Volume 3, Solids II* ed Y Horie (Berlin: Springer) pp 169–276

Grady D E 2010 Length scales and size distributions in dynamic fragmentation *Int. J. Fracture* **163** 85–99

Grady D E and Kipp M E 1979 Oil shale fracture and fragmentation at higher rates of loading *Proc. 20th U.S. Symp. on Rock Mechanics, (Austin TX, 4–6 June)* pp 403–06

Grady D E and Kipp M E 1980a Continuum modeling of explosive fracture in oil shale *Int. J. Rock. Mech. Mining Sci.* **17** 149–57

Grady D E and Lipkin J 1980b Criteria for impulsive rock fracture *Geophys. Res. Lett.* **7** 255–8

Grady D E and Benson D A 1983 Fragmentation of metal rings by electromagnetic loading *Exp. Mech.* **23**(4) 393–400

Grady D E, Kipp M E and Benson D A 1984 Energy and statistical effects in the dynamic fragmentation of metal rings, *Proc. Conf. of the Mechanical Properties of High Rates of Strain, (Oxford), Inst. Phys. Conf. Ser.* 70, 315–20

Grady D E and Kipp M E 1985 Geometric statistics and dynamic fragmentation *J. Appl. Phys.* **58** 1210–22

Grady D E and Kipp M E 1987 Dynamic rock fragmentation *Fracture Mechanics of Rock* (New York: Academic)

Grady D E and Hightower M M 1992 Natural fragmentation of exploding cylinders *Shock-wave and High-strain-rate Phenomena in Materials* ed M E Meyers, L E Murr and K P Staudhammer (Marcel Dekker) pp 713–21

Grady D E and Kipp M E 1993 Dynamic Fracture and Fragmentation *High-Pressure Shock Compression of Solids ed* J R Asay and M Shahinpoor (New York: Springer) pp 265–322

Grady D E, Parsons D A and Winfree N A 2000 Physics-based models and analysis tools for debris formation in meteoroid and orbital debris impact, *Applied Research Associates Report for National Aeronautics and Space Administration* March

Grady D E, Wilson L T, Reedal D R, Kuhns L D, Kipp M E and Black J W 2001 Comparing alternative approaches in the scaling of naturally fragmenting munitions *Proc. 19th Int. Symp. on Ballistics, (Interlaken, Switzerland, 7–11 May)* ed I R Crewther, pp 591–98

Grady D E and Winfree N A 2001 Impact fragmentation of high-velocity compact projectiles on thin plates: a physical and statistical characterization of fragment debris *J. Impact Eng.* **26** 249–62

Grady D E and Olsen M L 2003 A statistical and energy based theory of dynamic fragmentation *Int. J. Impact Eng.* **29** 293–306

Green S J and Perkins R D 1968 Uniaxial compression tests at varying strain rates on three geological materials, *Proc. 10th Symp. on Rock Mechanics*, K E Gray, pp 35–54.

Griffith L 1943 A theory of the size distribution of particles in a comminuted system *Can. J. Res.* A **21**(6) 57–64

Hahn G J and Shapiro S S 1967 *Statistical Models in Engineering* (New York: Wiley)

Held M 1990 Fragment mass distribution of HE projectiles *Propellants, Explosives, Pyrotechnics* **15** 254–60

Jeans J H 1921 *The Dynamic Theory of Gases* 3rd edn (Cambridge: Cambridge University Press)

Johnson W A and Mehl R F 1939 Reaction kinetics in processes of nucleation and growth *Trans. AIMME* **135** 414–58

Kanel G I 1982 The resistance of metals to spall fracture *Combustions, Explosions, Shock Waves* **18** 329–35

Kanel G I *et al* 1994 Spallation near the ultimate strength of solids *Shock Compression of Condensed Matter* ed S C Schmidt *et al* (Amsterdam: Elsevier) pp 1043–6

Kennedy J E 1997 The Gurney model for explosive output for driving metal *Explosive Effects and Applications* ed J A Zukas and W P Walters (Berlin: Springer) pp 221–58

Kerley G I 1992 CTH EOS package: porosity and reactive burn models, Sandia National *Laboratories Rept.* SAND92-0553, April.

Kiang T 1966 Mass distributions of asteroids, stars and galaxies *Z. Astrophys* **64** 426–32

Kiang T 1966 Random fragmentation in two and three dimensions *Z. Astrophys* **64** 433–9

Kick F 1885 *Das Gesetz der proportionalen Widerstand und Seine Anwendung* (Leipzig: Felix)

Kipp M E and Grady D E 1985 Dynamic fracture growth and interaction in one dimension *J. Mech. Phys. Solids* **33**(4) 399–415

Kipp M E and Grady D E 1986 Random flaw nucleation and interaction in one dimension *Metallurgical Applications of Shock-wave and High-strain-rate Phenomena* ed L E Murr, K P Staudhammer and M A Meyers (Marcel Dekker) pp 781–91

Kittel C 1971 *Introduction to Solid State Physics* (New York: Wiley)

Kolmogorov A N 1941 *Doklady Akad. Nauk., SSSR* **30** 9; **31** 99; **32** 16

Lawn B R and Wilshaw T R 1975 *Fracture of Brittle Solids* (London: Cambridge University Press)

Lankford J 1979 Uniaxial compressive damage in α-SiC at low homologous temperatures *J. Am. Cer. Soc.* **62** 310

Lee E H 1967 The continuum mechanics aspect of material properties determination *Energetics III* ed W Mueller and M Shaw (New York: Gordon and Breach) pp 85–122

Levesey J C 1980 Maxwellian distribution derived from *N*-body disintegration theory *Am. J. Phys.* **48** 222–5

Lienau C C 1936 Random fracture of a brittle solid *J. Franklin Inst.*, **221** 485–94, 674–86, 769–87

Lundberg B 1976 A split Hopkinson bar study of energy absorbtion in dynamic rock fragmentation *Int. J. Rock Mech. Min. Sci.* **13** 187–97

Mandelbrot B B 1982 *The Fractal Geometry of Nature* (San Fransisco, CA: Freeman)

Matthews J and Walker R L 1964 *Mathematical Methods of Physics* (New York: Benjamin) pp 228–34

Maxwell J C 1879 *Proc. Camb. Philos. Soc.* **12** 713

Mock W and Holt W H 1983 Fragmentation behavior of Armco iron and HF-1 steel explosive filled cylinders *J. Appl. Phys.* **54** 2344–51

Mott N F and Linfoot E H 1943 A theory of fragmentation, *United Kingdom Ministry of Supply* AC3348, February.

Mott N F 1943 Fragmentation of H E shells: a theoretical formula for the distribution of weights of fragments *United Kingdom Ministry of Supply* AC3642, March

Mott N F 1943 A theory of the fragmentation of shells and bombs *United Kingdom Ministry of Supply* AC4035, May

Mott N F 1944 A theory of fragmentation Application to wire wound bombs such as the American 20 lb. F, *United Kingdom Ministry of Supply*, AC6338, May

Mott N F 1947 Fragmentation of shell cases *Proc. R. Soc.* A **189** 300–8

Mott N F 1948 Brittle fracture in mild steel plates *Engineering* **165** 16

Piekutowski A J 1996 Formation and description of debris clouds produced by hypervelocity impact, *NASA Contractor Report*, 4704, February

Piekutowski A J 1997 Effects of scale on debris cloud formation *Int. J. Impact Eng.* **20** 639–50

Puri P S and Goldie C M 1979 Poisson mixtures and quasi-infinite divisibility of distributions *J. Appl. Prob.* **16** 138–53

Reedal D, Wilson L, Grady D, Chhabildas L and Reinhart W 1999 Impact and explosion induced failure and fragmentation studies on tungsten, *Proc. of the 15th US Army Symposium on Solid Mechanics (Myrtle Beach, SC, 12–14 April)* pp 569–85

Rinehart J S 1965 Dynamic fracture strengths of rocks, *Proc. Seventh Symp. Rock Mech.* (Pensylvania State University)

Romero L A 1989 The instability of rapidly stretching plastic jets *J. Appl. Phys.* **65** 3006–16

Romero L A 1991 The stability of stretching and accelerating plastic sheets I *J. Appl. Phys.* **69** 7474

Rosin P and Rammler E 1933 The laws governing the fineness of powdered coal *J. Inst. Fuel* **7** 29–36

Schuhmann R 1940 Principles of comminution I. Size distribution and surface calculations *AIME Tech. Publ* **1189** 1–11

Sharon E, Gross S P and Fineberg J 1996 Energy dissipation in dynamic fracture *Phys. Rev. Lett.* **76** 2117–120

Shih C J, Nesterenko V F and Meyers M A 2000 High-strain-rate deformation and comminution of silicon carbide *J. Appl. Phys.* **83** 4660–71

Skidmore I C 1965 *Appl. Mater. Res* **4** 131–47

Taylor G I 1963 *Scientific Papers of G I Taylor* vol III No 44 (Cambridge: Cambridge University Press)

Timoshenko S and Goodier J N 1951 *Theory of Elasticity* 2nd edn (New York: McGraw Hill) pp 372–84

Tonks D L, Vorthman J E, Hixson R S, Kelly A and Zurek A K 1999 Spall response of U-Nb (6%) alloy *Shock Compression of Condensed Matter* ed M D Furnish, L C Chhabildas and R S Hixson (New York: American Institute of Physics) pp 329–32

Tuler F R and Butcher B M 1968 A criterion for the time dependence of dynamic fracture *Int. J. Fracture Mech.* **4** 431–7

Turcotte D L 1986 Fractals and fragmentation *J. Geophys. Res.* **91** 1921–6

von Rittinger P R 1867 *Lehrbuch der Aufbereitungskunde* (Berlin: von Ernst & Korn)

Wang H and Ramash K T 1997 Dynamic strength and fragmentation of hot-pressed silicon carbide under uniaxial compression *Acta Met* **52** 355–67

Watson D F 1981 Computation of the *n*-dimensional Delaunay tessellation with application to Voronoi polytypes *Comput. J.* **24** 167–72

Waza T, Matsui T and Kani K 1983 Laboratory simulations of planetesimal collision II. Ejecta velocity distribution, *J. Geophys. Res.*

Weaire D and Rivier N 1984 Soap, cells and statistics—random patterns in two dimensions *Contemp. Phys.* **25**(1) 59–99

Wesenberg D L and Sagartz M J 1977 Dynamic fracture of 6061-T6 aluminum cylinders *J. Appl. Mech.* **44** 643–6

Zhou F, Molinari J F and Ramesh K T 2006 Effects of material properties and strain rate on the fragmentation of brittle materials *Appl. Phys. Lett* **88** 261918

Zurek A K, Hixson R S, Anderson W W, Vorthman J E, Gray G T and Tonks D L 2000 Elastic response of U-Nb (6%) alloy, *Proc. 6th Int. Conf. on Mechanical and Physical Behaviour of Materials under Dynamic Loading, (25–29 September)* EDP Sciences pp 677–82

IOP Publishing

Physics of Shock and Impact
Volume 1: Fundamentals and dynamic failure
Dennis Grady

Chapter 3

Applications in dynamic fragmentation—physics

The concepts of fragmentation developed in the previous chapter are pursued further and applied to a range of applications. Spall fragmentation is intimately related to the shock wave event. Calculations that illustrate the wave interactions leading to spall are explored within the context of several spall failure and fragmentation criteria. A contrasting analysis examines the impact failure and fragmentation of glass plate. Time-dependent failure criteria are commonly applied in the modeling of impact spall fracture and fragmentation. Tuler–Butcher fracture is one such criterion that has found wide application and is representative of this level of modeling.

Instability breakup and fragmentation of rapidly deforming ductile solid and liquid bodies offers an alternative application of dynamic fragmentation. One perspective is provided by the high-velocity impact breakup and fragmentation of polymeric liquids. The aerodynamic breakup of liquid metal droplets in an explosive gas environment is further illustrative.

Several untypical applications are explored that incorporated the principles of dynamic fragmentation. These topics include the breakup into a magnetic domain structure of shock compressed ferromagnetic material, spallation fragmentation in a high-energy nuclear collision and computational molecular dynamics (MD) simulations of fragmentation of a Lennard-Jones material. Formation of the early Universe into the present distribution of galaxies and breakup of the Shoemaker–Levi comet draw on issues of dynamic fragmentation.

3.1 Introduction

The present chapter continues the previous chapter's topic of dynamic fragmentation. Emphasis focuses on applications of the fundamental principles developed previously to a range of diverse topics. Spall is a process in which internal regions of a body are carried into rapid tension by the interaction of stress waves until failure through rupture of the material and the opening of internal voids occurs. Spall is

doi:10.1088/978-0-7503-1254-7ch3

fundamental to the shock physics of solids, and fragmentation is a natural conclusion of the spall process. Spall, and spall fragmentation, is addressed in this chapter on several levels. Several common spall criteria are described. Analysis of wave interactions leading to spall, and the spall fragmentation process, are pursued for several applications. The discussions offer perspective to the extensive spall wave data presented in chapter 1.

Dynamic failure and fragmentation, when free surfaces play a defining role in the wave interaction and fracture process, incur added complexity. The impact failure and fragmentation of brittle plates provides insights into both the ballistics issues and the brittle fragmentation aspects of such events. Issues relating to non-local and time-dependent factors governing the onset and progression of failure are brought to light that complicate computational simulations of the same applications. In a similar vein, complications are found in scaling consequences of dynamic fracture and fragmentation.

Bodies of ductile solids or liquids that are set into motion by impulsive forces that lead to stretching and thinning deformation undergo breakup and fragmentation through deformation instabilities. Examples include the fragmentation of rapidly stretching jets or filaments, and expanding shells or bubbles. Instability fragmentation represents a significant subclass of the broader topic of dynamic fragmentation. Also included within instability fragmentation is the aerodynamic breakup of high-velocity particles of ductile materials and liquids in an impeding atmosphere. Several topics in instability fragmentation are addressed in this chapter, including the impact-induced breakup of a polymeric liquid system, and the aerodynamic fragmentation of liquid metal droplets in an explosive gas environment. The internal instability partitioning of a rapidly deforming body through heterogeneous adiabatic shear-band deformation is a subject of dynamic fragmentation sufficiently extensive that chapter 5 is solely dedicated to the issue.

Dynamic fragmentation in a broader context has application to a number of quite diverse shock physics topics. Selected examples are addressed in the present chapter. Shocks in ferromagnetic materials lead to the demagnetization of the material through forces that realign the magnetic vector. The decomposition into a pattern of misaligned magnetic domains upon passage of the shock wave is a shock fragmentation event. Fragmentation in the shock demagnetization event is representative of a number of physical phase transformations that can occur under shock compression.

The relative high-speed impact of atomic nuclei and the subsequent breakup into smaller nuclear particles is an impact and shock fragmentation event on an entirely different scale. Such events nonetheless exhibit commonalities with other more familiar instances of dynamic fragmentation. Also informative is an example of dynamic fragmentation through application of smoothed-particle hydrodynamic computational simulations to the breakup of Lennard-Jones materials. Several cases of dynamic fragmentation are examined that are of astrophysical interest. The formation of the present distribution of galaxies in the Universe and breakup of the Shoemaker–Levi comet are considered in the context of dynamic fragmentation.

3.2 Fragmentation in the dynamic failure process

The fracture of condensed matter through spall is a process in which interaction of compression waves carries interior regions (away from free surfaces) into tension and fails the material if material strength is exceeded. Excepting the Hugoniot elastic limit, spall is probably the most familiar of failure processes to those in the shock compression community (e.g. Curran *et al* 1987). The subsequent fragmentation characteristics of spall are not commonly considered, however, most likely because the standard planar spall experiment and diagnostics do not prominently display the fragmentation features of the event; and further, it is not usually the topic of interest.

The processes of spall can result in intense fragmentation as the hypervelocity experiments of Piekutowski (1996, 1997) vividly illustrate. One radiograph from the hypervelocity impact experiments of Piekutowski is shown in figure 3.1. High-velocity normal impact of the aluminum sphere onto a thin aluminum plate results in a large-amplitude, short-duration attenuating pressure pulse which emerges at the back free surface of the sphere inducing intense consecutive multiple spall fragmentation. The complex sequence of wave interactions leads to the total disintegration and fragmentation of the projectile.

The simplest spall criterion is the prescription of a constant tensile stress or pressure that, when exceeded, prescribes the onset of fracture failure and internal spall damage of the body. In many applications, a constant spall stress that bounds the tension that a body can support adequately describes the requisite physics. A large body of experimental data, however, reveals that spall failure is a time-dependent phenomenon, and a number of time-dependent spall failure criteria have been proposed. A time-dependent spall criterion that has found wide application to both spall, and other related failure phenomena, is that due to

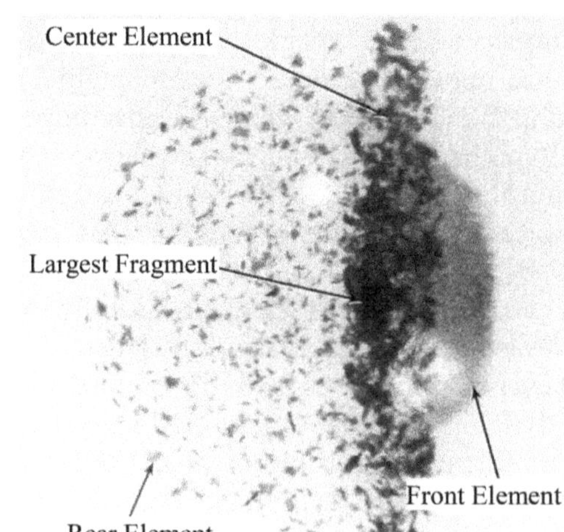

Figure 3.1. Normal impact of 12.7 mm diameter aluminum sphere on 0.59 mm thickness aluminum plate at 6.26 km^{-1}. Radiographic image at 46 μs after impact (Piekutowski 1996, 1997). Courtesy of NASA.

Tuler and Butcher (1968). The Tuler–Butcher spall criterion is applied in much of the analysis presented here.

In the present section, several analytic solutions of fragmentation in the spall process based on various spall failure criteria are examined to illustrate further issues of dynamic fragmentation and its relation to shock wave propagation in solids. These analytic examples provide insights into the fragmentation processes occurring in the Piekutowski experiments and the underlying intense spallation. The study also brings to light features of stress wave propagation and interaction key to the outcome of the spallation event. Related, but not strictly a spall event, is the impact failure and fragmentation of thin plates. Here, free surfaces play a primary role in the dynamic fracture and fragmentation process. The same type of time-dependent failure criterion, such as the Tuler–Butcher criterion, nonetheless sensibly captures observed failure behavior.

Correlation dynamics and the self-organization of the fracture fabric are central to the dynamic failure and fragmentation event. Issues underlying this dynamics are brought to light through exploration of the Mott crack (Mott 1948). The Mott crack is a physics-based analysis of an individual failure site that lends insights into the interaction dynamics establishing the failure fragmentation pattern.

3.2.1 Spall fragmentation from a pulse emerging at a free surface

Spall fragmentation is explored through investigation of the consequences of a saw-toothed compressive wave propagating toward the left and incident on a free surface as illustrated in figure 3.2. This wave adequately approximates, for example, the impact-induced shock waves emerging at the back surface of the Piekutowski sphere. The material will be linear elastic (or linear hydrodynamic) characterized by a density ρ and a wave speed c. If time $t = 0$ corresponds to emergence and reflection of the wave at the free surface, then the compressive stress profile at any time $t < 0$ is

$$\sigma(x,t) = P - \rho c^2 \dot{\varepsilon}(t + x/c)/2, \tag{3.1}$$

where P is the peak compressive stress and $\dot{\varepsilon}$ is the decompression strain rate (or the velocity gradient) behind the wave peak. The peak stress P is considered here to be

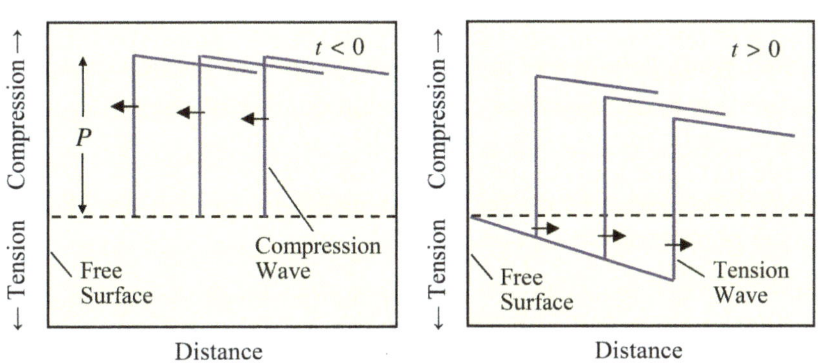

Figure 3.2. The compression wave (left) incident on the free surface and the reflected tension wave (right).

many times larger than the magnitude of the material tensile strength. (There will be occasional switches in sign convention for compression and tension in this section to forgo the awkward use of minus signs in the relations. The convention in use is evident from the text.)

The incident stress wave before reflection and the stress profile at some time after reflection are illustrated in figure 3.2. Ahead of the incident shock, material is initially at rest and stress free. Reflection of the wave results in a right-facing triangular wave as shown in the right-hand panel in figure 3.2. The stress field to the left of the leading reflected characteristic is a non-simple wave solution of the d'Alembert form (e.g. Courant and Hilbert 1962),

$$\sigma(x,t) = f(t - x/c) + g(t + x/c). \tag{3.2}$$

The stress-free boundary condition leads to the solution for the reflected tensile stress for time $t > 0$ as a function of distance x from the free surface and behind the right-facing decompression shock wave,

$$\sigma(x,t) = \rho c\dot{\varepsilon}x, \quad x - ct < 0. \tag{3.3}$$

The corresponding particle velocity behind the right-facing shock wave is

$$u(x,t) = (\rho c^2\dot{\varepsilon}t - 2P)/\rho c. \tag{3.4}$$

The parameter $\dot{\varepsilon}$ provides the rate at which the body is carried into tension and spall.

3.2.1.1 Spall and fragment size criteria

A criterion for spall is that of instantaneous fracture when a constant tensile stress of magnitude σ_s is achieved. Spall criteria, discussed in chapter 2, have also been proposed based on impulse or on work. A constant stress criterion has been by far the more common. For constant spall stress the distance $x = x_s$ to the spall plane is an estimate of the fragment size, and is calculated from equation (3.3) to be

$$x_s = \frac{\sigma_s}{\rho c\dot{\varepsilon}}. \tag{3.5}$$

A subsequent triangular tension wave originates at the spall plane free surface and propagates until the spall criterion is again achieved. A second spall fragment with size provided by equation (3.5) is again produced. The process repeats until the incident wave is exhausted, creating a fragment number of order P/σ_s.

A spall impulse failure criterion is derived by calculating the momentum imparted to the spalled plate and leads to the prediction of a fragment size and corresponding spall stress that is dependent on the critical spall impulse I_s, rather than a critical spall stress. The momentum expression

$$I(x_s) = \int_0^{x_s} \rho u d\xi = \int_0^{x_s} \rho\dot{\varepsilon}\xi d\xi = I_s, \tag{3.6}$$

when solved provides for the fragment size

$$x_s = (2I_s/\rho\dot{\varepsilon})^{1/2}, \tag{3.7}$$

and the spall stress at the spall plain

$$\sigma_s = (2\rho c^2 I_s \dot{\varepsilon})^{1/2}. \tag{3.8}$$

A strain-rate dependence of the spall stress based on a momentum criterion has been suggested by Skidmore (1965).

Alternatively, a spall energy criterion based on a critical work W_s follows from calculation of the elastic and kinetic energy imparted to the spall plate,

$$W(x_s) = \frac{1}{2}\int_0^{x_s} (\rho u^2 + \sigma^2/\rho c)d\xi = \frac{1}{2}\int_0^{x_s} 2\rho\dot{\varepsilon}^2\xi^2 d\xi = W_s. \tag{3.9}$$

For a critical work criterion, calculations yield for the fragment size

$$x_s = (3W_s/\rho\dot{\varepsilon}^2)^{1/3}, \tag{3.10}$$

and the spall stress

$$\sigma_s = (3\rho^2 c^3 W_s \dot{\varepsilon})^{1/3}. \tag{3.11}$$

Functional dependence on the energy and strain rate compares with the energy-horizon model described in chapter 2.

3.2.1.2 Tuler–Butcher spall

A spall stress that depends on details of the transient pressure load leading to spall has been observed in a number of materials (e.g. Kanel 1982, Grady and Kipp 1987). Various history-dependent spall criteria have been proposed. The relation of Tuler and Butcher (1968) has received wide application. Ignoring a threshold stress term the Tuler–Butcher fracture criterion can be written

$$I_n = \frac{1}{\rho c}\int \sigma^n(t)dt \leqslant K_n. \tag{3.12}$$

For the tensile stress field in equation (3.3) calculated for the saw-tooth wave incident on a free surface, the Tuler–Butcher integral can be explicitly solved for

$$I_n(x,t) = \frac{1}{\rho c}\int_{x/c}^{t} (\rho c\dot{\varepsilon}x)^n dt = \frac{1}{\rho c}(\rho c\dot{\varepsilon}x)^n(t - x/c). \tag{3.13}$$

The integral is calculated from the time of arrival $t = x/c$ of the tension wave at position x to any later time t. The Tuler–Butcher integral is illustrated in figure 3.3. Both ordinate and abscissa are normalized to provide amplitude of unity and a propagation distance of unity at spall criticality. On the left, the shape of the Tuler–Butcher integral is illustrated for the special cases of n = 1, 2 and 5. This plot

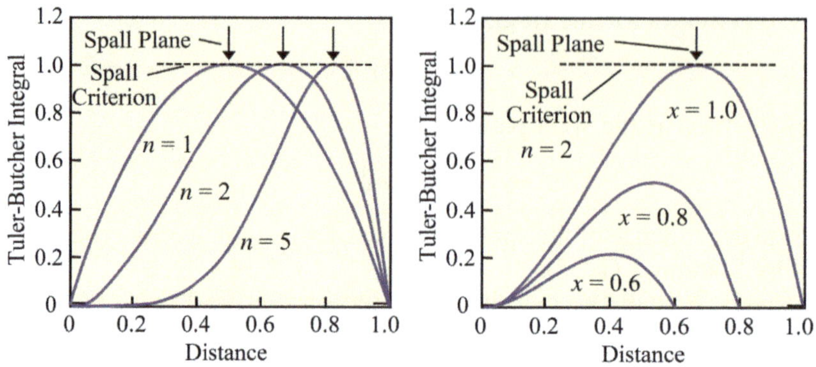

Figure 3.3. Tuler–Butcher integral for saw-tooth spall wave. The left plot is for selected values of the power n as the spall criterion is achieved. The right plot shows growth of the integral with propagation distance for the case of $n = 2$.

identifies the distance behind the tensile shock wave that the Tuler–Butcher integral grows the fastest. The position of the spall plane is identified for each of the cases, and is seen to approach the position of the front of the tensile stress wave as n becomes large. In the right-hand panel of figure 3.3, the propagation growth of the Tuler–Butcher integral is shown for the case of $n = 2$.

At any time t the Tuler–Butcher integral is a maximum at some interior point in the interval $0 < x < ct$. The Tuler–Butcher integral at this maximum point can be equated to the critical constant K_n to establish the distance to the fracture plane, and hence provide an estimate of the fragment size,

$$x_s = \left(\frac{n\rho c^2 K_n}{(\rho c\dot{\varepsilon})^n}\right)^{1/(n+1)}. \tag{3.14}$$

The tensile fracture stress, or spall stress, achieved at the fracture plane is

$$\sigma_s = \left(n\rho^2 c^3 K_n \dot{\varepsilon}\right)^{1/(n+1)}. \tag{3.15}$$

The Tuler–Butcher criterion for $n = 1$ retrieves the impulse criterion of equations (3.7) and (3.8) with $I_s = \rho c K_1/2$ the critical impulse. A value of $n = 2$ recovers the work criterion of equations (3.10) and (3.11) with $W_s = 2K_2/3$. As n becomes large the Tuler–Butcher spall criterion approaches a constant fracture stress and the fragment size dependence of equation (3.5).

Over a reasonable range of the Tuler–Butcher parameter n, the strain-rate power for the fragment size ranges from 1/2 for an impulse criterion, to 2/3 for a work criterion, to a value of 1 for a constant spall stress. In figure 3.4 the strain-rate power for fragment size determined for the aluminum sphere spall fragment size data of Piekutowski (1997) is shown to be in the neighborhood of 0.60–0.62. For the impact velocity range of the data, the assumption is made that strain rate scales with, and is reasonably approximated by, the ratio of the impact velocity to the sphere diameter as is plotted in the figure (Grady 2006). Comparison with the Tuler–Butcher criteria

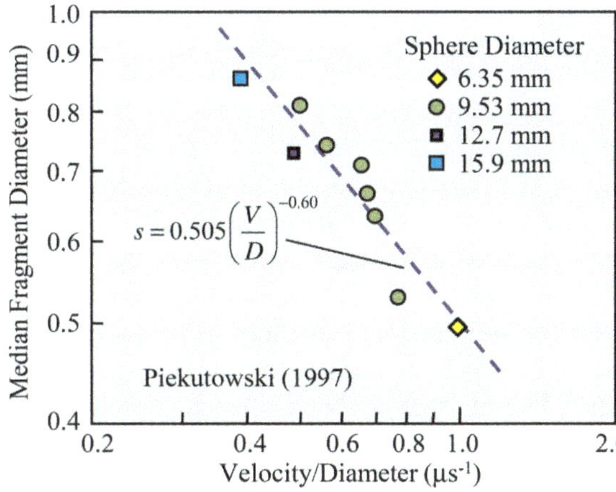

Figure 3.4. Spall fragment size data for aluminum sphere impact experiments of Piekutowski (1997) including both velocity and system scale variations.

suggests the data fall between a strain-rate power of 1/2 impulse criterion and a strain-rate power of 2/3 work criterion, perhaps closer to the latter.

The Tuler–Butcher criteria also predict a strain-rate dependence of the spall stress, and it is of interest to provide a comparison with available experimental data. The ultra-high strain-rate spall strength data for aluminum obtained by Kanel *et al* (1994) are plotted in figure 3.5 along with their companion data for molybdenum metal. The strain-rate power for aluminum and molybdenum are in the range of 0.23–0.27, somewhat below one-third for the work-based criterion and well below one-half for the impulse criterion. The data of Kanel *et al* (1994) are unique, however, and it is difficult to ascribe uncertainties to either the spall amplitude or the strain-rate dependence. Comparable spall-stress data at more modest strain rates are examined in chapter 1.

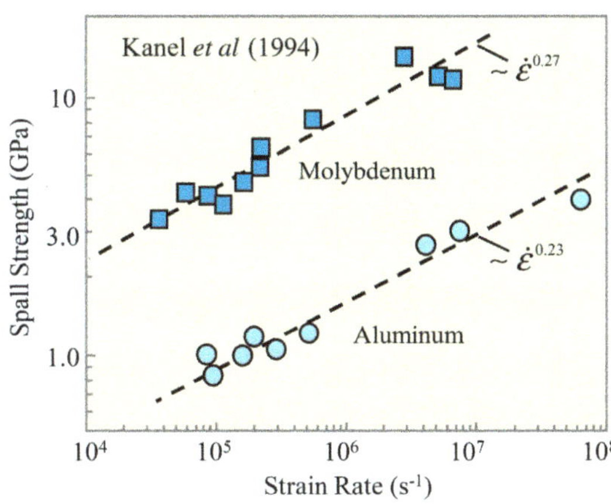

Figure 3.5. Ultra-high strain-rate spall strength data for aluminum and molybdenum from Kanel *et al* (1994).

3.2.2 Cohesive zone spall

Further analysis explores time-dependent features of tensile strength and fragment size in the spall fracture and fragmentation process. The preceding discussions reveal two critical features governing the time dependence in the dynamic tensile fracture of a solid. First, work must be performed to overcome the resistance to fracture of an opening crack and this energy must come from the local elastic or kinetic energy in the neighborhood of the crack. Second, momentum must be transported into the region of the crack in order to accommodate displacement of the opening crack. The properties influencing this process can be examined through the one-dimensional spall of a solid in which the material response is linear elastic for all regions of the body not specifically at the plane of fracture. The physics of crack opening is modeled separately.

Again, consider a linear elastic body initially subjected to an impact process leading to a one-dimensional compressive stress of magnitude P. The body is then carried into tension on the plane $x = 0$ by the symmetric interaction of two opposing release waves. The subsequent magnitude of tension is equal to P if the spall strength of the material is not exceeded. Select the coordinate system such that the center of the spall plane is at rest and remains at rest throughout the spall process. A tensile shock wave, illustrated in figure 3.6, propagates on the forward-facing characteristic $x = ct(x \geqslant 0)$. If the amplitude P exceeds the spall strength of the material $\sigma_s = \sigma_s(P)$, then an elastic shock wave with amplitude of the spall stress σ_s propagates outward from the plane of spall. A possible dependence of the spall stress on the initial shock amplitude P is assumed.

Within the x versus t domain enclosed by the spall plane $x = 0$, and the tensile shock wave $x = ct$, the d'Alembert solution again applies. The flow field on left-facing characteristics is constant. Consequently, stress and particle velocity, after properly matching jump conditions on the shock, have the simple wave solution

$$\sigma(x,t) = f(t - x/c), \tag{3.16}$$

$$u(x,t) = (P - f(t - x/c))/\rho c. \tag{3.17}$$

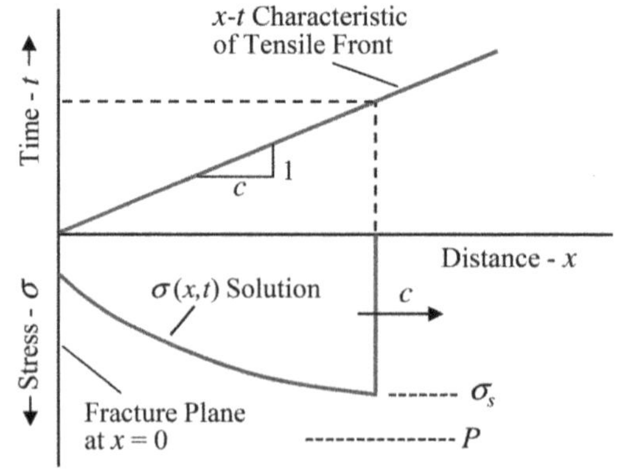

Figure 3.6. Distance versus time (upper plot) and tensile spall-stress profile (lower plot) emerging as time-dependent spall fracture proceeds on the $x = 0$ plane.

The solution is completed by applying the stress boundary condition at the spall plane $x = 0$. Consider a model for the cohesive zone spall crack-opening resistance of the form

$$f(t) = \sigma(0,t) = \sigma_s\left(1 - y(t)/y_c\right). \tag{3.18}$$

The tensile stress in the spall zone decreases linearly from the spall stress $\sigma_s(P)$ to zero over the crack opening displacement $0 \leqslant y(t) \leqslant y_c$. The distance y_c is the critical crack-opening displacement. Displacement is obtained from the particle velocity at $x = 0$,

$$y(t) = \int_0^t u(0,s)ds, \tag{3.19}$$

and leads to the differential solution for $f(t)$,

$$\frac{d}{dt}f(t) = -\frac{\sigma_s}{\rho c y_c}(P - f(t)). \tag{3.20}$$

Solutions for the stress and particle velocity field are

$$\sigma(x,t) = P - (P - \sigma_s)e^{\frac{\sigma_s}{\rho c^2 y_c}(ct-x)}, \tag{3.21}$$

$$u(x,t) = \frac{1}{\rho c}(P - \sigma_s)e^{\frac{\sigma_s}{\rho c^2 y_c}(ct-x)}. \tag{3.22}$$

In the present cohesive zone model for the spall resistance, the spall fracture energy is $\gamma_c = \sigma_s y_c/2$ and spall toughness can be defined through the expression $K_c^2 = 2\rho c^2 \gamma_c = \rho c^2 \sigma_s y_c$.

Spall strength is commonly found to depend on the shock strength P. (The amplitude of the elastic tension has magnitude P if spall does not occur). Dependence of the spall stress on the shock strength is included by assuming the form

$$\sigma_s(P) = \sigma_{so}\left(\frac{P}{\sigma_{so}}\right)^{m-1}, \quad 1 \leqslant m < 2, \tag{3.23}$$

where σ_{so} is a threshold spall stress (spall does not occur if $P < \sigma_{so} P < \sigma_{so}$). The spall stress is a constant σ_{so} for $m = 1$ and approaches P as $m \to 2$.

Let $R = P/\sigma_{so}$ be the ratio of the shock strength to the spall threshold stress and let $u_o = \sigma_{so}/\rho c$. Equation (3.21) provides the time-dependent stress at the spall plane

$$\sigma(t)/\sigma_{so} = R\left[1 - (1 - R^{m-2})e^{R^{2(m-1)}(ct/a_o)}\right], \tag{3.24}$$

where the length scale $a_o = (K_c/\sigma_{so})^2$. The stress and velocity history at any position x is provided by equations (3.21) and (3.22).

The time duration over which the resisting stress in the spall plane reduces to zero is calculated from equation (3.24),

$$-a_o R^{2(1-m)} \ln(1 - R^{m-2}) = ct_s = x_s/2. \tag{3.25}$$

The distance x_s is the minimum uninterrupted wave propagation distance required in order to transport the energy and momentum necessary to fracture completion at the spall plane. It is, for example, the minimum spacing allowing equally spaced fractures to achieve completion in a body uniformly loaded to a tension P.

Equation (3.25) then provides a reasonable measure of the nominal fragment size as a function of the tensile amplitude P as shown in figure 3.7. Fragment size is plotted as a function of $R = P/\sigma_{so}$ for both $m = 1$ and $m \to 2$.

Particle velocity histories at the spall plane are illustrated in figure 3.8 and approximate the pullback velocity signals measured in spall experiments. Again, for a value of $m = 1$ the spall strength measured is constant. Only the pulse duration reduces with increasing amplitude of the applied tensile stress. In contrast for larger values of m replicating spall strength dependence on the applied stress amplitude P through equation (3.23) (profiles for $m = 1.5$ are shown in the figure), both spall-stress amplitude and duration depend on the magnitude of $R = P/\sigma_{so}$.

For sensibly large R equation (3.25) reduces to the simpler form

$$2ct_s = x_s = 2a_o/R^m. \tag{3.26}$$

For the special case of $m = 1$ equation (3.26) provides

$$x_s = \frac{2\rho c^2}{P} y_c, \tag{3.27}$$

yielding a fragment size that depends only on the spall plane crack-opening displacement y_c, independent of the fracture energy, and with an inverse first-power dependence on the shock strength P. For $m \to 2$ the expression for fragment size becomes

$$x_s = \frac{4\rho c^2}{P^2} \gamma_c. \tag{3.28}$$

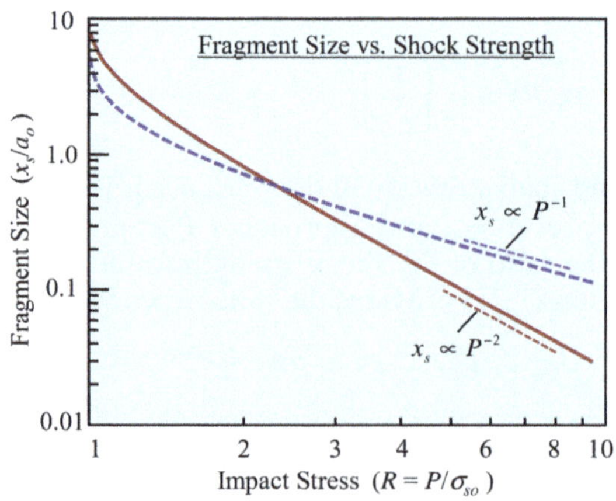

Figure 3.7. Minimum spall or fracture plane spacing as a function of the tensile stress amplitude.

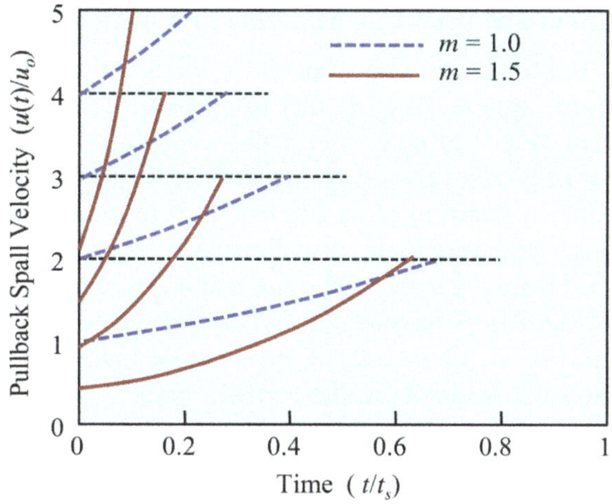

Figure 3.8. Calculated spall velocity profiles for tensile amplitudes of $P/\sigma_{so} = 2$, 3, 4 and 5.

The fragment size depends only on the fracture energy γ_c, and decreases as the second power of the shock strength P. A value of $m = 2$ in equation (3.23) leads to solution difficulties. This issue is troublesome and perhaps other functional forms for equation (3.23) would be more appropriate.

Lastly, the Tuler–Butcher criteria for the same two cases are

$$I_1 = \frac{1}{\rho c} P t_s = 2\gamma_c, \tag{3.29}$$

$$I_2 = \frac{1}{\rho c} P^2 t_s = 4\gamma_c. \tag{3.30}$$

The present cohesive spall zone model offers some insight into the several common forms of the Tuler–Butcher fracture criterion. For $m = 1$ the Tuler–Butcher criterion is an impulse criterion requiring a critical crack-opening displacement or, in one dimension, a fracture activation volume. When sufficiently above the threshold stress σ_{so}, the fracture criterion is independent of the threshold stress and the fracture energy, requiring only a sufficient transport of momentum into the spall zone to accommodate the requisite activation volume (crack-opening displacement). In contrast, for $m \to 2$ the Tuler–Butcher criterion approaches a critical work criterion requiring a critical fracture energy, or fracture activation energy, to complete the spall process. For a critical work criterion, both sufficient energy and momentum must be transported to the spall zone to accomplish fracture.

Such interpretations of the Tuler–Butcher integral are of interest because of its relatively wide use in modeling a range of critical-state activation processes under shock and high-rate applications. A form of the Tuler–Butcher integral is used, for example, in the HVRB reaction kinetics model of Kerley (1992).

3.2.3 Shock attenuation and spall fragmentation in a dissipative medium

A further analysis is explored. The analysis applies to the impact spall of the Piekutowski aluminum sphere. It also has applications to the attenuating shock and back surface spall of a ceramic or metal slab subject to a transient pressure load from an explosion or impact on the opposing surface, or to the spall of a metal plate subjected to the impulsive loading of an intense laser or electron beam. Here, shock wave load and spall in a concrete plate or wall provides the application. The situation is such that the lateral dimensions of the plate are large relative to the through-the-thickness dimensions and the analysis can be specialized to one-dimensional shock wave propagation and spall. This analysis encompasses issues of compressive shock propagation in dissipative media as well as tensile spall.

3.2.3.1 Momentum and energy coupling into a concrete barrier

The coupling of momentum and energy into a slab of material caused by the planar impact of a high-velocity plate or the detonation of an in-contact plane of explosive is readily estimated by analytic methods. The first is efficiently calculated through the appropriate application of pressure versus velocity diagrams (Boslough and Asay 1992), while the latter can be estimated with Gurney methods (Kennedy 1997). The pressure loading at the input side of the slab at $x = 0$ for time $t > 0$ is approximated with the exponential form

$$p(t) = p_o e^{-t/\tau_o}, \tag{3.31}$$

where the pulse amplitude and duration are captured by the two parameters p_o and τ_o. The corresponding particle velocity at the input interface is provided through the relation $p(t) = zu(t)$ where z is an appropriate constant shock impedance for the concrete at the prescribed input pressure. A constant relation between stress and particle velocity becomes increasingly approximate at high shock pressures. The total impulse provided by the load is

$$I_o = \int_0^\infty p_o e^{-t/\tau_o} dt = p_o \tau_o, \tag{3.32}$$

while the work applied and initial energy in the shock pulse is

$$E_o = \int_0^\infty pudt = \frac{1}{2z} p_o^2 \tau_o. \tag{3.33}$$

Let $E_x(x)$, $p_x(x)$ and $\tau_x(x)$ be the energy, pressure amplitude and pulse duration of the exponential shock pulse after propagating a distance x into the concrete barrier from the pressure loading surface, and such that $E_x(0) = E_o$, $p_x(0) = p_o$ and $\tau_x(0) = \tau_o$. The energy integral at position x corresponding to equation (3.33) yields

$$E_x = \frac{1}{2z} p_x^2 \tau_x. \tag{3.34}$$

Since momentum in the shock pulse is conserved ($I_o = p_o \tau_o = p_x \tau_x$), it follows that

$$E_x = \frac{I_o}{2z} p_x . \tag{3.35}$$

Attenuation of the shock energy and the peak pulse pressure with propagation distance are proportional and

$$\frac{dE_x}{dx} = \frac{I_o}{2z}\frac{dp_x}{dx} . \tag{3.36}$$

3.2.3.2 Dissipation in propagation of the shock pulse

Two forms of energy dissipation are accounted for in the analysis as the shock pulse propagates through the concrete slab. First, the material is assumed to exhibit a constant von Mises shear strength of Y. The shock compression and subsequent decompression will exhibit the stress versus strain response shown on the left side of figure 3.9. The energy dissipation is the enclosed area as shown and is written

$$Q_1 = \frac{4}{3} Y \varepsilon_h = \frac{4}{3}\frac{\rho_o Y}{z^2} p_x . \tag{3.37}$$

The second equality relates the strength dissipation in the concrete to the peak shock pressure of the pulse using the relation $p_x = (z^2/\rho_o)\varepsilon_h$.

The second source of dissipation is illustrated on the right of figure 3.9. The underlying physical causes are several, including rate-dependent plasticity and stress wave viscosity among others (Grady 1998). The dissipation energy is ultimately thermalized. This shock dissipation component can be written approximately as

$$Q_2 = \frac{1}{3}\rho_o C_o^2 S \varepsilon_h^3 = \frac{1}{3}\frac{\rho_o^2 S}{z^4} p_x^3 , \tag{3.38}$$

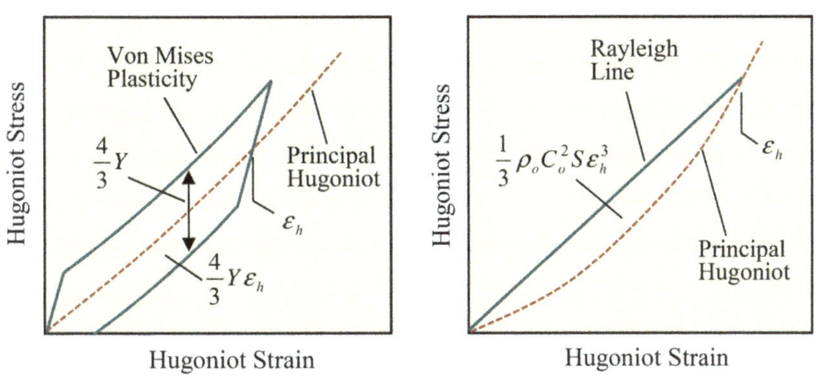

Figure 3.9. Illustrates mechanisms of dissipation through both von Mises strength and shock dissipation in the shock propagation process.

where C_o and S are the linear shock-velocity versus particle-velocity Hugoniot parameters for the material. The shock dissipation model becomes

$$\frac{dE_x}{dx} = -(Q_1 + Q_2) = -\frac{4}{3}\frac{\rho_o Y}{z^2}p_x - \frac{1}{3}\frac{\rho_o^2 S}{z^4}p_x^3. \tag{3.39}$$

Combining the previous equations provides a relation for attenuation of the peak shock pressure with propagation distance

$$I_o\frac{dp_x}{dx} = -\frac{8}{3}\frac{\rho_o Y}{z}p_x - \frac{2}{3}\frac{\rho_o^2 S}{z^3}p_x^3, \tag{3.40}$$

or

$$\frac{d\pi_x}{dx} = -\alpha_1\pi_x - \alpha_2{\pi_x}^3, \tag{3.41}$$

where $\pi_x = p_x/p_o$ and the coefficients α_1 and α_2 are identified from the previous equation. Equation (3.41) is solved for the attenuation of the peak pressure as a function of propagation distance into the concrete barrier. The solution can be performed separately for shock dissipation, which results in power-law attenuation, or for strength dissipation, leading to exponential attenuation. Generally, both dissipation mechanisms participate in the attenuation process. The calculated shock dissipation and attenuation are carried out for a reasonably well characterized concrete (Grady 1993). Hugoniot properties for the concrete are shown in figure 3.10.

The necessary strength and equation-of-state properties are the density, $\rho_o = 2300\ \mathrm{kgm^{-3}}$, the von Mises strength, $Y = 250\mathrm{MPa}$, shock-velocity intercept, $C_o = 2600\ \mathrm{ms^{-1}}$ and shock-velocity slope, $S = 1.48$. This concrete also exhibits some irreversible (pore) compaction over a portion of the Hugoniot curve that is not

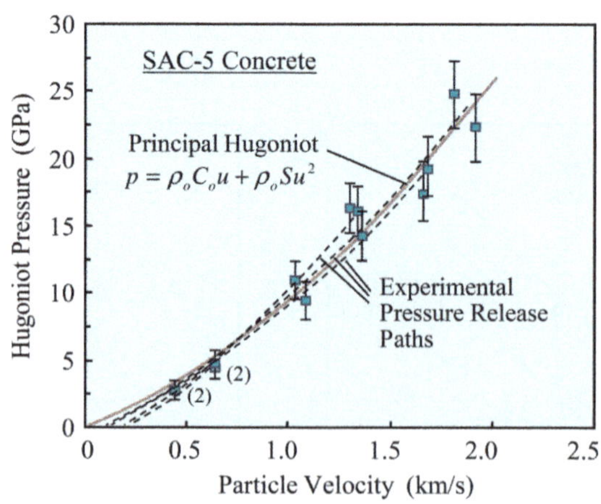

Figure 3.10. Shock Hugoniot data for SAC-5 concrete (Grady 1993) with a nominal initial density of 2300 kg m^{-3}. Two data points are indicated by (2). Solid curves provide the best quadratic fit to the Hugoniot data. Dashed curves are experimental stress unloading paths from the shock states.

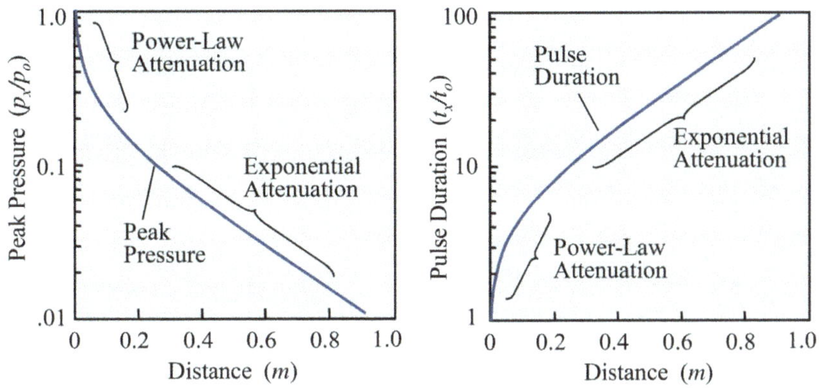

Figure 3.11. Attenuation of the peak pressure and growth of the pressure pulse duration in concrete due to both shock dissipation (power-law attenuation) and strength dissipation (exponential attenuation).

accounted for here. An additional dissipative term, say Q_3, could be determined from appropriate compaction modeling, however, and included in the dissipation and attenuation relations developed. Momentum conservation in the stress pulse becomes a less acceptable assumption, however, when shock wave crush becomes significant. A one-centimeter thickness TNT explosive detonated in contact with the concrete slab is assumed. Gurney equations provided an initial peak pressure and pulse duration of $p_o = 28.1$ GPa and $\tau_o = 1.46\ \mu$s.

Plots of the peak pressure attenuation and growth of the pressure pulse duration as a function of propagation distance into the concrete barrier are shown in figure 3.11. Calculations are based on the concrete and explosive properties provided. The calculated attenuation coefficients are $\alpha_1 = 3.34\ \text{m}^{-1}$ and $\alpha_2 = 71.8\ \text{m}^{-1}$, respectively. Shock dissipation and power-law attenuation dominates the early time decay of the pressure pulse within the first 15–20 cm of concrete. Strength dissipation and exponential attenuation govern later time decay. Peak pressure attenuates from 28 GPa to less than 0.2 GPa over approximately one meter of propagation in the concrete.

Pressure profiles within the two attenuation ranges are illustrated in figure 3.12. Noteworthy is the marked spreading of the pressure pulse necessary to conserving momentum as propagation and attenuation proceeds. Profiles on the left within the first tenth of a meter propagation distance are dominated by shock dissipation and power-law attenuation. Later time attenuation is controlled by dissipation due to the shear strength of the concrete.

3.2.3.3 Solution for the reflected tension wave

On emergence of the compressive pressure pulse at the back free surface of the concrete, subsequent wave reflection will carry the material into tension. When the tensile spall strength is exceeded, spall fracture will occur and spall debris will be ejected at velocities commensurate with the intensity of the incident pressure wave. Frequently more than one spall plane occurs, broadening the spread in the size of fragments produced and the velocity of the fragment ejecta. Multiple spall planes are

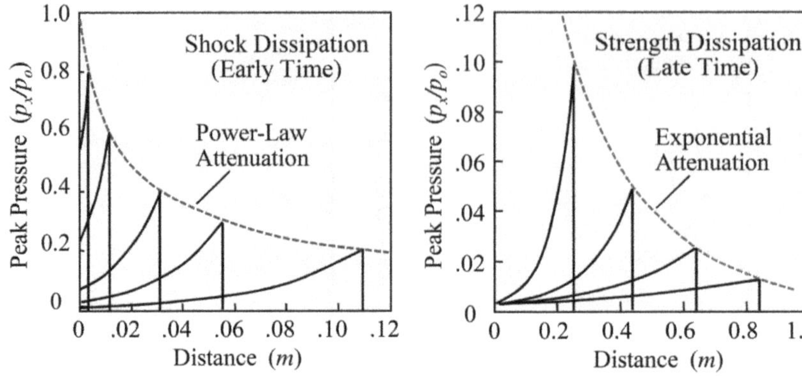

Figure 3.12. Attenuation and propagation of the shock pressure pulse in concrete. The initial values for the pressure pulse amplitude and duration are 28.1 GPa and 1.46 μs.

intrinsic to breakup of the Piekutowski sphere. The pressure wave approaching the back free surface of the concrete has the functional dependence

$$p(x,t) = p_x(x)e^{-(t-x/C_o)/\tau_x(x)}, \tag{3.42}$$

within the interval $x \leqslant C_o t \leqslant L$. On emergence at the back free surface following propagation through a thickness of concrete L, the pressure pulse has attenuated to

$$p(x,L/C_o) = p_x(L)e^{-(L-x)/C_o\tau_x(L)}. \tag{3.43}$$

Let $p_x(L) = p_L$ and $\tau_x(L) = \tau_L$, for times $t > L/C_o$ write

$$p(x,t) = p_L e^{-(C_o t-x)/C_o\tau_L}. \tag{3.44}$$

Subsequent wave propagation is reasonably approximated as linearly elastic with an initial condition at $t = L/C_o$ provided by equation (3.43).

In solving for the reflected tension wave it is convenient to introduce new distance and time variables $\bar{x} = L - x$ and $\bar{t} = t - L/C_o$, as illustrated in figure 3.13, providing from equation (3.44),

$$p(\bar{x},\bar{t}) = p_L e^{-(C_o\bar{t}+\bar{x})/C_o\tau_L}. \tag{3.45}$$

The D'Alembert solution to the elastic wave equation in domain III, behind the reflected tensile shock, for the pressure and particle velocity is

$$p(\bar{x},\bar{t}) = f(C_o\bar{t} + \bar{x}) + g(C_o\bar{t} - \bar{x}), \tag{3.46}$$

$$\rho C_o u(\bar{x},\bar{t}) = f(C_o\bar{t} + \bar{x}) - g(C_o\bar{t} - \bar{x}). \tag{3.47}$$

The pressure and particle velocity are continuous on characteristics $C_o\bar{t} + \bar{x} =$ constant, across the reflected tensile shock. Combining with the pressure free

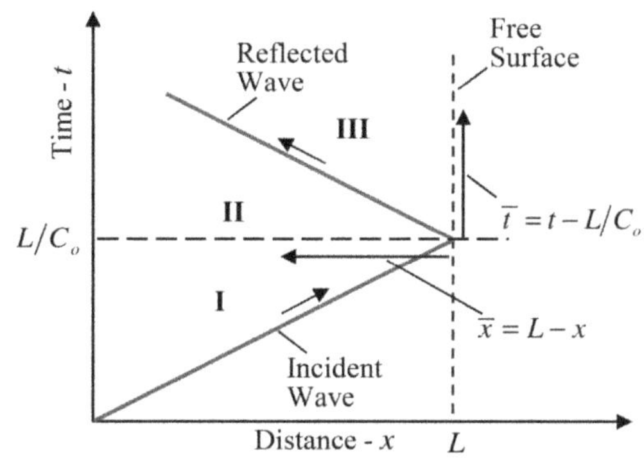

Figure 3.13. Solution domains for the incident pressure pulse and reflected tension wave in establishing wave interactions commensurate with spall failure.

boundary condition on $\bar{x} = 0$, equations (3.46) and (3.47) yield the solutions for tensile pressure and particle velocity in the reflected wave,

$$p(\bar{x},\bar{t}) = -2p_L e^{-\bar{t}/\tau_L} \sinh(\bar{x}/C_o\tau_L), \tag{3.48}$$

$$u(\bar{x},\bar{t}) = \frac{2p_L}{\rho C_o} e^{-\bar{t}/\tau_L} \cosh(\bar{x}/C_o\tau_L). \tag{3.49}$$

3.2.3.4 Spall criterion and fragment size

The reflected wave solution can be joined with various spall criteria to determine where spall planes are located as tension and failure occurs in the pulse loading event. A common criterion is that of abrupt spall fracture on planes where the tensile stress achieves a critical spall stress p_s. From equation (3.48) peak tension occurs on the characteristic $\bar{x} = C_o\bar{t}$ and the tension amplitude (a positive quantity) on this characteristic is

$$p(\bar{x}) = p_L[1 - e^{-2\bar{x}/C_o\tau_L}]. \tag{3.50}$$

Tension increases continuously with propagation distance $\bar{x}$ and spall occurs at $p(\bar{x}) = p_s$. Tensile stress drops to zero at this plane and the tension again grows with continued wave propagation into the material until p_s is again achieved at a second spall plane. The process continues until the tensile wave is exhausted or until the front surface of the wall is reached. Depths of the spall planes are readily determined from the relation

$$p(\bar{x}_n) = p_L[1 - e^{-2\bar{x}_n/C_o\tau_L}] = np_s, \tag{3.51}$$

or

$$\bar{x}_n = -\frac{1}{2}C_o\tau_L \ln\left(1 - n\frac{p_s}{p_L}\right). \tag{3.52}$$

Solving for n,

$$n(\bar{x}_n) = \frac{p_L}{p_s}[1 - e^{-2\bar{x}_n/C_o\tau_L}], \tag{3.53}$$

and spall occurs on planes at depths $\bar{x}_n$ that yield integer values of $n(\bar{x}_n)$. The total number of spall planes is given by

$$n_{tot} = \frac{p_L - \mathrm{mod}\,(p_L, p_s)}{p_s}. \tag{3.54}$$

3.2.3.5 Strain-rate-dependent fragment size

When a body subjected to stress wave loading is carried into tension and experiences spall fracture, a relation of the strain-rate-dependent form

$$s = \alpha\dot{\varepsilon}^{-m}, \tag{3.55}$$

is found to be appropriate for estimating the characteristic size of fragments or, for the present one-dimensional spall application, the characteristic spacing s of fractures. A relation of this form is discussed, for example, in the previous sections. The rate of strain as the body at a spall plane is carried into tension is $\dot{\varepsilon}$. The power parameter m is typically within the range $1/2 \leqslant m \leqslant 1$. In the present application, a calculation of the appropriate strain rate is determined from the elastic wave solution. The elastic pressure wave incident on the free surface is provided by equation (3.55), and the corresponding particle velocity field is

$$u(\bar{x},\bar{t}) = \frac{p_L}{\rho_o C_o} e^{-(C_o\bar{t}+\bar{x})/C_o\tau_L}. \tag{3.56}$$

The rate of strain is provided by the gradient of this expression on the characteristic $\bar{x} = C_o\bar{t}$,

$$\dot{\varepsilon} = \frac{\partial u}{\partial \bar{x}} = \frac{2p_L}{\rho_o C_o^2 \tau_L} e^{-2\bar{x}/C_o\tau_L}. \tag{3.57}$$

For the case of an inverse linear dependence of the fracture spacing on strain rate, or $m = 1$,

$$s = \frac{p_s}{\rho_o C_o \dot{\varepsilon}}, \tag{3.58}$$

where again p_s is the tensile spall strength, while ρ_o and C_o are the density and elastic wave speed. The equations yield for the fracture spacing

$$s(\bar{x}) = \frac{C_o\tau_L p_s}{2p_L} e^{2\bar{x}/C_o\tau_L}. \tag{3.59}$$

Location of the fractures can be determined from integer solutions of

$$n(\bar{x}) = \int_0^{\bar{x}} \frac{d\bar{x}}{s(\bar{x})}, \tag{3.60}$$

yielding the relation

$$n(\bar{x}) = \frac{p_L}{p_s}(1 - e^{-2\bar{x}/C_o\tau_L}), \tag{3.61}$$

or the same relation as equation (3.53) developed through the earlier analysis.

Alternatively, consider the fracture spacing dependence on strain rate,

$$s(\bar{x}) = \alpha\dot{\epsilon}^{-2/3}, \tag{3.62}$$

with $\alpha = (K_c/\rho_o C_o)^{2/3}$ where K_c is an appropriate spall fracture toughness for the material. The fracture spacing relation becomes

$$s(\bar{x}) = \left(\frac{K_c C_o \tau_L}{2p_L}\right)^{2/3} e^{4\bar{x}/3C_o\tau_L}, \tag{3.63}$$

and

$$n(\bar{x}) = \left(\frac{27}{16}\right)^{1/3}\left(\frac{p_L}{K_c/\sqrt{C_o\tau_L}}\right)^{2/3} e^{-4\bar{x}/3C_o\tau_L}. \tag{3.64}$$

For this case, and for any other case except for $m = 1$ in equation (3.55), the tensile stress achieved on a spall plane (the spall stress) is dependent on the strain rate at which the plane is carried into tension.

The spall relations provided in equations (3.61) and (3.64) both emphasize a dependence on two key parameters. The characteristic wavelength of the incident pressure pulse $C_o\tau_L$ provides the fracture spacing length scale. The dimensionless prefactor in each case determines the fracture intensity. Both equations (3.61) and (3.64) prefactors are a ratio of parameters with dimensions of stress. Fracture intensity increases with increased incident pressure amplitude p_L. Fracture intensity decreases when the fracture resistance, either p_s or $K_c/\sqrt{C_o\tau_L}$, increases. A spall fracture stress criterion resulting in equation (3.61) will exhibit replica scaling. That is, if in the present example the thickness of the concrete wall and explosive charge were scaled by some factor, the predicted spacing of spall fractures would correspondingly scale. An energy-based spall fracture criterion leading to equation (3.64) would not exhibit replica scaling.

3.2.3.6 Velocity of the spall ejecta

Emergence of the compressive wave at the free surface, and the subsequent reflected tension wave, leads to the development of successive planes of spall as described in previous paragraphs. Material ejected by the spall event (plates of various thicknesses in the present one-dimensional spall analysis) are expelled at velocities which are calculated though the present analysis. Referring to figure 3.13, the relevant relation from which subsequent motion of the spall ejecta can be calculated is that for the material velocity imparted by the initial compression wave in domain II and corresponding to the pressure relation in equation (3.45),

$$u(x,t) = \frac{p_L}{\rho_o C_o} e^{-(C_o t - x)/C_o \tau_L}. \tag{3.65}$$

Or, in a form simpler for the present analysis,

$$u(X,T) = u_L e^{-(T+X)}. \tag{3.66}$$

The dimensionless distance and time have been introduced,

$$X = \frac{L - x}{C_o \tau_L}, \quad T = \frac{C_o t - L}{C_o \tau_L}. \tag{3.67}$$

The elastic wave solution provides the particle velocity in domain III between the free surface and the first spall plane illustrated in figure 3.14, given in equation (3.49), and written here in the form,

$$u(X,T) = 2u_L e^{-T} \cosh(X). \tag{3.68}$$

The same elastic solution method provides the velocity

$$u_n(X,T) = 2u_L e^{-2X_n} e^{-(T-X_n)} \cosh(X - X_n), \tag{3.69}$$

in the region immediately behind the nth spall plane at position X_n. The free surface and first spall plate correspond to $n = 0$ and $X_0 = 0$.

The average velocity of successive spall plates is provided by calculation of the spall plate momentum at any convenient time, say, $T = X_{n+1}$,

$$I_n = \rho_o \delta_n \bar{u}_n = 2\rho_o C_o \tau_L u_L e^{-2X_n} e^{-(X_{n+1}-X_n)} \int_{X_n}^{X_{n+1}} \cosh(X - X_n) dX, \tag{3.70}$$

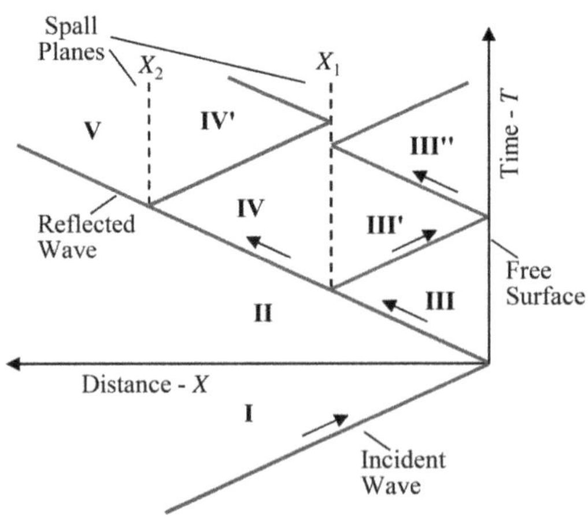

Figure 3.14. Arrival of the incident exponential pressure pulse at the free surface and the subsequent reflected tension and spall. Scaled distance and time (see the text) are on the abscissa and ordinate, respectively. Different domains are identified for which velocity solutions are obtained.

where $\delta_n = x_{n+1} - x_n$ is the spall plate thickness. The solution of equation (3.70) yields for the spall plate velocity

$$\bar{u}_n = 2u_L e^{-2\bar{x}_n/C_o\tau_L}\left(\frac{1 - e^{-2\delta_n/C_o\tau_L}}{2\delta_n/C_o\tau_L}\right), \tag{3.71}$$

where again $\bar{x}_n$ is the depth to the spall plane. When spall is intense ($p_s \ll p_L$) it follows that $\delta_n \ll C_o\tau_L$ over a reasonable range of n and

$$\bar{u}_n \simeq 2u_L e^{-2X_n}. \tag{3.72}$$

Equation (3.71) or (3.72) provides analytic illustrations of the velocity and momentum spread of spall ejecta. This outcome can be compared with statistical predictions of velocity and momentum dispersion of fragment debris when fewer details of the shock wave processes leading to fragmentation are known. Statistical aspects of velocity and momentum dispersion in the fragmentation event are discussed in chapter 2.

3.2.4 Impact breach of glass and ceramic plates

Failure and fragmentation through the processes of dynamic spall are of particular interest to shock wave scientists and address many issues of practical concern. Nonetheless, numerous dynamic fracture phenomena fall outside the province of spall failure. One example is that of impact failure and breach of brittle plates (Grady 2011).

The breach of a glass or ceramic plate by the impact of a comparable dimension chunky projectile provides a particularly challenging problem to computational solution methods. The reasons for this are several. Experimental results indicate that the properties of the plate material governing the strength and the resistance to breach are nonlocal and time-dependent. This behavior provides an immediate handicap to computational solutions based on local, time-independent strength and failure models. Further, the governing conditions at the onset of breach are statistical and provide additional complexity to a deterministic calculation of breach (e.g. Strack *et al* 2015).

In the present section, an analytic physics-based analysis of the onset of impact failure and breach of plates of brittle material is pursued that lends understanding to this dynamic fragmentation event. The model has application, for example, to the high-speed impact of solid fragments on transparent shielding, such as high-strength glass or possibly brittle polymeric materials such as Plexiglas (polymethyl methacrylate) or Lexan (polycarbonate). The model would be equally applicable to fragment or bullet impact onto barriers of ceramic plate. The emphasis of the model is on impact conditions near the threshold of the breach of the barrier.

The analysis makes use of data on the impact of steel spheres onto plates of float glass (Sun *et al* 2005) that nicely delineates the breach threshold. The data, when joined with the analytic model, clearly emphasize system scale dependence and time-dependent strength and failure features that are the challenge to a successful computational model.

3.2.4.1 The ballistic model

The ballistic event and the details of the physical model are illustrated in figure 3.15. A chunky projectile (fragment) of characteristic size a (a sphere is a reasonable but not necessary idealization) is incident on a plate of thickness b with normal velocity V. The velocity vector need not be normal to the barrier plate although the model is not appropriate for large angles of obliquity.

The transfer of momentum and energy from the projectile to the plate is assumed in the model to occur through the propagation of a radial elastic transverse shock wave. Details of the elastic stress wave propagation within the projectile and plate mass encompassed by the shock wave are ignored. The projectile, and the plate material encompassed by the shock wave, are assumed to be at the same momentum-averaged axial velocity $u(t)$. The projectile deceleration is calculated through a momentum balance between the projectile and the mass of the plate encompassed by the outwardly moving shock wave. The affected mass of the plate increases with time after impact according to $M(t) = \pi\rho b(ct)^2$, where M is the mass of the region, ρ is the plate density, b is the plate thickness and c is the elastic shear wave velocity of the plate material. If m is the mass of the projectile then momentum conservation requires

$$d(m + M)u = (m + M)du + udM = 0, \tag{3.73}$$

where $u(t)$ is the time-dependent velocity of the projectile and the plate mass $M(t)$, as illustrated in figure 3.15. Equation (3.73) integrates to

$$u = \frac{V}{1 + M/m}. \tag{3.74}$$

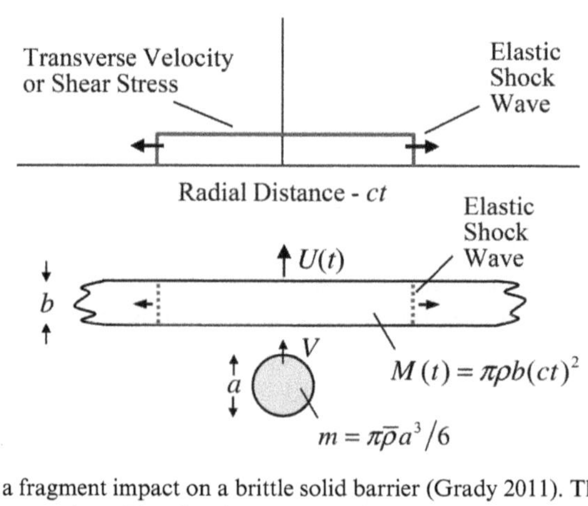

Figure 3.15. A model of a fragment impact on a brittle solid barrier (Grady 2011). The inset illustrates the size scales and initial relative motion. The drawing and graph show the motion and participating masses subsequent to impact at time t of the projectile and the disk portion of the plate of current radius ct.

By introducing the plate and projectile masses,

$$M = \pi\rho b(ct)^2, \quad m = \pi\bar{\rho}a^3/6, \tag{3.75}$$

equation (3.74) is rewritten in the time-dependent form,

$$u(t) = \frac{V}{1 + (t/\tau)^2}. \tag{3.76}$$

The time constant,

$$\tau = \sqrt{\frac{m}{\pi\rho c^2 b}} = \sqrt{\frac{1}{6}\frac{\bar{\rho}}{\rho}\frac{a^3}{c^2 b}}, \tag{3.77}$$

is a characteristic load transfer time that is dependent on both the dimensions and the properties of the plate and projectile material.

The decelerating force on the projectile is calculated from $F = -m du/dt$ and using $u(t)$ from equation (3.76),

$$F = \hat{F}\frac{t/\tau}{\left(1 + (t/\tau)^2\right)^2}, \tag{3.78}$$

with

$$\hat{F} = \frac{2mV}{\tau} = \left(\sqrt{\frac{2\pi^2}{3}\bar{\rho}\rho a^3 c^2 b}\right)V. \tag{3.79}$$

The time dependence of the decelerating force is shown in figure 3.16. A maximum force is achieved at $t/\tau = 1/\sqrt{3}$ with amplitude $F/\hat{F} = \sqrt{27}/16 \simeq 1/3$. The opposing shear stress $T(t)$ in the glass plate is calculated from the shock jump condition across the elastic shock wave,

$$T(t) = \rho c u(t). \tag{3.80}$$

From equation (3.76) this relation for the shear stress becomes

$$T(t) = \hat{T}\frac{1}{1 + (t/\tau)^2}, \tag{3.81}$$

where

$$\hat{T} = \rho c V. \tag{3.82}$$

The time dependence of the opposing shear stress in the plate is also shown in figure 3.16.

An interesting alternative calculation of the shear stress follows from energy considerations. Since momentum is conserved in the model, kinetic energy is not. The difference in kinetic energy of the plate and projectile at any time can be

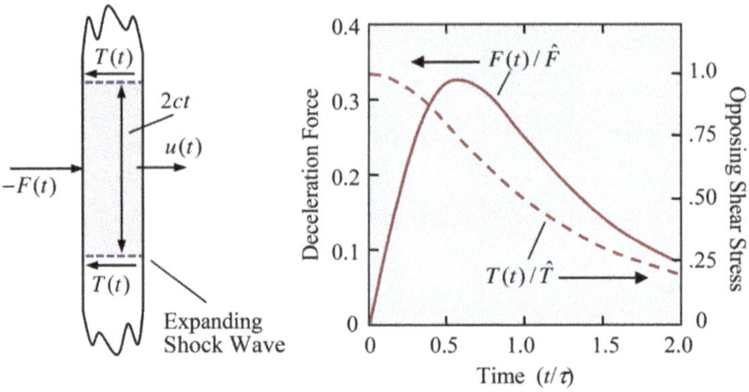

Figure 3.16. The decelerating force $F(t)$ and the opposing shear stress $T(t)$ in the responding portion of the barrier plate. Ordinates for $F(t)$ are on the left and for $T(t)$ on the right.

ascribed to strain energy in the plate. The elastic expression for strain energy results in the same measure of shear stress provided by the shock jump in equation (3.81).

Several assumptions are tacit in the present model. The several energy transfer processes that occur in the motions and deformations in the model, leading to arrest of the projectile or failure of the barrier plate, are decoupled. The kinetic energy of the incident projectile is assumed to transfer to kinetic energy of the impeding plate. Part of this kinetic energy converts in turn to elastic strain energy of the plate. The latter strain energy in not explicitly calculated or used in the analysis, however, and is only inferred from the shear-stress state.

Rigidity of the plate mass encompassed by the shock wave as implied by the uniform velocity $u(t)$ is certainly not correct. Spatial variations in this velocity as the elastic wave propagates outward from the point of impact would occur. The momentum averaged velocity $u(t)$ and the corresponding shear stress arising from the present model are assumed to adequately characterize the dynamics necessary for the prediction of failure in the present application. Failure is assumed to occur when and if some criterion resulting from this dynamics is exceeded during the process.

3.2.4.2 Failure criterion

A failure criterion appropriate to the impact event is now applied. The choice here is again the Tuler–Butcher failure criterion (Tuler and Butcher 1968) based on the time-dependent shear stress caused by the impact and stated in the form

$$I_n(t) = \int_0^t \left(\frac{T(t)}{\rho c} \right)^n dt \leqslant K_n. \tag{3.83}$$

Survival of the barrier is assured as long as the Tuler–Butcher integral $I_n(t)$ remains less than the material constant K_n. The barrier is breached when K_n is exceeded. Normalizing the stress by ρc is not necessary but is done here to simplify the expressions. In this form, the integral has dimensions of velocity to the power n multiplied by time.

This criterion is found to be quite general, and actually covers several specific criteria of particular interest. The parameter n in the Tuler–Butcher criterion can range from near unity to quite large, and the range $1 \leqslant n < \infty$ is considered. Within this range, a failure criterion governed by a critical impulse, a critical energy, or a critical stress is spanned. There are similarities with applications of the Tuler–Butcher criterion to spall earlier in the section. The Tuler–Butcher failure criterion, coupled with the present model, also brings to light issues of scale dependence observed in experiments, as well as some of the issues complicating computational simulation of the brittle failure event.

Introducing the expression for $T(t)$ from equation (3.81), the criterion can be written

$$I(t/\tau) = V^n \tau \int_0^{t/\tau} \left(\frac{1}{1+s^2} \right)^n ds \leqslant K_n. \tag{3.84}$$

The barrier is breached according to the present Tuler–Butcher criterion when the time-dependent integral on the left exceeds K_n. The time dependence of the integral is plotted as $I(t/\tau)/I(\infty)$ in figure 3.17 for values of $n = 1$, 2 and 10. Growth of the Tuler–Butcher integral is slowest for $n = 1$ (impulse criterion), more prompt for $n = 2$ (energy criterion) and most rapid for $n = 10$ (approaching a constant stress criterion).

The barrier is not breached if, as time becomes large,

$$\beta_n V^n \tau \leqslant K_n, \tag{3.85}$$

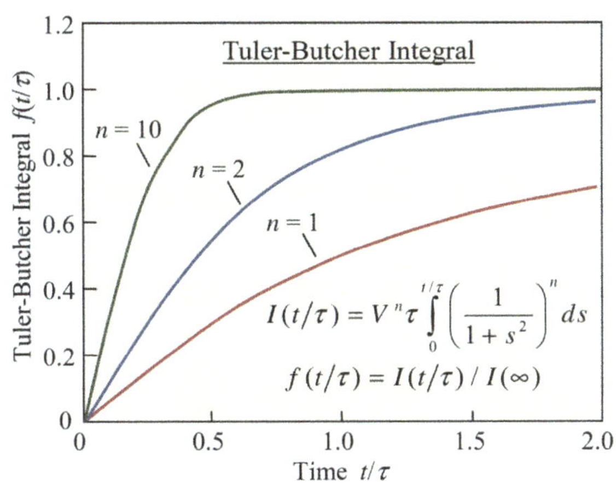

Figure 3.17. Time dependence of the Tuler–Butcher integral for breaching a glass barrier for selected values of the Tuler–Butcher power n.

where

$$\beta_n = \int_0^{\infty} \left(\frac{1}{1+s^2}\right)^n ds, \tag{3.86}$$

is the integral expression in equation (3.84) in the limit of large time. The integral in the Tuler–Butcher expression is analytic and equal to

$$\mathrm{atan}(t/\tau) \tag{3.87}$$

for $n = 1$,

$$\frac{1}{2}\left(\frac{t/\tau}{1+(t/\tau)^2} + \mathrm{atan}(t/\tau)\right) \tag{3.88}$$

for $n = 2$, and successively more complex functions of the inverse tangent for higher values of n. For equation (3.86) these relations provide $\beta_1 = \pi/2$ and $\beta_2 = \pi/4$.

The characteristic load-transfer time constant τ from equation (3.77) is written

$$\tau = \frac{a}{c}\sqrt{\frac{1}{6}\frac{\bar{\rho}}{\rho}\frac{a}{b}} = \frac{a}{c}R. \tag{3.89}$$

The parameter R is dimensionless and is constant when replica scaling of the system is considered (the projectile dimension a and the barrier thickness b are proportionally changed). Equation (3.85) is plotted as an equality for the selected values of n in figure 3.18 with $K_n/\beta_n = 1$. The time constant τ scales directly with the size a of the projectile, and hence the system size under replica scaling.

Increasing values on the abscissa, maintaining R constant, implies increasingly larger replica scaled projectile–barrier systems. The threshold velocity for onset of breach decreases with the system size for each of the three cases shown in figure 3.18. The scale dependence is most pronounced for the Tuler–Butcher impulse criterion

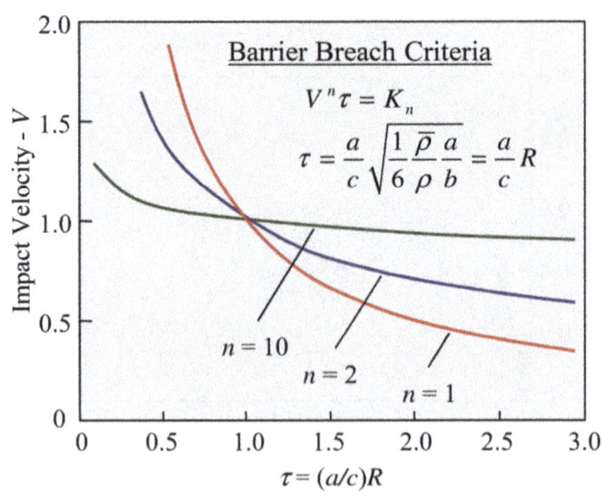

Figure 3.18. Barrier breach criteria for selected values of the Tuler–Butcher parameter n.

(n = 1), less so for the energy criterion (n = 2) and approaches independence of system scale for large n (approaching a failure stress criterion).

3.2.4.3 Experiments on impact breach of float glass

Experimental studies in which steel ball bearings impact plates of soda-lime glass (Sun *et al* 2005) provide appropriate data for exploring implications of the model descriptive of the impact breach process. In the tests, steel spheres are launched to velocities ranging over about 15–25 m s^{-1} and caused to undergo normal impact on glass plates. Steel spheres are either 1/4 inches (6.350 mm) or 3/8 inches (9.525 mm) diameter. Barrier plates are either 1.6 mm or 2.3 mm in thickness. The reported criterion for a breach of the glass plates was that they were either broken or unbroken.

The data of Sun *et al* (2005) are plotted from their report in figure 3.19. The ordinate is the impact velocity. The characteristic load transfer time provided by equation (3.77) or (3.89) is plotted on the abscissa. A density of 7840 kg m^{-3} is appropriate for the steel. A density of 2530 kg m^{-3} and shear wave speed of 3.47 km s^{-1} are used in the analysis for the soda-lime glass. Solid symbols identify the broken (breached) glass plates. Open symbols are for the unbroken plates. The breach and no-breach symbols at each impact condition are offset slightly in the figure for easier visualization.

Tuler–Butcher failure criteria for n = 1, 2 and 10 are independently compared with the data. Of the three, the data are best described with an n = 2, or an energy-based Tuler–Butcher criterion, although the significant scatter introduces some uncertainty to this observation. Particularly important is the lack of replica scaling of the breach strength data, as is most markedly shown by the two intermediate sets of data provided in figure 3.19.

A lack of replica scaling is most readily seen in figure 3.20 where the same data are plotted against the replica scale parameter R. Observe, in particular, the tests

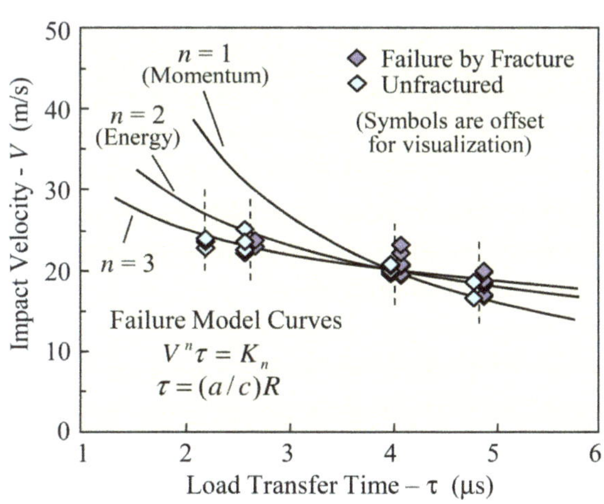

Figure 3.19. Breach data for steel sphere on glass plate experiments (Sun *et al* 2005). Comparison with selected Tuler–Butcher criteria (Grady 2011).

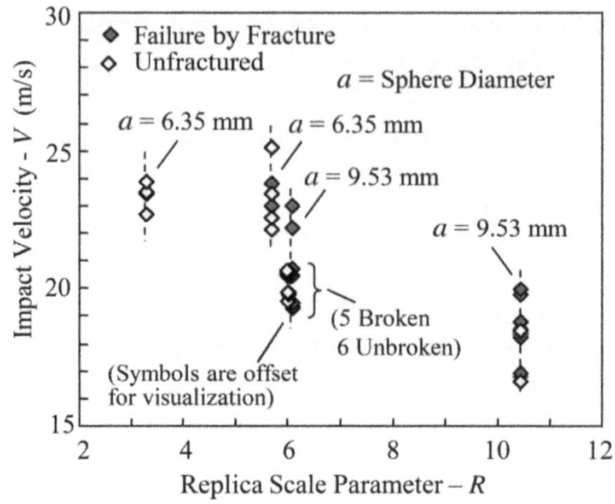

Figure 3.20. Breach data for steel sphere on glass plate experiments (Sun *et al* 2005). The abscissa is the replica scale parameter and is proportional to the ratio of the sphere diameter to plate thickness (Grady 2011).

with 6.35 mm steel spheres on 1.6 mm glass plates, and the 9.35 mm spheres on 2.3 mm plates. The replica scale parameter for the two differs by less than 6%, whereas system size differs by about 50%. The smaller scale test (a = 6.35 mm) exhibits a statistically higher breach strength.

3.2.4.4 Failure toughness of a glass plate

Working with the impact breach data of Sun *et al* (2005) and the Tuler–Butcher failure criterion, the data support an energy-based or $n = 2$ Tuler–Butcher failure criterion. Although there is no clear theoretical justification for an energy-governed criterion, there is a history of evidence that supports dynamic failure of brittle solids described by some form of an energy criterion.

For the present data on soda-lime glass, an energy-based Tuler–Butcher criterion best describes the breach data, as illustrated in figure 3.19. The value of the Tuler–Butcher constant K_2 determined from the fit to the glass failure data is $K_2 = 1.26 \times 10^{-3}$ m^2 s^{-1}. A failure energy is calculated through

$$\Gamma_f = \rho c K_2. \tag{3.90}$$

Using the density ρ and the shear wave speed c for soda-lime glass in equation (3.90), a value of $\Gamma_f = 11.0$ MJ m^{-2} is obtained. A property K_f with dimensions of fracture toughness is identified through the relation

$$K_f = \sqrt{2\rho c^2 \Gamma_f}\,. \tag{3.91}$$

The property K_f is here called the failure toughness of the glass plate. The Tuler–Butcher toughness may also be an appropriate naming. Carrying through the calculation yields a value of $K_f = 25.9$ MPa m$^{1/2}$. (This failure toughness property is markedly larger than the nominal fracture toughness for soda-lime glass which is of order $K_c = 0.5$–1.0 MPa m$^{1/2}$.)

The present failure toughness is not an intrinsic property of the glass, but may provide a sensible characterization of the impact failure of the glass plate. This strength property depends not only on the intrinsic strength of the glass but also on the flaw structure of the surface and interior of the plate.

3.2.4.5 Post-failure fracture damage

Failure of the glass plate upon impact based on the present model is achieved when the Tuler–Butcher integral exceeds the critical value. From the data of Sun *et al* (2005), failure of the present soda-lime glass plate is best described by a Tuler–Butcher energy criterion ($n = 2$) with a critical value of $K_2 = 1.26 \times 10^{-3}$ m^2 s^{-1}.

A measure of post-failure fracture damage can be calculated with the present model. The extent and the density of cracking resulting from the local impact is provided as well as the residual velocity of the impacting projectile and leading fragment debris. The model has validity to perhaps a few times the velocity at onset of failure.

In applying, the model considered the 3/8 inch steel sphere impact on 2.3 mm glass plate (Sun *et al* 2005) for which the failure threshold velocity is approximately 20 m s^{-1} as seen in figure 3.19 or calculated from the Tuler–Butcher criterion with a critical $K_2 = 1.26 \times 10^{-3}$ m^2 s^{-1}. The characteristic time constant from equation (3.77) for this problem parameter is approximately $\tau = 4 \times 10^{-6}$ s. The Tuler–Butcher energy integral from equation (3.84) or (3.88) is plotted on the left in figure 3.21 for selected velocities in excess of the failure threshold velocity. The time to failure t_f is calculated for each case in units of τ and is $t_f = 1.41, 0.623, 0.382$ and 0.202 times the constant τ for impact velocities of $V = 21, 25, 30$ and 40 m s^{-1}, respectively.

Post-failure fracture damage is estimated by calculating the characteristic fracture spacing at a radius r from the point of impact. An estimate of the circumferential fracture spacing is provided through a balance of the resisting fracture energy Γ and the driving expansion kinetic energy. This analysis provides the following relation for the number of fractures generated by the impact that persists out to a radius r,

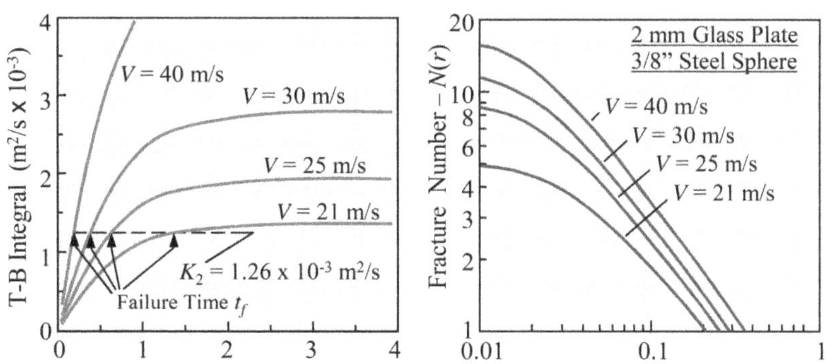

Figure 3.21. Tuler–Butcher energy integrals for impact velocities in excess of the threshold failure velocity are shown on the left. On the right are displayed plots of circumferential fracture number dependence on the radial distance from the impact point for steel sphere impact on a glass plate.

$$N(r) = 2\pi \sqrt[3]{\frac{\rho V^2}{48\Gamma}} r^{1/3} \left(1 + \frac{1}{\tau^2}\left(t_f + r/c_g\right)^2\right)^{-2/3}. \quad (3.92)$$

The fracture number from equation (3.92) for the selected impact velocities is shown on the right in figure 3.21. A fracture resistance energy of $\Gamma = 16.4 \text{ J m}^{-2}$ is used in equation (3.92) based on a fracture toughness K_c for glass of 1 MPa $\text{m}^{1/2}$ and the expression for $\Gamma = K_c^2/2\rho c^2$. The elastic shear wave speed for glass is assumed for the fracture growth velocity c_g in equation (3.92). The increased fracture number with increasing post-failure impact velocity and decreasing number with distance from the impact point is consistent with observation. It is notable that at large range the fracture number dependence on distance becomes power-law (fractal) with an $N(r) \sim 1/r$ dependence.

A pictorial representation of the same fracture data is shown on the left in figure 3.22. Fractures are distributed at random on the circumference through a Voronoi statistic process.

The fracture damage relation in equation (3.92) results from the following. At any radius r the average circumferential spacing L of fractures is determined from the energy-based fragment size relation developed in the previous chapter,

$$L = (48\Gamma/\rho\dot{\varepsilon}^2)^{1/3}. \quad (3.93)$$

The fracture number at a radius r is then

$$N(r) = 2\pi r/L = 2\pi r\left(\frac{\rho\dot{\varepsilon}^2}{48\Gamma}\right)^{1/3}. \quad (3.94)$$

The circumferential strain rate at r is provided by

$$\dot{\varepsilon} = \frac{u(r)}{r} = \frac{V}{r}\frac{1}{1 + (t/\tau)^2}, \quad (3.95)$$

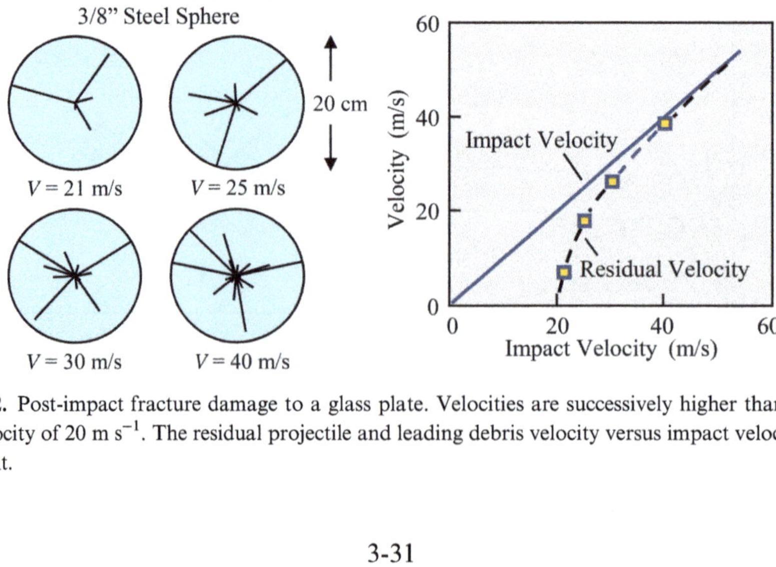

Figure 3.22. Post-impact fracture damage to a glass plate. Velocities are successively higher than the critical breach velocity of 20 m s^{-1}. The residual projectile and leading debris velocity versus impact velocity is shown on the right.

where the velocity $u(r)$ from equation (3.76) is applied. Equations (3.94) and (3.95) combine to give

$$N(r) = 2\pi\sqrt[3]{\frac{\rho V^2}{48\Gamma}}\, r^{1/3}\left(\frac{r}{(1 + (t/\tau)^2)^2}\right)^{1/3}. \tag{3.96}$$

A constant fracture growth velocity c_g is assumed such that $r = c_g(t - t_f)$ provides the radius of the fracture system at time t and is used to eliminate time from equation (3.96) providing the fracture damage relation in equation (3.92).

3.2.4.6 Residual projectile velocity

The post-failure residual velocity of the projectile and the leading fragment debris follows from the model. The time t_f at which the Tuler–Butcher failure criterion is exceeded and the failure (breach) of the plate achieved is calculated for any impact velocity. Time to failure for selected impact velocities are shown in figure 3.21. The velocity of the projectile at time t_f, and hence the residual projectile velocity, is provided by

$$u_f = \frac{V}{1 + \left(t_f/\tau\right)^2}. \tag{3.97}$$

A residual projectile velocity plot compared with the impact projectile velocity is shown on the right in figure 3.22. The selected impact velocity amplitude, and the projectile and plate dimensions, are the same as for the fracture damage calculation.

This and the preceding analytic solutions to idealized problems and data, interweaving the mechanics of dynamic fracture and fragmentation with solutions in shock wave propagation and interaction, ultimately have only modest application to real events. The value, however, rests in providing a background for the underlying physics and providing grounding points for larger-scale computational calculations. Also revealed are areas of difficulties for computational analysis in specific applications. Examples include the importance of non-local properties affecting the catastrophic failure and fragmentation of brittle solids and the limitations of geometric scaling in certain applications in dynamic fragmentation.

3.2.5 Failure dynamics of a crack

The subsection opens with resurrection of a calculation initially posed by Mott (1948) and pursued by others in which the dynamics of an individual crack is analyzed. Either a shear crack (adiabatic shear) or a tensile crack (spall) is envisioned, representative of a single local failure site in the pervasive dynamic failure of a solid. The crack length is identified as a generalized coordinate and Griffith energy conditions (Griffith 1924) are imposed. Expressions for the potential energy and the kinetic energy are obtained and the Mott crack is modelled as a conservative dynamic system. The Lagrange equation of motion provides the solution for the time history of the crack position and velocity in response to the applied far field stress.

Key to the failure dynamics of an individual Mott crack is the intensity of the radiated acoustic energy. The issue is investigated in the second part of the subsection through a simpler dynamic failure problem; namely, the one-dimensional dynamic shear or tensile failure at a planar interface in an otherwise linear elastic body. The failure problem as posed leads to informative analytic solutions. The principal result is that the amplitude of failure dissipation and the intensity of acoustic energy radiated from a site of failure are directly related. Such radiation energy is at the heart of the interplay of a system of cracks governing dynamic failure of a body. Through this interaction energy, correlation dynamics establishing the fabric of the crack pattern in failure is achieved.

3.2.5.1 The Mott dynamic crack

Failure in solids, in either tension or shear, when subject to sufficiently intense dynamic stress, entails the activation and the growth of a multiplicity of failure zones sufficient to alleviate the stress load exceeding the strength of the material. Mott, in probably his closing work on the topic of dynamic fracture and fragmentation (Mott 1948), considered the dynamics of failure in metals through a solution appropriate to the dynamic failure event. From this effort emerged the solution for the acceleration history of a fracture that has provided the backdrop for a number of subsequent analyses (e.g. Roberts and Wells 1954, Dulaney and Brace 1960) and addressed in fracture mechanics texts (e.g. Kanninen and Popelar 1985, Lawn 1993). The analysis leads to the velocity history of a crack of the form

$$\dot{a} \sim (1 - a_o/a)^n. \tag{3.98}$$

Mott arrived at equation (3.98) with $n = ½$. Later workers, in probably a more substantive analysis (Dulaney and Brace 1960), arrive at an exponent of $n = 1$, although the analysis of Mott is sufficiently terse that the arguments leading to his solution are not easily interpreted. In any case, in subsequent years more far-reaching solutions in the dynamics of a fracture have surpassed these earlier works. Although relegated to history, the physics underlying the analysis of Mott and later workers are sound and lend insights into the dynamics of fracture failure not afforded by later methods.

Here, the Mott fracture is resurrected for conceptual modeling of dynamic failure under shear or tension in the extreme shock environment. The Mott fracture replicates the dynamics of a single failure site, among many, in the high-rate failure of a solid. Either tensile failure (dynamic spall) or shear failure in the compression shock wave (adiabatic shear failure) is encompassed in the consideration.

The defect appropriate to the Mott analysis is that of a flat crack of initial half-length a_o subject to a normal tensile stress σ. The same crack is also considered subject to mode II loading under a shear stress σ. A circular, or for that matter, an elliptical crack could be considered, although not discussed here (see for example Wang and Lam 1995). Analysis is focused on the simpler problem of an infinite (two-dimensional) crack subject to plane-stress loading.

The mechanical potential energy assumed by Mott responsible for activation and subsequent dynamics is that provided by the Griffith energy (Griffith 1924)

$$V = 4a\Gamma - \pi a^2\sigma^2/E. \tag{3.99}$$

The first term on the right determines the fracture interface energy expended per unit length of growth of the crack. The energy Γ defines dissipation in the dynamic failure event. The constant E is the elastic modulus appropriate to the stress load. The second term provides the decrease in elastic strain energy with growth of the crack. Crack length $a \geqslant a_o$ applies at all times subsequent to onset of failure. Earlier analyses (Mott 1948, Roberts and Wells 1954, Dulaney and Brace 1960) initiate the solution at the Griffith instability state $a_o = a_g = 2\Gamma E/\pi\sigma^2$. Correspondingly, the Griffith stress σ_g is that just sufficient to achieve instability at initial crack length a_o. It is also appropriate to consider initial conditions where extreme rate loading achieves stress levels in excess of the Griffith stress at onset of crack growth.

Dynamic growth of fracture results in an elastic disturbance radiating outward from the propagating crack. As fracture proceeds, strain energy is transferred to both the fracture dissipation energy and kinetic energy of radiated disturbances in the proximity of the crack. Through dimensional arguments, Mott identified a local kinetic energy of the crack,

$$T = k\rho a^2\dot{a}^2\sigma^2/2E^2, \tag{3.100}$$

where k is an unspecified dimensionless constant determined by details of the kinetics of the dynamic strain field in the neighborhood of the crack.

The system as posed is conservative, and the total energy $H = T + V$, when equated to the Griffith energy at onset of crack motion, yields the history of the crack velocity. When the far field stress is equal to the Griffith stress, the solution for the crack length yields the relation of equation (3.98) with $n = 1$. When the far field stress is in excess of the Griffith stress the solution is more complicated but yields the same general behavior.

For a more complete characterization of the dynamics, the Lagrangian for the conservative system is $L = T - V$. The crack dynamics is then solved through the Lagrange equation

$$\frac{d}{dt}\frac{\partial L}{\partial \dot{a}} - \frac{\partial L}{\partial a} = 0. \tag{3.101}$$

The physics underlying dynamics of failure of the Mott crack is illustrated in figure 3.23. On the left is shown the respective surface and elastic energies combining to the Griffith energy. Competing tensions providing the Griffith force driving fracture growth are displayed on the right. The abscissa is normalized by the Griffith instability crack length $a_g = 2\Gamma E/\pi\sigma^2$.

Under the intense dynamic loading of a solid to tensile spall or to compressive shear failure, this representative Mott crack is, however, not alone. Many similar incipient failure sites are subjected to the same stress loading and experience similar crack activation, growth history and release of elastic disturbances. Thus, any

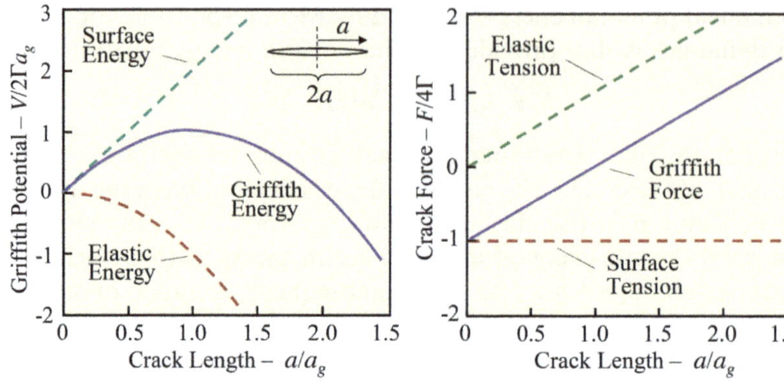

Figure 3.23. Energies and forces governing dynamics of the Mott crack.

particular crack is subjected to, not only the far field stress σ, but also to the random and chaotic force fields radiating from the many neighboring cracks experiencing similar failure dynamics. Such interaction and interplay resides at the heart of the correlation dynamics establishing the deformation fabric of the dynamic failure event; namely the characteristic spacing of the shear-band structure or the nominal size of fragments. Further perspective on this issue of failure in the shock environment is pursued in chapter 8.

Central to the underlying failure dynamics is the magnitude of the kinetic energy of the radiated disturbance from a single Mott crack. This issue is pursued further in the following subsection through a simpler elastic failure calculation.

3.2.5.2 Failure of the elastic planar dynamic crack

This subsection pursues an alternative and equally informative solution of the dynamics of failure at a single material site. Dynamic loads leading to inelastic failure of the material achieve stress levels at sites of weakness such that onset of failure and subsequent loss of load supporting ability occurs at the site. Consider an interior region of a material incurring increasing tensile stress at some dynamic rate of strain approaching the local spall threshold. Alternatively, consider an increasing shear stress until local shear failure and concomitant shear-stress release occurs. The energy needs at the site as failure ensues entail both dissipation at the site and the release of elastic strain energy.

Consider a process of one-dimensional failure on a planar interface subject to either a normal tensile stress or, alternatively, a parallel shear stress. A sketch illustrating the failure geometries is shown on the left in figure 3.24. This simplified model of failure has perhaps application to dynamics of the process zone of a propagating tensile crack. Under shear-stress loading the analysis is in line with previous solutions of adiabatic shear failure. The geometry is applicable to fracture failure in a rapidly stretching brittle wire. It is, in any case, a useful abstraction of essential features of the local dynamic failure event.

The following linear elastic one-dimensional wave propagation problem replicates the failure event. At time zero all space $x > 0$ is at constant tensile (or shear)

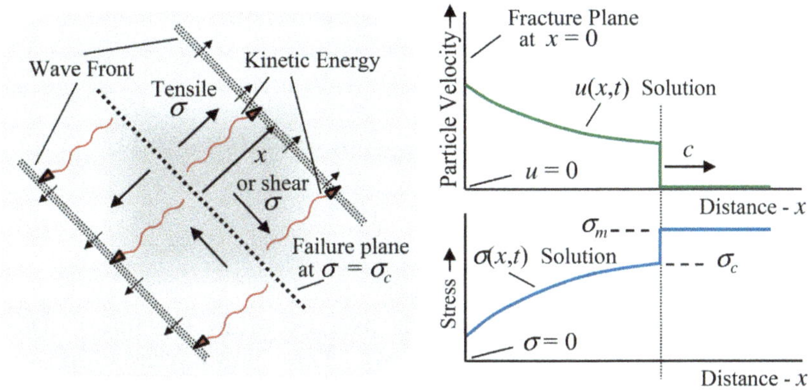

Figure 3.24. Material failure on the $x = 0$ plane initiates stress release wave propagation into the neighboring elastic medium.

stress σ_m achieved in the material through some dynamic process. At the failure plane $x = 0$ failure initiates at a stress $\sigma_c < \sigma_m$. Over subsequent time, stress at the failure plane reduces to zero in a prescribed manner replicating the failure process. The one-dimensional stress and particle velocity in the neighboring region satisfy the linear elastic wave equation. Wave propagation subsequent to failure is illustrated on the right in figure 3.24. Motions subsequent to failure are shown on the top as the velocity profile propagating into material at rest. The corresponding stress wave propagating into the medium initially at load stress σ_m is shown below.

Within the x versus t domain bounded at $x = 0$ by the failure plane and by the wave front $x = ct$, the d'Alembert solution to the elastic wave equation applies (e.g. Courant and Hilbert 1962). The flow field on left-facing characteristics is constant. Consequently, stress and particle velocity, after properly matching jump conditions on the characteristic $x = ct$, satisfy the simple wave solution,

$$\sigma(x,t) = \sigma_m - f(t - x/c), \tag{3.102}$$

$$\rho c u(x,t) = f(t - x/c). \tag{3.103}$$

The solution is completed by applying a boundary condition at the failure plane $x = 0$. Failure is described by the relation

$$f(t) = \sigma(0,t) = \sigma_c\left(1 - y(t)/y_c\right). \tag{3.104}$$

Namely, the stress in the failure zone decreases linearly from the failure stress σ_c to zero over the crack failure displacement, $0 \leqslant y(t) \leqslant y_c$. Displacement is calculated from the particle velocity at $x = 0$ through

$$y(t) = \int_0^t u(0,s)ds. \tag{3.105}$$

Equations (3.103)–(3.105) yield the differential equation

$$\frac{d}{dt}f(t) - \frac{\sigma_c}{\rho c y_c}f(t) = 0, \tag{3.106}$$

and solutions for the stress and particle velocity,

$$\sigma(x,t) = \sigma_m - (\sigma_m - \sigma_c)e^{\frac{\sigma_c}{\rho c y_c}(t-x/c)}, \tag{3.107}$$

$$\rho c u(x,t) = (\sigma_m - \sigma_c)e^{\frac{\sigma_c}{\rho c y_c}(t-x/c)}. \tag{3.108}$$

Energy dissipation in the failure zone is assessed from the linear stress versus displacement relation in equation (3.104). The failure dissipation energy is $\Gamma = \sigma_c y_c/2$. The interface dissipation energy Γ is a property of the material.

The far field stress σ_m is a tensile or transverse stress achieved through dynamic loading of the material. The rate of stress loading is on a time scale that is large compared to time scales of activation and growth at the site of failure. The stress σ_m is considered constant over the solution time. Onset of failure at the site of weakness is at stress $\sigma_c < \sigma_m$. The stress σ_m must necessarily exceed σ_c for failure at the site to ensue. The elastic solution is metastable otherwise. The stress difference $\sigma_m - \sigma_c$ can be identified with a hesitation period over which the nominal failure stress σ_c lingers during formation of the failure cohesive zone. This zone determines an activation strain or activation volume in the sense of the Dugdale–Leonov–Panasyuk cohesive zone model (Leonov and Panasyuk 1959, Dugdale 1960). A more extended model of the formation of failure activation is provided by the history-dependent viscoelastic treatment by Hakim and Mikhailov (2013).

With the foregoing calculation, it is convenient to simplify expressions in equations (3.107) and (3.108) solutions for the stress and particle velocity to

$$\sigma(x,t) = \sigma_m - \sigma_m(1 - r)e^{\frac{r^2}{\tau}(t-x/c)}, \tag{3.109}$$

$$\rho c u(x,t) = \sigma_m(1 - r)e^{\frac{r^2}{\tau}(t-x/c)}. \tag{3.110}$$

A characteristic failure time scale is $\tau = \rho c \sigma_c y_c/\sigma_m^2 = 2\rho c\Gamma/\sigma_m^2$. The arbitrary hesitation parameter $r = \sigma_c/\sigma_m$ is of order, but slightly less than, unity. The solution terminates when stress at the failure zone achieves zero, or equivalently, at the time $t = -\tau \ln(1 - r)/r^2$.

With the elastic solution for the one-dimensional failure event, the objectives are to calculate and assess the history of dissipation in the failure process zone and the history of kinetic energy radiated away from the failure zone. Results of this more basic solution can be compared with the partitioning of kinetic energy and failure dissipation in the Mott dynamic crack.

The kinetic energy in the elastic field neighboring the failure zone is provided by the relation

$$T(t) = \frac{1}{2}\rho\int_0^{ct} u^2 dx, \tag{3.111}$$

yielding

$$2T(t) = \Gamma\frac{(1-r)^2}{r^2}e^{\frac{2r^2}{\tau}t}. \tag{3.112}$$

Kinetic energy in the elastic field at any time in the failure process scales with the dissipation energy Γ.

Work and dissipation in the failure process zone at any time follows from

$$W(t) = \int_0^t \sigma u dt, \tag{3.113}$$

with solution

$$W(t) = 2\Gamma\frac{(1-r)}{r^2}(e^{r^2t/\tau} - 1) - 2T(t). \tag{3.114}$$

An energy of $2T$ is associated with the energy radiated from the actively failing zone. This includes both the kinetic energy and the elastic fluctuation from the far field stress in the outward direct wave.

Essence of the calculation is the direct relation between the radiation energy and the dissipation energy as captured by equation (3.112). Plots of the respective energies through the failure event illustrate the relationship. The time histories are shown in figure 3.25 for energy and stress for two values of the hesitation parameter $r = \sigma_c/\sigma_m$. Radiated kinetic energy and dissipation energy are normalized by the dissipation energy parameter Γ. Stress at the failure interface is plotted as the ratio $\sigma(t)/\sigma_m$. Dissipation precedes radiation in the time history. The two are comparable at failure completion. Implications of the calculation are that dissipation energy and radiation energy in propagation of the Mott crack are comparable in magnitude, and are governed by the same failure parameter Γ.

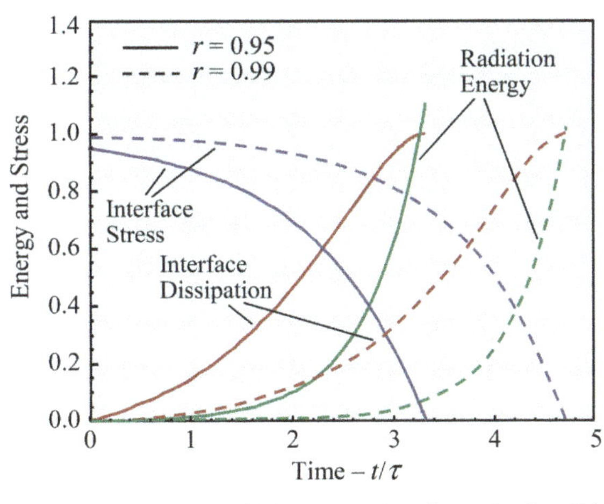

Figure 3.25. Time history of stress and energy of failure on a plane in an elastic solid. Radiated kinetic energy and interface dissipation energy is normalized by the dissipation parameter Γ. Stress is normalized by the far field stress σ_m.

3.2.5.3 Correlation dynamics and the failure event

The intensity of radiation energy induced in the failure of the Mott crack is central to interaction of the many active sites participating in the dynamic failure event. The deformation features and deformation fabric, and the evolution of such in the failure event, is intrinsic to the dynamic failure of solids. This dynamic applies to tensile spall as well as to failure in the shock wave under a state of confining pressure. A comparable correlation dynamic occurs within the delayed failure wave in brittle solids subject to shock, and in the shock-to-detonation response of reactive materials. The former is pursued further in chapter 6.

A deformation feature, depending on the mode of failure, is a tensile fracture or thin planar shear band whose inelastic dissipation is characterized by an interface dissipation energy Γ (units of J m^{-2}) that remains constant, or nearly constant, over a broad range of loading rates. The deformation fabric is the collection of deformation features characterized by a number density and spatial distribution necessary to accommodate the failure event. The deformation fabric can vary markedly as the range of loading rates is spanned. The deformation fabric is inherently a consequence of dynamics and the interaction of competing failure sites through the failure event. Correlation dynamics and dynamic self-organization subject to time, and consequently spatial, constraints are the determining causes.

This in general need not be. Deformation under quasi-static loading conditions also leads to a heterogeneous deformation fabric and deformation features. The underlying responsible physics is different, however. The transition from quasi-static-governed to dynamics-governed deformation fabric and the underlying responsible physics are not presently fully understood.

The correlation dynamics, dictated by the competition of energy and momentum needs of the heterogeneous failure features and the time-constrained redistribution of energy, lies at the heart of the dynamic failure event. In both spall failure and adiabatic shear failure, a balance is achieved between the number of deformation features necessary to the failure event and the energy/momentum redistribution length scale. In application, many more deformation features than necessary are seeded, but most drop out of the population as the energy balance is sought. This is very apparent in experimental testing involving the dynamic fragmentation of expanding metal rings or exploding metal shells where many fractures are activated, but few survive to complete the breakup process. A less apparent, but similar failure dynamics occurs in brittle materials. Here an initially sparse activation of fractures is further seeded through the multiple branching of existing fractures. This failure process cascades until the fracture density is consistent with the load intensity and dissipation requirements of the failure event. Similar multiplication of failure features through branching activation occurs in the adiabatic shear failure of metals subject to high-velocity plugging. These issues of dynamic adiabatic shear failure are revisited in chapter 5.

3.3 Fragmentation through thinning and stretching instability

A number of intense dynamic events lead to the high-rate deformation of solids or liquids where materials stretch and thin until instabilities in the deformation cause breakup and fragmentation. Common applications include the fragmentation of

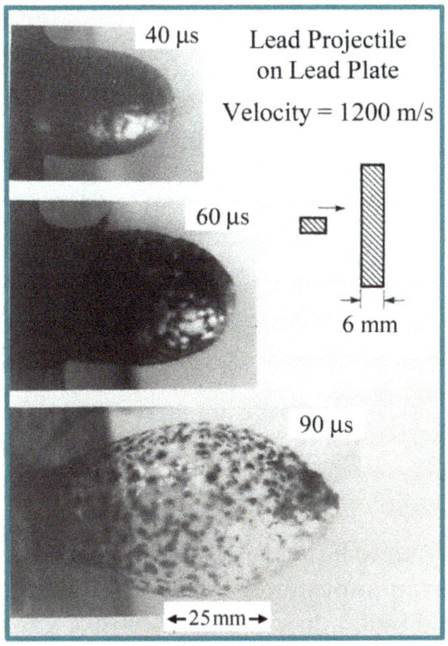

Figure 3.26. High-speed impact of a lead projectile onto a lead plate illustrating the deformation bubble formation leading to thinning instability and fragmentation. Reproduced with permission from Grady *et al* (1985), Grady and Passman (1990). Copyright 1985, AIP Publishing LLC.

shaped-charge metal jets, the explosive acceleration and breakup of metal shells, and the fracture of magnetic expanding rings.

Thinning and instability breakup is illustrated, for example, by the high-velocity impact of a solid metal lead projectile onto a stationary lead plate shown in the sequence of high-speed photographs in figure 3.26. The high ductility and inertial stability of the impact-induced dynamic deformation leads to the generation of an expanding bubble of ductile metal. Thinning of the bubble wall proceeds until thinning instabilities ensue. Onset of instability is governed by the flow strength of the lead and the inertial energy of the thinning body. The event culminates in dynamic fragmentation into wafer-shaped fragments injected at high down-stream velocity.

A further interesting application of thinning instability and fragmentation is shown in figure 3.27, where selected sketches illustrate the dynamic response of certain liquid polymers when expelled in a thin jet (De Kee and Wissbrun, 1998). Stretching and thinning is induced though the elongating motion imparted to the jet. Geometric softening in the deformation thinning process leads to onset of breakup through necking at statistical sites along the jet. Subsequent orientation hardening of the polymer microstructure within the necking region causes restabilization of the flow, leading to continued elongation in necking regions, inhibiting complete fragmentation until substantially later in the process. Experimental images illustrating this aspect of dynamic fragmentation in polymer materials are seen in the works of De Kee and Wissbrun (1998).

Selected topics on dynamic instability fragmentation of liquids or liquid-like matter are addressed in the present chapter. Calculations are described of the impact breakup

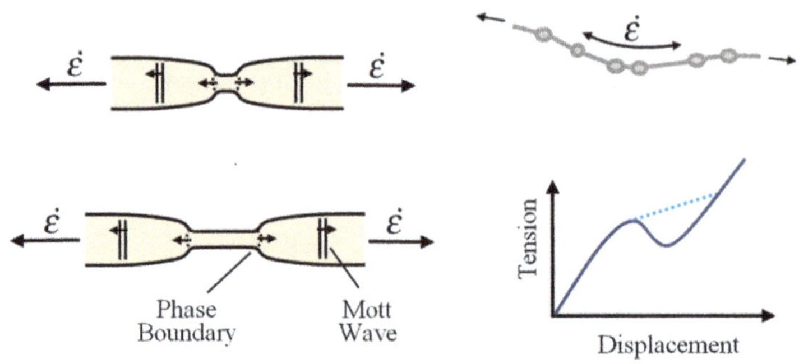

Figure 3.27. Stretching and thinning instability, and subsequently fragmentation in a liquid polymer.

and fragmentation of polymeric liquid systems. The phenomenon of the aerodynamic breakup of liquid droplets in atmosphere is addressed and applied to the fragmentation of high-velocity liquid metal drops in an explosive gas environment.

3.3.1 High velocity impact fragmentation of a liquid system

The impact of a high-velocity projectile onto a metal container filled with liquid will catastrophically disrupt and disperse the system materials. In particular, the liquid component is violently fragmented into a distribution of small liquid particles. The resulting particles are ejected outward at a comparably high velocity and at trajectories distribution over the full 4π solid angle. Here, calculations are pursued that expand on the earlier dynamic fragmentation principles. Focus is on the onset of breakup and fragmentation of the rapidly dispersing liquid. The calculations are compared with impact experiments performed on containers filled with liquid tributyl phosphate (TBP) (Borg *et al* 2001).

3.3.1.1 System configuration and fragmentation model

Impact experiments were performed on cylindrical containers of tributyl phosphate (TBP) in both an unthickened form (density 970 kg m^{-3} and viscosity 0.0033 Pa s), and also in a thickened form. Through addition of about 5% of a copolymer liquid, the viscosity of the thickened form was increased to 1.1 Pa s. The liquid–container systems are impacted at high velocity (2–6 km s^{-1}) by an appreciably smaller solid projectile along a trajectory that passes reasonably close to the geometric center of the liquid-bearing container. In the center-of-mass frame materials are subjected to an outward impulse and a residual outwardly directed kinetic energy. Given an initial relative impact velocity and the masses of both projectile and target, the radial outwardly directed velocity u can be estimated through one of several models. The velocity u is determined from measurement in the present test series. Calculation assumes an outwardly directed average constant specific kinetic energy $\tau = u^2/2$ where u is the outwardly directed velocity of the liquid. A characteristic initial

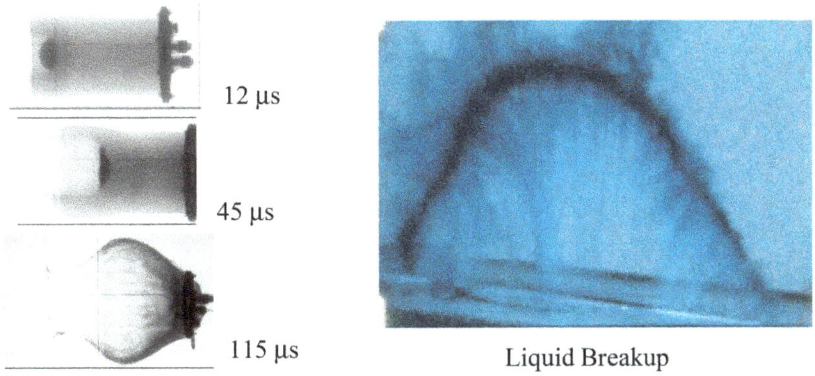

Figure 3.28. The sequence on the left provides radiographic images of the impact-induced fragmentation of the system steel container and liquid. An image of the expanding liquid expelled through a stationary slit in a barrier plate is shown on the right. Reprinted from Borg *et al* (2001). Copyright 2001, with permission from Elsevier.

dimension b_o is ascribed to the container such that the initial volume is approximately b_o^3.

Radiographic images of a representative test are shown in figure 3.28. A sequence of three images on the left illustrates the expansion and fragmentation of the steel container. From this test at an impact of 2.4 km s^{-1} an expansion velocity of approximately 500 m s^{-1} is measured. On the right is shown the expansion and breakup of the liquid as imaged after passage through a slit in a stationary barrier somewhat later in time.

The thin steel container undergoes fragmentation early in the expansion event and probably plays some role in effecting the liquid fragmentation. Container breakup is ignored in the subsequent analysis. Attention focuses on the physics governing instability onset and breakup of the expanding liquid.

Additionally, the expanding liquid body is modeled as an expanding spherical shell with inner and outer radii at any time t of $a(t)$ and $b(t)$, such that

$$b^3(t) - a^3(t) = b_0^3, \tag{3.115}$$

as illustrated in figure 3.29. The liquid is modeled as an incompressible viscous fluid.

3.3.2 High-velocity impact fragmentation of a liquid system

The impact of a high-velocity projectile onto a metal container filled with liquid will catastrophically disrupt and disperse the system materials. In particular, the liquid component is violently fragmented into a distribution of small liquid particles. The resulting particles are ejected outward at a comparably high velocity and in trajectories distributed over the full 4π solid angle. Here, calculations are pursued that expand on the earlier dynamic fragmentation principles. The focus is on the onset of breakup and fragmentation of the rapidly dispersing liquid. The

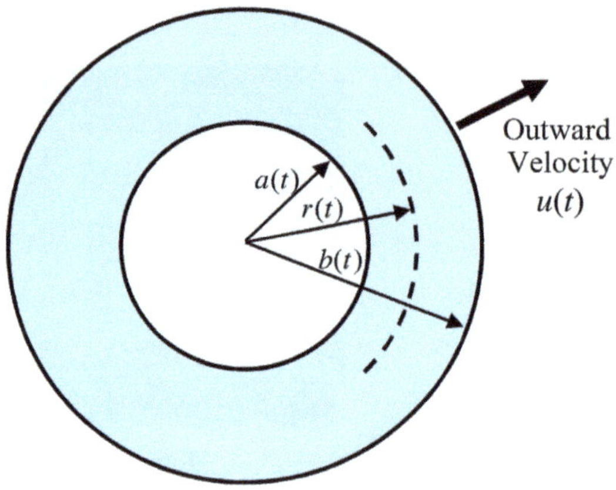

Figure 3.29. Arbitrary configuration of an expanding spherical shell fluid body with inner radius a, and outer radius b. The coordinate r references a spatial point $a \leqslant r \leqslant b$.

calculations are compared to impact experiments performed on the containers filled with liquid TBP (Borg *et al* 2001).

3.3.2.1 Motion and pressure field

For the assumed motions, equations of continuity and momentum conservation are applied to an incompressible non-viscous fluid body to calculate an inertial pressure field existing within the freely expanding spherical body of the form (Grady and Passman 1990, Grady *et al* 1990)

$$p(\xi,\beta) = \rho\tau\frac{\beta^3 - 1}{3\beta^3}\left(\frac{\beta}{\xi} - 1\right)\left[(\beta^2 + 2\beta + 3) - \left(\frac{\frac{\beta}{\xi} - 1}{\beta - 1}\right)\left(\frac{\beta^2}{\xi^2} + 2\frac{\beta}{\xi} + 3\right)\right], \quad (3.116)$$

where

$$\beta = \frac{b}{a} \quad \text{and} \quad \xi = \frac{r}{a}. \quad (3.117)$$

This expression for the pressure simplifies considerably when the shell thickness h becomes small relative to the expansion radius a. Letting

$$\beta = 1 + \frac{h}{a}, \quad (3.118)$$

and

$$\xi = 1 + \frac{z}{a}, \quad (o \leqslant z \leqslant h), \quad (3.119)$$

equation (3.116) for the pressure for $h \ll a$ becomes

$$p(\xi,\beta) = 6\rho\tau(\xi - 1)(\beta - \xi) = 6\rho\tau\frac{z(h - z)}{a^2}, \tag{3.120}$$

by series expansion and retention of lowest-order terms.

To account for viscous flow, a characteristic viscous flow stress Y is included in the analysis such that the radial and circumferential stress components become

$$\sigma_r = -p - \frac{2}{\sqrt{3}}Y, \tag{3.121}$$

$$\sigma_\theta = -p + \frac{1}{\sqrt{3}}Y. \tag{3.122}$$

A mean stress comparable to the pressure in equation (3.120) is

$$p = -\frac{1}{3}(\sigma_r + 2\sigma_\theta) = 6\rho\tau(\xi - 1)(\beta - \xi) - \frac{2Y}{\sqrt{3}}. \tag{3.123}$$

The large inertial compressive pressure and the accompanying pressure gradients at the inner and outer surfaces of the expanding spherical fluid shell account for the stability of the surfaces to Rayleigh–Taylor perturbations during much of the expansion (Romero 1991).

3.3.2.2 Instability leading to breakup

To predict the onset of instability of the expanding fluid shell, a criterion for instability is shown (Romero 1989) to reasonably emulate more rigorous stability analysis. The method is viewed as an application of Le Chatelier's principle, which states that for a system to be stable any deviation from equilibrium must bring about forces that tend to restore equilibrium (Reif 1965). A similar criterion has been stated by Considère (1885) concerning the instability of stretching plastic bodies.

To obtain the circumferential tension (or compression) in the expanding shell the circumferential stress from equation (3.122) is integrated through the thickness h of the shell,

$$T = \int_0^h \sigma_\theta dz = a\int_1^\beta \sigma_\theta d\xi, \tag{3.124}$$

yielding

$$\frac{T}{a} = \sqrt{3}Y(\beta - 1) - \rho\tau(\beta - 1)^3. \tag{3.125}$$

The tension T is plotted as a function of $a/h = 1/(\beta - 1)$ in figure 3.30. Instability onset is identified as the point at which $dT/d\beta = 0$ giving

$$\beta_c - 1 = \frac{h_c}{a_c} = \sqrt{\frac{Y}{\sqrt{3}\rho\tau}}, \tag{3.126}$$

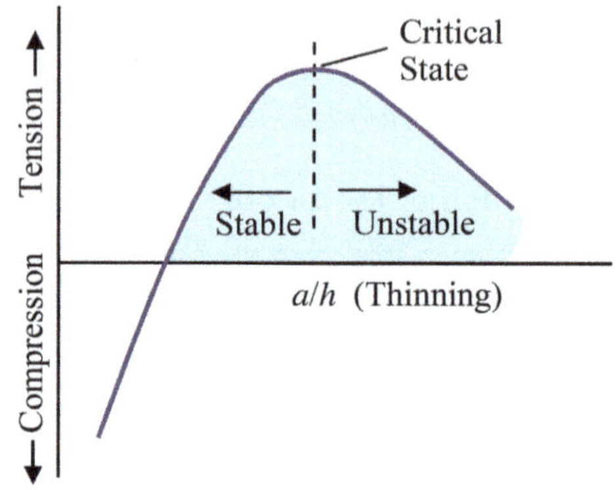

Figure 3.30. Illustrates the stress state as function of configuration of the expanding fluid shell and the point of onset of instability and fragmentation.

where, using equation (3.115),

$$a_c = \left(\frac{1}{3}\frac{b_o^3}{h_c/a_c}\right)^{1/3}. \tag{3.127}$$

3.3.2.3 Viscous liquid relations

Although it is expected that the complex molecular liquids of concern, both neat (unthickened) or thickened with a polymer agent, would be nonlinear in viscous behavior, a linear relation is assumed of the form,

$$Y = \lambda\dot{\varepsilon}. \tag{3.128}$$

The strain rate is $\dot{\varepsilon}$ and λ is an appropriate dynamic viscosity coefficient.

Using equation (3.128) in equations (3.126) and (3.127), approximating the strain rate by $\dot{\varepsilon} \sim u/a_c$, and ignoring the numerical terms of order unity provides

$$\frac{a_c}{b_o} = \left(\frac{\lambda}{\rho u b_o}\right)^{-1/5}, \tag{3.129}$$

and

$$\frac{h_c}{b_o} = \left(\frac{\lambda}{\rho u b_o}\right)^{2/5} \tag{3.130}$$

for the extent of expansion and thickness of the shell at onset of instability.

3.3.2.4 Expansion to instability

Equations (3.129) and (3.130) provide, within the constraints of the approximations and model, both a qualitative picture and a quantitative prediction of the radius a_c

Table 3.1. Parameters representative of test series and predicted instability threshold conditions

Property	Symbol	Neat (unthickened)	Thickened
Initial size (radius)	b_o	0.037 m	0.037 m
TBP liquid density	ρ	970 kg m^{-3}	970 kg m^{-3}
TBP liquid viscosity	λ	0.0033 Pa s	1.1 Pa s
Expansion velocity	u	500 m s^{-1}	500 m s^{-1}
Instability radius	a_c	0.83 m	0.26 m
Instability thickness	$h_{.c.}$	0.12 mm	0.77 mm

and the thickness h_c at which instability breakup proceeds in an expanding shell of viscous fluid. It is the expansion point at which the healing forces of inertial pressure are exceeded by the destroying forces of viscous resistance.

Approximate test conditions for the experiments of Borg *et al* (2001) are used to estimate threshold instability conditions and assess the nature of equations (3.129) and (3.130). Table 3.1 lists the test conditions and the estimated instability radius and instability thickness as predicted from equations (3.129) and (3.130).

As provided in the table, equation (3.129) predicts an expansion breakup radius of about 0.83 m for the neat liquid and 0.23 m for the thickened liquid. Equation (3.130) predicts the liquid shell thickness at the breakup radius to be approximately 0.12 mm and 0.77 mm for the neat and thickened liquids, respectively.

The essence of the calculation establishes that when impulsive loads on a liquid system are sufficiently intense, extensive contiguous expansion stabilized by inertia precedes ultimate breakup. In the present analysis, viscosity is the dissipative mechanism defining onset of instability. If viscosity is sufficiently low, the liquid surface tension is the force determining instability onset. The consequences of the calculation are in sensible accord with experimental observations in the tests illustrated in figure 3.28.

3.3.2.5 Instability fragmentation

The onset of instability is the precursor to instability growth, breakup and fragmentation of the expanding body. Changes in the physical parameters on the right in equations (3.129) and (3.130) that govern the onset of instability lead to self-similar changes in the observed breakup and fragmentation. That is, changes in the density, the expansion velocity or the initial system size cause corresponding changes in the shell thickness and radius at instability onset. The details of breakup, such as instability growth, fragment size and fragment shape, would—except for length scale changes—look similar.

Self-similar behavior is not necessarily anticipated for changes in viscosity. Since viscosity variations in the application discussed are brought about by addition of polymeric thickener, significant changes in the mechanisms of instability growth and fragmentation are expected. Polymer alignment and possible orientation hardening would fundamentally change the nature of the growth and breakup process. This orientation-hardening effect is illustrated by the sketches shown in figure 3.27 and was also observed in the test series on thickened TBP material (Borg *et al* 2001).

An estimate of particle size following fragmentation is provided by the shell thickness h_c at onset of instability. Determination of a dominant lateral instability wavelength is difficult due to the time-varying nature of the base state. Analysis, however, suggests a relatively flat dominant growth rate over wavelengths of order of the shell thickness h_c to wavelengths of the order of a dimensionless parameter times h_c. The ratio of inertial to viscous forces determines this parameter (Romero 1991), but is of order ten. Hence, a lateral characteristic fragment size of a few h_c is probably reasonable. The fragment size predictions based on the energy balance methods outlined in the previous chapter assuming viscous-dissipation-governed fragmentation, yield similar fragment size results (see also Grady 1988). Neither method would be satisfactory for complex viscous effects in polymer-thickened liquids when orientation hardening leads to stringer formation in the fragmentation process.

3.3.2.6 Fragmentation of viscous liquids

In summary, an approximate analysis investigates the onset of instability leading to fragmentation of a rapidly expanding shell of liquid. The simplifying geometry assumed is intended to model the key parameters governing liquid breakup occurring in the kinetic energy impact of a liquid–container system. The essential physics pursued is the stabilizing nature of inertial forces brought about by the expansion dynamics in competition with destabilizing effects due to viscous properties of the liquid medium.

An instability criterion based on a balance of these opposing forces predicts instability onset for kinematic and material properties representative of the experiments of Borg *et al* (2001). The extensive expansion prior to fragmentation that was observed in the experiments, and was also predicted by the theory, lends support to this underlying physics.

The present analysis barely probes the complex phenomena of liquid fragmentation under the dynamic conditions of interest and a number of physical features were not touched on. For example surface tension, which is an important first-order effect in many liquid fragmentation events, is judged to be of second-order in the present application. For example, surface tension is fundamental to the fragmentation of liquid metals (Grady 1995). In addition, in the present analysis the liquid shell is assumed to be freely expanding rather than opposed by an exterior gas pressure, as was the case in selected experiments in the present program. With the complication of exterior gas pressure, Rayleigh–Taylor instability plays a role in the onset of instability and fragmentation. Finally, the analysis focuses principally on the onset of instability, and the issues of instability growth and fragmentation in complex molecular liquids is only briefly pursued (e.g. Romero 1991).

3.3.3 Hydrodynamic erosion of high-velocity fragments

A further example of dynamic fragmentation through deformation instability mechanisms is provided by the breakup of liquid (or solid) particles that are

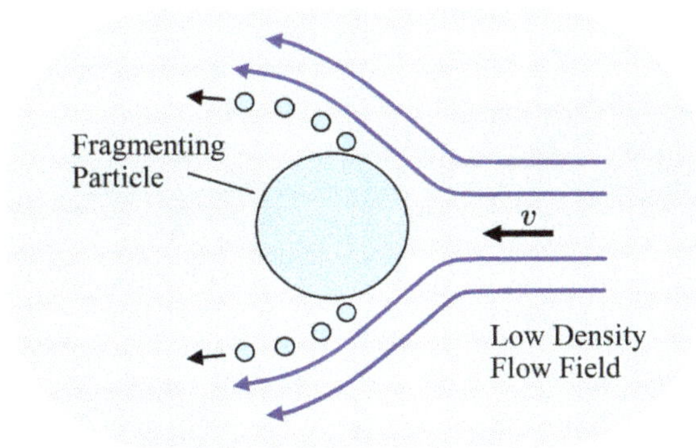

Figure 3.31. Hydrodynamic erosion fragmentation of a liquid or solid particle subject to an intense high-velocity low-density flow field.

introduced into a gaseous medium at a high relative velocity. The particle–flow interaction and hydro-erosion fragmentation are illustrated in figure 3.31. Although a spherical body of liquid undergoing breakup due to forces brought about by the high-velocity flow is here pursued further, the concepts are equally applicable to a solid body when inertial forces of the flow are large compared to the deformation strengths of the solid. The high-velocity entry of meteoroids into the Earth's upper atmosphere is one example.

The present example of hydrodynamic fragmentation is motivated by the consequences of the high-explosive implosion of a spherical or cylindrical metal shell. Metals in the shell are both intensely shocked and subsequently fragmented by the explosive event. High-temperature liquid and hot solid metal fragments are ejected at high relative velocity into the gaseous explosive products and neighboring air environment. This outcome sets up conditions for further fragmentation through hydrodynamic erosion of the metal fragments, as illustrated in figure 3.31.

An example of such an event is shown in figure 3.32, where the consequences of the high-explosive implosion of a hollow uranium metal cylinder are shown in a late-time photograph (Grady 1982b, 2010). Much of the metal cylinder is consumed by primary and secondary fragmentation mechanisms within the explosive fireball. Random larger high-velocity fragments escape the fireball in this test, providing the dramatic streamers exhibiting further ablative breakup and burning in the air environment.

The present analysis of the problem exploits a model more commonly applied to the mechanics of penetration into solid bodies. The model is applied to the present application and joined with the previous fragmentation theories to assess hydro-erosion fragmentation.

3.3.3.1 Specifics of the processes leading to hydro-erosion fragmentation

Computation and analysis of the motions and breakup of imploding metal have resulted in a reasonably clear picture of the phenomena (see the discussion of Gurney analysis in chapter 4). The sequence of events is illustrated in figure 3.33. Explosive detonation leads to rapid inward acceleration of shell metals with maximum inward

Figure 3.32. Fireball and streamers from the implosion of a hollow cylinder of a depleted uranium metal. Copyright 2010 with permission of Springer.

velocities $v_{in}^{\max}$ reaching values in the range of about 1–3 km s^{-1}. Collapse of the metal shell structure, such as in the event depicted in figure 3.32, is achieved within 30–100 μs, resulting in extremely high transient pressures. The concomitant entropy production during the shock process leaves the metal at elevated temperatures. After near-adiabatic pressure release, metal will persist in a high-temperature solid or liquid state. Several percent levels of metal vapor are also calculated. The phase fractions and temperatures depend on the initial configuration of the explosive–metal system. The collapse of the metal shell and high transient pressures near the metal core results in an outward-directed shock wave within the compressed metal. The shock wave arrests the inward motion and reverses the direction of the velocity vector. Peak outward velocities achieve an appreciable fraction of the maximum inward velocity. Scaling studies have shown the approximate relationship

$$v_{\text{out}}^{\max} = \alpha\sqrt{v_{\text{in}}^{\max}}, \tag{3.131}$$

where α is a constant (Grady *et al* 1991). This relation between the maximum outward velocity and the maximum inward velocity is appropriate over a sensibly broad range of system parameters. The sequence of events leading to explosive shock acceleration and subsequent fragmentation are illustrated in figure 3.33

The outwardly directed shock wave in the compressed metal is reflected at the metal–explosive interface. Interior regions of the metal are carried into tension and spall fragmentation of the shell metal initiates. Spall strain rates lead to fragmentation into the tens to hundreds of micrometers particle size range. Collapse irregularities and shadow regions result in some fraction of larger fragments. Spall fragmentation proceeds through much of the interior of the metal. Due, however, to confinement of

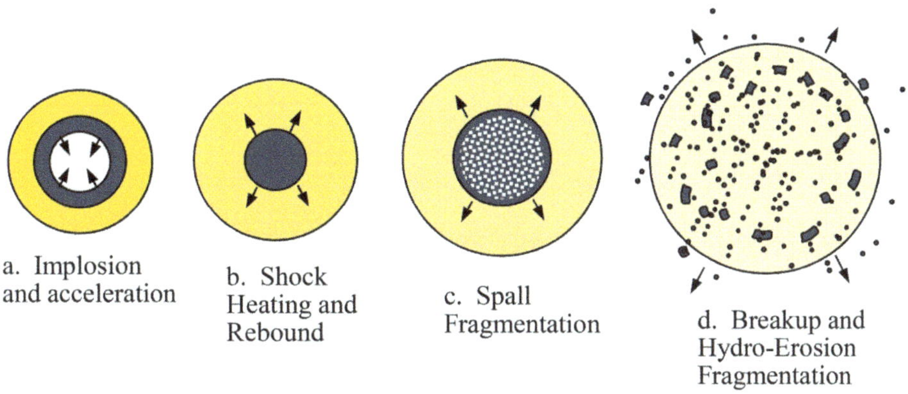

Figure 3.33. Sequence of events leading to implosion-induced hydro-erosion fragmentation.

the explosive products, the propagating spall front is quenched before reaching the metal–explosive interface. A thin contiguous metal shell (of order several millimeters in thickness) continues to expand outward, driving the explosive products and preventing mixing of these products with the interior metal spall fragments.

This state of the system is transient. Expansion and thinning of the shell continues until a critical configuration is achieved at which the metal shell undergoes fragmentation through ductile instability breakup (Grady *et al* 1990). Breakup permits the first mix of explosive products and metal fragments. Material in this outer metal shell is at the lowest temperatures as shock pressures are least intense at this radius. The material is solid (possibly liquid in very intense implosion configurations). This shell constitutes 20–40% of the initial shell metal.

Material in the expanding shell has the highest outward velocities and after breakup is injected into the explosive products, achieving a high relative velocity (of order 0.5–1.5 km s^{-1}). Explosive products, at this time in the problem (approximately 200 μs), have expanded to densities of order 1–10% of the initial density (several tens to a hundred kg m^{-3}).

At the high relative velocity of the fragmented metal particles in the high-density explosive products, further ablative and erosive fragmentation occurs. Whether metal fragments escape the near-region explosive products environment to become far-ranging fragment streamers, or whether they undergo erosion fragmentation and arrest within the near field fireball environment, depends on the initial conditions of the motion and fragmentation resistance properties of the high-velocity metal particles. Both the intensity of the initial explosive loading and the configuration of the shell metal (the thickness of the initial shell wall and extent of the shell cavity) play a key role in determining these governing conditions and outcomes of the event.

3.3.3.2 Aerodynamic breakup of liquid particles

Before embarking on a specific model of hydro-erosion fragmentation, a brief background is warranted on the topic of the aerodynamic breakup of liquid particles (or drops). This discussion draws heavily on the survey of the topic provided by Pilch

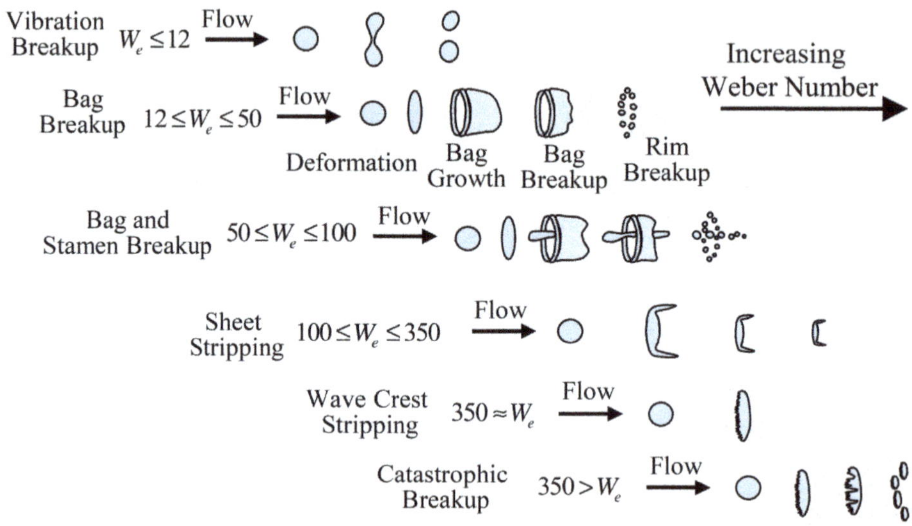

Figure 3.34. Aerodynamic fragmentation mechanisms of liquid particles (Pilch and Erdman 1987).

and Erdman (1987). Experimental observations of the breakup of liquid drops have revealed a rich display of fragmentation mechanisms. A systematic organization of the observed fragmentation dynamics is provided in this earlier study. The display of breakup mechanisms shown in figure 3.34 is a redrawing of a figure from Pilch and Erdman (1987).

The various breakup mechanisms from top to bottom reveal the effects of the increasing relative velocity of the disrupting flow, and concurrently, the increasing intensity of the fragmentation phenomena. The relative flow and breakup intensity is quantified through the magnitude of a dimensionless Weber number W_e. The Weber number is provided by a ratio of the measure of dynamic pressure ρV^2 imparted by the flow and responsible for disruption and breakup of the liquid drop, and a corresponding measure of the strength γ/D resisting the breakup process. In the dynamic pressure expression, ρ is the density of the flow media and V the relative velocity. The surface tension of the liquid is γ while D is the diameter of the drop. The Weber number is then

$$W_e = \frac{\rho V^2 D}{\gamma}. \tag{3.132}$$

It is sensible to explore comparable consequences of other governing properties of the drop material. For example, if the liquid is strongly viscid while surface tension is of little consequence, the corresponding strength of the drop is reasonably expected to be of order $\eta V/D$ where a measure of the Newtonian material viscosity is η. The ratio of the dynamic pressure to strength of the drop is then the dimensionless Reynolds number,

$$R_e = \frac{\rho V D}{\eta}. \tag{3.133}$$

The relative effects of liquid surface tension and viscosity are uncovered by joining the Weber number and Reynolds number eliminating flow velocity,

$$\sqrt{W_e}/R_e = \sqrt{\frac{\rho_d}{\rho}}\frac{\eta}{\sqrt{\rho_d \gamma D}} = \sqrt{\frac{\rho_d}{\rho}}O_n, \tag{3.134}$$

where, in these relations, ρ_d and ρ are the respective densities of the drop material and flow field material. The dimensionless Ohnesorge number, $O_n = \eta/\sqrt{\rho_d \gamma D}$, is dependent only on properties of the drop liquid, and is found to capture the transition from surface tension to viscosity dominated fragmentation. Plots of the critical Weber number governing onset of drop fragmentation are observed to exhibit marked upturn as the Ohnesorge number achieves and exceeds unity.

Similar considerations can be applied to solid particles subjected to comparable flow field conditions. The plastic yield strength Y of a ductile metal, depending on both material type and ambient temperature, can range over about 10^2–10^3 MPa. A corresponding ratio of the dynamic pressure to strength yields a dimensionless number,

$$H_Y = \frac{\rho V^2}{Y} = \frac{\rho}{\rho_d}\frac{V^2}{Y/\rho_d}. \tag{3.135}$$

Again, aero- or hydro-dynamic breakup is expected when the number H_Y achieves some critical value. The relation $\sqrt{Y/\rho_d}$ is identified as the plastic wave velocity, and ranges, for the material strengths noted above, over about 100–500 m s^{-1}.

Brittle solids can also experience catastrophic fragmentation in an intense flow environment. A rocky meteoroid entering the upper Earth atmosphere is a case in point. For example, a competent basaltic rock exhibits a tensile strength T of several hundred megapascals, and a number corresponding to equation (3.135),

$$H_T = \frac{\rho V^2}{T}, \tag{3.136}$$

would have a similar interpretation regarding onset of breakup.

It is notable that the numbers in either equation (3.135) or (3.136), in contrast to the corresponding dimensionless numbers for liquid breakup, are independent of the size scale of the particle. This scale independence would disappear if some deformation-rate or length-scale dependence was ascribed to the solid material strength.

Pilch and Erdman (1987) have described in detail the various breakup modes pictorially identified for liquid particle breakup in figure 3.34. At the lowest Weber numbers of order $W_e = 12$ or less, onset of breakup is characterized by vibrational instability and separation into several large fragments. As Weber number increases, $10 \leqslant W_e \leqslant 100$, several modes of deformation identified as bag breakup are observed. The disrupting flow initiates an expanding and thinning bag or film of liquid supported by a more substantial liquid ring. Ultimately the bag bursts, subsequently followed by breakup of the rim. In the higher flow-velocity range a stamen will form and contribute to the fragmentation complexity. At Weber numbers above about 100 a markedly difference breakup regime is entered. At the lower range a process described as sheet stripping occurs where a thin membrane of liquid is drawn to the particle periphery. The sheet continues to thin on the down-wind side of the particle and ultimately particulates. The breakup process is self-similar, scaling to smaller size as the particle is exhausted. At Weber numbers of about 350 the dynamic pressure activates the growth of Rayleigh–Taylor instability on the windward side of the particle. Crests of the instabilities are eroded and contribute to particulation by the mechanism known as wave-crest stripping. At increasingly higher Weber numbers the large-amplitude growth of Rayleigh–Taylor instabilities will lead to catastrophic particle breakup before wave stripping can significantly reduce the mass. In this intense Weber number range it is expected that the breakup will become independent of the initial particle size, undergoing repeated catastrophic breakup, cascading through the size scales until the disrupting forces are exhausted. With similarities to hydrodynamic turbulence, power-law characteristics are expected in both the breakup dynamics and the ultimate fragmentation distribution.

3.3.3.3 Hydrodynamic erosion theory of Alekseevskii and Tate

Pilch and Erdman (1987) note that two aspects of particle dynamics must be addressed in assessing aero-ballistic breakup. First, is the deceleration of the particle through the duration of breakup. Second, is the mechanics of the fragmentation process. These issues are particularly true for the problem under consideration; namely, the high-velocity insertion of hot solid and liquid metal particles into the gaseous environment of the confining explosive products. The present analysis pursues applicability to the current problem of an alternative hydro-erosion model, which has found broad application in the field of high-velocity penetration mechanics. The approach is also found useful in the prediction of secondary fragment size in the present erosion breakup of the high-velocity particles.

Birkhoff *et al* (1948) identified the fundamental ideas of the high-velocity hydrodynamic erosion of a penetrating rod of metal. They considered a rod of an incompressible inviscid liquid of density ρ_p and current length l undergoing steady-state normal penetration into an equivalent liquid of density ρ_t. By applying Bernoulli's equation,

$$p + \frac{1}{2}\rho v^2 = \text{const.}, \tag{3.137}$$

to the central streamline in both projectile and target material, they obtain the penetration and erosion Bernoulli relation,

$$\frac{1}{2}\rho_p(v - u)^2 = \frac{1}{2}\rho_t u^2. \tag{3.138}$$

The velocity of the penetrating projectile is v, whereas u is the progression velocity of the eroding interface between the target and projectile material.

Alekseevskii (1966) and Tate (1967), to account for the strength of a projectile material, modified the model of Birkhoff *et al* (1948). A strength term σ with the dimensions of stress is added to the left side of equation (3.138), providing the modified Bernoulli equation

$$\frac{1}{2}\rho_p(v - u)^2 + \sigma = \frac{1}{2}\rho_t u^2. \tag{3.139}$$

Equation (3.139) is supplemented with an equation of erosion,

$$\frac{dl}{dt} = u - v \text{ (conservation of mass)}, \tag{3.140}$$

and an equation of motion of the eroding projectile,

$$\rho_p l \frac{dv}{dt} = -\sigma \text{ (conservation of momentum)}. \tag{3.141}$$

3.3.3.4 Application to high-velocity fragment erosion

Erosion and arrest of high-velocity solid (or liquid) metal fragments undergoing relative motion in the reasonably dense high-explosive products is the application of interest. The modified hydrodynamic theory of Alekseevskii and Tate is applied to this problem. Penetrating and eroding fragments are assumed spherical with diameter l. The erosion rate dl/dt is the rate of reduction of the diameter of the eroding spherical fragment.

A model of the event is illustrated in figure 3.35. Motion from the perspective of a laboratory reference frame on the left depicts a spherical fragment of radius $r = l/2$ moving at a velocity v into an incompressible fluid at rest (the gas is certainly not incompressible but this aspect is readily modified with a compressible fluid model). Due to erosion, the front interface of the particle in contact with the fluid moves at a velocity $u \leqslant v$. It is also useful to view the process in a reference frame in which the eroding interface is at rest (see figure 3.35). In this frame, the fragment is moving at a velocity of $v - u$, while the fluid approaches from the right at a velocity u.

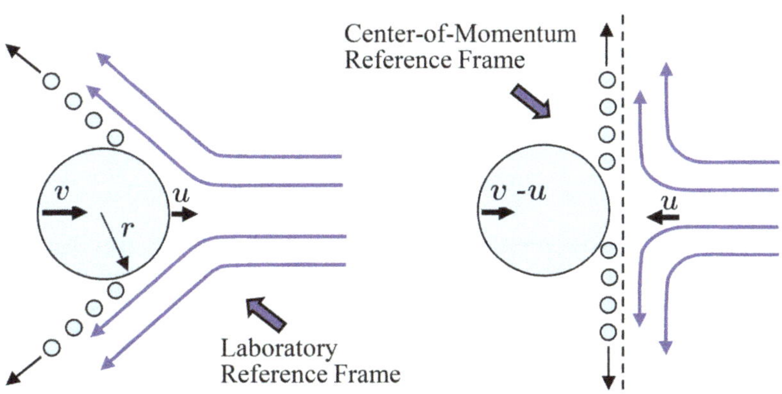

Figure 3.35. Aerodynamic fragmentation in laboratory and center-of-momentum reference frames.

The elimination of time between equations (3.140) and (3.141) provides the following differential expression for the rate of erosion,

$$\frac{dl}{l} = \frac{\rho_p}{\sigma}(v - u)dv. \tag{3.142}$$

From equation (3.139), the velocity of the stagnation point u *is* solved for

$$u = \frac{v - \mu\sqrt{v^2 - \dfrac{\sigma}{\rho_t}(1 - \mu^2)}}{1 - \mu^2}, \tag{3.143}$$

where $\mu = \sqrt{\rho_t/\rho_p}$. For $\mu^2 \ll 1$, the simplification

$$v - u = \mu\sqrt{v^2 - \frac{\sigma}{\rho_t}}, \tag{3.144}$$

is appropriate. Substitution into equation (3.142) provides an explicit differential equation,

$$\frac{dl}{l} = \frac{\rho_p}{\sigma}\mu\sqrt{\left(v^2 - \frac{\sigma}{\rho_t}\right)}dv, \tag{3.145}$$

which can, in certain cases, be solved analytically. The erosion rate goes to zero when $v = \sqrt{\sigma/\rho_t}$. The current depth of penetration d is determined from the integration

$$d = \int_0^t v dt, \tag{3.146}$$

which can be transformed to the velocity integral

$$d = -\frac{\rho_p}{\sigma}\int_{v_0}^{v} l(v) v dv. \tag{3.147}$$

The solution of equation (3.145) provides $l(v)$, and for certain limiting cases, equation (3.147) can also be solved analytically. For example, in the case of a projectile material without strength ($\sigma = 0$) complete erosion occurs with a depth of penetration $d = l_o/\mu$.

The present model of penetration is completed with a specification of the dependence of the fragment strength properties on the conditions of the problem. For a solid fragment, the fragment strength σ is approximately equal to the metal flow stress Y. The flow stress Y is known to elevate somewhat at high flowing rates so a dynamic value of Y commensurate with the high velocity rates of deformation is appropriate. Also, Y is known to be a sensitive function of temperature, dropping dramatically as temperatures approaching melt are achieved. (Here consideration is with the injection temperatures of the metal fragments incurred during shock heating prior to the prompt fragmentation process and not additional ablative or frictional heating.) If the injected metal fragment is in the liquid state, then surface tension γ provides the principal resistance to erosion breakup. It is reasonable to propose a strength term of $\sigma = \gamma / l$ to complete the erosion and penetration analysis through equations (3.145) and (3.147).

Following the physics of erosion resistance somewhat further, it is reasonable to consider possible reductions in strength of the particle surface due to friction or through heat conduction due to intimate contact with the high-temperature explosive products. The effect of surface layer heating would be to lower the flow stress and enhance erosion fragmentation.

Predictions of secondary fragment size associated with the erosion products follow from the present erosion model. As is shown in the center-of-momentum reference frame in figure 3.35, erosion fragments are ejected along radial streamlines. For velocities reasonably high relative to σ / ρ_t, kinetic energy is conserved on these streamlines and the divergent strain rate associated with erosion fragmentation is

$$\dot{\varepsilon} \approx \mu \frac{v}{r}, \tag{3.148}$$

where r is the current radius of the particle. In this reference frame, an average fragment size associated with the erosion fragments is calculated from the energy balance relation developed in the previous chapter,

$$s \approx \left(\frac{\Gamma}{\rho_p \dot{\varepsilon}^2} \right)^{1/3}. \tag{3.149}$$

The fracture surface energy of the metal particle is Γ and is determined from either the flow stress of the solid metal or the surface tension of the liquid metal.

The present model of erosion fragmentation has not been widely applied. Some experiments were conducted by the author in which small metal spheres were accelerated to between 1 and 3 km s^{-1} and injected into low-density ($\sim$100 kg m^{-3}) polyurethane foams. Flash radiography examined penetration and erosion. The cursory results of the study were sensibly consistent with the present model of fragment deceleration and erosion.

3.4 Selected applications in dynamic fragmentation

The catastrophic disruption and breakup of solid structures brought about by an abrupt intensely energetic event such as an explosion or high-velocity impact has dominated the discussions of dynamic fragmentation to this point. Further examples of dynamic fragmentation are considered in this section that reach into areas of physics that are not typical of the earlier studies. One purpose is to show that aspects of the energy principles introduced earlier have application to physical phenomena unrelated to mechanical fracture. Physical transformation brought about by a transient stimulus (a shock wave for example) is typically heterogeneous, exhibiting a granular or cellular structure with a dominant length scale. Examples can include polymorphic or metallurgic transformation, inelastic deformation such as adiabatic shear localization and energetic reaction, among others. One example described here will be that of a transformation in the domain structure of a ferromagnetic ceramic when subjected to an intense shock wave. Energy principles found appropriate in this early study of shock demagnetization provided the stimulus for the application of energy balance concepts to the dynamic fragmentation of solids (Grady 1982a). These and other diverse applications of the theoretical dynamic fragmentation concepts are explored in the present section.

3.4.1 Fragmentation in the shock demagnetization event

Shock demagnetization is a striking example of an alternative physical phenomenon where fragmentation plays an important role in the shock wave event (Shaner and Royce 1968, Grady *et al* 1972). The shock demagnetization event is illustrated in figure 3.36. A ferrimagnetic material is subjected to an external applied magnetic field H of strength sufficient to saturate the material to its characteristic magnetization strength M_s. (In contrast to ferromagnetism typical of metals, the more complex exchange interaction in certain ceramics is identified as ferrimagnetism.) The material is further subjected to a shock wave with propagation direction orthogonal to the applied field and magnetization direction. Magnetization of the ferrimagnetic material is reduced behind the shock wave by an amount dependent on both the strength of the shock wave and the amplitude of the external magnetic field.

The physics of the shock demagnetization process is illustrated in figure 3.37. On the left, the external magnetic field has oriented the magnetic moment in the crystal lattice along the direction of the field. On the right the uniaxial elastic strain of the crystal lattice brought about by the shock wave induces magneto-elastic forces that compete with the external field and rotate the magnetic unit toward alignment with the shock propagation direction. The process is not so simply illustrated when the

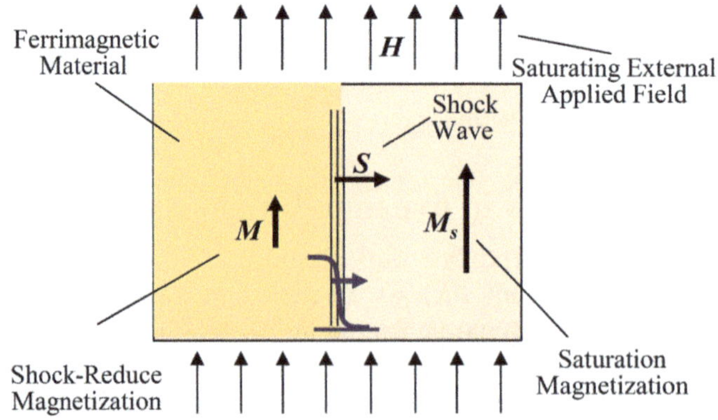

Figure 3.36. Shock demagnetization, or the reduction in magnetization produced by a shock wave normal to the applied field and directed magnetization.

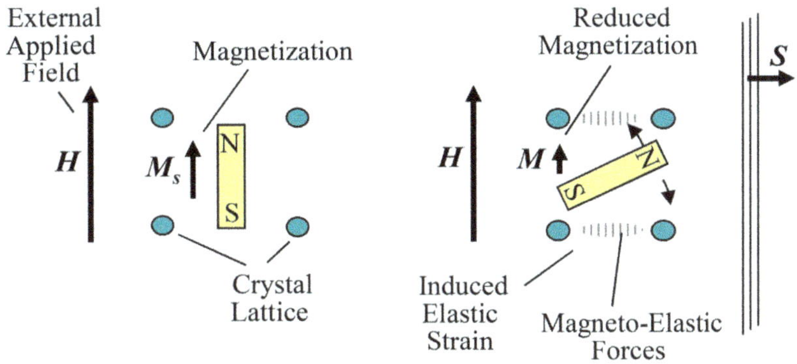

Figure 3.37. Magneto-elastic forces brought about by elastic strain behind the shock wave rotate the magnetic moment out of alignment with the applied field and reduce the directed magnetization.

crystal lattice is arbitrarily oriented relative to the applied field and shock propagation direction but the same physics nonetheless applies.

Rotation of the magnetic moment toward alignment with the shock direction does not occur uniformly over the entire body. Rather, select regions rotate with the north pole oriented towards the shock direction, while other regions rotate in the opposite direction, as illustrated in figure 3.38. Magnetic regions that rotate in synchronism are called magnetic domains and are separated by a thin boundary layer of magnetic material identified as domain walls.

The breakup into ferrimagnetic domains during the shock demagnetization event occurs to lower the self-energy of the magnet (also called the demagnetization energy). The smaller the domain size, as quantified by the domain width d identified in figure 3.38, the lower the magnetic self-energy. The decomposition into ferrimagnetic domains has its own energy expense, however. Disorientation of the magnetic vector through the domain wall, which occurs over a few tens to a few hundred lattice spacings, is accompanied with an increase in a quantum-based magnetic exchange energy. The exchange energy is quantified as domain wall energy (energy per unit area). Thus, reduction of the domain size leads to an increase in the magnetic domain wall interface area, and energy.

A theory of the shock demagnetization process based on energy methods has been developed that provides a sensible description of the phenomenon (Grady 1972a, Grady *et al* 1972). The theory in turn is based on a more general description of energy-based principles in the magnetic materials (Kittel 1949). An energy expression that depends on the magnetic rotation θ and the domain wall spacing d is provided by

$$U(\theta,d) = -M_s H \cos\theta + b\varepsilon \sin^2\theta + \frac{d}{L} M_s^2 \sin\theta + \frac{8}{d}\sqrt{Ab\varepsilon}\,\sin^2\theta. \qquad (3.150)$$

On the right the respective terms are the energy of the external magnetic field, the magneto-elastic energy, the magnetic self-energy and the ferrimagnetic exchange (domain wall) energy.

The energy, actually a Legendre transform energy, is a function of the magnetic rotation and the domain wall spacing. The magnitudes of the demagnetization

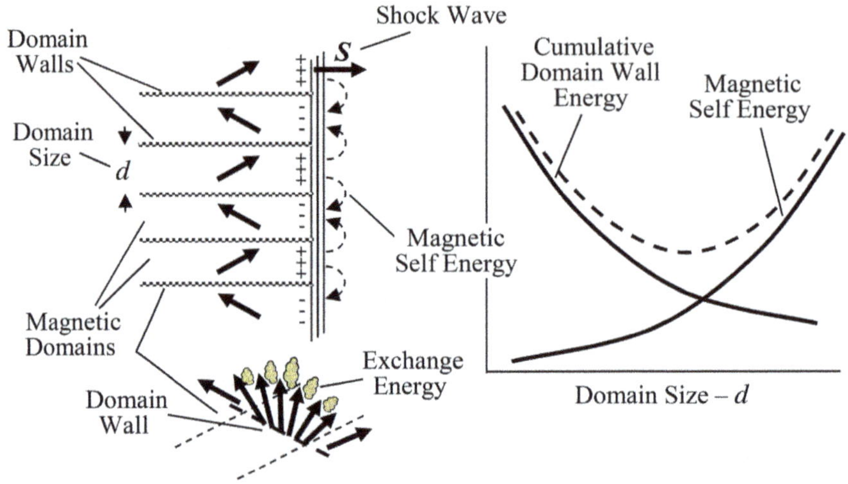

Figure 3.38. Shock rotation of the magnetic moment and decomposition (fragmentation) into ferrimagnetic domains is illustrated on the left. On the right is shown the competing energy terms that determine the characteristic domain size (Grady 1972a, Grady *et al* 1972).

(magnetic rotation) and the domain wall spacing are predicted through minimization of the magnetic energy expression in equation (3.150) at specified values of the applied field H and the amplitude of the shock strain ε. Specific magnetic properties are provided by the saturation magnetization M_s, the magneto-elastic constant b, the magnet thickness L and the exchange energy constant A.

Equation (3.150) has strict application only to single-crystal ferrimagnets with shock-strain oriented along the symmetry [100] or [111] axis where the magneto-elastic constant b takes the cubic crystal values of b_1 or b_2, respectively. Complications arise for polycrystalline ferrimagnets where suitable averages based on simplifying assumptions are required. Such treatment of the physical properties of polycrystalline material is common to other physical phenomena—the Voigt and the Ruess bounds for the elastic constants of polycrystalline solids being a familiar one.

A theory of demagnetization of polycrystalline ferrimagnets based on a sensible averaging of the energy expression in equation (3.150) is found in accordance with experimental shock demagnetization data (Grady 1972b, Grady *et al* 1972). Experimental results based on the test method described, and performed on the ferrimagnetic yttrium iron garnet (YIG) ceramic, are shown in figure 3.39 to be in sensible agreement with the theory. Porosity inherent to ceramics further complicates the shock wave response through a component of non-uniaxial strain in the neighborhood of pores. The demagnetizing contribution due to porosity (Grady 1971) is shown to be small through the range of shock states tested in the reported study.

The nominal spacing of ferrimagnetic domains (fragmentation) achieved in the shock compression event as assessed from the energy-based theory is also shown in the figure. Although this aspect of the theory was not tested, it is integral to the development and illustrative of dynamic fragmentation in other areas of shock compression science.

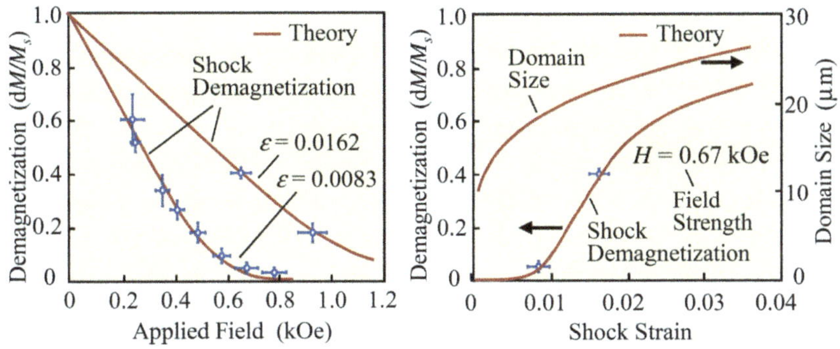

Figure 3.39. Experimental shock demagnetization results for ferrimagnetic YIG ceramic and comparison with an energy-based prediction of demagnetization and domain size (Grady *et al* 1972).

3.4.2 Fragmentation in a nuclear collision

The consequences of a nuclear fragmentation event brought about by the 70 MeV per nucleon collision of a carbon nucleus (atomic mass 12) with a silver nucleus (atomic mass 107) is shown in figure 3.40 (Greiner and Stocker 1985). In this event, 16 electrically charged fragments were produced ranging from deuterium (two nucleons) to beryllium (nine nucleons). The fragment number distribution for this nuclear fragmentation experiment is compared with the statistical and energy-based principles of dynamic fragmentation described in the present and preceding chapters (Grady 1990).

First, a cumulative Mott plot of the fragment number against nucleon mass shown in figure 3.40 demonstrates that the data are not inconsistent with an exponential distribution, as would be surmised from a random, or Poisson, partitioning of the nuclear mass into fragments. Second, an estimate of the fragment number generated in the high-energy impact event is provided through application of the energy balance fragment size prediction in the form

$$s \approx \left(\frac{\Gamma}{\rho \dot{\varepsilon}^2}\right)^{1/3}, \tag{3.151}$$

where s is a measure of the fragment size, ρ is the mass density, Γ is the fracture surface energy and $\dot{\varepsilon}$ is the expansion rate. Fragmentation analysis makes use of a liquid-drop model of the atomic nucleus introduced by Bohr (e.g. Evans 1955), although an approximate expression relating surface energy to the nuclear binding energy is used here. The leading constant of order unity in equation (3.151) is not included because of the elementary nature of the analysis. The number of fragments produced in the collision is then

$$N \approx \frac{R^3}{s^3} \approx \frac{\rho R^3 \dot{\varepsilon}^2}{\Gamma}, \tag{3.152}$$

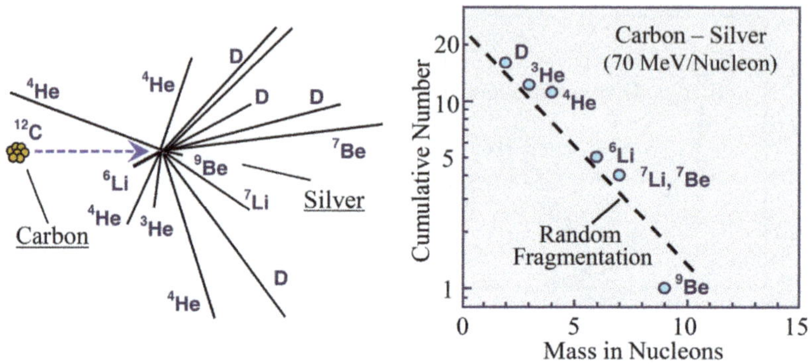

Figure 3.40. Left: the fragmentation results recorded in a photographic emulsion caused by the 70 MeV per nucleon relative collision of a carbon and a silver nucleus. From Greiner and Stocker (1985), attributed therein to Bo Jakobsson and colleagues at the University of Lund. Right: comparison of cumulative fragment number distribution and random fragmentation (Grady 1990).

where R is a measure of the nuclear radius prior to fragmentation. The expansion strain rate is determined from the impact kinetic energy through

$$\dot{\varepsilon}^2 \approx \frac{\alpha V^2/2}{R^2}. \tag{3.153}$$

The impact specific kinetic energy is $E_k = V^2/2$, whereas αE_k is the subsequent expansion kinetic energy driving the breakup process. The factor α accounts for a modest reduction from the impact kinetic energy required by momentum conservation. For the carbon-silver system α is about 0.9. Equation (3.152) becomes

$$N \approx \frac{\alpha R E_k}{\Gamma/\rho}. \tag{3.154}$$

Breakup of the nucleus is resisted by the nuclear binding energy Γ_b. The nuclear binding energy is an energy per unit mass, and is related to the surface energy in equation (3.154) through $\Gamma/\rho \simeq \Gamma_b s$. Equation (3.154) becomes

$$N \approx \frac{R}{s}\frac{\alpha E_k}{\Gamma_b}. \tag{3.155}$$

The ratio $R/s = N^{1/3}$ from equation (3.152) yields the relation,

$$N \approx \left(\frac{\alpha E_k}{\Gamma_b}\right)^{3/2}, \tag{3.156}$$

for the number of nuclear fragments produced in the collision fragmentation event. Using $E_k = 70$ MeV/nucleon, $\alpha = 0.9$ and $\Gamma_b = 8.5$ MeV/nucleon (the nuclear binding energy of the elements (e.g. Evans 1955) is shown in figure 3.41), a fragment number of approximately $N = 20$ is calculated from equation (3.156). The calculated number is remarkably close to the intercept of the linear fit to the cumulative fragment

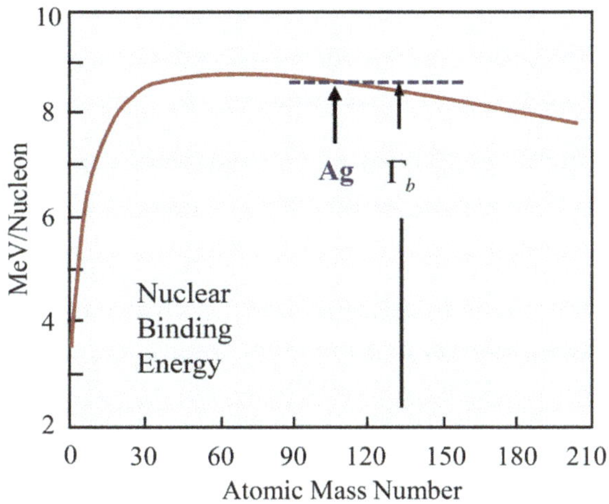

Figure 3.41. Nuclear binding energy of the elements with that for silver identified.

number distribution of the test data provided in figure 3.40. Sixteen fragments were observed in the test.

3.4.3 Lennard-Jones fragmentation

The random fragmentation of an infinite line provides a geometric statistics example whose fragmentation is in reasonable accord with experiment in certain fragmentation applications (e.g. Lienau 1936). In two and three dimensions, different geometric constructions lead to different fragment distributions. Because of inherent test complexities, laboratory fragmentation experiments are seldom able to differentiate among the various statistical theories. Computer simulations of fragmentation, although fraught with their own list of problems, do not suffer from the practical limitations of laboratory testing or the necessity of constructing statistical fragmentation algorithms.

A series of computational fragmentation experiments (Holian and Grady 1988) are particularly revealing of the dynamic fragmentation phenomena. In these calculations, MD techniques are used to subject condensed matter to adiabatic expansion and homogeneous fragmentation. Because the kinematic conditions imposed on the body approximate the uniform Euclidian expansion (the Hubble constant) associated with the 'Big Bang' model for the formation and subsequent evolution of the Universe, the present calculations were referred to in the paper as a microscopic big bang. The analysis parallels MD simulations of fragmentation of liquids by Blink and Hoover (1985) induced by rapid heating.

Informative features of the dynamic fragmentation process are uncovered in the computer simulations. The theory of energy balance between the kinetic energy of expansion and the potential energy of breaking molecular bounds are found to be in reasonable agreement with the calculations. Furthermore, statistical size distributions are in sensible accord with a Poisson nature of the breakup process.

3.4.3.1 Specifics of the molecular dynamics fragmentation calculation

The computational material selected for study is a two-dimensional fluid using a Lennard-Jones pair potential (inverse-twelfth-power repulsion and inverse-sixth-power attraction) for the intermolecular force. The Lennard-Jones units are the molecular mass m, the characteristic length scale σ at the pair potential crossing point and the well depth energy ε, as illustrated in figure 3.42. The initial state of the fluid is at a density of $\rho\sigma^2/m = 0.75$ and a temperature of $kT/\varepsilon = 0.6$.

The two-dimensional fluid consisted of 4200 particles subjected to an initial condition of uniform expansion characterized by an expansion rate $\dot{\eta}$. The initial velocity of each particle consists of an average separation velocity from its nearest neighbors proportional to $\sigma\dot{\eta}$ and a random velocity component associated with its current thermal state. The finite region is subjected to periodic boundary conditions, thus reasonably approximating an infinite region experiencing homogeneous expansion. Five subsonic expansion rates ranging from $\dot{\eta}\tau = 0.0535$ to $\dot{\eta}\tau = 0.856$ are used in the series of fragmentation experiments where $\tau = (\sigma^2 m/\varepsilon)^{1/2}$ is the characteristic Lennard-Jones time.

The system is initially well within the fluid state. Expansion is nearly isentropic and fragmentation initiates near the spinodal of the liquid–gas transition where insipient cavities form due to thermal fluctuations. This initial cavitation is illustrated in the earlier time image shown in figure 3.42. The termination of the calculation is illustrated in the later image. The state of the material at this later time is non-equilibrium as most of the fragments are far from spherical and the forces of surface (line) tension have not yet brought about spherical shape. Furthermore, the kinetic temperature indicates that at this state most of the material should be vaporized. Hence continued calculation would lead to additional boiling (detachment) of molecules from the larger clusters until equilibrium is achieved.

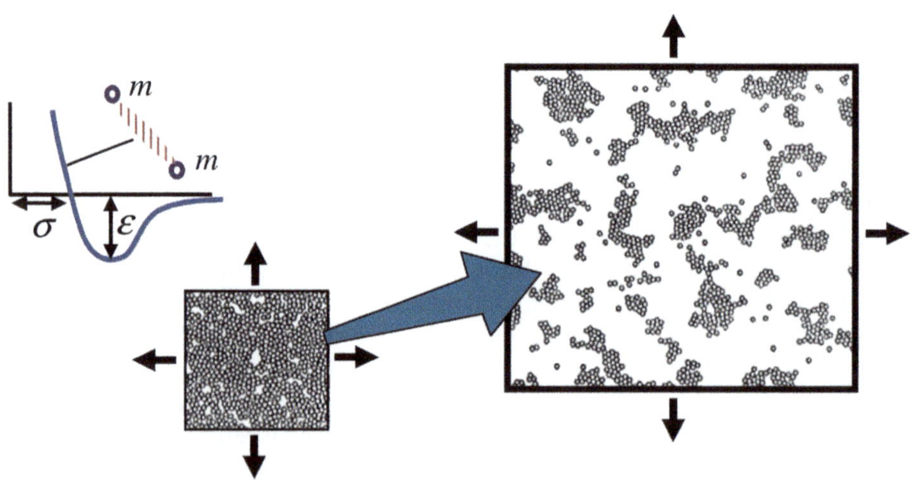

Figure 3.42. Fragmentation is brought about by the uniform expansion of a two-dimensional fluid composed of particles exhibiting two-body Lennard-Jones bonding (Holian and Grady 1988).

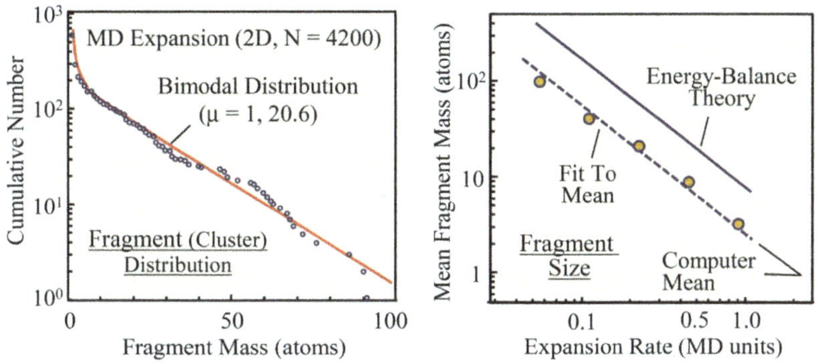

Figure 3.43. Computed average fragment size (mass) is compared with the energy-based prediction on the right. The distribution of fragment masses for one calculation is shown on the left with a bimodal (two exponentials) fit to the computed distribution (Holian and Grady 1988).

At termination of the five calculations at five different rates of expansion, the number and mass of clusters resulting from the fragmentation event are calculated based on a systematic algorithm that accounts for all cluster connectivity. The distribution resulting from one such calculation at an expansion rate of $\dot{\eta}\tau = 0.21$ is shown on the left in figure 3.43. A semi-logarithmic or Mott plot of the cumulative number distribution against the cluster mass is shown. The preponderance of the mass of the body is well described by an exponential distribution with average mass $\mu = 20.6$. The very small fragments (single-molecule, two-molecule and three-molecule clusters) show an upturn, which probably represents a transient state associated with vaporization kinetics. The solid curve is a bimodal exponential fit with exponential scale parameters of 1.0 and 20.6. An exponential distribution of fragments resulting from the MD fragmentation simulation is consistent with fragmentation governed by a random Poisson process with a uniform hazard function (Grady and Kipp 1985). This computer simulation rules out other random geometric constructions such as the Voronoi fragmentation, which results in a markedly different distribution shape. In particular, the observed exponential distribution identifies a fragment size scale for the fragmentation process. The distribution does not exhibit a power-law behavior, which would imply a fractal nature of the fragment size and no governing length scale.

3.4.3.2 Energy-based predictions of fragment mass

The average fragment mass (the larger exponential scale parameter) for each of the five MD simulations are plotted against the expansion rate on the right in figure 3.43. This aspect of the computational fragmentation calculation is compared with the earlier kinetic-energy-based theory of dynamic fragmentation (Grady 1982a), and found to be in reasonable accord (Holian and Grady 1988) with the theory. An average fragment mass is provided through $\mu \simeq \rho s^2$, where the fragment size $s \simeq (\gamma/\rho\dot{\eta}^2)^{1/3}$ results from the kinetic energy theory of fragmentation. The surface energy γ for the Lennard-Jones liquid is calculated from the molecular cohesive energy through $\gamma \simeq \rho E_{coh}\sigma$, where the cohesive energy is $E_{coh} \simeq 3\varepsilon/m$. An average fragment mass,

$$\mu \simeq 3\pi\rho\left(\frac{\sigma\varepsilon}{m\dot{\eta}^2}\right)^{2/3}, \tag{3.157}$$

is calculated for a Lennard-Jones liquid undergoing homogeneous fragmentation at an expansion rate of $\dot{\eta}$.

The mass predicted from the energy-based theory is compared with the computed average fragment sizes in figure 3.43. Prediction of fragment mass decreasing as the 4/3 power of the increasing expansion rate is in accord with the computed result. The quantitative prediction is approximately a factor of three higher than the MD calculations. Arguments were offered that the fractal character of the cluster connectivity could account in part for the discrepancy (Holian and Grady 1988). Alternatively, a later formulation of the energy theory (Grady 1988), which includes elastic strain energy, brings the theoretical predictions and computational results into markedly closer agreement (e.g. Ashurst and Holian 1999).

3.4.4 Cosmic fragmentation

There is no grander stage for dramatic examples of dynamic fragmentation than the Universe beyond. Examples abound in the Solar System, in the Milky Way Galaxy, and in the mist of the galactic expanse. Several cosmic events are explored that are indicative of dynamic fragmentation in the formation process. The topics are that of the breakup of the Shoemaker–Levi comet on entry to the Jovian gravity field, the birth of stars in the NGC 3603 Carina stellar cluster and the size distribution of galaxies consequent to early-Universe formation. Much of the discussion is comparable to more familiar dynamic fragmentation events introduced earlier in this chapter and in chapter 2. The physics, where applied, is sensible, but should be regarded as speculative.

3.4.4.1 Galactic fragmentation

Of the many dramatic fragmentation phenomena that can occur in the reaches of space outside of this planet, the event that reigns supreme is that of the creation of the known Universe. The theory of this creation most commonly accepted is that of a cataclysmic release of energy (often referred to as the ‘Big Bang’) that resulted in the globally expanding collection of galactic matter viewed today. This creation event is illustrated in figure 3.44.

The Universe is observed to be reasonably isotropic and to undergo uniform expansion, when viewed on an adequately large length scale, with a velocity of recession V that is proportional to the distance r,

$$V = Hr, \tag{3.158}$$

where H is the Hubble constant. The distance r is measured from any point selected as the origin of the coordinate system. A classical continuum mechanics description

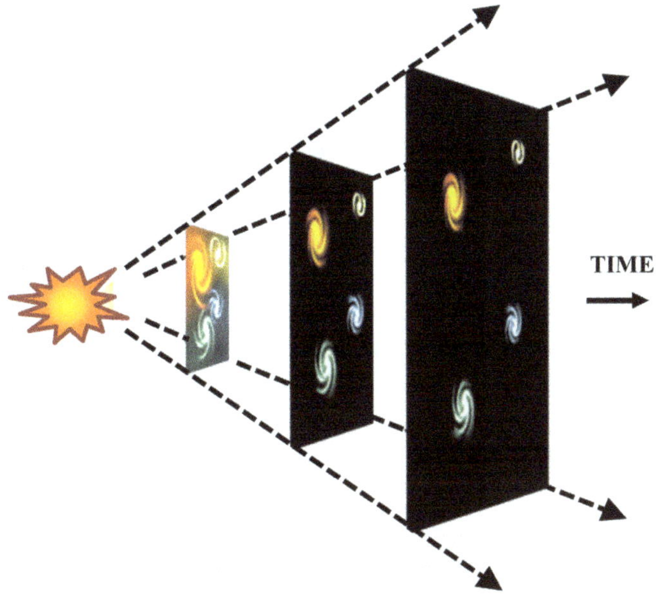

Figure 3.44. The singularity illustrates the time at zero origin of the Universe while fragmentation and subsequent expansion is pictured at the later times. This image has been modified by the author(s) from the Wikimedia website [https://commons.wikimedia.org/wiki/File:Universe_expansion2.png], where it is stated to have been released into the public domain. It is included within this article on that basis.

of the dynamic process at any time provides a uniform dilational strain rate $\dot{\rho}/\rho$ in terms of the current density ρ, given by

$$\dot{\rho}/\rho = -\text{div}\,V = -3H. \tag{3.159}$$

It is a reasonable assumption to extrapolate this observed dynamics of the present Universe back to earlier times in the creation, and attempt an understanding of galaxy formation from the physics of more common fragmentation phenomena. For example experiments conducted to cause the dynamic fragmentation of rapidly expanding metal rings and shells lead to collections of fragments whose sizes and distributions are reasonably well understood in terms of material properties and energetics of the breakup event. The latter energetics is commonly quantified through an expansion rate much like equation (3.159).

In the dynamic breakup of metal rings and shells the metal ductility and fracture toughness provide the forces resisting breakup, and counterbalance the expansion energy in achieving a characteristic fragment size and distribution. If such a correspondence with more common fragmentation makes sense, then the corresponding resisting forces at the critical time of fragmentation in the Big Bang are not as well understood, although gravity probably played a significant role. Further, more recent astronomical theories and observations postulate an inflationary period in the early formation, and that the measured Hubble constant is not constant, but is increasing as the Universe expands (e.g. Schilling 2014). Thus, whether the current Hubble constant, as couched in equation (3.159), has any relevance to the expansion energetics at the time of fragmentation is open to question. It nevertheless remains a useful exercise to consider the consequences of this original fragmentation event in terms of classical theories of dynamic fragmentation.

The observable consequences of this fragmentation event are the vast number of galaxies currently viewed in the visible Universe. (More recent evidence for an appreciable dark matter component to the Universe complicates this observation.) These galaxies are possibly the fragments of the original breakup event, and their stochastic distributions in size and spacing are suggestive of this premise. The premise of an initial fragmentation genesis, of course, can also be questioned. With the ever-increasing observational resolution of galactic dynamics, the collisions of neighboring galaxies and the subsequent disruption show, to some degree, a continuing modification of the primordial galactic size distribution. Whether the original distribution has been significantly modified is not known. Further, with recent imaging of the Hubble telescope, the distribution of galaxies throughout a vast lacelike network speaks perhaps to a hierarchy of fragmentation events accounting for the present state of the Universe.

Still, an investigation of the galactic size distribution in the framework typical of more common fragmentation phenomena is of interest. Frequently, fragment distributions are observed to be exponential, or exponential-like (Voronoi, Weibull), indicative of a characteristic length scale governing the fragmentation physics. Alternatively, and common to certain material types, power-law distributions of fragments are observed, attesting to the lack of a governing length scale and the fractal nature of the breakup phenomena.

A study of this type is described here that was carried out over 30 years ago (Brown *et al* 1983), and is principally the efforts of the first author of this publication. Certainly, with the revolution in galactic observational tools over the past several decades, a more thorough study could be conducted today. The results of this early study are nonetheless of interest, and may encourage a more extensive analysis.

3.4.4.2 The galactic number distribution

This assessment of galactic fragmentation started with the data available in the *Second Reference Catalogue of Bright Galaxies* (de Vaucouleurs et al 1976), which contains information on 4364 galaxies of which 1011 galaxies contained sufficient information for the fragmentation distribution analysis. The data allow calculation of the absolute luminosity, which is assumed to be proportional to the galactic (fragment) mass.

Various plots of the galactic distribution data were attempted. The most successful was a semi-logarithmic plot (a Mott plot) of the cumulative number of galaxies versus the absolute galactic luminosity shown on the left in figure 3.45. In particular, these data were not well described by a power-law cumulative number distribution. Separate plots were made of the different galaxy types (elliptical, spiral, and lenticular). These distributions looked the same as the collective distribution.

It was recognized in the study that increasingly distant galaxies would be increasingly difficult to observe, and the smaller, fainter galaxies would be missed more readily. An algorithm was developed to account for missing galaxies by

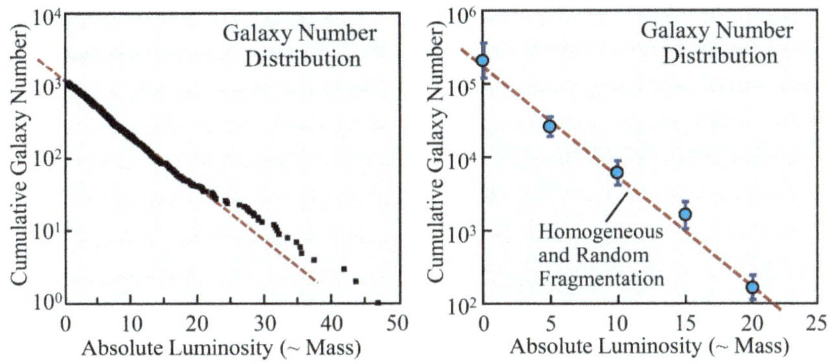

Figure 3.45. Left: cumulative galaxy number versus absolute luminosity, assumed proportional to galactic mass, is shown for 1011 galaxies from the *Second Reference Catalogue of Bright Galaxies* (de Vaucouleurs *et al* 1976). Right: the corrected galactic number distribution over a representative volume. The distribution is corrected based on assumption of universal homogeneity (Brown *et al* 1983).

assuming universal homogeneity over the representation volume of galaxies. This corrected representation of the galactic distribution is displayed on the right in figure 3.45. Again, the distribution exhibits the near-exponential character observed in the earlier figure. The uncertainties shown are a consequence of the processes used to account for universal homogeneity (Brown *et al* 1983).

3.4.4.3 Comparisons with conventional fragmentation

Although the representation of the distribution may still suffer from selective loss of galaxies toward the smaller, fainter end, the apparent exponential character of the distribution has interesting implications. Abrupt, dynamic fragmentation events are typically observed to be exponential in character. This is observed in both experiment and computation. Computer simulations of condensed matter subjected to uniform expansion to fragmentation show an exponential distribution of the resulting fragments over mass (Holian and Grady 1988). Further, the most probable distribution of mass over a fixed number of fragments, ignoring any altering physics, is exponential (Grady and Kipp 1985).

In contrast, it is observed that when material is subjected to fragmentation through multiple breakup processes—commonly called continued comminution—the resulting number distribution is invariably power-law. For example, when a brittle solid is reduced to powder in a ball mill, the final particle distribution is power-law. The crushing of coal leads to a power-law distribution (Bennett 1936). The shatter of spheres of glass and brittle minerals pressed between platens until failure provides power-law distributions of the resulting fragment debris. The distributions of sizes of asteroids in solar orbit, which are believed to have evolved through repetitive collisions over an extended time span, are reported to be power-law (e.g. Donnison and Sugden 1984).

The apparent exponential character of the galactic distribution inferred from the investigation of the available data, when compared with the nature of the size distributions resulting from more conventional fragmentation phenomena, suggests that the current distribution of galaxies are the result of the original primordial fragmentation event. Modifications of the distribution due to subsequent galactic interactions would appear to be reasonably minor.

3.4.4.4 Further features of the galactic distribution

The observation of galactic clustering is a feature that does not have a counterpart in more common fragmentation applications. In fact, in the statistical fragmentation of a rapidly expanding ductile metal ring, the distribution in spacing between fragments is the same as that of the distribution in fragment lengths (Grady 2006). The statistical spread in both fragment length and spacing is described by a Weibull distribution, which is a particular exponential representation. Further, the distribution in spacing retains the same shape; only the scale parameter evolves in time.

It is still an open question as to whether or not the observed clustering of the galaxies is some remnant signature of the original fragmentation event—perhaps a hint of the turbulence associated with the dynamics of clumping leading to fragmentation. Alternatively, it is reasonable to speculate that clustering is a post-fragmentation feature of the galactic distribution. The long-range forces of gravity have continued to influence the spatial separation of galaxies, leading to galactic interaction and the slow dissipation of the local kinetic energy of separation of select groups of galaxies. Such a process could lead to local collections of gravitationally bound galaxies and the observed clustering.

One further point concerning the exponential distribution is of interest. It would not have been surprising if the number distribution plotted in figure 3.45 was found to be power-law. There are many examples in nature where such distributions in size are observed to be fractal. The governing physics leading to the power-law, and fractal, character of such distributions would not contain a characteristic size scale over the span of the distribution observed. In contrast, the distribution of the galaxies observed within the present limited data, if indeed exponential, provides a characteristic size scale, where again proportionality is assumed between galactic mass and luminosity. In every day fragmentation phenomena, this length scale relates directly to the energy resisting the breakup.

3.4.4.5 Fragmentation of the Shoemaker–Levi comet

One of the more dramatic examples of an extra-terrestrial fragmentation event is that of the breakup of the Shoemaker–Levi comet prior to impact on the Jovian planetary surface. The well-known image of the disrupted comet is shown in figure 3.46. Fragmentation of the comet was most likely a consequence of tidal forces brought about by entry into the gravitational field of Jupiter. It is instructive to explore details of the comet fragmentation process through physical models that have proven productive in the understanding of more conventional fragmentation phenomena.

The original size of the Shoemaker–Levi comet is not well constrained. Initial diameters ranging from about 4 or 5 km to as much as 15 km are reported. The competence of the initial constituent material is also not well known. The comet may have been reasonably solid with the constituent material exhibiting a reasonable

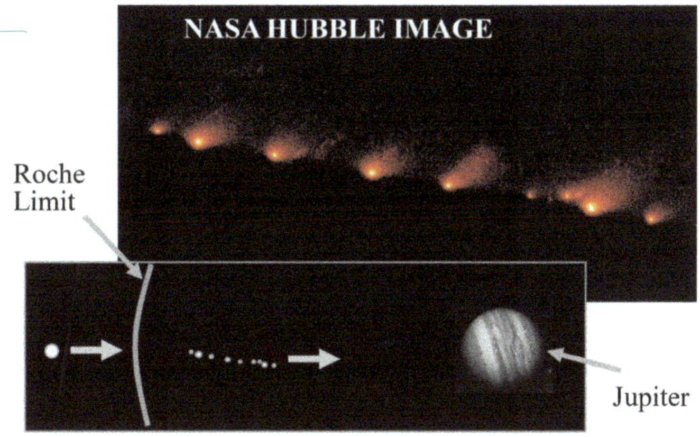

Figure 3.46. Image of the Shoemaker–Levi comet breakup on entry into the Jovian gravitational field. The lower inset illustrates a probable cause for the fragmentation event. Courtesy of NASA.

measure of strength. Alternatively, the comet may have been a gravitationally bound aggregate of loose material that would have exhibited no tensile strength and minimal shear strength. More on this issue shortly.

Disruption and breakup of the comet is brought about by tidal forces on entry into the gravitational field of Jupiter. This point is quantified as the Roche limit (e.g. Gu *et al* 2003) for the distance from Jupiter at which the gravitational gradient exceeds the self-gravitational strength of the body, and the potential for breakup is achieved. This event is illustrated qualitatively within the inset in figure 3.46. It is useful to provide a rough development of the Roche limit for purposes of assessing the fragmentation process.

The force exerted on the comet by the gravitational field of Jupiter is

$$F = G\frac{mM_J}{r^2}, \tag{3.160}$$

where m and M_J are the mass of the comet and of Jupiter, respectively, r is the distance from the center of Jupiter to the comet and G is the gravitational constant. The force differential over the diameter of the comet d is approximately

$$\delta F \simeq G\frac{mM_J}{r^3}d. \tag{3.161}$$

Thus, a measure of the tension capable of disrupting the comet is $T \sim \delta F/d^2$, or

$$T \sim G\frac{mM_J}{r^3 d}. \tag{3.162}$$

The self-gravitational energy of the comet is of the order $U \sim Gm^2/d$ and the pressure $P \sim U/d^3$, or

$$P \sim G\frac{m^2}{d^4}. \tag{3.163}$$

The Roche limit is achieved when the tidal forces characterized by the tension T are comparable to the gravitational pressure P. Equating equations (3.162) and (3.163) yields

$$r_l \sim R_J(\rho_J/\rho)^{1/3}, \tag{3.164}$$

where R_J is the radius of Jupiter, while ρ_J and ρ are the nominal mass density of Jupiter and the comet, respectively. A more exacting analysis yields

$$r_l \sim 2.456R_J(\rho_J/\rho)^{1/3} \tag{3.165}$$

for the Roche limit.

A display of the Shoemaker–Levi comet fragment distribution is provided in figure 3.47, which shows a Mott distribution plot (cumulative number larger versus fragment mass) based on the assumption that the luminosity of individual fragments is proportional to the fragment mass. The state of the comet material is not known. This leaves open the exploration of several suppositions that would play a role in the mechanisms leading to the observed fragmentation.

Following passage of the Roche limit on the inward trajectory of the asteroid, tidal forces brought about by the gravitational pull of Jupiter would increasingly distort and ultimately disrupt the body. Details of the processes of deformation would depend on the specific makeup of the asteroid. Two more or less bounding assumptions regarding the constitution of the asteroid can be considered. First, the asteroid might be composed of reasonably competent solid rock. Tidal forces would increasingly distort the asteroid until the elastic strain energy achieved a level at which fracture would activate and propagate through the body, breaking the asteroid into a collection of fragments. The number of fragments produced would depend on the magnitude of the elastic strain energy at the onset of fracture.

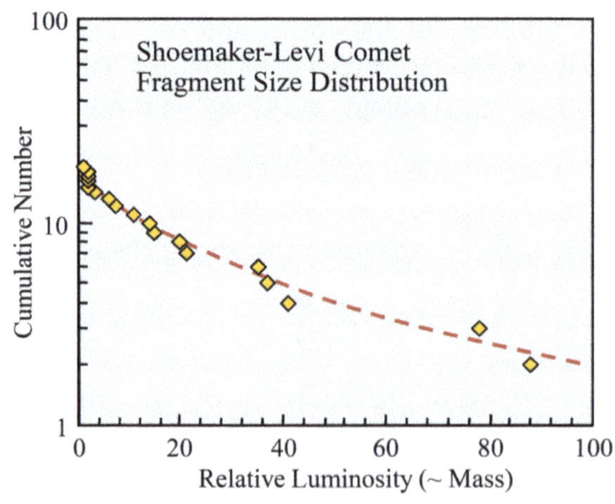

Figure 3.47. Mott plot fragment size distribution of the Shoemaker–Levi comet breakup based on the assumption that fragment mass is proportional to the relative luminosity

A second condition of the asteroid that is reasonable to consider is that of a gravitationally bound aggregate of loose, gravel-like material. Under the distortional loads provided by both the self-gravity, and the gravitational gradient of Jupiter, this material would flow in shear, although probably with a modest level of resisting frictional stress. The nature of the gravitational loads after passage of the Roche limit would lead to gradual elongation of the body along the axis of Jovian gravitational force. At some point in the dynamics of this process, instability forces would lead to the disruption and fragmentation of the asteroid.

3.4.4.6 Birth of stars

A more recent cosmic event, on a galactic time scale, has occurred that is evocative of dynamic fragmentation in the present context. In a region of the Milky Way, at a distance of about 20 000 light years from Earth, an enormous body of dust and gas imploded under the forces of gravitational collapse. The subsequent rebound resulted in an expanding profusion of fragments, which soon ignited into new-born stars. This star cluster in the Carina spiral arm, identified as NGC 3603, was first observed by John Herschel in 1834. Stars in this young cluster, shown in figure 3.48, span the size gamut, with some having masses hundreds of times that of the Sun. The largest of these stars resides closest to the center of the cluster.

Within the present theme of dynamic fragmentation, it is difficult to resist comparing this dramatic cosmic event to the imploding uranium metal shell depicted in figure 3.32 of the present chapter. The two events have outward similarities. There are also noteworthy differences. The uranium metal is imparted inward momentum through explosive energy. The uranium properties of surface tension and fracture toughness provide the balancing forces that dictate the sizes and distribution

Figure 3.48. Hubble Space Telescope image of the NGC-3603 open star cluster in the constellation Carina. Courtesy of NASA.

of resulting fragments in the outward disruption of the body. In the stellar event gravitation forces presumably play both roles. Gravitation collapse provides the inward rush of matter. One can speculate that gravity substitutes for surface tension or fracture toughness in determining the sizes of fragments in the outward breakup of the body.

A further correspondence is worthy of note. The random distribution of velocities and kinetic energies of fragments resulting from a centered exploding fragmentation event is explored in the statistical section of chapter 2. Given sufficiently random conditions leading to breakup, the statistical distribution of kinetic energy over the fragments is determined to be Maxwellian. One consequence is that larger fragments have slower velocities and remain probabilistically closer to the fragmentation origin. This is in keeping with observations of the largest stars statistically positioned nearer to the center of the NGC-3603 stellar cluster.

3.4.4.7 Galactic penetration

In the neighborhood of Milky Way exists a swarm of globular clusters and dwarf galaxies. These galactic bodies wander the reaches of space near the Milky Way and the two other major galaxies that constitute the local galactic group. On a grander scale these clusters and miniature galaxies can be likened to the collection of asteroids that roam the Solar System. In the galactic past chaotic forces have aligned to redirect one or more of these small galactic bodies into a collision course with the Milky Way. The consequences of such an event provides an interesting study in fragmentation and impact penetration. The penetration physics of Bernoulli and the theories of Alekseevski and Tate outlined earlier in the chapter come to mind.

To envision the deformation and fragmentation event, consider a liquid drop model of the small galactic body. (Think of the liquid drop model of the atomic nucleus on a vastly difference length scale.) Self-gravity of the galactic body plays the role of surface tension. As the redirected smaller body approaches the Milky Way the accelerating gravity forces and gravitational gradient overpower the self-gravity. The body elongates into an extended streamer or perhaps a collection of galactic fragments. The event can be likened, on a grander scale, to dynamics of the Shoemaker–Levi comet illustrated in figure 3.46. On impact of this galactic jet a colossal shock wave and subsequent penetration of the jet ensues. To the solid materials shock physicist, analogy with mechanics of the long-rod penetrator and the shaped-charge jet is brought to mind.

Consequences of the collision depend on the impact conditions. In the language of terminal ballistics, the angle of attack is determined by that angle between the incoming trajectory and the axis of elongation. If the angle of attack is large then the impact of the elongated body is oblique. Depending on the magnitude of the attack angle, sequential contact of the body will distribute and extend shock over some span of the outer reaches of the impacted galaxy. If the angle of attack is small then the penetration mechanics of Bernoulli comes into play.

Remnant of past such galactic penetration events are identified as streamers; threads of stars through the Milky Way moving at vast relative speeds. Notable is the Arcturus stream, possibly the vestiges of an earlier impact of a small dwarf galaxy (Schilling 2014). A colossal gas cloud, the Smith Cloud, is currently on a collision course with the Milky Way (e.g. Lockman *et al* 2008). Elongation from gravitational forces of the Milky Way is underway. The pending impact will entail cosmic shock physics on the grandest of scales.

References

Alekseevskii V P 1966 Penetration of a rod into a target at high velocity *Fiz. Goren. Vzryva* **2** 99–106

Ashurst W T and Holian B L 1999 Droplet formation from rapid expansion of a liquid *Phys. Rev.* E **59** 6742–52

Bennett J G 1936 Broken coal *J. Inst. Fuel* **10** 22–39

Birkhoff G D, MacDougall D, Pugh E and Taylor G 1948 Explosives with lined cavities *J. Appl. Phys.* **19** 563–82

Blink J A and Hoover W G 1985 Fragmentation of suddenly heated liquids *Phys. Rev.* A **32** 1027–35

Borg J P, Grady D E and Cogar J R 2001 Instability and fragmentation of expanding liquid systems *Int. J. Impact Eng.* **26** 65–76

Bosłough M B and Asay J R 1992 Basic principles of shock compression *High-pressure Shock Compression of Solids* ed J R Asay and M Shahinpoor (Berlin: Springer) pp 7–42

Brown W K, Karpp R R and Grady D E 1983 Fragmentation of the Universe *Astrophys. Space Sci.* **94** 401–12

Considère A 1885 Mémoire sur l'emploi du fer et de l'acier dans les construction *Ann. Ponts Chaussees* **9** 574–775

Courant R and Hilbert D 1962 *Methods of Mathematical Physics* vol 2 (New York: Interscience)

Curran D R, Seaman L and Shockey D A 1987 Dynamic failure of solids *Phys. Rep.* **147** 253–88

De Kee D and Wissbrun K F 1998 Polymer rheology *Phys. Today* **51**(6) 24–9

De Vaucouleurs G, De Vaucouleurs A and Corwin H G 1976 *Second Reference Catalogue of Bright Galaxies* (Austin, TX: University of Texas Press)

Donnison J R and Sugden R A 1984 The distribution of asteroid diameters *Mon. Not. Roy. Astron. Soc.* 210 673–82

Dugdale D S 1960 Yielding of steel sheets containing slits *J. Mech. Phys. Solids* **8** 100–4

Dulaney E N and Brace W F 1960 Velocity behavior of a growing crack *J. Appl. Phys.* **31** 2233–6

Evans R D 1955 *The Atomic Nucleus* (New York: McGraw-Hill)

Grady D E 1971 Origin of the linear term in the expression for the approach to saturation in ferromagnetic materials *Phys. Rev.* B **4**(11) 3982

Grady D E 1972a Shock-induced anisotropy in ferromagnetic materials I. Domain-theory analysis of single-crystal behavior *J. Appl. Phys.* **43**(4) 1942

Grady D E 1972b Method for shock wave investigation of magnetic materials *Rev. Sci. Instrum.* **43**(5) 800

Grady D E 1982a Local inertial effects in dynamic fragmentation *J. Appl. Phys.* **53** 322–5

Grady D E 1982b Analysis of prompt fragmentation in explosively-loaded uranium cylindrical shells *Sandia National Laboratories Technical Report* SAND82-0140 February

Grady D E 1987 Fragmentation of rapidly expanding jets and sheets *Int. J. Impact Eng.* **5** 285–92

Grady D E 1988 Spall strength of condensed matter *J. Mech. Phys. Solids* **36**(3) 353–84

Grady D E 1990 Particle size statistics in dynamic fragmentation *J. Appl. Phys.* **68**(12) 6099–105

Grady D E 1993 Impact compression properties of concrete *Proc. Sixth Int. Symp. on Interactions of Nonnuclear Munitions with Structures (Panama City Beach, FL, 3–7 May)*

Grady D E 1995 Spall and fragmentation in high-temperature metals *High-pressure Shock Compression of Solids II—Dynamic Fracture and Fragmentation* pp 219–36

Grady D E 1998 Scattering as a mechanism for structured shock waves in metals *J. Mech. Phys. Solids* **46** 2017–32

Grady D E 2006 Comparison of hypervelocity fragmentation and spall experiments with Tuler–Butcher spall and fragment size criteria *Int. J. Impact Eng.* **33** 305–15

Grady D E 2010 Length scales and size distributions in dynamic fragmentation *Int. J. Fracture* **163** 85–99

Grady D E 2011 Impact breach and fragmentation of glass plate *Int. J. Impact Eng.* **38** 446–50

Grady D E, Duvall G E and Royce E B 1972 Shock-induced anisotropy in ferromagnetic materials II. Polycrystalline behavior and experimental results for YIG *J. Appl. Phys.* **43**(4) 1948

Grady D D, Bergstresser T K and Taylor 1985 Impact fragmentation of lead and uranium plates *Sandia National Laboratories Technical Report* SAND84-1545 November

Grady D E and Kipp M E 1985 Geometric statistics and dynamic fragmentation *J. Appl. Phys.* **58** 1210–22

Grady D E and Kipp M E 1987 Dynamic rock fragmentation *Fracture Mechanics of Rock* (New York: Academic)

Grady D E, Dunn J E, Wise J L and Passman S L 1990 Analysis of prompt fragmentation, *Sandia National Laboratories Technical Report* SAND90-2051 August

Grady D E and Passman S L 1990 Stability and fragmentation of ejecta in hypervelocity impact *Int. J. Impact Eng.* **10** 197–212

Grady D E, Swegle J W and Ang J A 1991 Analysis of prompt fragmentation: October 1989 through September 1990 *Sandia National Laboratories Report* SAND91-0483 June

Grady D E, Hall C A and Reinhart W D 1996 Perforation and fragmentation under high velocity oblique impact *Sandia National Laboratories Technical Memorandum* April, unpublished

Greiner W and Stocker H 1985 *Sci. Am.* **252** 76

Griffith A A 1924 Theory of rupture *Proc. First Int. Congr. on Applied Mechanics (Delft)* pp 55–63

Gu P G, Lin D N C and Bodenheimer P H 2003 The effect of tidal inflation instability on the mass and dynamical evolution of extrasolar planets with ultrashort periods *Astrophys. J.* **588** 509–34

Hakim L and Mikhailov S 2013 Integral equations in cohesive zones modelling of fracture in history dependent materials *Proc. World Congr. on Engineering* Newswood Limited, International Association of Engineers pp 226–31

Holian B L and Grady D E 1988 Fragmentation by molecular dynamics: the microscopic Big Bang *Phys. Rev. Lett.* **60** 1355–58

Kanel G I 1982 The resistance of metals to spall fracture *Combustions, Explosions, Shock Waves* **18** 329–35

Kanel G I *et al* 1994 Spallation near the ultimate strength of solids Shock *Compression of Condensed Matter* ed S C Smith *et al* (Elsevier) pp 1043–46

Kanninen M F and Popelar C H 1985 *Advanced Fracture Mechanics* (Oxford: Oxford University Press)

Kennedy J E 1997 The Gurney model for explosive output for driving metal *Explosive Effects and Applications* ed J A Zukas and W P Walters (Berlin: Springer) pp 221–58

Kerley G I 1992 CTH EOS package: porosity and reactive burn models Sandia National Laboratories Report SAND92-0553

Kipp M E and Grady D E 1995 Experimental and numerical studies of high-velocity fragmentation *High-Pressure Shock Compression of Solids II* ed L Davison, D E Grady and M Shahinpoor (New York: Springer) pp 282–339

Kittel C 1949 Physical theory of ferromagnetic domains *Rev. Mod. Phys.* **21** 541

Lawn B R 1993 *Fracture of Brittle Solids* 2nd edn (Cambridge: Cambridge University Press)

Leonov M Y and Panasyuk V V 1959 Development of the smallest cracks in the solid *Appl. Mech. (Prikl. Mekh.)* **5**(4) 391–40

Lienau C C 1936 Random fracture of a brittle solid *J. Franklin Inst.* **221** (485–494, 674–686) 769–87

Lockman F J, Benjamin R A, Heroux A J and Langston G I 2008 The Smith cloud: a high-velocity cloud colliding with the Milky Way *Astrophys. J.* **679** (1) L21

Mott N F 1948 Brittle fracture in mild steel plates *Engineering* **165** 16

Piekutowski A J 1996 Formation and description of debris clouds produced by hypervelocity impact *NASA Contractor Report* **4704** (February)

Piekutowski A J 1997 Effects of scale on debris cloud formation *Int. J. Impact Eng.* **20** 639–50

Pilch M and Erdman C A 1987 Use of breakup time data and velocity history data to predict the maximum size of stable fragments for acceleration-induced breakup of a liquid drop *Int. J. Multiphase Flow* **13** 741–57

Reif F 1965 *Fundamentals of Statistical and Thermal Physics* (New York: McGraw-Hill)

Roberts D K and Wells A A 1954 The velocity of brittle fracture *Engineering* **178** 820–1

Romero L A 1989 The instability of rapidly stretching plastic jets *J. Appl. Phys.* **65** 3006–16

Romero L A 1991 The stability of stretching and accelerating plastic sheets. I *J. Appl. Phys.* **69** 7474

Romero L A 1991 The stability of stretching and accelerating plastic sheets. II *J. Appl. Phys.* **69** 7487

Schilling G 2014 *Deep Space: Beyond the Solar System to the Edge of the Universe and the Beginning of Time* (New York: Black Dog & Leventhal)

Shaner J W and Royce E B 1968 Shock-induced demagnetization of YIG *J. Appl. Phys.* **39** 492

Skidmore I C 1965 *Appl. Mater. Res.* **4** 131–47

Steverding B and Lehnigk S H 1976 The fracture penetration depth of stress pulses *Int. J. Rock Mech. Min. Sci.* **13** 75–80

Strack O E, Leavy R B and Brannon R M 2015 Aleatory uncertainty and scale effects in computational damage models for failure and fragmentation *Int. J. Numer. Methods Eng.* **102** 468–95

Sun X, Khaleel M A and Davies R W 2005 Modeling of stone-impact resistance of monolithic glass ply using continuum damage mechanics *Int. J. Damage Mech.* **14** 165–78

Tate A 1967 A theory of the deceleration of long rods after impact *J. Mech. Phys. Solids* **15** 387

Tuler F R and Butcher B M 1968 A criterion for the time dependence of dynamic fracture *Int. J. Fracture Mech.* **4** 431–7

Wang Z and Lam K Y 1995 Evolution of microcracks in brittle solids under intense dynamic loading *J. Appl. Phys.* **77** 1–5

Chapter 4

Applications in dynamic fragmentation—engineering

This chapter closes the topics of dynamic fragmentation with several examples of engineering analysis. Applications are explored of ballistic impact leading to dynamic fragmentation. Spall fragmentation is considered resulting from the impact of chunky projectiles onto thick barrier plates. Conversely, fragmentation is examined under conditions of extreme velocity impact of projectiles onto thin plates. The former is relevant to conventional armor applications, whereas the latter approximates conditions of space debris protection technology. Additionally, theories of R W Gurney that address the explosive acceleration of metal shells leading to dynamic fragmentation are developed and applied to relevant fragmentation problems. The Gurney model is contrasted with contemporaneous studies of G I Taylor and N F Mott.

4.1 Introduction

Several theoretical arguments including carefully posed dimensional considerations lead to a length scale characterizing the nominal fragment size in many dynamic fragmentation events of the form

$$s = \left(\frac{48\Gamma}{\rho\dot{\varepsilon}^2}\right)^{1/3}. \tag{4.1}$$

The numerical factor in this relation results from one development of the fragmentation theory and can take different values for alternative developments and different applications. The density of the material is ρ and is that of condensed matter—either solid or liquid. The kinematic parameter $\dot{\varepsilon}$ is some measure of the deformation rate, or rate of strain, characterizing the dynamics of failure through fracture and fragmentation. Alternatively, dynamics can be cast as an instantaneous Heaviside load, or a constant input of energy (power). The same theoretical energy

doi:10.1088/978-0-7503-1254-7ch4

arguments lead to fragment size relations comparable to equation (4.1). The energy Γ (dimensions of energy per unit area) is a measure of the resistance to fracture and fragmentation, and can take different forms dependent on the nature of fracture and fragmentation of the material. Even in more brittle dynamic fragmentation events, where resulting distributions tend to exhibit power-law behavior spanning one or more decades in fragment size, equation (4.1) plays a role in identifying a measure of the distribution center bounded by the upper and lower size limits of power-law behavior.

Equation (4.1), when combined with some formulation of a statistical distribution of fragment sizes, has been successfully applied to a range of dynamic fragmentation problems over the past several decades. Generalizations of equation (4.1) have been implemented into computational models and joined with physics-based computer codes (e.g. Kipp and Grady 1995) to investigate fragmentation in complex dynamic events. Applications of such computational fragmentation calculations to a broad range of dynamic fragmentation events have not proceeded without difficulties, however. A reasonably clear picture of the failure and fragmentation physics specific to the event is usually necessary to a successful calculation. Commonly, a joining of an engineering analysis of the problem with supporting computational simulations leads to a more confident calculation of the fragmentation outcome.

In this chapter, engineering models and analysis of several dynamic fragmentation problems are explored that emphasize the underlying fragmentation physics and that are constructively improved on through supporting computational analysis. Archetypal of such engineering models with application to dynamic fragmentation is the theory of Gurney (1943). The Gurney model provides transparent relations for accessing the rates of expansion of exploding shells and allows for ready estimates of fragmentation. The Gurney theory is examined in the chapter along with the concurrent work of Taylor (1941) and Mott (1943).

Comparable engineering models have also emerged to assess the effects of ballistic impact in various geometries. Selected ballistic models and resulting consequences leading to fragmentation are pursued. When intensities of impact achieve critical levels, the thermophysical properties of materials enter the analysis of the fragmentation event. Elevated temperature states, melting and even vaporization accompany the fragmentation consequences. An equivalent Hugoniot model provides sensible engineering analysis of such extreme-energy impact fragmentation events.

4.2 Impact spall fragmentation

The fragmentation event explored here is that of a chunky projectile such as an explosively formed fragment undergoing impact on a metal plate at velocities of a few hundred to perhaps a few thousand meters per second. The thickness of the plate is considered to be several times to several tens of times larger than the dimension of the projectile. For simplicity, analysis is restricted to normal or near-normal impact. The consequences of such an impact are arrest of the projectile, propagation of a compact stress wave and, lastly, spall fracture at the opposite free surface with ejection of the spall fragment debris. The objectives of the analysis will be the assessment of the impact-induced stress wave propagation and the subsequent spall fragmentation. The topic of higher-velocity impact of projectiles onto thin plates that results in intense fragmentation of the projectile is considered later in the chapter.

4.2.1 Stress wave propagation

The illustration in figure 4.1 shows the initial consequences of the impact. Impact and arrest of the projectile induces a stress wave that propagates through the plate and emerges at the back surface. The axial extent or thickness (in the direction of propagation) of the stress wave remains nominally constant during propagation through the plate thickness, and is of the order of the dimension a of the projectile.

The wave spreads laterally during propagation to the back surface as is illustrated by the hatched region in figure 4.1. A parabolic description of the lateral amplitude of the compressive stress in the wave is provided by

$$\sigma(r) = \sigma_o(1 - r^2/d^2). \tag{4.2}$$

The stress amplitude shown in the figure on the right is a reasonable approximation of the maximum stress in the wave as a function of the lateral extent $0 \leqslant r \leqslant d$ as the wave emerges at the back surface. The strength of the stress wave arriving at the back surface will determine the intensity of spall fragmentation. In an engineering analysis of the wave strength, an energy assessment could be applied. There are, however, dissipative processes and energy losses in the wave during stress compression and decompression that are not easily accounted for. Momentum conservation more readily yields a measure of the wave strength in the present problem.

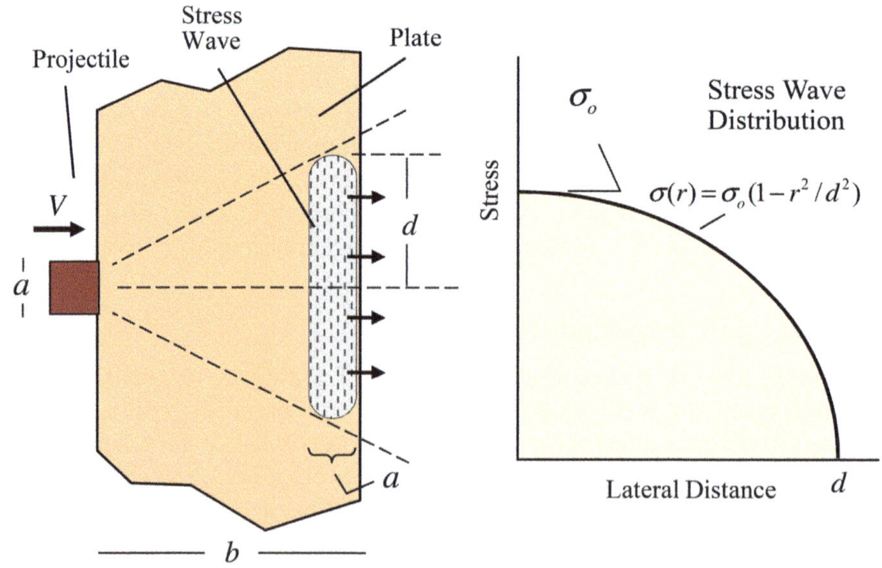

Figure 4.1. Left: impact-induced stress wave emerging at the back free surface. Right: the lateral stress distribution in the wave.

From equation (4.2), particle velocity in the wave is provided by

$$u(r) = \frac{\sigma(r)}{\rho c} = \frac{\sigma_o}{\rho c}(1 - r^2/d^2), \tag{4.3}$$

where ρc is the elastic impedance of the plate material. An appropriate mass element of the material through the thickness a of the stress wave is

$$dm = 2\pi\rho a r dr, \tag{4.4}$$

and integration provides the total wave momentum

$$\int_0^d u(r)dm = \frac{\pi}{2} a d^2 \sigma_o / c. \tag{4.5}$$

Equate equation (4.5) to the initial momentum of the impacting projectile,

$$\frac{\pi}{4}\rho_p a^3 V, \tag{4.6}$$

where an equal dimension cylinder is assumed and ρ_p is the density of the projectile material. Solving for σ_o yields

$$\sigma_o = \frac{1}{2}\rho_p c \frac{a^2}{d^2} V. \tag{4.7}$$

It is reasonable to presume that the propagation velocity of the wave and the speed of the lateral spread of the wave are comparable and assume that $d = b$, the thickness of the plate. The centerline stress amplitude emerging at the free surface is

$$\sigma_o = \frac{1}{2}\rho_p c \frac{a^2}{b^2} V. \tag{4.8}$$

4.2.2 Extent of spall

The extent of the spall region is calculated through comparison of the stress in the wave responsible for spall to the spall strength σ_s of the material, as illustrated in figure 4.2. Using equation (4.2), the radius of the spall disc is calculated from

$$\sigma_s = \sigma_o(1 - d_s^2/d^2). \tag{4.9}$$

The depth from the free surface to the spall plane is approximately $a/2$, where a is the dimension of the impacting projectile. This estimate is consistent with a wave of thickness a incident on the free surface.

4.2.3 Fracture criterion

A further feature of the deformation mechanics leading to failure and fragmentation of the spall material is particularly pertinent to metals. The spall disc is ejected at an

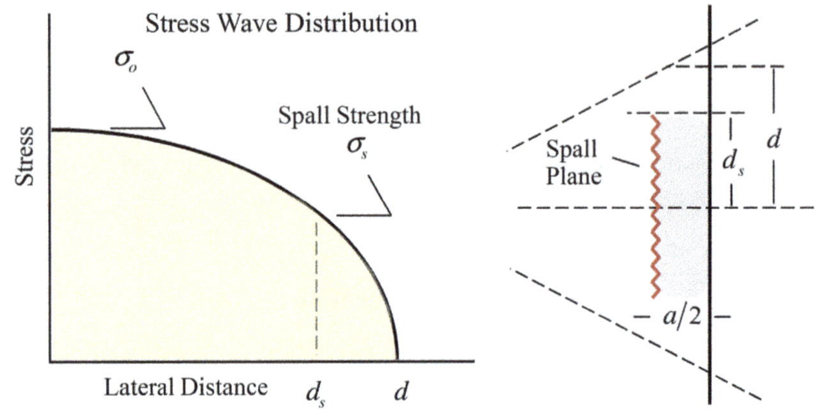

Figure 4.2. Assessment of the lateral extent of the spall zone through comparison of the spall strength of the material to the lateral stress of the wave incident on the free surface.

initial central velocity of $u_o = 2\sigma_o/\rho c$ (see equation (4.3)). Divergent motion leads to plastic deformation at the tensile flow stress Y of the metal until a strain to failure ε_f is achieved, as illustrated in figure 4.3. The specific kinetic energy of the spall disc is reduced according to

$$\frac{1}{2}u^2 = \frac{1}{2}u_o^2 - \frac{Y}{\rho}\varepsilon_f. \tag{4.10}$$

The velocity of the spall disc at onset of fracture is

$$u = u_o\left(1 - \frac{Y\varepsilon_f}{\frac{1}{2}\rho u_o^2}\right)^{1/2}. \tag{4.11}$$

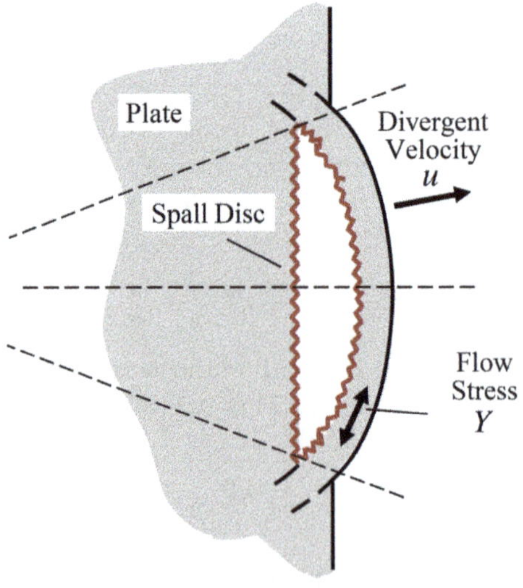

Figure 4.3. Divergent motion and plastic flow subsequent to spall and preceding fragmentation.

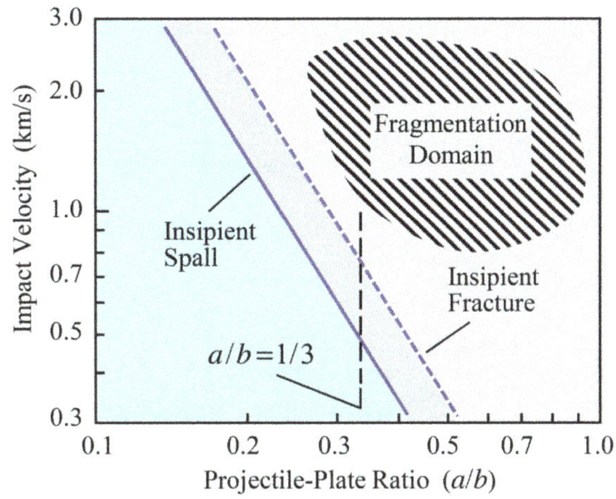

Figure 4.4. Prediction of insipient spall and insipient fracture for normal impact of a tungsten carbide projectile on a steel plate. The projectile–plate dimension ratio is identified for the example calculation in the text.

If the plastic dissipation term exceeds the initial specific kinetic energy, then the spall disc arrests without further fragmentation.

The model of the spall process provides for estimates of critical states in the dynamic event. Equating equation (4.8) to the spall stress σ_s provides the conditions for insipient spall. Equation (4.11), when the argument on the right is equal to zero, determines insipient fracture and subsequent fragmentation of the spall disc.

An example is provided by the impact of a tungsten carbide projectile with ρ_p = 15 000 kg m^{-3} and a = 10 mm onto a steel plate with ρ_p = 8000 kg m^{-3} and b = 30 mm. Representative properties for the steel are: for the spall stress σ_s = 2 GPa, for the yield stress Y = 1 GPa, for the fracture toughness K_c = 60 MPa m$^{1/2}$ and for the strain to failure ε_f = 0.1. For the present materials a plot of the impact velocity versus the ratio of the projectile dimension to the plate thickness a/b is shown in figure 4.4. The graph displays curves of insipient spall and insipient fracture. The present example of a/b = 1/3 is identified in the plot with respective insipient impact velocities of 480 m s^{-1} for spall and 760 m s^{-1} for fracture and onset of fragmentation.

4.2.4 Fragmentation

Fragmentation intensity is calculated for the spalled material. The energy-based model described in chapter 2 identifies a fragment-size length scale of

$$\lambda \simeq \left(\frac{5K_c}{\rho c\dot{\varepsilon}}\right)^{2/3}. \tag{4.12}$$

The divergent strain rate of the ejected spall disc is provided by $\dot{\varepsilon} \simeq u/b$ with the velocity u from equation (4.11). The length scale λ determines a characteristic spacing of fractures in the breakup of the spall disc. A nominal fragment mass is then

$$m = \rho\lambda^2\frac{a}{2} \simeq \rho\frac{a}{2}\left(\frac{5K_c}{\rho cu/b}\right)^{4/3}, \tag{4.13}$$

where $a/2$ is the thickness of the spall disc. Normalizing by the projectile mass $m_p = \pi\rho_p a^3/4$ yields

$$m/m_p \simeq 5\frac{\rho}{\rho_p}\frac{b^2}{a^2}\left(\frac{K_c}{\rho cub^{1/2}}\right)^{4/3}. \tag{4.14}$$

The total mass of ejected spall debris is also relevant to the calculation and is estimated from the extent of the spall disc provided in equation (4.9).

The total spall mass is

$$M = \pi\rho\frac{a}{2}d_s^2 = \frac{\pi}{2}\rho ab^2\left(1 - \frac{\sigma_s}{\sigma_o}\right), \tag{4.15}$$

where σ_o from equation (4.8) is dependent on the impact velocity. Again, normalizing by the projectile mass,

$$M/m_p = 2\frac{\rho}{\rho_p}\frac{b^2}{a^2}\left(1 - \frac{\sigma_s}{\sigma_o}\right). \tag{4.16}$$

Lastly, the average number of fragments in the ejected spall debris is provided through the ratio $N = M/m$. For the present example of a tungsten carbide impact onto a steel plate, the calculated mass of spall debris, nominal fragment mass, fragment number and ejecta velocity as a function of impact velocity are shown in figure 4.5. The ejecta velocity is normalized by the impact velocity.

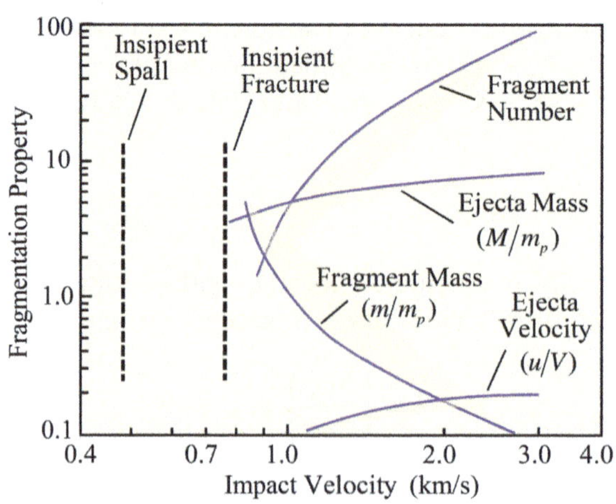

Figure 4.5. Calculation for spall fragmentation dependence on impact velocity for a 10 mm tungsten carbide projectile onto a 30 mm steel plate.

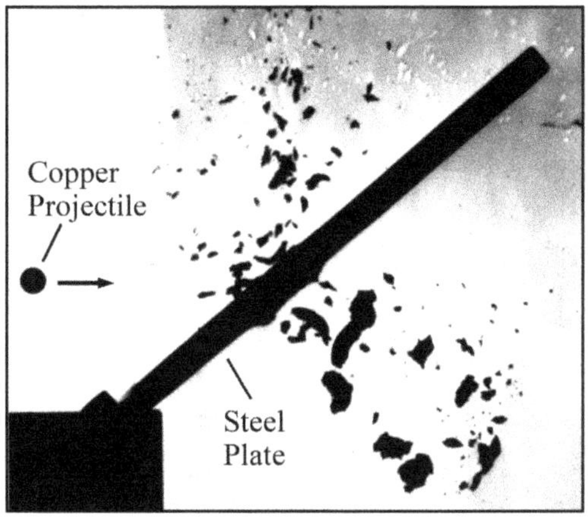

Figure 4.6. A flash radiograph of fragmentation from a copper projectile impact on 304 stainless steel at 3.09 km s^{-1} and an obliquity angle of 50 degrees (Grady 2010). Copyright 2010 With permission of Springer.

The model, although approximate, is consistent with experimental observation and captures the essential physics. Improvements to the model can be made by tuning to the results of several relevant experiments, and with wave-code calculations of the impact shock propagation. The nominal fragment mass calculated from equation (4.13) is expected to provide the length (mass) scale in an appropriate fragment distribution relevant to the impact spall event.

A flash radiograph of fragment ejecta is shown in figure 4.6 that resulted from the oblique impact of a 6.35 mm diameter copper spherical projectile onto a 6 mm 304 stainless steel target plate (Grady 2010). Although impact conditions differ somewhat from the model assumptions, the results are informative. (When the dimensions of the projectile and plate thickness are comparable, spall is not likely to play a significant role in the breakup.)

Cumulative fragment mass distributions resulting from two such tests are shown in figure 4.7, and are in reasonable accord with an exponential distribution in fragment number of the form

$$N(m) = N_o e^{-m/\mu}. \tag{4.17}$$

The calculated fragment mass from equation (4.13) provides a measure of the distribution scale parameter μ in equation (4.17).

4.3 Equivalent Hugoniot fragmentation

The concept of equivalent Hugoniot fragmentation refers to an analytic model that has application to a number of intense energy fragmentation problems. The problems are those for which the energy available to the fragmentation event can be readily quantified before the event. Usually this refers to the available center-of-mass kinetic energy prior to the high-velocity collision of two bodies. The general modeling method does not preclude, however, determination of the relevant chemical energy or

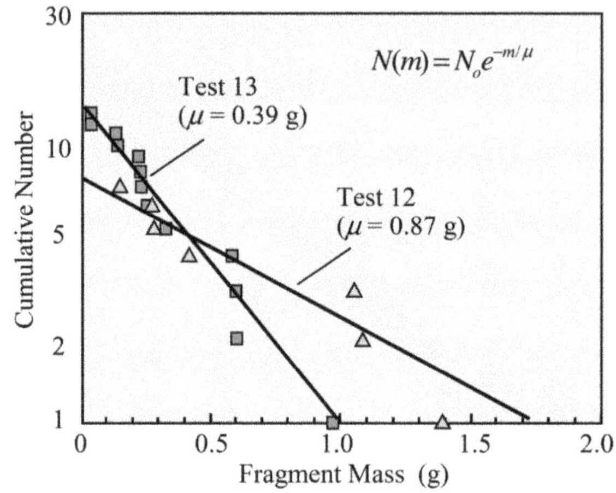

Figure 4.7. Cumulative fragment distributions for behind-target ejecta for two copper projectile impact tests on 304 stainless steel (Grady 2010).

radiation energy causing the fragmentation event, if the fragmentation problem is amenable to such an approach. For example, the use of a Gurney characterization of the explosive energy (Gurney 1943, Kennedy 1997) resulting in an intense energy fragmentation event is quite appropriate, as will be discussed in one of the later sections.

This section begins with discussion and analysis of the impact of two parallel plates. The simple example illustrates how the initial kinetic energy of the system is partitioned through the shock Hugoniot process into dissipation energy and recovered kinetic energy, with the potential to cause failure and fragmentation of the plate material. Next, the essence of the example is broadened to the equivalent Hugoniot model capable of providing approximate solutions to the shock impact and subsequent fragmentation of a wide range of impact problems. Three problems are pursued. First, the high-velocity impact of a projectile on a thin plate representative of the functionality of space debris shielding is examined. Second, the implosion of a metal cylinder that gives rise to the intense fragmentation of uranium shells described in earlier chapters is described. Lastly, the intense energy impact of a projection on a structured system is explored through the equivalent Hugoniot model. The fragmentation examples are interrupted later in the chapter by an extended discussion of the Gurney solution methods, including the related contemporaneous works of Taylor and Mott.

4.3.1 Plate impact

As a starting point, consider the symmetric collision of the two flat plates illustrated in figure 4.8. Equal mass plates collide with equal and opposing velocities, and on impact come to rest on the center plane. From symmetry, the equivalent problem is that of one plate undergoing impact on a perfectly rigid wall. Calculations supporting this discussion will make use of the latter equivalent representation.

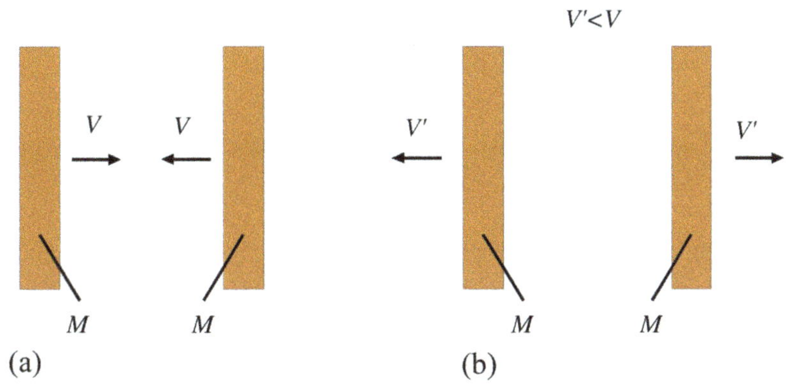

Figure 4.8. Impacting plates with equal masses at velocities V will rebound with velocities V' less than V due to shock dissipation in the impact event.

Prior to impact the total energy of the system resides in the kinetic energy of the plate traveling at velocity V. Upon impact, the entire plate comes to rest in a time equal to T/U where T is the thickness of the plate and U is the velocity of the backward-facing shock wave originating at the plane of collision. During passage of the shock wave, the initial kinetic energy converts to lattice strain energy and energy dissipated in the shock compression process. The latter dissipation energy may accommodate some measure of non-recoverable defect or other non-thermal energy. Nonetheless, this energy is predominantly heat energy, giving rise to temperature increase of the material. On emergence of the shock wave at the free surface, the recoverable lattice strain energy accelerates the plate material through an isentropic decompression wave. Energy is reverted back to kinetic energy and the plate rebounds at some velocity $V' < V$. Complications in assigning a single velocity to the rebound plate will emerge in the discussion. The rebound velocity and corresponding kinetic energy are always less than the initial velocity and kinetic energy because of the energy loss in dissipation. Thus, energy balance requires

$$E_k = \frac{1}{2}V^2 - E_d, \tag{4.18}$$

with E_k and E_d the residual specific kinetic energy of the rebounded plate and shock dissipation energy, respectively, following the impact event. (Throughout the present section, the symbol E may refer to either total energy or energy per unit mass. The choice is clear from the context.)

Processes of shock dissipation can take various forms. For the present example, a plate material described by a linear shock velocity versus particle velocity $U = C_o + Su$ and a von Mises yield stress Y is assumed. As the shock wave passes through the material, two mechanisms of shock dissipation are participating in the event, as illustrated in figure 4.9.

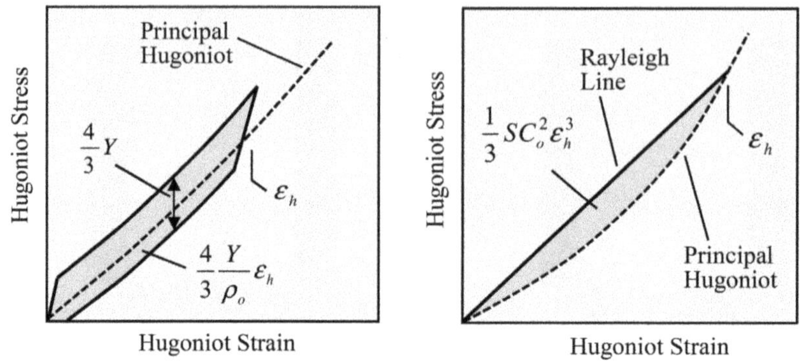

Figure 4.9. Mechanisms of shock dissipation through both von Mises plasticity (left) and shock entropy production (right) in the impact event.

The first dissipation mechanism illustrated on the left is deformation plasticity that, for shock compression to and decompression from a Hugoniot strain ε_h, is provided by

$$W_p = \frac{4}{3}\frac{Y}{\rho_o}\varepsilon_h. \tag{4.19}$$

The second is a consequence of entropy production and temperature rise in the nonlinear shock compression event. Specific dissipation energy in the shock, as illustrated in the figure, is provided to a reasonable first order as the area enclosed by the Hugoniot and the Rayleigh line to the final Hugoniot strain state ε_h,

$$W_s = \frac{1}{3}SC_o^2\varepsilon_h^3. \tag{4.20}$$

Equation (4.20) would not apply for example if the material exhibited phase transformation on the Hugoniot, or for excessive Hugoniot pressures when the decompression isentrope resides substantially above the Hugoniot. Relations more generally applicable for equation (4.20) are developed later in the section. This particular shock dissipation model is applied in earlier chapters to examine the nature of shock attenuation and spall in solid matter.

To lowest order the Hugoniot strain and impact velocity are related through

$$\varepsilon_h \simeq V/C_o, \tag{4.21}$$

and equations (4.18)–(4.21) join to provide the ratio of the residual to the initial kinetic energy,

$$\frac{E_k}{V^2/2} = 1 - \frac{8}{3}\frac{Y/\rho_o}{C_o}\frac{1}{V} - \frac{2}{3}\frac{S}{C_o}V. \tag{4.22}$$

This rebound kinetic energy ratio, accounting for shock dissipation, is illustrated for aluminum in figure 4.10. Shock Hugoniot and strength properties are appropriate for aluminum tested in the impact fragmentation experiments of Piekutowski (1996)

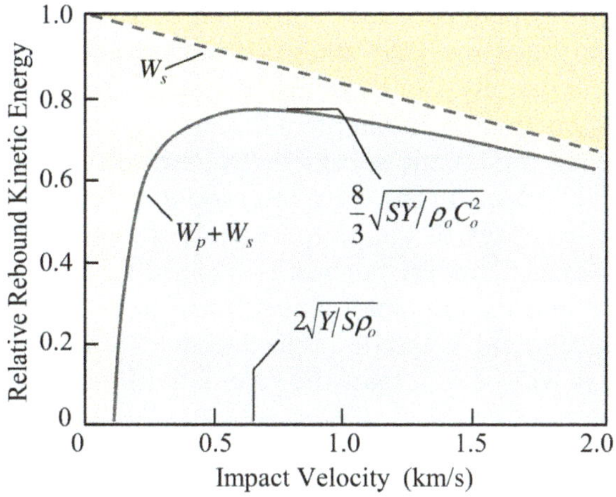

Figure 4.10. Rebound kinetic energy of the plate relative to the incident kinetic energy resulting from shock entropy dissipation alone, and from both dynamic plasticity and shock entropy dissipation.

referred to in the previous chapter and considered further later in this section. Plastic deformation in the shock wave dominates dissipation at low impact velocity. Shock entropy production and material heating take precedence as impact velocities increase. The relative dissipation (one minus the ratio calculated in equation (4.22)) is identified by the amplitude above the curves in figure 4.10. That above the dashed line is from entropy production alone. Both entropy production and plastic dissipation contribute to the amplitude above the solid curve. Relative dissipation is a minimum at a velocity of $2\sqrt{Y/S\rho_o}$ with a minimum of $8\sqrt{SY/\rho_o C_o^2}/3$. Namely, for the example calculated, a maximum rebound kinetic energy of nearly 80% and dissipation of 20% is achieved at an impact velocity of about 700 m s^{-1}. Shock Hugoniot and strength properties for aluminum in the studies of Piekutowski are $\rho_o = 2800$ kg m^{-3}, $C_o = 5.3$ km s^{-1}, $S = 1.35$ and $Y = 0.45$ GPa.

Closer examination of the simple impact example posed in figure 4.8 reveals further complexities. Real materials are nonlinear, as is captured in the model by the linear shock velocity versus particle velocity description of the material. When the impact-induced shock wave reaches the back of the plate, a centered rarefaction wave emerges from the free surface. Centered nonlinear waves are discussed in chapter 1. Consider sufficiently strong overdriven shock waves so that elastic–plastic shock structure is not an issue. The decompression wave spreads as it propagates. For a linear shock velocity versus particle velocity description, the leading head of the wave propagates at a velocity of approximately $C_h \simeq C_o + 2Su_h$, and the tail at a velocity of C_o. On propagating through the thickness L of the plate, the ramp-shaped decompression wave, incident on the impact interface, will have a thickness of $\lambda \simeq L(1 - C_o/C_h)$. On acceleration of the plate during rebound, a region of the plate at the impact interface of approximate thickness of $\lambda/2$ constitutes the wave interaction region. Material in the interaction region experiences tension during acceleration. Excessive tension leads to spall. Strain rates carrying material into tension and spall are approximately $\dot{\varepsilon} \simeq 2V'/\lambda$. The number and spacing of spall planes are calculated through the several methods introduced in chapters 2 and 3.

Based on the model, the kinetic energy recovered on rebound is that plotted in figure 4.10. Because of effects in the wave interaction zone, not all of this energy is center-of-mass kinetic energy of the plate. A modest portion is expended in spall fracture energy. A further modest portion is trapped within the spall plates. The principal plate body and trailing spall plates from the wave interaction zone will exhibit a spread in rebound velocities. The rebound kinetic energy of the plate debris center of mass will be less than the calculated kinetic energy shown in the figure.

4.3.2 Colliding bodies and the equivalent Hugoniot

Applications of interest in the analysis developed in this section are abstracted to the illustration shown in figure 4.11. The relative closing velocity and masses of the two bodies are ascribed. The picture is not intended to convey any particular geometry of the bodies. The collision of bodies in Earth orbit is certainly an apparent application. The model will also encompass the impact of a projectile on an armor plate, such as space debris shields or other more conventional armor applications. When applied to armor fragmentation events the participating portion of the armor plate must be identified and specified as one of the masses.

First, the two-body system is transformed to a coordinate system in which the total momentum is zero. In this center-of-momentum coordinate system mass M travels at velocity V_c and mass m at velocity $V - V_c$ along the path of closing trajectory, such that

$$m(V - V_c) = MV_c. \tag{4.23}$$

The total kinetic energy in center-of-momentum coordinates is

$$\frac{1}{2}m(V - V_c)^2 + \frac{1}{2}MV_c^2. \tag{4.24}$$

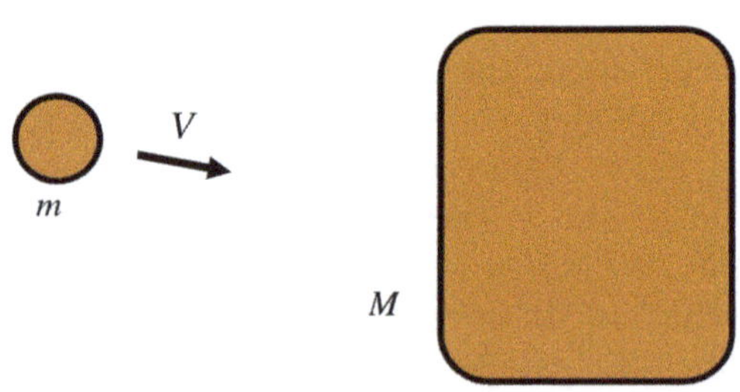

Figure 4.11. Two colliding bodies that will lead to impact shock effects with subsequent fragmentation and dispersion of the resulting debris.

Eliminating V_c between equations (4.23) and (4.24), and dividing by the total system mass $m + M$, yields a specific kinetic energy (energy per unit mass) of the system before impact of

$$E_x = \frac{1}{2}(\mu V)^2, \quad \text{where } \mu = \sqrt{\frac{mM}{(m+M)^2}}. \tag{4.25}$$

Development of a model to predict the consequences of the high-velocity impact event requires attention to characteristics of the impact-induced shock compression processes. On impact of the colliding bodies, shock compression and the attendant shock dissipation is intense. Shock intensity and dissipation are not uniform throughout the participating bodies. Shock intensity is highest at the initial point of impact. Shocks will attenuate rapidly with propagation distance from the point of contact, and accompanying dissipation will correspondingly lessen. For given geometries of the participating bodies the extent and location of dissipation will depend on details of the shape and orientation.

The equivalent Hugoniot model assumes an energy-averaged Hugoniot and decompression dissipation process in the impact shock event. This model provides a reasonable prediction of shock dissipation in the impact event when details of the colliding bodies are only approximately known.

The equivalent Hugoniot model assumes that on impact of the two-body system the center-of-momentum specific kinetic energy achieves a transient energy-averaged Hugoniot state, such that

$$E_x = E_h. \tag{4.26}$$

The equivalent Hugoniot-state energy in turn determines on decompression of the system the partitioning into energy of dissipation and recovered kinetic energy. Namely, $E_h = E_k + E_d$. The equivalent Hugoniot model provides a relation between the closing velocity V of a two-body system and the partitioning of the available energy into dissipation and kinetic energy subsequent to impact,

$$\frac{1}{2}(\mu V)^2 = E_x \rightarrow E_h \rightarrow E_k + E_d. \tag{4.27}$$

Arrows in the relation displayed in equation (4.27) identify the path of energy through the impact event. Specifically, the initial closing kinetic energy transitions to an equivalent transient shock Hugoniot state energy on impact, and then to the residual center-of-momentum kinetic energy and dissipation energy on decompression from the shock Hugoniot state.

The present development will focus on dissipation due to Hugoniot entropy production within the intense shock compression resulting from the high-velocity impact event. Dynamic plasticity dissipation, discussed in the previous subsection, becomes significant only at the lower impact velocities. Plastic dissipation when warranted can be included through the methods discussed earlier. Further, a level of porosity in the colliding system can be accounted for when justified.

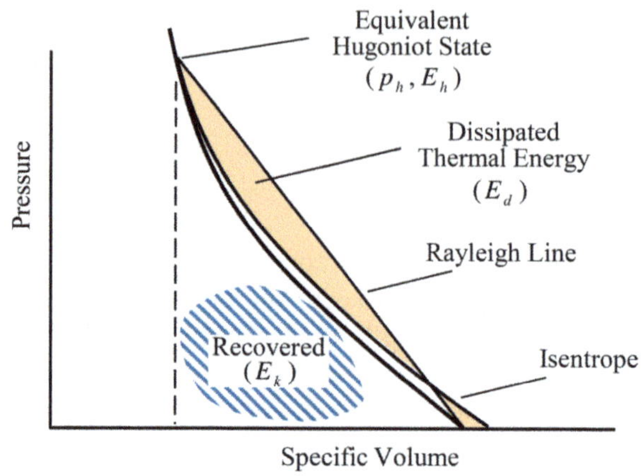

Figure 4.12. Equivalent Hugoniot state and partitioning of dissipated and recovered kinetic energy.

The Hugoniot and respective compression and release paths governing the shock compression to the equivalent Hugoniot state are illustrated in figure 4.12. Shock compression follows the Rayleigh line from the initial state to the equivalent Hugoniot state. Subsequent decompression follows an isentrope from the Hugoniot state to the ambient (near zero) pressure state. Dissipation in the shock compression and subsequent decompression process is provided by

$$E_d = \int_0^{p_h} p_R(v)dv + \int_{p_h}^{0} p_S(v)dv, \tag{4.28}$$

where $p_R(v)$ and $p_S(v)$ specify the pressure–volume relations for compression along the Rayleigh line and the decompression isentrope, respectively. Concurrently, the decompression isentrope determines the recovered kinetic energy. Namely, the area under the isentrope shown in figure 4.12.

Specification of the thermodynamic equation-of-state (EOS) of the material is needed to provide quantitative measures of the dissipated energy and recovered kinetic energy. For the present purposes it is assumed that the colliding bodies are the same material and are adequately described by a Mie–Gruneisen EOS with a linear shock velocity versus particle velocity Hugoniot relation $U = C_o + Su$. Analysis would proceed accordingly for more complex EOSs, and if colliding materials are dissimilar.

Shock compression to an equivalent Hugoniot state yields a Hugoniot specific energy,

$$E_h = \frac{1}{2}\frac{p_h \varepsilon_h}{\rho_o} = \frac{1}{2}\frac{C_o^2 \varepsilon_h^2}{(1 - S\varepsilon_h)^2} = \frac{1}{2S^2}\frac{C_o^2 (S\varepsilon_h)^2}{(1 - S\varepsilon_h)^2}, \tag{4.29}$$

where $\varepsilon_h = 1 - v_h/v_o$ is a measure of the volume strain to the Hugoniot state. Integration along the decompression isentrope determines the recovered lattice strain energy and center-of-momentum kinetic energy of the colliding system. Integration along the decompression isentrope requires appropriate implementation of the Mie–Gruneisen EOS (see chapter 10). For the elementary case of the

Gruneisen gamma equal to zero, the decompression isentrope coincides with the Hugoniot and the integral is

$$E_k = \int_{\varepsilon_h}^{0} \frac{C_o^2 \varepsilon}{(1 - S\varepsilon)^2} d\varepsilon = \frac{C_o^2}{S^2}\left[\frac{S\varepsilon_h}{1 - S\varepsilon_h} + \ln(1 - S\varepsilon_h)\right]. \tag{4.30}$$

An approximate accounting for a finite Gruneisen gamma is introduced later in the section.

The difference of equations (4.29) and (4.30), $E_d = E_h - E_k$, corresponds to equation (4.38) and provides the entropy production dissipation in the equivalent Hugoniot shock compression event. A Taylor expansion in powers of $S\varepsilon_h$ results in

$$E_d = E_h - E_k = \frac{1}{3}\frac{C_o^2}{S^2}\left((S\varepsilon_h)^3 + \frac{9}{4}(S\varepsilon_h)^4 + \frac{18}{5}(S\varepsilon_h)^5 + \cdots\right). \tag{4.31}$$

4.3.3 High-velocity projectile on plate

The shatter of a chunky projectile incident at high velocity onto a barrier plate provides an exemplary example of colliding bodies and the equivalent Hugoniot model. The problem provides a unique interplay of impact mechanics, shock thermodynamics and fragmentation statistics. The event has relevance to conventional armor applications. It is perhaps most notably relevant to the functionality of protective shielding from meteorite and orbital debris in space assets applications.

In figure 4.13 are shown the consequences of two such fragmentation events that lends support to the present solution methods. The first illustrates the consequences of a lead projectile normally incident on a lead plate at a typical ballistic velocity. The second is an aluminum sabot that is representative of a piece of space debris, incident at high velocity onto a thin (approximately one millimeter) zinc metal protective spacecraft plate. The intense fragmentation and spread of the fragment debris is representative of the objective of spacecraft shielding.

4.3.4 Momentum and energy constraints

For the purposes of the present study, the problem is abstracted to a spherical projectile of diameter D incident on a plate of thickness T, and with the projectile dimension substantially larger than the plate thickness. Issues of non-sphericity of the projectile are discussed where appropriate. Normality of impact is not assumed, although extreme obliquity is not considered and is usually treated with different solution methods (e.g. Grady *et al* 2000). The projectile and plate collide at a relative velocity V. The impact geometry and consequences are illustrated in figure 4.14.

A general representation of conservation of momentum and conservation of energy through the impact event is expressed as

$$m_p \vec{V} = m'_p \vec{V}_r + \vec{I}_{\text{int}}, \tag{4.32}$$

$$\frac{1}{2} m_p V^2 = \frac{1}{2} m'_p V_r^2 + E_{\text{int}} + E_x. \tag{4.33}$$

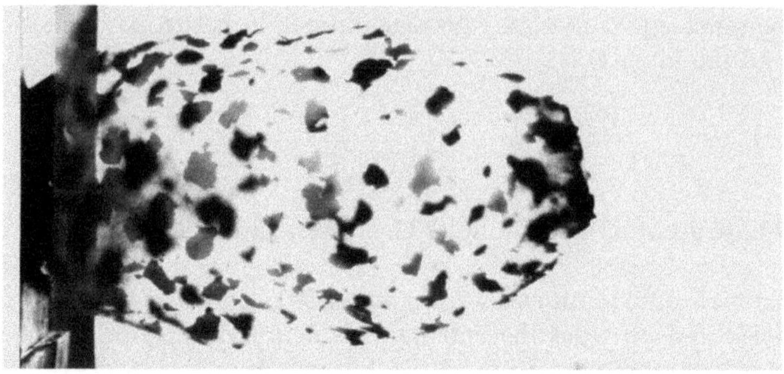

a.) Lead cylinder projectile on lead plate.

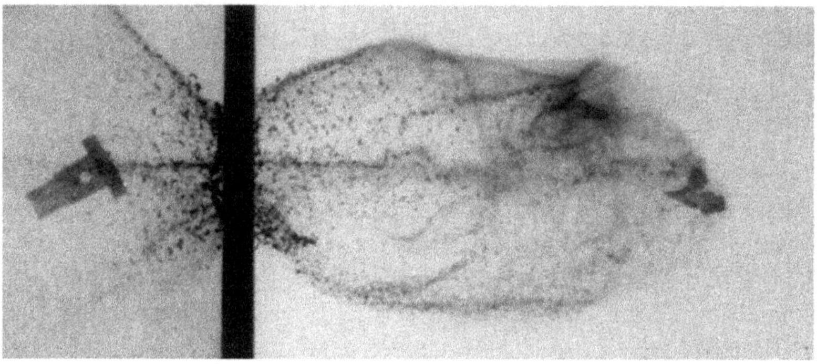

b.) Aluminum sabot on zinc metal plate.

Figure 4.13. Impact shatter and fragmentation of (a) a post-impact photograph of a 6 mm × 6 mm lead projectile on a 6 mm thickness lead plate. Normal impact at approximately 1500 m s^{-1}. Image 175 μs after impact (Grady *et al* 1985) and (b) successive pre- and post-impact radiographs of an aluminum sabot on a 1 mm thick zinc metal plate at 5000 m s^{-1} (Piekutowski 1996). Courtesy of NASA.

In these relations m_p is the initial projectile mass while m_p' reflects a possible loss of projectile mass due to back splash or erosion. The vector velocities $\vec{V}$ and $\vec{V}_r$ are, respectively, the velocity of the incident projectile and the center-of-mass velocity of the projectile fragment debris. The vector impulse $\vec{I}_{\text{int}}$ accounts for interaction with the barrier in the impact event. Also included in this term are residual momenta retained in the barrier material or eroded projectile material. Correspondingly, the interaction-energy term E_{int} accounts for work on the barrier plate during perforation and any dissipative energy or kinetic energy retained by barrier or eroded plate material. In subsequent subsections upper case E will have dimensions of energy rather than specific energy.

The excess energy E_x as previously discussed is that remaining energy available for thermomechanical heating and velocity dispersion of the projectile debris. Equations (4.32) and (4.33) can be solved for the excess energy

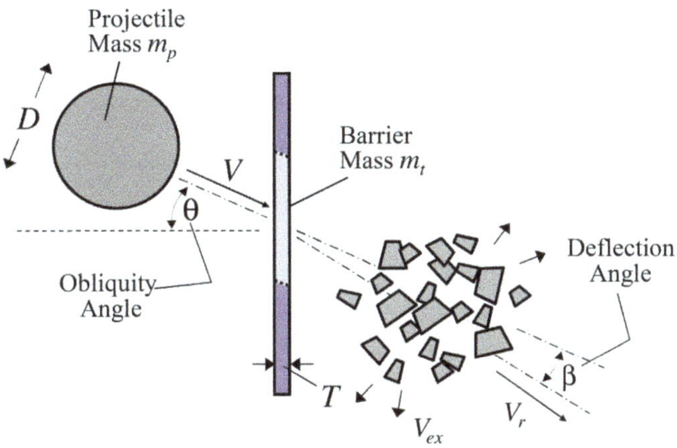

Figure 4.14. Geometry for projectile on plate impact analysis (Grady *et al* 2000).

E_x and the residual center-of-mass velocity of the secondary projectile fragment ejecta $\vec{V}_r$,

$$E_x = \frac{m_p}{m_p'}\vec{V} \cdot \vec{I}_{\text{int}} - \frac{1}{2}\frac{1}{m_p'}I_{\text{int}}^2 - \frac{1}{2}m_p V^2\left(\frac{m_p}{m_p'} - 1\right) - E_{\text{int}}, \tag{4.34}$$

$$\vec{V}_r = \frac{1}{m_p'}\left(m_p\vec{V} - \vec{I}_{\text{int}}\right). \tag{4.35}$$

Equations (4.34) and (4.35) are reasonably rigorous. Further approximations of the impact and interaction physics are necessary to yield tractable predictive equations.

4.3.4.1 Swift plate impact model

Swift *et al* (1982, 1983), following earlier analytic studies of Richardson (1969), developed a predictive model for high-velocity projectile impact on barrier plates. The geometry in figure 4.14 is restricted to normal impact in the Swift model. From experimental observations of bubble-shaped clouds of ejecta debris following high-velocity impact, a spherical shell of debris ejecta is assumed in the model with center-of-mass velocity V_r and radial expansion velocity of V_{ex}. Through sensible approximations of the equations (4.32) and (4.33) relations, the model yields

$$E_x = \frac{1}{2}m_p V^2\left[\frac{\alpha\rho_t T/\rho_p D}{1 + \alpha\rho_t T/\rho_p D}\right], \tag{4.36}$$

$$V_{ex} = \frac{\sqrt{\alpha q\rho_t T/\rho_p D}}{1 + \alpha\rho_t T/\rho_p D}V, \tag{4.37}$$

$$V_r = \frac{V}{1 + \alpha\rho_t T/\rho_p D}. \tag{4.38}$$

The barrier plate and projectile densities are ρ_t and ρ_p. The parameter α accounts for uncertainty in the participating plate mass. Evidence suggests that $\alpha \simeq 1$. An undetermined fraction q accounts for shock dissipation in the Swift model.

4.3.4.1 Recht plate impact model

A similar modeling effort was pursued by Recht and co-workers (Recht and Ipson 1963, Recht and Yatteau 1983, Recht 1990). Its development contrasts with the Swift model in addressing impact obliquity and attending to subsonic aspects of the impact mechanics. The author's approximations of the conservation of momentum and energy relation from equations (4.32) and (4.33) result in the relations

$$E_x = \frac{1}{2} m_p' V^2 \left[\frac{m_t + m_p' \sin^2 \beta}{m_t + m_p'} \right], \tag{4.39}$$

$$V_r = \frac{\cos \beta \sqrt{V^2 - V_{\text{int}}^2}}{1 + m_t / m_p'}, \tag{4.40}$$

with a plate mass that participates in the impact interaction of

$$m_t = \rho_t A_p T \sec\theta. \tag{4.41}$$

The presented plate area is A_p, while θ and β are the impact obliquity and angle of deflection identified in figure 4.14. The initial projectile mass typically replaces m_p' in the equations. Analytic relations are also developed for the deflection angle β but for initial impact velocities in excess of 150% of the ballistic limit, β seldom exceeds 10 degrees and $\cos \beta$ in equation (4.40) is replaced by unity.

4.3.5 The thin plate model

A further projectile-on-plate analysis, identified as the thin plate model (Grady *et al* 2000, Grady and Winfree 2000, 2001), emphasizes spacecraft protective shielding. In the model equations (4.32) and (4.33) reduce to

$$m_t^* V = \left(m_t^* + m_p \right) V_r, \tag{4.42}$$

$$\frac{1}{2} m_t^* V^2 = \frac{1}{2} \left(m_t^* + m_p \right) V_r^2 + E_x. \tag{4.43}$$

Interaction momentum I_{int} and energy E_{int} are accounted for through impact interaction with the effective plate mass m_t^*. For analytic convenience, energies and momenta are referenced to a coordinate system in which the impacting projectile is at rest. Both the impact velocity (of the plate element) and the residual velocity (projectile and plate element) components lie along the same impact trajectory. The plate mass m_t identified in figure 4.14 is that silhouetted by the impacting projectile and is experimentally observed to be approximately equal to the plate mass ejected in high-velocity impact on thin plates. The mass m_t^* is the effective plate mass that achieves the appropriate momentum and energy balance in equations (4.42) and (4.43). The relation between m_t and m_t^* follows from further analysis.

Equations (4.42) and (4.43) combine to yield

$$E_x = \frac{1}{2} m_t^* V^2 \left[\frac{1}{1 + m_t^*/m_p} \right], \tag{4.44}$$

$$E_r = \frac{1}{2} m_t^* V^2 \left[\frac{m_t^*}{m_p} \frac{1}{1 + m_t^*/m_p} \right], \tag{4.45}$$

where E_r is the residual center-of-mass kinetic energy following impact. Written this way it is seen that in the thin plate limit with $m_t^* \ll m_p$ a large portion of the initial plate mass kinetic energy converts, in the impact process, into potentially destructive and dispersive excess energy E_x. Only a small portion is required by the dictates of momentum conservation to remain as translational kinetic energy. Thus, in the thin plate limit the following approximations are assumed,

$$E_x = \frac{1}{2} m_t^* V^2, \tag{4.46}$$

$$V_r = \frac{m_t^*}{m_p} V. \tag{4.47}$$

On returning to the reference frame with the barrier plate at rest, the residual center-of-mass velocity of the projectile debris is $V - V_r$.

4.3.5.1 The effective plate mass

The ellipsoidal silhouette plate mass for the oblique impact of a spherical projectile illustrated on the left in figure 4.15 is

$$m_t = \pi (D/2)^2 T / \cos\theta. \tag{4.48}$$

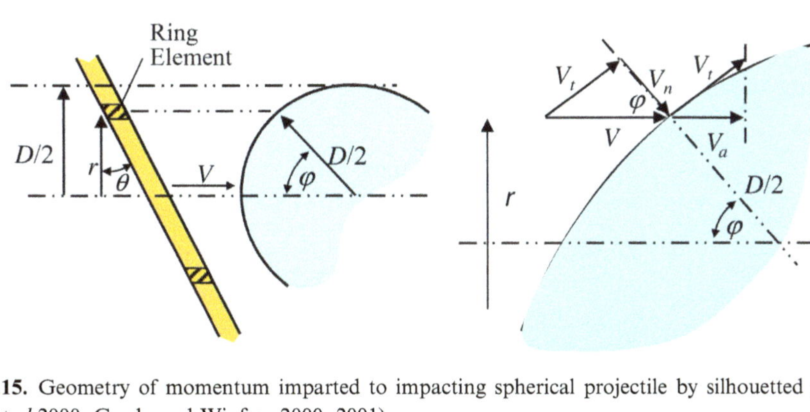

Figure 4.15. Geometry of momentum imparted to impacting spherical projectile by silhouetted plate mass (Grady *et al* 2000, Grady and Winfree 2000, 2001).

This mass does not fully transfer momentum and kinetic energy to the projectile in the impact process, however. The effective plate mass m_t^* is assessed through momentum considerations of the impact event. The interaction is illustrated in figure 4.15 where attention is focused on a skewed elliptical annular mass element of the plate that is incident at a radius r from the sphere centerline trajectory. The impact angle of obliquity is θ. The ring of material impacts the sphere at a position determined by an angle φ also referenced from the centerline trajectory. The impulse imparted to the sphere by this ring element is integrated over the sphere. The temporal difference in impact of separate elements on the ring will not lead to impulse imbalance and lateral deflection of the sphere provided the impact is at sufficiently high velocity. This criterion relates to the supersonic versus subsonic nature of the sphere and plate contact point.

The illustration on the right identifies the vector velocities relevant to the impacting annular element. At velocities of interest, shearing forces will be negligible and only the normal component V_n of the initial impact velocity V will contribute to the impulse. Thus, the component of velocity tangent to the impact surface of the sphere will remain unchanged through the impact process. Two approximations are implicit. First, that to first-order the sphere remains unmoved during the impact event. This approximation is reasonable in the thin plate limit of $m_t \ll m_p$. Second, that the normal component of velocity is zero after impact. That is, the impacting plate element does not 'bounce'. Shock analysis shows that this would be true if shock impedance of the plate material equals or exceeds that of the sphere. Lower impedance plate material would exhibit some rebound thus increasing the impulse on the sphere. Increase of a factor of two would be an upper limit.

The geometry illustrated in the figure yields a post-impact tangent velocity component of $V_t = V \sin\varphi$ and an axial component of $V_a = V_t \sin\varphi = V \sin^2\varphi$. The change in the axial velocity of the ring element is, accordingly,

$$\Delta V = V - V_a = V\cos^2\varphi, \tag{4.49}$$

and therefore the increment of impulse imparted to the sphere is

$$\delta\Delta p = \delta m_t \Delta V = \frac{2\pi r \delta r T}{\cos\theta} V \cos^2\varphi. \tag{4.50}$$

Integrating to the radius of the sphere,

$$\Delta p = \frac{2\pi\rho T V}{\cos\theta}\int_0^{D/2}\cos^2\varphi r dr, \tag{4.51}$$

yields

$$\Delta p = \frac{\pi}{2\cos\theta}\rho T (D/2)^2 V = \frac{1}{2} m_t V, \tag{4.52}$$

with m_t provided by equation (4.48).

Thus, for the oblique impact of a sphere on a thin plate, the effective plate mass and projected plate mass are related through

$$m_t^* = \alpha m_t = \frac{1}{2} m_t, \tag{4.53}$$

and, from equation (4.47), the residual velocity of the sphere (in the reference frame with the sphere initially at rest) is

$$V_r = \frac{1}{2}\frac{m_t}{m_p} V = \frac{3}{4\cos\theta}\frac{\rho_t T}{\rho_p D} V. \tag{4.54}$$

Predictions of the residual velocity (Grady *et al* 2000, Grady and Winfree 2000, 2001) are compared with the normal impact data of Piekutowski (1996) in figure 4.16. These data show both the dependence of the reduction of the center-of-mass velocity of the projectile with impact velocity at a specified ratio T/D, and variations in T/D at a specified impact velocity. Agreement with the model is reasonably good for a T/D ratio less than about 0.1.

With effective and projected plate mass related through $m_t^* = m_t/2$, excess energy is calculated from equation (4.46),

$$E_x = \frac{1}{4} m_t V^2. \tag{4.55}$$

Thus in the thin plate model exactly one-half of the projected plate mass kinetic energy is predicted to participate in the destruction and dispersion of the incident spherical projectile. The other half remains as the outwardly imparted kinetic energy of plate debris resulting from the impact process. Such a halo of fine particulate plate debris is observed in radiographs of Piekutowski (1996). The same analysis can be performed for non-spherical projectile geometries and different ratios between

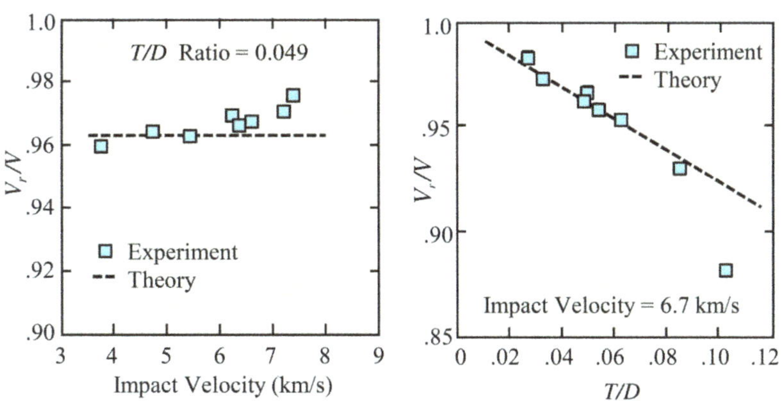

Figure 4.16. Velocity of fragment debris center-of-mass relative to impact velocity at a specified value of T/D (right) and impact velocity (left). The normal impact of aluminum spheres onto aluminum plates. Data from Piekutowski (1996).

effective and silhouette plate mass are found. For example, normal impact of a cubic projectile, face-on, edge-on or point-on, is yielded through the same analysis $\alpha = 1$, 1/2 and 2/3, respectively, in equation (4.53) (Grady and Winfree 2000, 2001).

4.3.5.2 Supersonic contact

In the thin plate model, independent shock impacts of elemental rings of the plate are integrated to assess the impulse and energy imparted to the projectile. A sufficient condition for independent shock coupling is that the interaction contact point between the plate and projectile moves at a velocity that is supersonic with respect to the sound speeds in either material. The sketch on the left in figure 4.17 illustrates the relative impact of a thin plate and a surface of the projectile at velocity V, and is characterized by an impact angle of obliquity and an angle of incidence between the plate and projectile surfaces. Velocity of the contact point can be determined from the geometry. The contact velocities relative to a frame of reference at rest with respect to the target plate and with respect to the projectile are, respectively,

$$V_{ct} = \left(\frac{\cos\theta}{\tan\varphi} + \sin\theta\right)V, \qquad V_{cp} = \frac{\cos\theta}{\sin\varphi}V. \tag{4.56}$$

Various effects are apparent. First, contact velocities scale with the impact velocity, and under reasonably normal and reasonably planar impacts will significantly exceed the impact velocity. Sound speeds in most metals are of order 4–7 km s^{-1}. When the projectile impact velocities of concern range from this order to higher, contact velocities are supersonic and conditions are met for the hypervelocity impulse analysis. For normal impact the contact velocity in the projectile exceeds that in the plate for all angles of incidence.

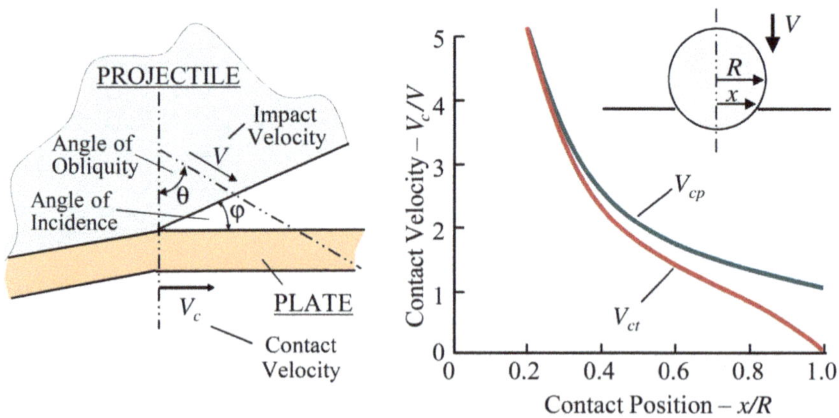

Figure 4.17. Left: projectile impacting plate at velocity V. The angle of obliquity and angle of incidence determine the velocity of contact between projectile and plate surfaces. Right: relative contact velocities for normal impact of a sphere onto a plate relative to reference frames within the plate V_{ct} and within the sphere V_{cp}, respectively (Grady *et al* 2000).

It is of interest to examine the contact velocities in terms relevant to an impacting spherical projectile. Rewriting the equations in terms of the sphere radius $R = D/2$ and the distance X from the centerline of the sphere to the contact point in the plane of obliquity yields

$$V_{ct} = \left(\frac{\sqrt{R^2 - X^2}}{X} \cos\theta + \sin\theta\right)V, \quad V_{cp} = \left(\frac{R}{X}\cos\theta\right)V. \tag{4.57}$$

Note that the same expressions apply with $\theta = 0$ for contact velocities normal to the plane of obliquity. Contact velocities are shown on the right in figure 4.17 as a function of the contact radius for normal impact of a spherical projectile onto a plate. For a projectile and plate of the same material, transition to subsonic motion will occur first in the plate. The transition radius will depend on the impact velocity. Although transition to subsonic contact always occurs in the normal and oblique impact of a sphere, for reasonable impact velocities and for less than extreme obliquities, the transition occurs near the outer radius of the sphere in plate material that does not contribute significantly to the sphere impulse.

4.3.5.3 Fragment debris expansion

Based on the thin plate model the residual kinetic energy following the shock impact event of a spherical projectile undergoing normal impact on a plate is

$$E_k = E_x - E_d = m_p\left(\frac{1}{4}\frac{m_t}{m_p}V^2 - e_d\right), \tag{4.58}$$

where E_x is eliminated through equation (4.55) and $e_d = E_d/m_p$ is the energy per unit mass (specific energy) expended in shock dissipation in the projectile.

Normal impact experiments of aluminum spheres onto thin aluminum plates (Piekutowski 1996) can be compared with the kinetic energy calculated from equation (4.58). The experimental radiographs (see figure 4.20) reveal features that are quantified in the study. The preponderance of sphere mass resides within the central expanding lens of material. The velocities of radial expansion of this region are reported in the experimental study. A relation between the expansion kinetic energy in equation (4.68) and the experimental radial expansion velocity is written as

$$E_k = \frac{1}{g}m_p V_k^2, \tag{4.59}$$

where g is an undetermined constant because the radial distribution of mass within this disc region is not known. For example, if all the mass is in the outer perimeter, then $g = 2$. If mass and velocity are uniformly distributed throughout the disc, then $g = 4$ is more appropriate. Joining equations (4.58) and (4.59) yields

$$\frac{V_k}{V} = \sqrt{g\left(\frac{3}{8}\frac{T}{D} - \frac{e_d}{V^2}\right)}, \tag{4.60}$$

where plate and sphere masses and corresponding dimensions are related through $m_t/m_p = 3T/2D$.

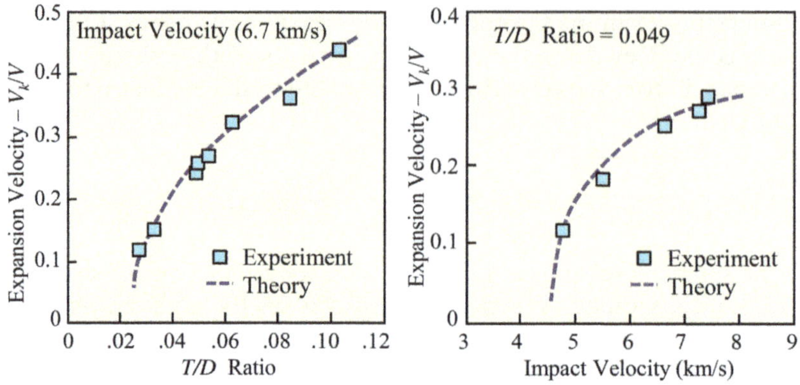

Figure 4.18. Expansion velocity as a function of impact velocity for normal impact of aluminum spheres on aluminum plates (Grady *et al* 2000, Grady and Winfree 2000). The T/D ratio is varied at a specified impact velocity shown on the left. Variations in impact velocity at a fixed T/D ratio are presented on the right. Data from Piekutowski (1996).

Dissipation in the impact event is due to shock entropy production and plastic work. Shock entropy production increases as some power of the impact velocity. Plastic work will depend on material strength and strain to failure. A more detailed discussion of e_d is pursued later. Here a constant value of e_d is assumed over the relatively narrow range of the impact velocities ($\sim$3.5–7.5 km s^{-1}) and equation (4.60) provides a two-parameter equation for the expansion velocity. Values of $e_d = 370$ J g^{-1} and $g = 6.7$ provide the quite satisfactory fit of equation (4.60) to the experimental data shown in figure 4.18. The reasonably large value of g (see equation (4.59)) suggests an expanding debris mass and kinetic energy are weighted towards the central region of the expanding fragment debris. This result is consistent with the experimental observation of survival of a large central fragment even at the more intense impact velocities (Piekutowski 1996). At a midpoint impact velocity of about 5.5 km s^{-1}, the Hugoniot volume strain in aluminum is approximately $\varepsilon_h = 0.3$. Equations (4.19) and (4.20) provide estimates of approximately 65 J g^{-1} and 360 J g^{-1} for plastic dissipation and shock entropy production, respectively. The latter is in sensible agreement with $e_d = 370$ J g^{-1}, inferred from the fit of the model to the data.

4.3.5.4 Shock dissipation

The thin plate model resulting in equation (4.60) provides a reasonable description of the impact data of Piekutowski. Further, the energy dissipation and debris dispersion inferred from the model are in sensible agreement with observation and shock properties of aluminum. To assess a broader range of impact conditions, however, the constant dissipation energy parameter assumed in equation (4.60) is not appropriate. The analysis is broadened to more general applications with the equivalent Hugoniot assumption introduced in section 4.3.2.

To the lowest order in the Hugoniot strain, the specific energy dissipated in the shock event from equation (4.31) is provided by

$$E_d = \frac{1}{3} C_o^2 S \varepsilon_h{}^3. \tag{4.61}$$

Other dissipation mechanisms, such as von Mises plasticity considered in the introduction to this section, are ignored here but can be included when warranted. Introducing the Hugoniot energy, again to lowest order, $E_h = C_o^2 \varepsilon_h{}^2/2$, provides the ratio of the shock dissipation energy to the Hugoniot energy,

$$\frac{E_d}{E_h} = \sqrt{\frac{8}{9}} S \left(\frac{E_h}{C_o^2} \right)^{1/2}. \tag{4.62}$$

Thus, for modest shock compression, the relative dissipation increases as the square root of the Hugoniot energy, or linear with the Hugoniot compression strain.

For increasing shock compression equation (4.62) becomes inappropriate and a more careful accounting of shock dissipation must be undertaken. As noted in section 4.3.2, application of the full Mie–Gruneisen EOS, or whatever EOS is deemed appropriate for description of the shock response, is required for this analysis. If the Gruneisen response of the material is ignored ($\Gamma = 0$), the decompression isentrope coincides with the Hugoniot and, for a linear shock-velocity versus particle-velocity response, is provided by the integral in equation (4.30). The Hugoniot energy is

$$E_h = \frac{1}{2S^2} \frac{C_o^2 (S\varepsilon_h)^2}{(1 - S\varepsilon_h)^2}. \tag{4.63}$$

The recovered strain energy (the area under the Hugoniot if Gruneisen gamma is zero) on isentropic decompression is

$$E_k = \frac{C_o^2}{S^2} \left[\frac{S\varepsilon_h}{1 - S\varepsilon_h} + \ln(1 - S\varepsilon_h) \right], \tag{4.64}$$

and the shock dissipation energy is provided by $E_d = E_h - E_k$.

Accounting for Gruneisen properties of the material without resorting to full application of the Mie–Gruneisen EOS can be accomplished through a reasonably straightforward approximation. At the Hugoniot state, the pressure difference between the isentrope and the Hugoniot is provided by

$$\Delta P \simeq \rho_o \Gamma_o E_d, \tag{4.65}$$

where Γ_o is the ambient Gruneisen parameter and the sensible assumption of $\rho\Gamma$ equal to a constant is used. Through inspection of figure 4.12, one observes that dissipation differs from the difference between equations (4.63) and (4.64) by the area between the Hugoniot and the decompression isentrope. This area is approximated by

$$\rho_o \Delta E_d \cong \frac{1}{2} \Delta P \varepsilon_h. \tag{4.66}$$

Joining equations (4.65) and (4.66) yields

$$E_d = (E_h - E_k)\left(1 - \frac{1}{2}\Gamma_o \varepsilon_h\right). \tag{4.67}$$

Equation (4.67) recovers the previous approximation for the dissipation when $\Gamma_o = 0$.

The several relations for shock dissipation on the Hugoniot are illustrated for aluminum in figure 4.19. On the left, the relative dissipation (the ratio of the dissipation energy to the shock Hugoniot energy) is plotted against the Hugoniot energy. The shock Hugoniot dissipation is plotted on the right in figure 4.19 as a function of the impact velocity of an aluminum plate onto an aluminum plate that would achieve the corresponding Hugoniot energy. The simplest approximation is provided by equation (4.62), in which the relative dissipation is proportional to the square root of the Hugoniot energy, is perhaps useful to impact velocities approaching 2 or 3 km s^{-1}. The approximation of estimating the shock dissipation from the area between the Rayleigh line and the Hugoniot (equivalent to assuming $\Gamma_o = 0$) also rapidly becomes a poor approximation over the range of interest. Equation (4.67), although also an approximation to the application of a complete EOS, provides a quite acceptable calculation of the shock dissipation for aluminum. Over most of the range of concern, the shock dissipation energy is between about 20% and 30% of the shock Hugoniot energy. With a melt enthalpy of 0.4 kJ g^{-1}, aluminum at ambient barometric conditions subsequent to intense shock

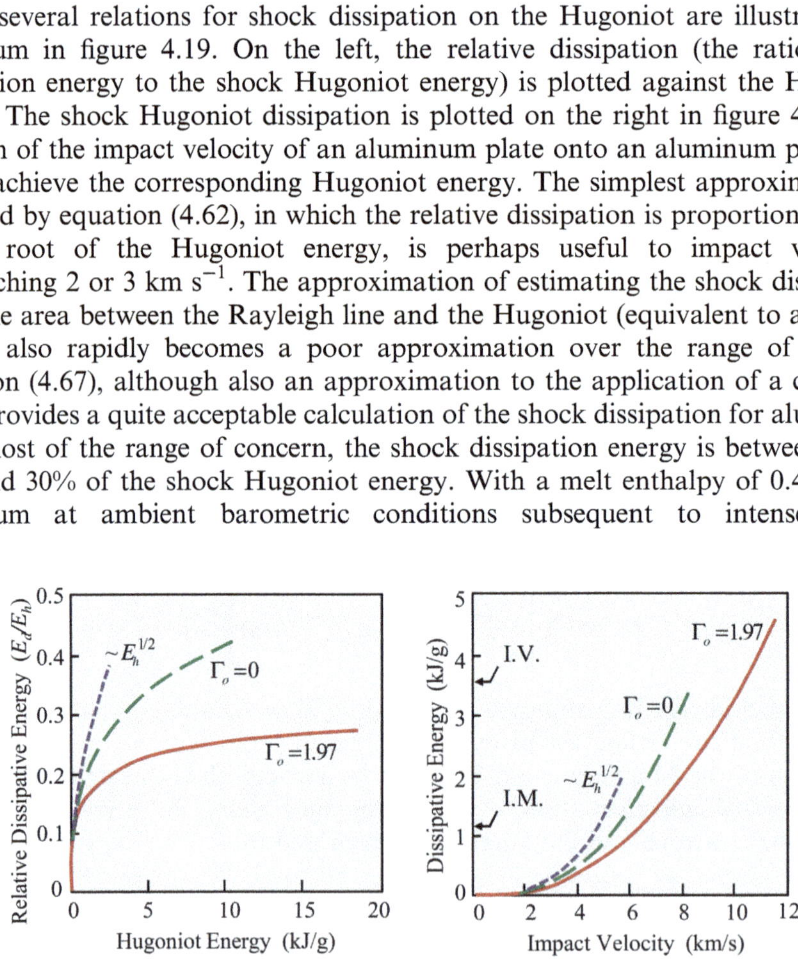

Figure 4.19. Left: relative shock dissipation as a function of the Hugoniot energy. Right: shock dissipation versus the impact velocity of aluminum on aluminum that would achieve the corresponding Hugoniot energy. Also shown on the right are energy levels corresponding to insipient melt (I.M.) and insipient vaporization (I.V.) subsequent to shock compression (Grady *et al* 2000).

compression melts over the range of approximately 1.2–1.6 kJ g^{-1}, corresponding to aluminum on aluminum impact velocities between about 6–8 km s^{-1}.

The present calculations of shock dissipation on the Hugoniot are informative. A complication to note, however, regards the relatively simple and well-behaved assumption of the Gruneisen properties of the material. As shock dissipation becomes excessive and the thermal energy becomes comparable to the binding energy of the lattice and higher, approximations of a temperature-independent Gruneisen parameter and Mie–Gruneisen EOS behavior become increasingly poor. Levels of molecular ionization are incurred. Shock thermal states enter a span of material behavior where even detailed physics-based theories are challenged.

4.3.5.5 Application to plate impact fragmentation

A notable study of impact-induced fragmentation that has provided the experimental backdrop for the previous several subsections is provided by the experimental simulation of orbital debris impact on spacecraft bumper shields performed by Piekutowski (1996, 1997). In that work aluminum spheres were accelerated to a high velocity with a two-stage light gas gun and caused to undergo normal impact onto thin aluminum plates. The x-ray image of the fragment debris resulting from one such test is shown on the left in figure 4.20. In one study of Piekutowski the ratio of the bumper plate thickness T to projectile sphere diameter D was maintained constant at $T/D = 0.049$, and geometric scaling of the fragmentation event explored. Maintaining this geometric ratio constant over the series of experiments, impact velocity was varied from about 4.5–7.5 km s^{-1} and the sphere diameter was varied from 6.35 to 15.9 mm.

Piekutowski outlined underlying causes for most of the details observed in the fragment debris pattern pictured in the figure. Quantitative analysis was focused on two key characteristics of the fragment debris. The first was the size of the largest

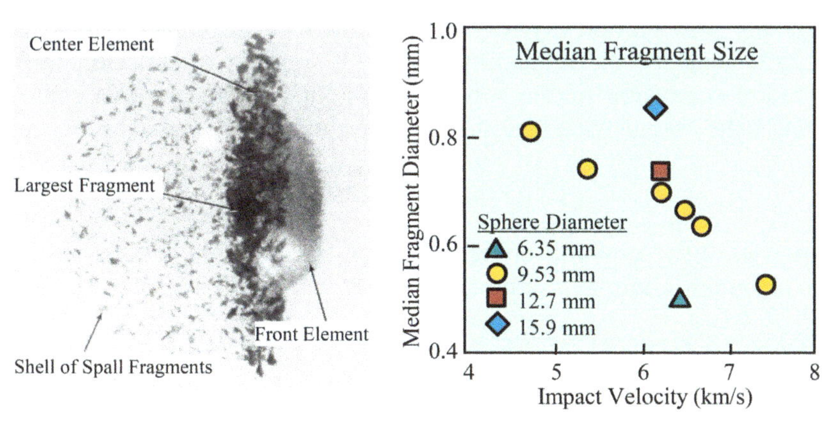

Figure 4.20. Left: test radiograph of an aluminum spherical projectile on aluminum-plate impact-induced fragmentation experiment used to quantify breakup and expansion characteristics (Piekutowski 1996, 1997). Right: spall fragment size in the aluminum sphere impact experiments including both velocity and system size scaling. Courtesy of NASA.

central fragment identified in the forward portion of the radiographic image. The second was an average measure of the fine fragments resulting from shock-induced spall in the aluminum sphere and residing in the trailing spherical portion of the fragment cloud. For this latter measure Piekutowski chose the median Martin diameter of the fragment distribution.

The median Martin diameter for all of the impact experiments is plotted on the right in figure 4.20. Several features are observed in the plot. First, a series of experiments was performed on spheres at a single 9.53 mm diameter over an impact velocity range of approximately 4.5–7.5 km s^{-1}. Second, a series of four tests was performed on four different diameter aluminum spheres at nominally the same intermediate impact velocity (about 6.3 km s^{-1}). Again, the ratio T/D was maintained constant over all tests plotted in figure 4.20. Each series separately, and both collectively, elucidate scaling of fragmentation in the experimental impact fragmentation study.

A test of both the equivalent Hugoniot model and the fragmentation theories explored in the present and previous chapters is provided by these unique impact fragmentation data. The equivalent Hugoniot theory has found application to a number of complex impact and explosive load events. The theory emerged, however, in efforts to develop analytic models to evaluate the effects of the very high-velocity impact of chunky projectiles onto thin barrier plates for assessing the shielding of spacecraft from meteor and orbital debris (Grady *et al* 2000). Accordingly, description of the application of the equivalent Hugoniot model to fragmentation in the Piekutowski impact experiments is warranted.

Application of the equivalent Hugoniot model yields a residual specific kinetic energy E_k that is directed outward from the center of momentum of the impacting bodies, and is responsible for fragmentation and dispersion of the participating material. For the Piekutowski experiments the contributing plate mass is small compared to that of the impacting sphere and can reasonably be ignored in the fragmentation calculation.

Recognizing the complexity of the statistical distribution of E_k over the fragmentation mass of the impacting sphere, a simplest assumption is that of a uniform radial expansion of the sphere consistent with the kinetic energy E_k. A calculation of the volumetric expansion rate of a uniformly expanding sphere yields

$$\dot{\varepsilon} = d \ln v/dt = \sqrt{120E_k/D^2}\,. \tag{4.68}$$

Earlier analysis of the available energy resulting from the impact of a sphere onto a thin plate of similar material at a normal impact velocity V is, from equation (4.55),

$$E_x = \frac{3}{8}\frac{T}{D}V^2. \tag{4.69}$$

Here E_x is a specific energy. The equivalent Hugoniot theory equates E_x to $E_h = E_k + E_d$, yielding

$$E_k = E_h - E_d = E_h(1 - E_d/E_h), \tag{4.70}$$

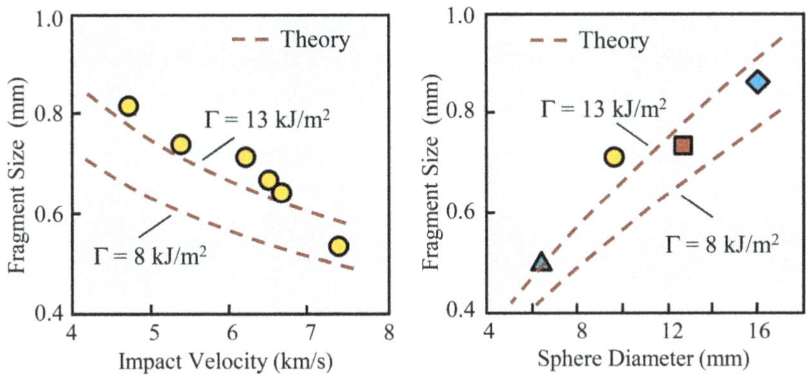

Figure 4.21. Fragment size measurements for normal impact of aluminum spheres on thin aluminum plates (right) (Piekutowski 1996, 1997) (right) and comparison with model predictions (Grady 2006b) (left).

and a relation for the expansion strain rate,

$$\dot{\varepsilon} = \sqrt{45\frac{T}{D}}(1 - E_d/E_h)^{1/2}\frac{V}{D}. \tag{4.71}$$

The relation for average fragment size from the energy-horizon model of fragmentation developed in chapter 2 is

$$s = \left(\frac{48\Gamma}{\rho\dot{\varepsilon}^2}\right)^{1/3}. \tag{4.72}$$

Using strain rates calculated from equation (4.71) and $\Gamma = 8$ kJ m^{-2}, representative of the more modest temperature fragmentation energy for aluminum plotted in figure 4.23, yields the lower curves plotted with the data in figure 4.21. An energy of $\Gamma = 13$ kJ m^{-2} provides the curves more representative of the data in the figure. The two predictions differ by less than a factor of two in the average fragment mass, and reasonably capture the trend with both impact velocity and size scale. The prediction is remarkable considering approximations in both the theory and in estimating parameters representative of the experimental conditions.

4.3.5.6 Energy distribution

The equivalent Hugoniot model provides a method for assessing the partitioning of kinetic energy and dissipation energy in a complex impact event. The model does not specify how that energy is distributed over the participating bodies. As discussed in the preceding subsections, the high-velocity impact of a projectile mass onto a stationary structure leads to a complexity of shock-induced acceleration and dissipation throughout the participating masses. The distribution of thermal and kinetic energy is not uniform because of rapid shock attenuation from the point or points of impact to extremities in projectile and target bodies.

Consequently, in developing physics-based equations describing the distribution of thermal energy over the affected matter due to the impact shock process, it is reasonable to rely on principles of statistical mechanics. Given the complexity of the impact shock process, combined with the uncertainty of the impact configuration, it is reasonable to propose that the thermal energy is distributed over the available debris mass in a statistically probable manner.

When a specified energy E_d is distributed randomly over a large number of material elements, classical statistical mechanics leads to a very tractable relation for the most probable distribution of the mass m over the specific energy E. Namely,

$$m(E) = Me^{-\alpha E}, \tag{4.73}$$

where M is the total mass over which the energy is distributed and the constant α is determined from the constraint

$$\int_0^\infty E dm = ME_d. \tag{4.74}$$

The average specific dissipated thermal energy E_d is that determined from the equivalent Hugoniot model. Carrying out the integral yields $\alpha = 1/E_d$. Combining the equations yields

$$m(E) = Me^{-E/E_d}, \tag{4.75}$$

where $m(E)$ is the cumulative mass constituting material with specific thermal energy greater than E. In an intense impact event, the equivalent Hugoniot theory provides the specific dissipation energy E_d. Equation (4.75) is used to calculate the mass fractions of solid, melt and vapor in the shock event. Let

$$x(E) = e^{-E/E_d} \tag{4.76}$$

be the mass fraction of material in excess of thermal energy E. Given the critical state energies, E_{m1}, E_{m2}, E_{v1} and E_{v2}, corresponding to incipient melt, complete melt, incipient vaporization and complete vaporization, respectively, define

$$x_{m1} = e^{-E_{m1}/E_d}, \quad x_{m1} = e^{-E_{m2}/E_d}, \quad x_{m1} = e^{-E_{v1}/E_d}, \quad x_{m1} = e^{-E_{v2}/E_d}, \tag{4.77}$$

as mass fractions of material with thermal energies greater than the respective critical states. The mass fraction of vapor is then

$$x_v = x_{v2} + \frac{1}{2}(x_{v1} - x_{v2}) = \frac{1}{2}(x_{v1} + x_{v2}). \tag{4.78}$$

Similarly the mass fraction of melt is provided by

$$x_m = \frac{1}{2}(x_{m1} + x_{m2}) - x_v, \tag{4.79}$$

while the mass fraction of solid is

$$x_s = 1 - x_m - x_v. \tag{4.80}$$

Melt and vaporization energies are extracted from appropriate material EOSs, or are calculated from available specific heat data, and melt and vaporization enthalpies.

4.3.6 Hydrocode simulations

The equivalent Hugoniot theory has been compared with computer simulations of intense impact events (Grady *et al* 2000, Grady and Winfree 2001). In the computer calculations, the CTH wave code (McGlaun *et al* 1990, Bell *et al* 2000) is used to simulate the normal impact of a solid aluminum sphere 10 mm in diameter onto an aluminum plate 1 mm in thickness at velocities of 6, 9 and 12 km s^{-1}. A full three-phase EOS for aluminum is used to describe the shock response of projectile and target material. For each simulation, the internal energy of the spherical projectile is interrogated on a cell-by-cell basis at a time sufficiently late in the calculation, such that all internal energy can be assumed to be thermal. The results are assembled as a cumulative mass fraction of the internal energy and are plotted for each of the three simulations in figure 4.22. Internal energy levels sufficient for incipient melt and incipient vaporization at ambient barometric conditions are identified on the plot. Theoretical curves based on the equivalent Hugoniot theory through equation (4.75) are compared with the numerical predictions. Both theory and numerical simulation capture the same trends, although differences are noted.

Although reasonable agreement between computational simulation and analytic theory are observed in the figure, it is important to note that such a comparison is not the final arbiter in validation of the analytic theory. The full physics of the impact event are not necessarily captured in the numerical simulation. Phenomena relating to material heterogeneities and to dynamic instabilities of material interfaces, which contribute substantially to the randomization of thermal energy, are not satisfactorily treated in discretized, continuum hydrocodes. Total dissipation energy based on the equivalent Hugoniot model is compared on the right in figure 4.22, and is found to be modestly less than that of the hydrocode simulation.

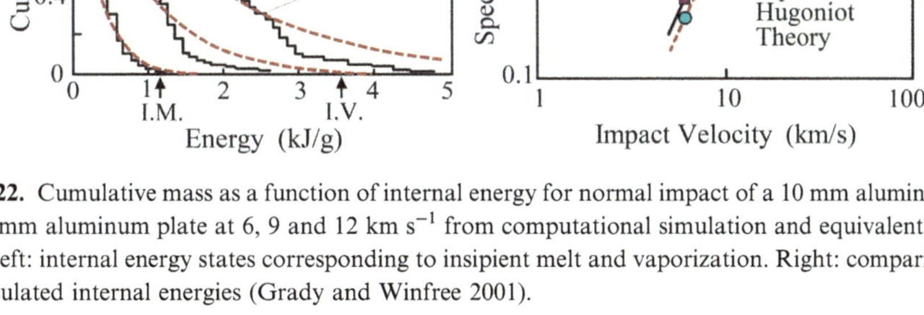

Figure 4.22. Cumulative mass as a function of internal energy for normal impact of a 10 mm aluminum sphere onto a 1 mm aluminum plate at 6, 9 and 12 km s^{-1} from computational simulation and equivalent Hugoniot theory. Left: internal energy states corresponding to insipient melt and vaporization. Right: comparison of the total calculated internal energies (Grady and Winfree 2001).

4.3.7 Fragmentation

In intense impact events such as the examples referred to in the preceding subsections, the equivalent Hugoniot theory provides a method for assessing the partitioning of available energy into shock dissipation energy and kinetic energy. The energy not absorbed in the shock dissipation process, remains as a disruptive kinetic energy, and is responsible for the forces causing fragmentation and dispersion of the participating bodies. Like the dissipation energy, the kinetic energy is also distributed randomly over the disrupted mass. Simplifying assumptions are similarly required to assess the intensity of fragmentation. For example, in the case of the Piekutowski aluminum sphere impact experiments where the sphere constitutes most of the mass, the assumption of a uniform divergent strain rate descriptive of the disruptive dynamics leads to $\dot{\varepsilon} \sim \sqrt{E_k/D^2}$, where E_k is the specific kinetic energy assessed from the equivalent Hugoniot theory and D is the sphere diameter. The constant of proportionality is of order unity and is calculated by assuming some distribution of the kinetic energy divergent from the center of momentum and carrying out the analysis. With the strain rate and some measure of the fracture resistance of the participating materials, characteristic fragment size can be estimated by the methods described in chapters 2 and 3.

The issue of the thermal state of the fragmenting material brought about by shock dissipation in the impact event must also be considered. Shock dissipation determines both the temperature and the physical state of the participating material. Intense shock can lead to a mix of solid, liquid and vapor matter as the calculations displayed in figure 4.22 illustrate. Fragmentation of the condensed matter (solid and liquid) through the theoretical approaches discussed in chapters 2 and 3 require specification of the fragmentation energy Γ resisting the fragmentation process. This energy will in general depend on both the temperature and the physical state of the material.

The fragmentation resistance energy for aluminum shown in figure 4.23 is illustrative. This plot of the fragmentation energy is constructed from experimental spall data on ambient and elevated temperature samples, and from surface tension data on liquid aluminum. A comparison is shown with fracture energy that is based on ambient fracture toughness properties of a typical structural aluminum. High-temperature spall properties of aluminum are from the measurements of Kanel *et al* (1996) discussed in chapter 1. The onset of melt and vaporization are also identified on the abscissa.

4.3.8 Projectile on plate impact

The section continues with several further examples of spherical projectile impact onto stationary plates. The experimental tests lead to intense deformation, failure and fragmentation of the participating materials. Through selective radiographic diagnostics of such experiments, data are acquired for testing computational codes, and associated failure and fragmentation models.

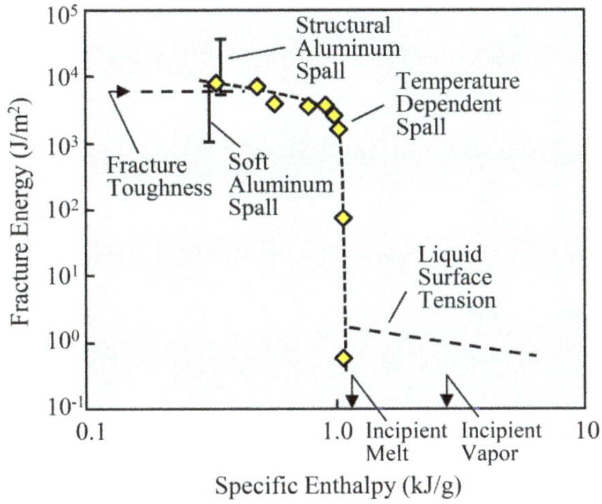

Figure 4.23. Fracture energy appropriate to the fragmentation resistance properties of aluminum metal (Grady and Winfree 2001).

4.3.8.1 Copper on steel

An illustrative impact fragmentation test is provided by experiments in which copper spheres impact steel plates of comparable dimensions (Grady and Kipp 1994). In the experiments, pure copper spherical projectiles 6.35 mm in diameter impact 4340 steel plates 5.6 mm in thickness at impact velocities ranging over 3.5–5.0 km s^{-1}. Two tests were performed under normal impact conditions. A 30 degree oblique impact was achieved in the third. Flash radiography 11.6 μs after impact in two of the tests and 131 μs in one test provide orthogonal shadowgraphs of the impact debris configuration and fragmentation. Radiograph images of two tests are shown in figure 4.24. The third test is shown in figure 4.25.

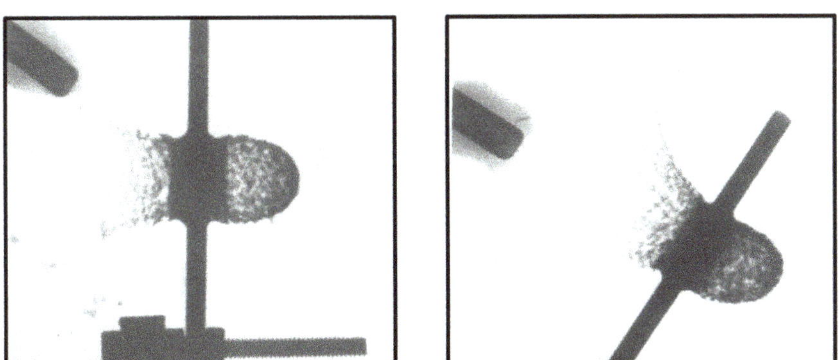

Figure 4.24. Normal and oblique impact of a copper sphere onto a 4340 steel plate Reprinted from Grady and Kipp (1994). Copyright 1994, with permission from Elsevier. Radiographic images at 11.6 μs after impact, with impact velocities of 4.96 km s^{-1} and 4.52 km s^{-1} for the left and right panels, respectively. Plate thickness = 5.6 mm, sphere diameter = 6.35 mm.

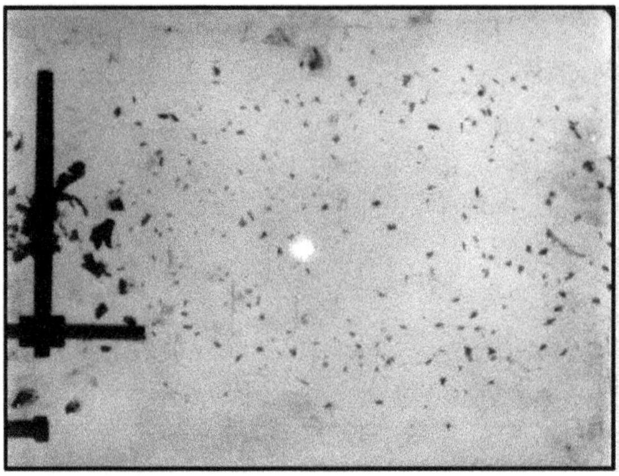

Figure 4.25. Normal impact of copper sphere onto 4340 steel plate imaged at 131.4 μs after impact. Impact velocity is 3.43 km s^{-1}. Reprinted from Grady and Kipp (1994). Copyright 1994 with permission from Elsevier. Plate thickness = 5.6 mm, sphere diameter = 6.35 mm.

The radiograph at 11.6 μs after normal impact at 4.96 km s^{-1} is shown in the left-hand image. The leading edge of the behind-target debris is moving at a velocity of approximately 2.3 km s^{-1}, slightly less than half of the initial impact velocity. Velocities in excess of 3 km s^{-1} are associated with the highest velocity fragments in the back-splash debris. In contrast, large fragments associated with the lip of the spall ring (both front and back) emerging from the peripheral region of the penetration spall crater are moving at velocities of about 0.4 km s^{-1}. At the time of imaging, fragmentation appears complete although high fragment density precludes resolution of individual particles.

The radiograph resulting from the oblique impact experiments at 4.52 km s^{-1} is shown on the right in the figure. (The trajectory of the copper projectile in the figure is horizontal.) This image is also 11.6 μs after impact. The leading edge of the behind-target debris is moving at approximately 1.8 km s^{-1}. There is a slight asymmetry of the envelope of the behind-target ejecta debris due to the oblique impact. There is more material in the upper side of the cloud and fragment separation is large. Computer simulations (figure 4.26) reveal that much of the projectile copper resides in this region laminating the inside of the steel debris cloud. Details of the back-splash debris associated with jetting from the contact interface between the target and projectile are particularly apparent. The trajectories of particles released at successive times in the back-splash jetting process determine the curvature shown in the debris. The leading fragments in this back-splash are traveling at velocities in excess of 6 km s^{-1}. A ray pattern in the back-splash debris is associated with circumferential instability in the initial breakup.

The radiographic image in figure 4.25 shows the behind-target debris for a similar normal impact event (although somewhat lower impact velocity) late in time. The low spatial density of material at this time (131.6 μs after impact) allows for near full resolution of the fragments. The impact velocity for this test was 3.43 km s^{-1}. The velocity of the leading edge of the debris determined from the radiograph is

Figure 4.26. Numerical simulation of an oblique impact experiment of a copper projectile onto a steel plate with the CTH shock physics code. Reprinted from Grady and Kipp (1994). Copyright 1994, with permission from Elsevier.

approximately 1.6 km s^{-1}. The majority of the observed fragments are uniformly distributed through an elliptical cloud. They have an average projected area fragment size very close to one square millimeter with a remarkably narrow size distribution. These fragments constitute about 2 g of the approximately 12 g of target ejecta and projectile mass. The remaining mass is composed of some back-splash material, but primarily the large spall ring fragments residing close to the target plate in the figure. Details of the impact crater geometry are described in Grady and Kipp (1993).

Detailed three-dimensional numerical simulations of the impact experiments were undertaken using the Eulerian wave propagation code CTH (McGlaun *et al* 1990, Bell *et al* 2000). Calculations focused primarily on the oblique impact experiment shown in figure 4.24. Both copper and steel are described by a Mie–Gruneisen EOS including the α to ϵ solid–solid phase transformation at 13 GPa in the steel. Fracture is treated in the computational analysis by introducing void into the computational cell when a principal stress component exceeds the specified tensile fracture strength.

Computer simulations at 11.6 μs after impact, shown in figure 4.26, provide comparisons with the experimental radiograph in figure 4.24. Removal of material in front of the center plane of symmetry reveals that all material (both copper projectile and target ejecta) reside in the outer periphery of the ejecta bubble. The color on the right confirms that projectile copper laminates the inner surface of the expanding steel. Fragmentation of the ejecta bubble begins about 5 μs after impact. Onset of fragmentation is not a consequence of the tensile fracture model. Rather, breakup originates from dynamic instabilities in the rapidly thinning shell of material. Fragmentation of the copper occurs later in time than the steel, a consequence of the lower plastic flow stress of the copper and the influence of flow stress on instability fragmentation (Grady and Passman 1990, Romero 1991).

The thermodynamic state of the materials and breakup modes inferred from the numerical calculation reveals that projectile copper disintegrates into extremely fine particulate. The preponderance of fragments imaged in figure 4.25 are from the target steel. Analysis of the radiograph and recovered ejecta fragments finds that 90% of the fragment mass resides between about 0.5–3.5 mm. Computer simulations or approximate analysis of the experimental radiographs provide expansion strain rates of the expanding ejecta shell of about 0.5–1.0 × 10^5 s^{-1}. At these strain rates, and with properties for 4340 steel of $\rho = 7850$ kg m^{-3}, $c = 5.0$ km s^{-1} and $K_c =$ 50 MPa $m^{1/2}$, the fragment size relation from chapter 2,

$$s = \left(\frac{48\Gamma}{\rho\dot{\varepsilon}^2}\right)^{1/3} = \left(\frac{\sqrt{24}\,K_c}{\rho c\dot{\varepsilon}}\right)^{2/3}, \tag{4.81}$$

predicts a fragment size of 1.5–2.5 mm, in remarkably good agreement with experimental measurements.

4.3.8.2 Experimental fragmentation from projectile impact

An alternative sphere-on-plate impact experiment has found application in testing dynamic fragmentation properties of material. In such tests, the spherical projectile is the material of interest and the target plate provides a specified impulsive load on the sphere that leads to failure and fragmentation. Flash radiography is the principal diagnostic in the experiment and a target material is selected that is reasonably transparent to x-rays so that only the fragmented sphere material is imaged. A test setup is illustrated in figure 4.27 where steel is the test material of interest and a PMMA plate provides the impulsive load leading to fragmentation. From the extent of fragmentation and the velocity of fragment debris expansion determined from the test, the dynamic fragmentation properties of the test material are inferred.

This impact test method has been used to determine the fragmentation properties of a number of metals including aluminum, copper, tantalum, titanium and various steels (Grady and Kipp 1997). Discussion of the method here is restricted to a

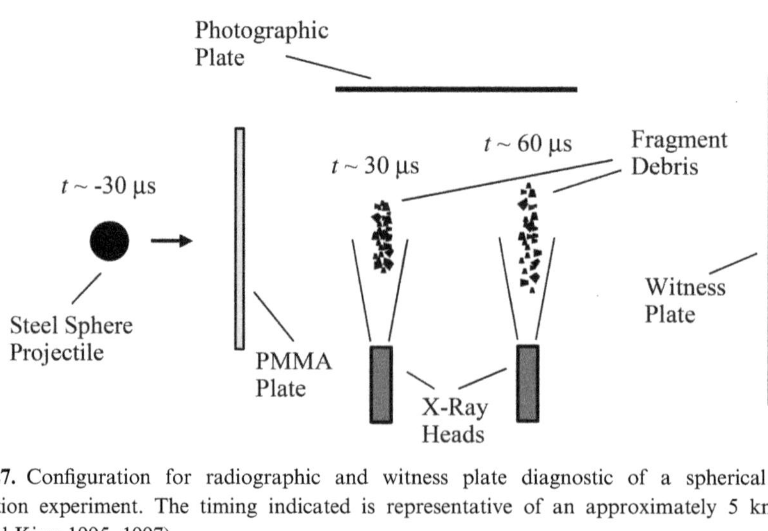

Figure 4.27. Configuration for radiographic and witness plate diagnostic of a spherical test sample fragmentation experiment. The timing indicated is representative of an approximately 5 km s^{-1} impact (Grady and Kipp 1995, 1997).

specific steel, namely AISI E52100 high-carbon chromium steel heat-treated to a Rockwell-C hardness of 60–67. The static yield strength of this steel is 2.0 GPa and the fracture toughness is 30–40 MPa $m^{1/2}$. In the tests, as illustrated in the figure, one-quarter inch spheres of the test steel are launched with a light gas gun to velocities in the range of 3–5 km s^{-1}. The spheres are cause to undergo normal impact onto plates of PMMA of selected thickness (1–12 mm thicknesses with most tests in the 2–6 mm thickness range). Two orthogonal radiographs image the fragmented spherical sample at approximately 15 and 30 cm downstream from the point of impact. Additionally, a witness plate intercepts the fragment debris late in time.

A set of experimental radiographs for one test is shown on the left in figure 4.28. The impact velocity for this test is 4.6 km s^{-1}. The velocity of the fragment debris subsequent to impact is reduced by 270 m s^{-1} because of impulse imparted by the impact. The disc-shaped distribution of the fragment debris in the figure is representative of all experiments with PMMA plate thickness ranging over about 3–6 mm. Thicker plates lead to spread of the fragment debris caused by drag on downstream plate material. For thinner plates a distinct separation into two regions indicative of a spall plane is observed. For yet thinner plates, the shock impulse is not sufficient to cause fragmentation of the spherical sample within the velocity range tested. As seen in the radiograph, the steel fragments remain well grouped and continue to move along the original trajectory at a velocity slightly less than the initial impact velocity. Axial dispersion of the fragment debris is quite limited, with

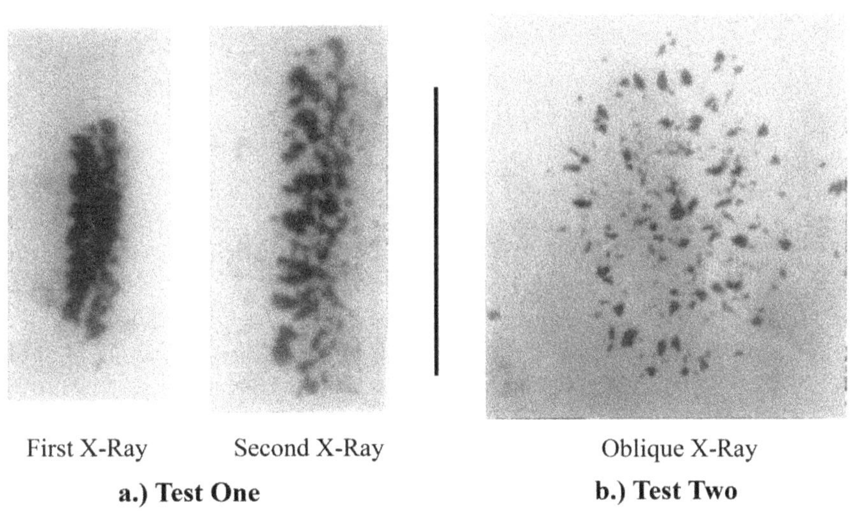

Figure 4.28. Two fragmentation experiments of 1/4 inch steel test sphere impact on PMMA plate. Reprinted from Grady and Kipp (1995). Copyright 1995, with permission from Elsevier. (a) Two images from normal impact at 4.6 km s^{-1} onto a 3.4 mm thick PMMA plate. The time between the two radiographs is 33 μs. The diameter of the disc of debris is 28 mm and 46 mm in the first and second image, respectively. (b) Oblique radiograph from normal impact at 4.5 km s^{-1} onto a 3.3 mm thick PMMA plate. Radiographic imaging is canted 30 degrees from the orthogonal. Vertical extent of the fragment debris is about 45 mm.

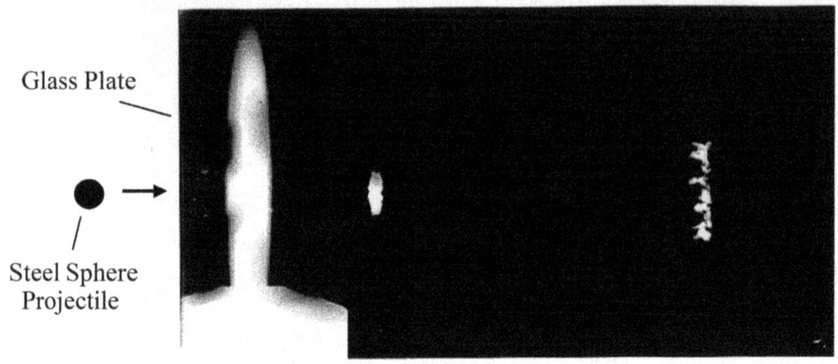

Figure 4.29. Composite radiograph revealing target plate (3 mm thick glass) and projectile debris at two positions. The projectile is a 1/4 inch sphere of 304 stainless steel. The impact velocity is 2.82 km s^{-1}. Reprinted from Grady and Kipp (1997). Copyright 1997, with permission from Elsevier.

reasonably well defined and planar boundaries. Radial dispersion of the debris is substantial, however.

An off-normal radiograph of a comparable experiment is shown in the right-hand image in figure 4.28 and reveals that the debris is sensibly uniformly distributed in both fragment size and fragment density throughout the disc region. Thus, to a reasonable approximation the experiment creates a fragmentation event in which breakup is characterized by a single radial expansion strain rate. This expansion strain rate is calculated directly from the ratio of the radial expansion velocity of the debris obtained from experimental measurement and the initial radius of the spherical sample.

A further example of the experimental method is shown in figure 4.29 where a composite radiographic image reveals the consequences of a 1/4 inch 304 stainless steel sphere impact on a 3 mm thick glass plate at 2.82 km s^{-1}. Although imaged somewhat askew by the radiograph, the glass plate is observed in the radiograph in addition to the fragment debris at two positions downstream subsequent to impact. The expansion velocity and strain rate are assessed from the temporal change in diameter of the debris disc at the two positions.

Some understanding of the physics governing the sphere-on-plate normal impact event is achieved through consideration of the pressure history on the centerline at the first point of contact. A qualitative sketch of the pressure is shown on the left in figure 4.30. The initial pressure p_h is determined by the velocity of impact, and by the Hugoniot properties of the projectile and target plate materials. Hugoniot pressure is sustained until a shock decoupling time t_h. Subsequent interface pressure will drop to a level p_b characterized by hydrodynamic Bernoulli resistance of the material. This pressure is sustained to a time t_b determined by the thickness of the target plate.

Hugoniot pressure is calculated from

$$p_h = \frac{z_S z_T}{z_S + z_T} V, \tag{4.82}$$

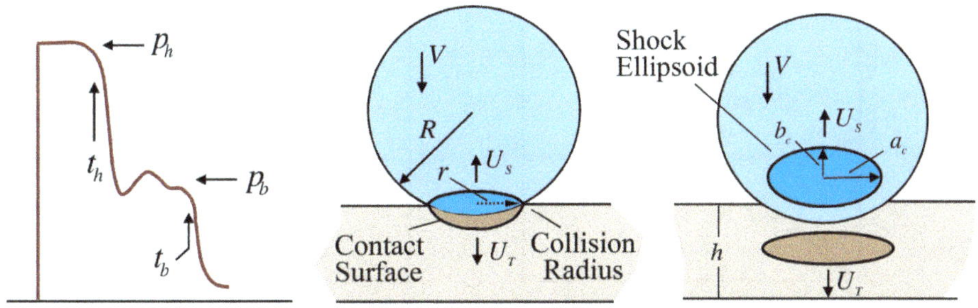

Figure 4.30. Time history of the pressure at the centerline contact point illustrating transition from shock Hugoniot to hydrodynamic Bernoulli pressure. The pictures on the right illustrate coupling of shock energy into the spherical projectile (Grady and Kipp 1995).

where V is the impact velocity while z_s and z_t are the shock impedance of the projectile sample material and target material, respectively. The time t_h over which the Hugoniot pressure is sustained is determined by both the time over which the projectile–target contact radius is supersonic, as described in section 4.3.5, plus the time of transit back to the center point of a sonic release wave from the supersonic-to-subsonic transition radius.

The hydrodynamic Bernoulli pressure dominates momentum transfer to the projectile and is reasonably approximated with the relation (Backman and Goldsmith 1978),

$$m\frac{dV}{dt} = -\frac{1}{2}\rho V^2 A, \tag{4.83}$$

with $p_b = \rho V^2/2$ and where m and A are the projectile mass and projected area, respectively. For relatively modest reductions from the initial impact velocity, equation (4.83) yields a velocity reduction for a plate of thickness h of

$$\Delta V = \frac{\rho A}{2m} hV. \tag{4.84}$$

Comparison of this estimate of velocity decrement from normal impact experiments of steel spheres onto PMMA plates (Grady and Kipp 1995) are shown on the left in figure 4.31. The reasonable agreement confirms that hydrodynamic momentum transfer accounts for the preponderance of velocity reduction of the projectile material.

Energy coupled into the spherical projectile during the initial shock, on the other hand, imparts entirely different motions. Evidence indicates that this energy is largely responsible for the lateral impulse that induces the diametral expansion and fragmentation of the projectile. The extent and magnitude of the impact-induced shock energy can be estimated (Ang 1990, Grady and Kipp 1997) and the process is illustrated by the pictures on the right in figure 4.30. The analysis proceeds by calculating the lateral radius $r \leqslant a_c$ on the contact surface at which the speed of the

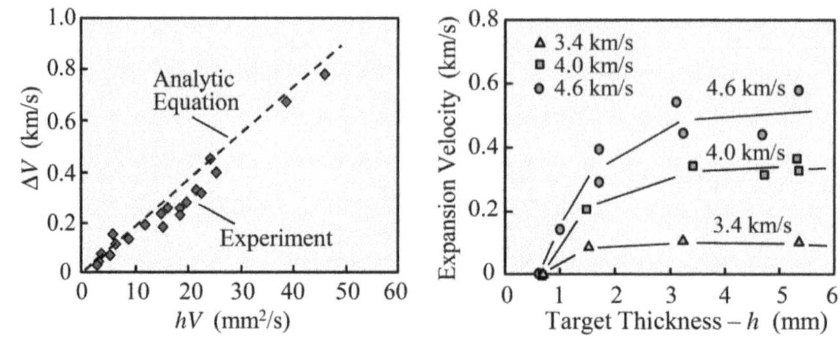

Figure 4.31. Experimental data for normal impact of steel 1/4 inch spheres onto PMMA target plates. Left: reduction in fragment debris velocity from projectile impact velocity. Right: expansion velocity as a function of PMMA target plate thickness for selected impact velocities (Grady and Kipp 1995).

contact radius transitions from supersonic to subsonic. The normal extent of the shock b_c is determined by the shock speed and subsequent release speed of the sphere material. The total shock energy coupled into the projectile is provided by the double integral

$$E_s = 2\pi \int^{a_c} \int^{t(r)} \frac{p_h^2}{\rho c} r dr dt, \tag{4.85}$$

where ρ is the density and c is a measure of the elastic sound speed of the projectile material. The Hugoniot pressure is provided by equation (4.82). Approximating the shock volume as an ellipsoid as illustrated in the figure yields

$$E_s = \frac{4\pi}{3} a_c^2 b_c \frac{p_h^2}{\rho c^2}, \tag{4.86}$$

and is shown to provide a reasonable measure of the shock wave energy coupled into the sphere (Kipp *et al* 1993).

The consequences of the shock energy on fragmentation and expansion of the fragment debris cloud are illustrated by the data plotted on the right in figure 4.31. For PMMA target plate thicknesses in excess of about 3 mm, the shock energy as provided by equation (4.86), and illustrated by the ellipsoidal region on the right in figure 4.30, is full coupled into the sphere. Plate thicknesses in excess of 3 mm do not further contribute to the shock energy or the debris expansion velocity. Conversely, for target plates of thicknesses less than about 3 mm, the shock energy for the specified impact velocity is foreshortened by release from the target back surface. The shock energy and the imparted expansion velocity is proportionally less. The data for steel projectile impact on PMMA plates displayed in figure 4.31 clearly illustrate the separate consequences of hydrodynamic forces on projectile deceleration and the impact-shock-caused lateral impulse, leading to projectile failure and fragment debris expansion.

4.3.8.3 Fragment size from experimental radiographs

The impact fragmentation experiments illustrated by figures 4.27–4.29, when properly configured, lead to an experimental fragmentation event that is reasonably characterized by a single kinematic parameter; namely the radial expansion strain rate. The strain rate is readily determined from the measured radial expansion velocity of the fragment debris and the initial dimension of the spherical sample. If the characteristic size of the resulting fragments can be assessed from the radiographic measurement, then the single fragmentation resistance property intrinsic to the energy-based fragmentation theory can be determined. Namely,

$$K_c = \frac{\rho c \dot{\varepsilon} s^{3/2}}{\sqrt{24}}, \tag{4.87}$$

where K_c is a measure of the dynamic toughness or fragmentation toughness of the test material.

The second image on the left in figure 4.28 is representative of the experimental radiographic measurement used to assess the fragment size created in the breakup event. The radiograph on the right is perhaps more informative in this regard. Image timing and equipment configuration renders this test more difficult to perform. The image of concern clearly reveals that the body is fully fragmented, and reasonably uniform. Assessment of a consistent and systematic fragment size characteristic of the test debris is made difficult, however, by the substantial shadowing of overlapping fragments along the trajectory of the x-ray image. A method to determine a consistent and unique fragment size from such radiographic fragmentation data has been developed (Grady and Kipp 1995, 1997, Grady *et al* 1999).

A method for extracting fragment number and average fragment size from radiographic measurements of reasonably tightly packed fragment debris was initially based on a statistical analysis of crystal growth put forth by Johnson and Mehl (1939). A sounder mathematical framework is provided by the model of Getis and Jackson (1971) developed to describe area patterns randomly dispersed on a plane. The present fragmentation radiographic measurement is illustrated in figure 4.32 where the shadowing of fragments is accounted for by statistical methods in assessment of the fragment size and fragment number.

The statistical issue is conceptualized by imagining coins of various sizes that are tossed and land at random positions on a surface. After N coins are tossed, the total area of interest A is the sum of the areas of the N coins. The area of the surface covered, however, will be less because of overlapping due to the random disposition of the coins. If the area of the surface covered (the projected or observed area of either of the debris images in the figure) is identified as A_{obs}, then statistical theory leads to a relation between the observed area A_{obs} and the actual area A through the expression

$$A = A_{obs} \ln (1 - f)^{-1/f}, \tag{4.88}$$

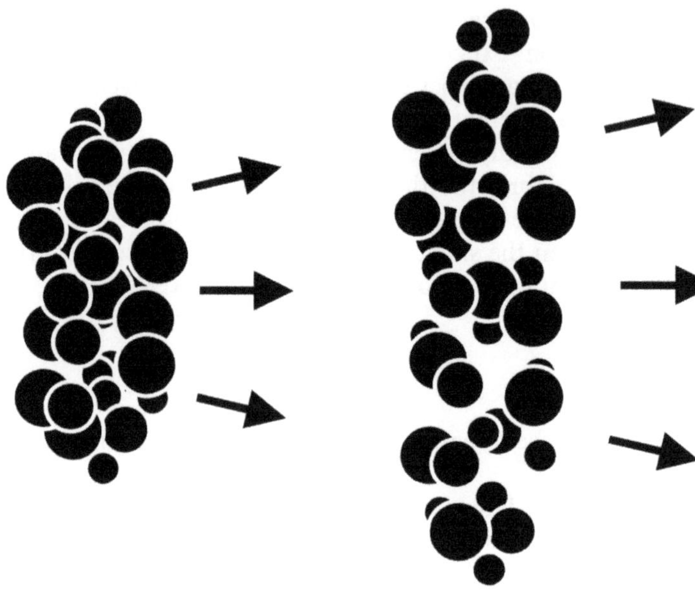

Figure 4.32. Illustration of a radiograph image of fragment debris elucidating fragment shadowing, which is accounted for in statistical assessment of fragment size and number.

where f is the fraction of the surface covered. A Taylor expansion of the relation,

$$A = A_{obs}\left(1 + \frac{f}{2} + \frac{f^2}{3} + \cdots\right), \tag{4.89}$$

reveals that A is always greater than A_{obs} and converges to A_{obs} as f becomes small. Equation (4.88) can be applied to either image of the fragments in figure 4.32. The total projected area A_{obs} is determined and a fractional measure f of the covered area estimated by an appropriate image-processing algorithm. Either image should in principal yield the same value for A. Uncertainties in the fraction f, however, would render the second image with smaller f less prone to uncertainties in A.

The Getis–Jackson model describes area patterns by modeling the number of circles that covers a particular point on the surface. The distribution of the center points of the circles are assumed to be Poisson distributed with spatial frequency λ. The radius of the circles is assumed to come from a distribution of circles with mean $E(r^2)$. Consider a region of the surface A_{reg} over which the circles are randomly distributed. The probability that k circles cover a particular point in the plane is then

$$P(k) = \frac{\theta^k e^{-\theta}}{k!}, \tag{4.90}$$

where

$$\theta = A/A_{reg} = \pi E(r^2)\lambda, \tag{4.91}$$

and with A the total area of circles within the region A_{reg}. The fraction of the region covered is then

$$f = A_{obs}/A_{reg} = P(k \geqslant 1) = 1 - P(k = 0) = 1 - e^{-\theta}. \tag{4.92}$$

Joining equations (4.91) and (4.92) yields equation (4.88).

Knowing the total volume V of the fragmented material (the volume of the spherical sample) within the debris cloud illustrated in figure 4.32, the number of fragments N and the average fragment size s are related through $V = Ns^3$. Correspondingly, the total fragment area calculated from equation (4.88) satisfies $A = Ns^2$. The experimental average fragment size and total fragment number are then provided by

$$s = V/A \text{ and } N = A^3/V^2. \tag{4.93}$$

An alternative analysis assuming an exponential distribution over fragment mass (Grady and Kipp 1997) provides a characteristic fragment size of $s = 3V/2A$.

Fragment size and strain rate are then used with equation (4.87) to calculate a dynamic toughness governing fragmentation. Experimental results are shown in figure 4.33 for E52100 steel and a tungsten heavy alloy. Fragmentation tests for the steel are illustrated in figure 4.29. On the left in the figure are shown average fragment sizes for the six experiments calculated from equations (4.88) and (4.93). Strain rated is calculated from the measured expansion velocity. The dynamic toughness of the respective metals calculated from equation (4.87) is plotted on the right.

Calculated toughness governing dynamic fragmentation is compared with the quasi-static fracture toughness of the respective metals. Static and dynamic toughness differ by about a factor of three. Physical arguments regarding the constrained fracture process zone in the extreme rates of the impact fragmentation event can be offered for the reduced dynamic toughness. Alternatively, the starkly different test methods probably contribute to the difference.

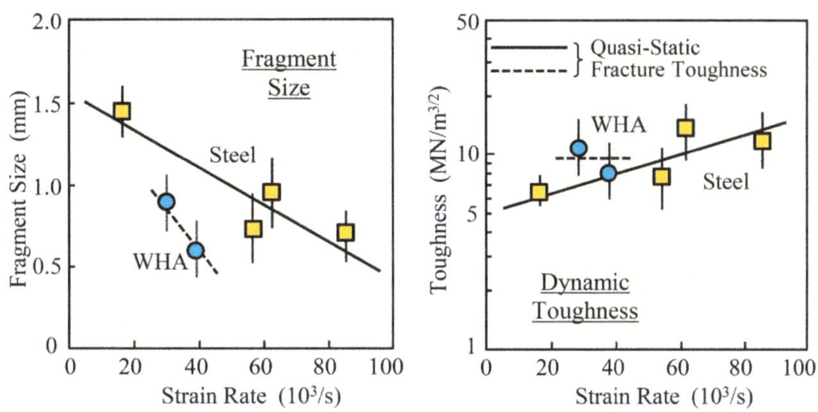

Figure 4.33. Experimental fragment size (left) and dynamic toughness (right) for E52100 steel and a tungsten heavy alloy (Grady *et al* 1999) obtained from sphere impact fragmentation tests.

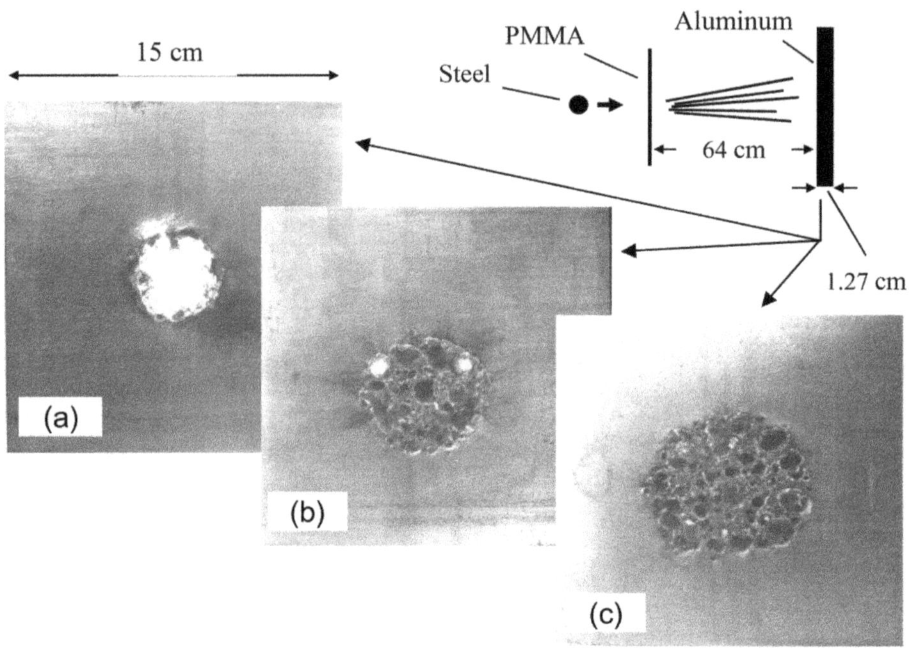

Figure 4.34. Aluminum plate damage from steel sphere impact fragment debris. 6061-T6 aluminum plates 6 inches square by one-half inch thick, 25 inches behind the PMMA impact plate (Kipp *et al* 1993). (a) 0.99 mm thick PMMA and an impact velocity of 4.7 km s^{-1}. (b) 4.75 mm thick PMMA and an impact velocity of 4.04 km s^{-1}. (c) 4.71 mm thick PMMA and an impact velocity of 4.75 km s^{-1}. Courtesy of Sandia National Laboratories.

4.3.8.4 Witness plate effects

Fragment debris spread and damage from impact fragmentation experiments are illustrated by the images shown in figure 4.34. In the tests, 1/4 inch diameter steel spheres are fragmented by high-velocity normal impact onto thin PMMA plates. Aluminum plates six inches square and 1/2 inch in thickness are placed 25 inches behind the impact point. The fragment debris shown in the radiographs in figure 4.28 are illustrative of the debris incident on the aluminum plates.

Selected impact velocities and PMMA plate thicknesses account for differences in the fragment debris spread observed in the figure. The several experiments illustrate the significance of enlarging the impulse footprint in enhancing structure survival to high-velocity impact.

4.3.9 Projectile on wall impact and rebound

Chunky metal projectiles incident on stationary barriers at ballistic velocities undergo a critical thermomechanical change of state as a consequence of the considerable relative kinetic energy at impact. Elevated temperature rise and change of phase, including melt and vaporization, can occur as a consequence of extreme rate plastic work and shock entropy production. The divergent motion associated with rebound flow causes fragmentation and particulation of the body. Post-impact energetic reaction can proceed as a consequence of newly exposed surface area and Weber motions imparted to impact generated particulate debris.

The present subsection addresses issues of impact-induced partitioning of energy into heating and fragmentation, rebound motion of particulate debris and fragmentation statistic of the breakup event. The effort is weighted towards analytic modelling of chunky projectile impact on stationary walls with the intent of identifying relevant underlying physics and developing scaling relations. For purposes of quantitative discussion, aluminum metal is assumed for the projectile material. Iron, and in some cases, other high-strength metals provide the barrier or wall material. The analytic models are independent of material, however, and are readily directed to other applications.

The present application of projectile impact on an impenetrable wall offers a further opportunity to demonstrate effective use of the equivalent Hugoniot model in assessing pertinent features of the impact consequences and displaying the relevant impact physics. The study first examines the near normal impact of a chunky projectile onto a rigid wall to identify the essential features of the analysis. The subsection concludes with the more general case of projectile impact on a non-rigid wall.

4.3.9.1 Projectile on wall and the equivalent Hugoniot model

Development of a model to predict the consequences of the projectile-on-wall impact event requires attention to characteristics of the impact-induced shock compression processes. On impact with the wall, shock compression and the attendant shock dissipation in the projectile are intense. Shock intensity and dissipation are not uniform throughout the projectile. Shock intensity is highest at the initial point of impact. Shocks will attenuate rapidly with propagation distance from the point of contact, and accompanying dissipation will correspondingly lessen. For given geometries of the projectile and impact, the extent and location of dissipation will depend on details of the shape, orientation and impact speed.

The projectile impact and rebound event is again appropriate to approximate solution through the equivalent Hugoniot model. Consider the impact of a chunky projectile with a rigid wall as depicted in figure 4.35. As displayed, the figure illustrates the three states of the system interaction identified in equation (4.27). First, the total energy of the system resides in the kinetic energy of the incident projectile. Second, the transient impact state in which the projectile comes to rest and projectile kinetic energy is assumed to be converted in total to the Hugoniot state energy of the projectile material. Lastly, material rebound in which Hugoniot state energy is partitioned into the dissipation and rebound kinetic energy of the projectile.

Application of the equivalent Hugoniot model to the projectile-on-wall impact event will restrict attention to dissipation through plastic deformation and shock entropy production, approximated as illustrated in figure 4.9 and as expressed by the relations in equations (4.19) and (4.20). Equation (4.22) applies directly for the ratio

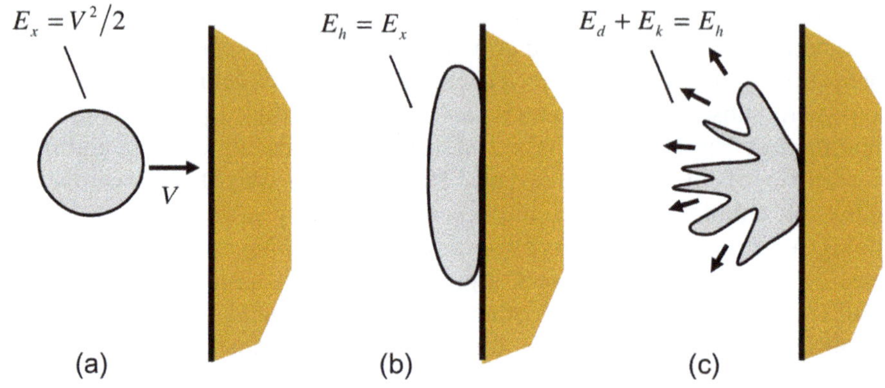

Figure 4.35. Application of the equivalent Hugoniot model to the projectile-on-wall impact event.

of the residual rebound kinetic energy to the incident kinetic energy based on the equivalent Hugoniot impact prediction. Namely,

$$\frac{E_k}{V^2/2} = 1 - \frac{8}{3}\frac{Y/\rho_o}{C_o}\frac{1}{V} - \frac{2}{3}\frac{S}{C_o}V \tag{4.94}$$

relates the rebound kinetic energy to the initial impact velocity V of the projectile. Again, consider the 2017 aluminum used in the study of Piekutowski (1996, 1997) with shock Hugoniot and plasticity properties $\rho_o = 2800$ kg m^{-3}, $C_o = 5.3$ km s^{-1}, $S = 1.35$ and $Y = 0.45$ GPa. The plot in figure 4.10 applies directly to the present impact event. The model estimates no rebound for impact velocities below approximately 90 m s^{-1}. Maximum relative rebound kinetic energy occurs at an impact velocity of about 700 m s^{-1}. Dissipation through shock entropy production increasingly dominates at velocities in excess of 1000 m s^{-1}, accommodating excessive temperature rise in the rebound fragment debris.

The equivalent Hugoniot model provides a method for assessing the partitioning of kinetic energy and dissipation energy in the complex impact event. The model does not specify how that energy is distributed over the participating material. The high-velocity impact of a projectile mass onto a stationary wall leads to a complexity of shock-induced rebound acceleration and dissipation. The distribution of thermal and kinetic energy is not uniform because of rapid shock attenuation from the point or points of impact to regions well removed in the projectile body.

Consequently, in developing physics-based equations describing the distribution of thermal and kinetic energy over the affected material due to the impact shock event, it is appropriate to draw on principles of statistical physics. A sensible statistical description of kinetic energy over the rebound projectile material is appropriately treated with the Maxwell velocity statistics developed in chapter 2. Fragmentation intensity is then addressed through application of the same energy-based calculations leading to equations (4.70)–(4.72). The dissipation and thermal energy governing material heating, including melting and vaporization at the higher impact velocities, is sensibly addressed through the Boltzmann statistics methods described earlier in this chapter.

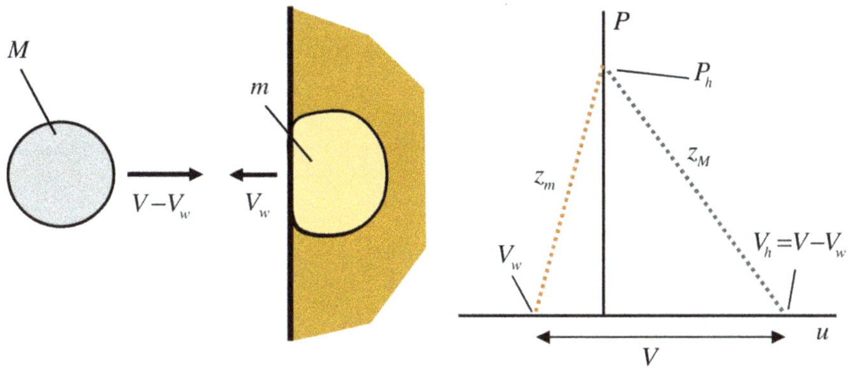

Figure 4.36. Projectile-on-wall impact in center-of-momentum coordinates. The pressure versus particle velocity plot illustrates the path to the equivalent Hugoniot state achieved on impact.

4.3.9.2 Rebound on a non-rigid wall

Projectile impact and rebound is complicated by the finite strength and impact impedance of the wall material. Both dissipation and kinetic energy are coupled to the wall material on impact, diminishing that allocated to the rebound projectile material. Analysis of the impact and rebound of a projectile onto a non-rigid wall can be accomplished through an extended application of the equivalent Hugoniot model. Development of the analysis is predicated on the mechanics illustrated in figure 4.36, which parallels that of figure 4.35 for a rigid wall.

As before, application of the model identifies the kinetic energy of the incident projectile prior to impact, a transient state in which the incident kinetic energy transitions to and momentarily resides as energy at the Hugoniot state of the impacting bodies, and lastly the terminal state in which the Hugoniot energy transitions to rebound kinetic energy and dissipation in both projectile and wall material. In the illustration on the left in figure 4.36, an unspecified mass m of the wall material is identified such that, in the impact and momentum coupling phase, projectile mass M and wall mass m come to rest in a center-of-momentum coordinate system. The center-of-momentum coordinate system is identified through a wall velocity V_w and a reduced projectile velocity $V - V_w$. In the center-of-momentum coordinate system,

$$M(V - V_w) = mV_w, \tag{4.95}$$

and the respective specific kinetic energies (energy per unit mass) of the projectile mass M and the wall mass m are

$$E_M = \frac{1}{2}(V - V_w)^2, \quad E_m = \frac{1}{2}V_w^2. \tag{4.96}$$

Assume at the transient impact state that projectile mass M and the wall mass m come to rest at the equivalent Hugoniot state as illustrated in the pressure versus particle velocity plot on the right in figure 4.36. Equating incident kinetic energies from equation (4.96) to respective transient Hugoniot state energies yields

$$E_M = \frac{1}{2}\frac{P_h^2}{z_M^2} = \frac{1}{2}(V - V_w)^2 \tag{4.97}$$

for the projectile material and

$$E_m = \frac{1}{2}\frac{P_h^2}{z_m^2} = \frac{1}{2}V_w^2 \tag{4.98}$$

for that of the wall material.

The shock impedance of the projectile and wall material are z_M and z_m, respectively. The relations between Hugoniot energy and Hugoniot pressure in equations (4.97) and (4.98) follow from the Hugoniot conservation relations. Solve for V_w and P_h,

$$V_w = \frac{z_M}{z_M + z_m}V, \quad P_h = \frac{z_M z_m}{z_M + z_m}V. \tag{4.99}$$

Using the relations in equation (4.99), the equivalent Hugoniot energy for the projectile mass from equation (4.97) is calculated,

$$E_M = \frac{1}{2}\frac{P_h^2}{z_M^2} = \frac{1}{2}\frac{z_m^2}{(z_m + z_M)^2}V^2 = \frac{1}{2}(\alpha V)^2 = \frac{1}{2}(V_h)^2. \tag{4.100}$$

The velocity $V_h = V - V_w = \alpha V$, with $\alpha = z_m/(z_M + z_m)$, is the particle velocity jump to the Hugoniot state of the projectile material. Ambient impedance properties are appropriate for the present approximate analysis. The solution approaches that of a rigid wall for z_M significantly greater than z_m as α approaches unity.

Assuming shock dissipation in the projectile material through plastic work and entropy production, the kinetic energy of the projectile material following rebound is provided by

$$E_k = E_M - E_d = \frac{1}{2}(V_h)^2 - \frac{4}{3}\frac{Y}{\rho_o}(V_h/C_o) - \frac{1}{3}SC_o^2(V_h/C_o)^3. \tag{4.101}$$

where $V_h = \alpha V$ relates the equivalent Hugoniot particle velocity and the incident projectile velocity. As before, kinetic energy E_k and dissipation energy E_d are applied to, respectively, estimate the intensity of the fragmentation and thermodynamic state of the rebound material subsequent to impact, where appropriate statistical methods distribute energy over the rebound mass. The rebound kinetic energy is directly related to the incident impact velocity through

$$\frac{E_k}{V^2/2} = \alpha^2\left[1 - \frac{8}{3}\frac{Y}{\rho_o C_o}\frac{1}{\alpha V} - \frac{2}{3}\frac{S}{C_o}\alpha V\right]. \tag{4.102}$$

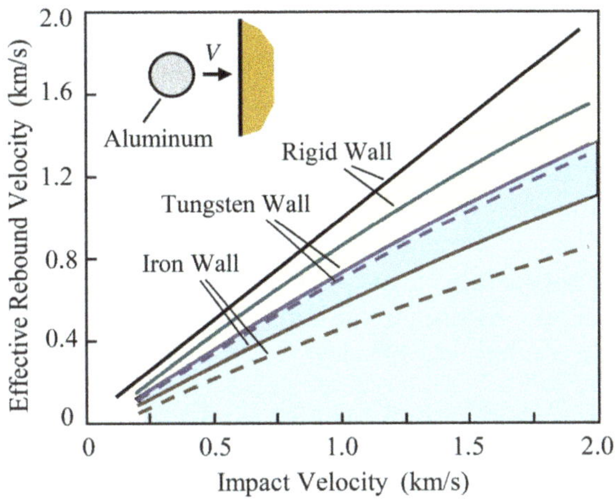

Figure 4.37. Calculations of effective rebound velocity of a chunky aluminum projectile subsequent to normal impact on a stationary wall. The shock impedances used are 14.8 kg s^{-1} m^{-2} for aluminum, 28.1 kg s^{-1} m^{-2} for iron and 77.4 kg s^{-1} m^{-2} for tungsten.

The kinetic energy E_k is referenced to the center-of-momentum coordinate system identified in figure 4.36. A theorem of classical mechanics identifies decomposition of the kinetic energy of a many-body system (the rebound fragment debris) into the kinetic energy of the center of mass plus the kinetic energy about the center of mass (e.g. Goldstein 1965). Thus the kinetic energy of the rebound debris in the initial (rest) frame becomes,

$$E_k' = E_k - \frac{1}{2}V_w^2. \tag{4.103}$$

An effective rebound velocity V_k *is* defined through $E_k = V_k^2/2$. The rebound velocity V_k (or V_k') is not a vector property and is not indicative of a rebound speed normal to the impacted plate. It is a velocity measure of the rebound kinetic energy ascribed by the model to the projectile material. Under certain impact conditions the dominant direction of residual motion is actually parallel to the plate surface.

Effective rebound velocity as a function of normal impact velocity is illustrated in figure 4.37 for selected stationary barrier conditions. The one-to-one line corresponds to perfect rebound from a rigid wall with no dissipation in projectile material. The next curve corresponds to a rigid wall rebound accounting for plastic work and shock dissipation in the 2017 aluminum projectile material. The following two solid curves correspond to tungsten and iron metal wall barriers, respectively. The dashed curves correspond to correction to the rest frame of reference through equation (4.103).

4.4 Gurney equations and dynamic fragmentation

In a time without computers R W Gurney (1943) developed a set of solutions for estimating the motion induced by the explosive loading of metal shells and plates.

More importantly, his analysis revealed the important dimensional relations and provided a method for scaling specific explosive systems over different materials and geometries. Gurney provided, rather than a set of solutions, a solution concept for addressing a broad class of problems. Rather than rote application of his several relations, the Gurney solution is most profitably used in its fundamental form with a certain level of ingenuity to specific problems of interest. Kennedy (1997) provides an overview of the Gurney method and subsequent developments.

Concurrently, in another laboratory, G I Taylor (1941) and N F Mott (1943) were pursuing related studies. Although perhaps less well known, these latter works expand sizably on the underlying physics.

4.4.1 The classic Gurney solution

In the early report of Gurney, specific objectives relating to the explosive acceleration of metal cases were of paramount interest. These goals dictated the initial geometries considered by Gurney. It is noteworthy that Taylor and Mott were pursuing similar analysis contemporary with the efforts of Gurney. In his seminal paper (Gurney 1943), Gurney references a report of Taylor on the subject.

4.4.1.1 The Gurney method

The elementary problems first treated by Gurney are illustrated in figure 4.38. The picture on the left depicts an explosive charge set against a rigid wall with total mass C per unit area. The detonated charge will accelerate the metal plate of mass M per unit area to some terminal velocity V_g. Calculation of the velocity V_g and the scaling of V_g over different material and system parameters resides at the heart of the Gurney analysis. The initial thickness of the charge and Lagrangian position of the metal plate is X_o. The Lagrangian position at some interior point of the charge is X

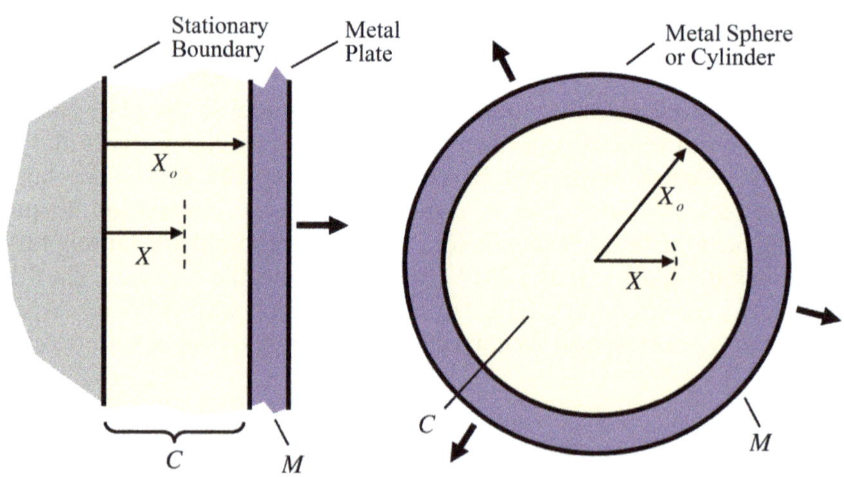

Figure 4.38. Initial configurations for an explosive accelerated plate on the left and an expanding cylinder or sphere on the right.

and the material coordinate is the cumulative mass at the position X or $m = (X/X_o)C$. The mass density for the planar charge is $\rho_c = C/X_o = m/X$.

On the right side of figure 4.38 is depicted either a spherical metal shell of mass M containing a charge of mass C, or a cylinder of mass M and charge C, each of unit length. The coordinates X_o and X are the respective radii while the cumulative mass coordinate is $m = (X/X_o)^2C$ or $m = (X/X_o)^3C$ for the cylinder or spherical system, respectively.

For any of the three configurations conservation of energy provides the relation

$$CE = \frac{1}{2}MV_g^2 + \frac{1}{2}\int_0^C u^2(m)dm. \tag{4.104}$$

On the left side of equation (4.104) is the initial potential energy of the explosive charge, whereas on the right is the terminal kinetic energy of the metal and the explosive products, respectively.

Several assumptions are implicit in this relation. Most fundamental is the assumption of a constant E providing the potential per unit mass of the explosive for doing mechanical work independent of the size and configuration of the charge. Gurney himself had concerns with this assumption and pursued efforts to justify it. Wide use by many others over the intervening years has demonstrated that this assumption is reasonable over practical variations in the system parameters.

The specific potential energy E can be estimated if the Chapman–Jouguet properties and release isentrope properties of the explosive are known. It has been more common to determine E for a particular explosive empirically through several experiments designed for this purpose. Other authors commonly identify $\sqrt{2E}$ as the Gurney velocity and a property of the material. This convention is not used here.

Also tacitly assumed in equation (4.104) is the lack of any further dissipation processes providing additional energy sinks for the explosive potential energy. Perhaps most obvious would be the plastic work in expanding the metal cylinder or sphere to the point of breakup and fragmentation. Such plastic work would vary markedly from metal to metal. In most applications, plastic dissipation is found to be small. If, however, one pursues a wider range of applications for the model, this and other additional energy sinks should be considered. The contribution of plastic work is assessed in the study of Mott (1943).

A second assumption key to the Gurney model is that the distribution of the velocity through the explosive charge is linear. That is

$$u = V_g\left(\frac{X}{X_o}\right) = V_g\left(\frac{m}{C}\right)^{1/n}, \tag{4.105}$$

where, in the second step, n = 1, 2 or 3 for the planar, cylindrical or spherical configuration, respectively. The assumption of velocity linearity (a constant velocity gradient) along the axis of motion is quite reasonable, particularly in the terminal phases of the explosive acceleration process. Motion details imparted by the initial detonation-induced shocks will of course not be revealed with the model.

Equation (4.104), after substitution of equation (4.105) and an integral transformation, becomes

$$CE = \frac{1}{2}MV_g^2 + \frac{1}{2}CV_g^2 \int_0^1 y^{2/n} dy. \tag{4.106}$$

Solving for V_g yields

$$V_g = \frac{\sqrt{2E}}{(M/C + k_n)^{1/2}}, \tag{4.107}$$

where $k_n = 1/3$, $1/2$ or $3/5$, for $n = 1$, 2 or 3, respectively. The basic relations of Gurney provided by equation (4.107) have found wide engineering application over the many years since the model was developed.

4.4.1.2 The Taylor and Mott analysis of exploding cylinders

The seminal report of Gurney (1943) noted an earlier study of Taylor (1941) on the explosion-driven expansion of metal cylindrical cases. Gurney suggested some restrictions to the Taylor analysis that are probably not warranted. In fairness, the Gurney model provides engineering level scaling relations that have admirably served the munitions community to the present day.

The Taylor analysis, however, is in a class of its own. The analysis of the explosive expansion of a long cylindrical shell by Taylor entails physical principles and analytic tools that uncover a wide range of understanding of the explosive interaction and case expansion process. Taylor's analysis was integral to the efforts of Mott (1943) in modeling explosive fragmentation of metal cases. The analysis of Mott more closely parallels the application-oriented effort of Gurney. The works of both Taylor and Mott precede that of Gurney, and warrant closer examination.

Taylor analysis

Taylor considered a long cylinder such that detonation initiated at one end achieves steady state. Further, analysis of the cylinder and explosive system is undertaken in a coordinate system in which the detonation wave is stationary with respect to the observer. Accordingly, cylinder material and unreacted explosive enter from the left at a velocity D as illustrated in the left-hand sketch in figure 4.39. A distance x along the cylinder axis originating at the detonation front identifies the position into the detonation products determining the pressure driving cylinder expansion. The function $r(x)$ provides the radial expansion of the cylinder wall as a function of the distance behind the detonation front. In this coordinate system, the profile shown in the figure is unvarying in time throughout the calculation. (The plot can be likened to a streak camera image of one position of the cylinder through the detonation expansion event.)

At any position x the pressure $p(x)$ applies force to the cylinder at right angles to the direction of motion in the selected coordinate system. Consequently, kinetic energy of the cylinder remains unchanged and the cylinder velocity D at any mass element is constant through the problem. The radial acceleration of the explosive

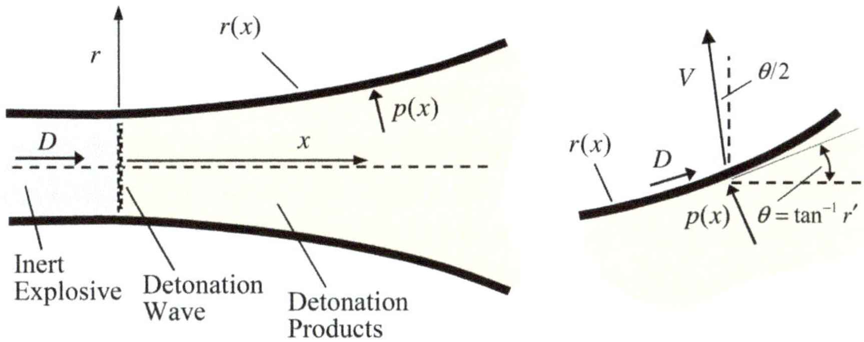

Figure 4.39. Left: profile of explosive-driven expansion of a cylindrical case in a reference frame in which the detonation wave is stationary (Taylor 1941). Right: the expanded image identifies the normal pressure at position x inducing centrifugal acceleration of the cylinder element.

products is assumed to be sufficiently small that radial gradients in pressure can be ignored. The same assumption is implicit in the Gurney model of the same event. The centripetal force applied by the pressure is balanced by the centrifugal acceleration D^2/R_c, where R_c is the radius of curvature of the case at the position x. Accordingly, the equation of motion is provided by

$$\frac{M}{2\pi r}\frac{D^2}{R_c} = p(x), \tag{4.108}$$

where M is the mass per unit length of the metal cylinder. The radius of curvature at x is,

$$R_c = \frac{(1 + r'^2)^{3/2}}{r''} \simeq \frac{1}{r''}. \tag{4.109}$$

The speed D relative to the outward motion renders the first derivative of $r(x)$ small and reasonably ignored in the radius of curvature expression. Equation (4.108) then yields the differential equation for the expanding case profile $r(x)$,

$$\frac{MD^2}{2\pi r}\frac{d^2r}{dx^2} = p(x). \tag{4.110}$$

Solution closure is achieved through characterization of the detonation gas dynamics. For barotropic flow, conservation of momentum is expressed through the Bernoulli equation,

$$\int_p \frac{dp}{\rho} + \frac{u^2}{2} = f(p) + \frac{u^2}{2} = C_1, \tag{4.111}$$

while conservation of mass yields

$$\rho u r^2 = C_2. \tag{4.112}$$

The integral in equation (4.111) yields $f(p)$ through an adiabatic EOS $\rho(p)$ for expansion of the detonation gases. The constants C_1 and C_2 are determined by conditions proceeding, and just after, the detonation wave. Namely,

$$C_1 = f(p_j) + \frac{u_j^2}{2},\ C_2 = \rho_o D r_o^2. \tag{4.113}$$

In these constants, p_j and u_j are for example the Chapman–Jouguet pressure and particle velocity behind the detonation wave for an ideal detonation. The properties ρ_o and r_o are the density of the unreacted explosive and initial radius of the cylinder, respectively.

From this point, the solution is straightforward. Particle velocity can be eliminated between equations (4.111) and (4.112), and pressure solved for as a function of the radius r. Equation (4.110) becomes a second-order differential equation that is readily solvable through numerical methods. Taylor, not having the computer power of today, proceeded further with the solution through analytic means. The following parallels the analysis of Taylor.

In equation (4.110) define $v = dr/dt$ and replace x with Dt yielding the system of two first-order differential equations,

$$M\frac{dv}{dt} = 2\pi r p(r), \tag{4.114}$$

$$\frac{dr}{dt} = v. \tag{4.115}$$

That equation (4.114) has the appearance of a straightforward application of Newton's law is misleading. The velocity v is the rate of expansion of the cylinder radius at a fixed axial position, and not the path of a material element. Equation (4.114) transforms to

$$M\frac{dv}{dt} = M\frac{dv^2/2}{dr} = 2\pi r p(r), \tag{4.116}$$

and integration yields

$$\frac{v^2}{2} = \frac{2\pi}{M}\int_{r_o}^{r} r p(r) dr = \frac{1}{M}\int_{a_o}^{a} p(a) da, \tag{4.117}$$

where the change of variable $a = \pi r^2$ has been introduced. Integration by parts gives

$$M\frac{v^2}{2} = ap(a) - a_o p(a_o) - \int_{p_j}^{p} a dp. \tag{4.118}$$

The mass conservation relation from equation (4.112) is used to replace a and a_o, providing

$$M\frac{v^2}{2} = \pi r_o^2 \rho_o D\left(\frac{p}{\rho u} - \frac{p_j}{\rho_j u_j} - \int_{p_j}^{p} \frac{dp}{\rho u}\right). \tag{4.119}$$

Note that equation (4.112), when applied to the detonation wave, yields the shock mass conservation relation $\rho_o D = \rho_j u_j$ applied in the second term in the parentheses. From the Bernoulli equation (4.111) $dp/\rho = -udu$ and equation (4.119) becomes

$$M\frac{v^2}{2} = \pi r_o^2 \rho_o D\left(\left(\frac{p}{\rho u} + u\right) - \left(\frac{p_j}{\rho_j u_j} + u_j\right)\right). \tag{4.120}$$

Application of the shock jump conditions for conservation of mass and momentum through the detonation wave gives

$$\frac{p_j}{\rho_j u_j} + u_j = D. \tag{4.121}$$

The mass of explosive per unit length in the Gurney notation is $C = \pi r_o^2 \rho_o$. Equation (4.120) is rewritten as

$$v = \frac{\sqrt{2D^2((p/\rho u D + u/D) - 1)}}{\sqrt{M/C}}. \tag{4.122}$$

Equation (4.122) provides the Taylor solution for the velocity history of the case expansion. In this form, the relation can be compared with the Gurney solution. The term in the numerator in the limiting expansion velocity provides a measure of the Gurney energy. The denominator differs from the Gurney relation only when M/C becomes small, a condition also known to be problematic for the Gurney equation (Kennedy 1997). Equation (4.122) fails when fragmentation of the case occurs.

Equation (4.122) provides a measure of the case expansion radial velocity. The relation does not provide velocity and directionality of mass elements, and ultimately fragments, from the ruptured case. Referring to the illustration on the right-hand side of figure 4.39, the magnitude of the velocity of any mass element in the case in the reference frame of the analysis (the detonation wave stationary) is equal to the detonation velocity D. At a position x along the axis, the velocity is directed at an angle θ relative to the axis of the cylinder. Returning to the reference frame with the cylinder at rest, the material element exhibits a velocity magnitude in the direction of the detonation along the cylinder axis of $D(1 - \cos\theta)$ and perpendicular to the axis of $D \sin\theta$. The velocity of the fragments is then

$$D\left((1 - \cos\theta)^2 + \sin^2\theta\right)^{1/2} = 2D\sin\theta/2, \tag{4.123}$$

at an angle of $\theta/2$ relative to the normal, and where

$$\theta = \tan^{-1} dr/dx = \tan^{-1} v/D. \tag{4.124}$$

On completion of the analysis, Taylor undertook a specific calculation for the explosive TNT using appropriate adiabatic expansion data and numerically calculating the Bernoulli relation in equation (4.111). Although known to be less than adequate for describing the expansion properties of detonation gases, a gamma-law gas nonetheless yields an analytic relation for the Bernoulli equation,

$$\frac{\gamma}{\gamma-1}\frac{p}{\rho}+\frac{u^2}{2}=\frac{\gamma}{\gamma-1}\frac{p_j}{\rho_j}+\frac{u_j^2}{2}, \tag{4.125}$$

where

$$p/\rho^{\gamma}=p_j/\rho_j^{\gamma}, \tag{4.126}$$

and markedly simplifies the analysis.

Mott analysis

The development of Mott that incorporates analysis of the explosion-induced expansion of cylindrical cases is included in his investigation of case fragmentation (Mott 1943). The full analysis including formulation of Gurney-like methods for predicting case expansion velocities is instructive.

Mott in 1943 developed two relations that are intended for scaling exploding high-explosive shells with respect to the statistical distribution in the sizes of metal fragments resulting from the explosion (see also Grady 2006a). They were, in his notation,

$$N(m)dm=C_1e^{-M/M_A}dM, \tag{4.127}$$

$$M_A=C_2t^{5/6}d_2^{1/3}(1+t/d_2). \tag{4.128}$$

In equations (4.127) and (4.128), C_1 and C_2 are constants while $M=m^{1/2}$ where m is a fragment mass. Cylinder dimensions are provided by d_2 and t. The geometry is illustrated in figure 4.40.

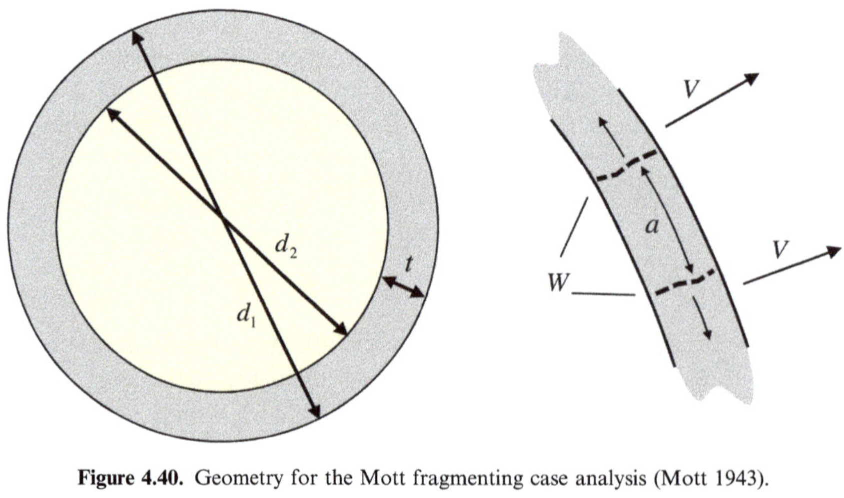

Figure 4.40. Geometry for the Mott fragmenting case analysis (Mott 1943).

Equation (4.127) is a description of the statistical fragment size distribution. Equation (4.128) provides a scaling relation for the characteristic fragment size. These relations continue to be used, frequently inappropriately, up to the present day. In equation (4.128) M_A, with dimensions of $m^{1/2}$, provides the scale parameter in the equation (4.127) statistical fragment distribution and can be written

$$M_A = \sqrt{\rho tal}. \tag{4.129}$$

The characteristic length and width of the average fragment are l and a, while t is the through-the-case thickness of a fragment. Mott assumed initially, and in a later report provided analytic support for, a constant of proportionality between fragment length and width $l = \alpha a$, where the calculated fragment aspect ratio α is, for example, approximately 3.5 for the breakup of steel cases. The mass scale is then

$$M_A = \sqrt{\alpha \rho t a^2}. \tag{4.130}$$

Mott pursued two theories in establishing a predictive scaling relation for the mean fragment size or, equivalently, the circumferential fracture spacing a. First was an energy balance wherein a resisting fracture energy is equated to a driving inertial energy in the expanding cylinder breakup. The second was the tensile stress activating statistically distributed fracture sites. He clearly favored his second analytic pursuit and in later discussions offered arguments that downplayed the applicability of the former.

It was, however, the former energy balance calculation that led to the precise form of the oft-quoted fragment-size scaling rule. To calculate the characteristic fracture spacing a, illustrated on the right in the figure, Mott equated the work W required to open a single fracture with the stretching kinetic energy on both sides of the fracture. This calculation provided the following relation for the characteristic fragment width,

$$a = \left[\frac{24r^2 W}{\rho V^2}\right]^{1/3}, \tag{4.131}$$

with V the velocity of expansion at the radius of breakup. Combining material parameters into a constant of proportionality $C_3 = 24^{2/3}\alpha^{1/2}\rho^{-1/6}W^{2/3}$, equations (4.130) and (4.131) yield

$$M_A = C_3 r^{2/3} t^{1/2} V^{-2/3}. \tag{4.132}$$

The dimensions r and t are values for the radius and thickness of the cylinder at the time of fracture. Since the product rt is the constant cross-sectioned area (volume) of the expanding cylindrical shell, it is a reasonable approximation, as noted by Mott, to replace these geometry parameters with their initial values in equation (4.132).

Mott's attention at this point in the analysis turned to addressing the velocity V at the time of rupture and fragmentation of the case. In words, Mott proposed '... V^2 to be proportional, not to the charge-weight ratio of the whole shell, but to the charge-weight ratio of a cross-section of it ...'. That is, V^2 is proportional to C/M

where C and M are the cylinder charge and metal case masses per unit length. Comparison to the Gurney relation for explosive expansion of a cylinder from equation (4.107),

$$V^2 = \frac{2E}{M/C + 1/2}, \tag{4.133}$$

reveals disparity only in the factor of 1/2 in the Gurney solution. This factor, resulting from accounting for acceleration of the explosive mass in the Gurney analysis, is a negligible contribution unless the ratio M/C is inordinately low. Kennedy (1997) has noted difficulties with the Gurney relation in this limit. Some contributing problems to this difficulty are considered later in this section.

Mott (1943), in his third technical report on case fragmentation, also considered the reduction in case velocity because of plastic work incurred during expansion. The analysis resulted in an estimate of velocity reduction of

$$\Delta V = \frac{Y}{\rho V}\varepsilon, \tag{4.134}$$

where V is the calculated velocity at the expansion state ignoring plastic work, Y is a measure of the unconfined tensile strength of the metal, ρ is the metal density and ε is the plastic strain at breakup. For applications of concern at the time, his estimates showed reductions of order 10% or less. Conclusions were that unless the charge to metal weight ratio was unusually low, velocity reduction due to plastic work is not an important issue.

Mott posed the velocity relation in the form

$$RV^2 = \text{const.} \tag{4.135}$$

with

$$R = \frac{d_1^2 - d_2^2}{d_2^2}, \tag{4.136}$$

choosing to keep the charge and metal density ratio in the constant to obtain a fragment size relation dependent strictly on the geometry. Letting $r = (d_1 + d_2)/4$ and combining with equation (4.132) yields Mott's relation for the size scale parameter,

$$M_A = C_2 t^{5/6} d_2^{1/3}(1 + t/d_2). \tag{4.137}$$

Comparison of equations (4.127) and (4.128) with ordinance fragmentation data available at the time showed good agreement.

The analyses of detonating expansion and fragmentation of metal cases in the early 1940s during the height of the war years by both Taylor and Mott are amazement undertakings. It is difficult to imagine the daily laboratory culture in which these theoretical developments and supporting experimental studies emerged.

4.4.2 Alternative Gurney geometries

Various extensions of the Gurney solution method to other system geometries have been pursued that have application to explosive engineering. Two of these Gurney solutions are considered. Others are readily formulated to address the many explosive–structure interaction applications. The power of the Gurney model is the ready application of the solution method to diverse problems of interest.

4.4.2.1 Unequal parallel plates

The geometry shown in figure 4.41 is that of two flat and parallel metal plates of in general unequal masses M_A and M_B per unit area separated by a charge of mass C per unit area. Detonation of the explosive charge leads to opposing accelerations of the plates. This Gurney problem is treated by Jones *et al* (1980). A more transparent development is provided here.

A stationary plane can be identified within the charge that separates charge C_A and the mass M_A accelerated in the minus X direction from charge C_B and the mass M_B accelerated in the plus X direction, where $C_A + C_B = C$. The linear velocity relations in the two regions (positive quantities) are

$$V(m) = \frac{m}{C_A} V_A, \text{ and, } \quad V(m) = \frac{m}{C_B} V_B, \tag{4.138}$$

where V_A and V_B are the respective terminal velocities of the two plates. The coordinate m ranges over $0 \leqslant m \leqslant C_A$ or $0 \leqslant m \leqslant C_B$, respectively, for the opposing sides.

Momentum conservation is written

$$M_A V_A + \int_0^{C_A} \frac{m}{C_A} V_A dm = M_B V_B + \int_0^{C_B} \frac{m}{C_B} V_B dm, \tag{4.139}$$

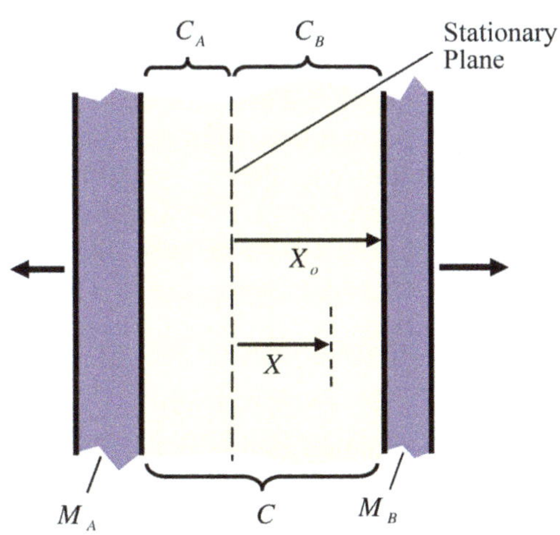

Figure 4.41. Plates of dissimilar masses M_A and M_B accelerated in opposing directions by explosive charge C.

or

$$(M_A + C_A/2)V_A = (M_B + C_B/2)V_B. \tag{4.140}$$

The assumption of velocity linearity through the charge provides the additional relation

$$V_A/C_A = V_B/C_B. \tag{4.141}$$

Combining equations (4.140) and (4.141) provides expressions for the opposing charge proportions separated by the stationary plane,

$$C_A = \frac{C}{2}\frac{2M_B + C}{M_A + M_B + C} \tag{4.142}$$

and

$$C_B = \frac{C}{2}\frac{2M_A + C}{M_A + M_B + C}. \tag{4.143}$$

For each side of the stationary plane the system is equivalent to the rigid boundary geometry pictured on the left side of figure 4.38 and solutions for Gurney velocities are immediately written from equation (4.107) as,

$$V_A = \frac{\sqrt{2E}}{(M_A/C_A + 1/3)^{1/2}} \tag{4.144}$$

and

$$V_B = \frac{\sqrt{2E}}{(M_B/C_B + 1/3)^{1/2}}, \tag{4.145}$$

where C_A and C_B are provided by equations (4.142) and (4.143).

4.4.2.2 Compound shells

The following analysis has a number of applications. The general analysis, when separated into its component parts allows, for example, Gurney acceleration solutions for such curvilinear geometries as linear or concentric shaped-charge jets, explosively formed projectiles or explosive preparation of powder compacts. The solution addresses both inward and outward explosive acceleration of metal component parts. Aspects of the solution have been addressed in reports of Chou *et al* (1981, 1983), and in the review of Kennedy (1997). The present solution is based on work reported in Grady (2001).

4.4.2.3 Nested shells

The exploding shell geometry is illustrated in figure 4.42. The symmetrically nested metal shells may be spheres, cylinders or partial sections of curvilinear geometries reasonably approximated by a sphere or a cylinder. The region between the inner and outer shell is filled with explosive. A cavity resides within

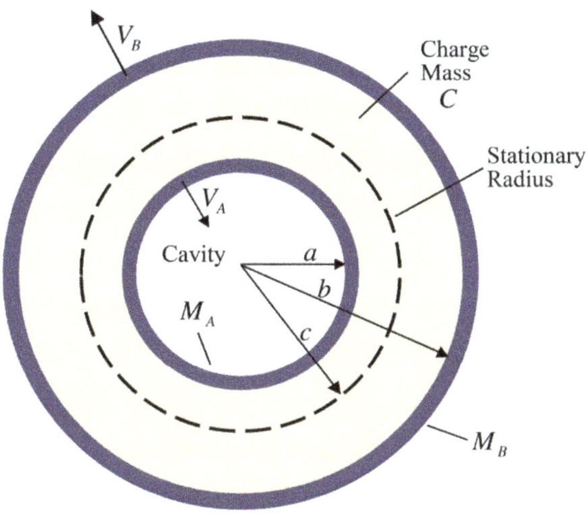

Figure 4.42. Nested spherical or cylindrical shells with intervening explosive charge. Explosive acceleration leads to outward velocity V_B of metal mass M_B and inward velocity V_A of metal mass M_B.

the inner shell. The inner and outer shells have mass M_A and M_B. For the situation of cylindrical shells either a mass per unit length or the total mass of a sufficiently long cylinder can be considered. The mass of the charge will be identified by the traditional symbol C.

Detonation of the explosive charge is assumed to occur in a time that is short when compared to time scales of the subsequent motions. This is known to be a rough approximation when a system such as that illustrated in the figure is detonated at a single point. The model system better replicates a system that has multiple ignition points achieving closer to simultaneous detonation of the explosive. Metal velocities induced by the charge detonation are not strongly affected by the non-simultaneity of the detonation except in regions of colliding detonation waves where significant jetting and early shell rupture may occur.

Detonation in the model system leads to rapid radial (either spherical or cylindrical) acceleration of metal shells to terminal velocities, namely, an outwardly directed velocity V_B for metal mass M_B and an inwardly directed velocity V_A for the inner shell of mass M_A. Possible complications of early rupture and fragmentation are ignored.

The explosive mass C is also accelerated by the explosive energy. A fundamental Gurney assumption is that the explosion products expand uniformly following the acceleration process. A linear gradient in the radial velocity of the explosive from an inward directed V_A at the inner shell to an outward directed V_B at the outer shell is a consequence. (A rigorous sign convention is not used here and scalar velocities are regarded as positive except in specific calculations where more careful attention to sign is necessary.) Metal masses during and following acceleration to terminal velocities are regarded as lumped and any internal wave features during coupling of the explosive energy are ignored.

The linear gradient in the radial velocity of the expanding explosive from an outward V_B to an inward V_A requires that there is a radius c at which the explosive products remain stationary during the system acceleration. Explosive products and metal mass exterior to the stationary radius c expand outward in the system acceleration process, whereas interior to the radius, metal and explosive products accelerate inward. This assumption provides an important decoupling of solutions in the analytic development as was introduced in the solution of the acceleration of unequal plates in the previous section. If the stationary radius is known then this radius can be regarded as a rigid spherical or cylindrical boundary. The solution for acceleration and velocities for explosive and metal exterior to this boundary can be developed separately. Similarly, the solution can be separately developed for materials inside this rigid boundary. Such solutions are also appropriated for the explosive launch from a curvilinear surface that remains stationary. An example would be that of a solid metal sphere or cylinder. Assumption of a stationary radius for the curvilinear geometries is approximate. Further discussion is provided later in the section.

In the subsequent analysis first assume that this stationary radius is known. Solutions are developed independently for both the inward and the outward motions of explosive charge and metal materials. Calculation of the position of the stationary radius is then undertaken.

Exploding shell

The configuration for the outward explosive acceleration of a spherical or cylindrical metal shell is illustrated on the left in figure 4.43. The stationary radius within the explosive charge is represented as a rigid boundary at a radius c. The metal mass M resides at a radius b while the enclosed explosive charge mass is C.

The solution proceeds by developing an expression for conservation of energy. At the time of interest, detonation has occurred and the metal shell is assumed to have

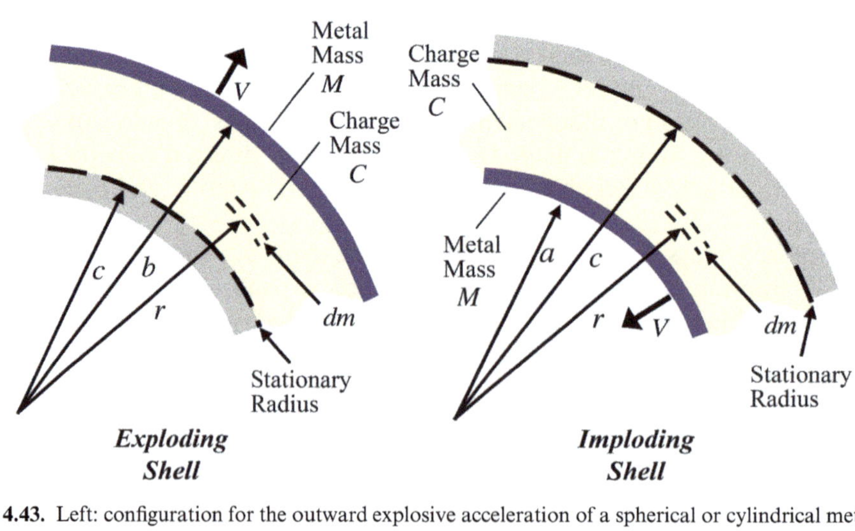

Figure 4.43. Left: configuration for the outward explosive acceleration of a spherical or cylindrical metal shell. Right: the corresponding configuration for an imploding metal shell.

accelerated to its limiting velocity V_g. It would be equally appropriate to query the solution at any velocity $V < V_g$ at a specified radius of expansion as was done by Taylor (see section 4.4.1.2). This would require a more detailed characterization of the energy of the explosive products as a function of expansion. A later section addresses this extension of the Gurney solution method.

Conservation of energy is written as

$$\frac{1}{2}MV_g^2 + \frac{1}{2}\int_c^b u^2(r)dm = CE. \tag{4.146}$$

The terms on the left are, respectively, the kinetic energy of the metal shell and explosive charge at velocity V_g. The radius r is the Lagrangian (material) coordinate over $c \leqslant r \leqslant b$ while $u(r)$ is a functional expression for the velocity of the explosive products over this radial coordinate. The total kinetic energy of the explosive products results from integration over the charge mass.

Ignored on the left-hand side are other energy sinks, the dominant one probably being the work of plastic distortion in stretching the spherical or cylindrical shell during outward expansion. This is usually considered small and ignored (see analysis of Mott in the preceding section).

On the right of equation (4.146) is the explosive energy source. Total charge mass is C, while E is a measure of the explosive specific energy available to do expansion work. Commonly called the Gurney energy, it is usually determined experimentally.

Equation (4.146) can be solved for the velocity V_g and written in a form which is canonically similar for most of the Gurney solutions,

$$V_g = \frac{\sqrt{2E}}{\left(\dfrac{M}{C} + \dfrac{1}{CV_g^2}\displaystyle\int_c^b u^2(r)dm\right)^{1/2}}. \tag{4.147}$$

The Gurney assumption of a constant velocity gradient with $u(c) = 0$ and $u(b) = V_g$ requires that

$$u(r) = \frac{r - c}{b - c}V_g. \tag{4.148}$$

The curvilinear mass element at a radius r is

$$dm = \frac{nr^{n-1}dr}{b^n - c^n}C, \tag{4.149}$$

where $n = 2$ or $n = 3$ corresponds to either cylindrical or spherical symmetry in figure 4.43. At no additional cost, $n = 1$ will provide the planar geometry solved earlier. The integral expression within the radical expression in equation (1.147) then becomes

$$k_n(y) = \frac{1}{CV_g^2}\int_c^b u^2(r)dm = \int_1^y \left(\frac{\xi - y}{1 - y}\right)^2 \frac{n\xi^{n-1}}{1 - y^n}d\xi, \tag{4.150}$$

and equation (4.147) is written

$$V_g = \frac{\sqrt{2E}}{(M/C + k_n(y))^{1/2}}, \tag{4.151}$$

where again $n = 1$, 2 or 3 corresponds to planar, cylindrical or spherical symmetry, respectively. Dimensionless parameters $y = c/b$ and $\xi = r/b$ are used in the transformed integral in equation (4.150). The integral in equation (4.150) is readily solved yielding

$$k_1(y) = 1/3 \qquad \text{(planar)} \tag{4.152}$$

$$k_2(y) = \frac{1}{6}\frac{y+3}{y+1} \qquad \text{(cylindrical)} \tag{4.153}$$

$$k_3(y) = \frac{1}{10}\frac{y^2+3y+6}{y^2+y+1} \qquad \text{(spherical)} \tag{4.154}$$

over the range $0 < y < 1$.

Imploding shell

The contrasting configuration for the imploding system is illustrated on the right in figure 4.43. Again the stationary radius is identified as a rigid boundary at a radius c. Metal mass M at radius r is accelerated radially inward to limiting velocity V_g by the intervening concentric explosive charge C. Similar energy arguments lead to an expression for the limiting velocity formally the same as equation (1.147),

$$V_g = \frac{\sqrt{2E}}{\left(\frac{M}{C} + \frac{1}{CV_g^2}\int_a^c u^2(r)dm\right)^{1/2}}. \tag{4.155}$$

Equation (4.155) differs only in the limits of the integral expression within the radical. Similar substitutions provide the same integrals,

$$k_n(y) = \int_1^y \left(\frac{\xi - y}{1 - y}\right)^2 \frac{n\xi^{n-1}}{1 - y^n} d\xi, \tag{4.156}$$

with variables $y = c/a$ and $\xi = r/a$. Solutions of equation (4.156) are provided for the imploding shell by equations (4.152)–(4.154) with the parameter $y > 1$.

Difficulties with the applicability of the approximate Gurney solution to the inward implosion of a metal shell is immediately apparent. It has application only to reasonably thin shells and becomes increasingly problematic as the shell thickness is increased.

The solutions to the Gurney velocity integral appropriate to equations (4.147) and (4.155), and provided in equations (4.152)–(4.154), are plotted in figure 4.44. The solutions exhibit the appropriate limiting values of 3/5 and 1/2 as $y \to 0$ for a fully

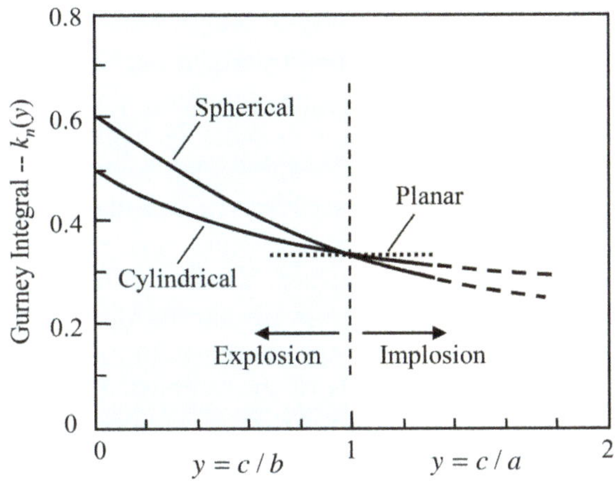

Figure 4.44. Gurney velocity integral for explosion (outward) and implosion (inward) accelerations of cylindrical or spherical shells. The abscissa is the ratio of the stationary radius to the metal shell radius.

loaded spherical or cylindrical shell and 1/3 for $y \to 1$ as either curvilinear geometry approaches planarity.

Gurney velocity of nested shells

The more general Gurney problem illustrated in figure 4.42 of the simultaneous inward and outward acceleration of metal shells by an intervening explosive charge provides a challenge to the Gurney solution method. The geometry encounters physics not occurring in the simple Gurney geometries and becomes sufficiently complex that pursuit of approximate engineering solutions and scaling relations can be questioned.

The effort nonetheless uncovers important phenomena in the explosive–metal interaction event and lends worthwhile insight into the underlying physics. Decomposition of the previous planar explosive acceleration of unequal plates (figure 4.41) into two separate Gurney solutions by identification of a stationary plane provides both clarity and simplification of the solution. A similar approach if reasonable would also simplify the curvilinear problem. If a stationary radius could be identified separating metal and explosive products impelled outward from the metal and explosive products impelled inward, then the individual explosion or implosion solutions for cylindrical or spherical geometries developed in the previous section can be employed.

Problems arise, however. In the unequal plates solution momentum balance is additionally required to determine the stationary plane. Momentum balance is also required to determine the stationary radius in the curvilinear system. The curvilinear geometry, however, brings about an impulse to the system during explosive acceleration that contributes to the momentum balance and complicates the solution. The same impulse is orthogonal to the momentum vector in the planar system and does not enter into the analysis.

The present first-order solution analysis solving for the stationary radius ignores this impulse. This simplification is justified by assuming that the impulse amplitude and duration are sufficiently small, or that geometry curvilinearity is sufficiently small, such that the impulse contribution is sensibly negligible compared to other momentum contributions.

In this approximation, it is then necessary to calculate the inwardly directed momentum and the outwardly directed momentum and balance the two to determine the radius of the stationary surface. This scalar measure of the momentum is calculated through integral expressions similar to the kinetic energy relations obtained in the previous section. Using the geometry in figure 4.42 with a and b the radii of the inner and outer metal shells of the cylindrical or spherical geometries, and c that of the stationary radius, the inward directed scalar momentum is then

$$M_A V_A + \int_a^c u(r) dm, \tag{4.157}$$

and similarly for the outwardly directed momentum,

$$M_B V_B + \int_c^b u(r) dm. \tag{4.158}$$

Both momentum relations can be solved and equated to give the stationary radius c. The Gurney assumption of a uniform radial velocity gradient that provided the additional relation in equation (4.141) for unequal parallel plates also applies to the present problem. The relation is more complex, however, as curvilinearity of the geometry no longer yields a linear relation between mass and velocity. Attempting a full analytic solution is probably not profitable. A more appropriate approach is to simply adjust the stationary radius c until equality between equations (4.157) and (4.158) is achieved. Approximate measures of the inward and outward Gurney velocities are then provided by equation (4.151) using appropriate values for $k_n(y)$.

4.4.3 Gurney acceleration

An estimate of the acceleration of the system is also possible within the model framework developed by Gurney. This solution is useful when characteristic times and motions to the terminal Gurney velocity are of interest. It is also useful for applications in which impeding media exterior to the system lead to late time deceleration. (The imploding metal shell falls into this category.) The Gurney acceleration solution is described by Jones *et al* (1980) and the development outlined here roughly follows their solution. It was shown by M McGlaun at Sandia Laboratories (see also Jones *et al* 1980) that the equilibrium conditions assumed in the solution are not consistent with continuum laws of motion. If warranted, higher-order solutions can be pursued using the McGlaun observation.

4.4.3.1 Equilibrium acceleration

As before, a uniform velocity gradient within the explosive products consistent with the problem symmetry is assumed. It is further assumed that the explosive products are at a uniform pressure P that provides the acceleration force on the metal mass M. For the geometries in figure 4.38 conservation of energy during the acceleration process is written

$$E = U(v/v_o) + \frac{1}{2}\left(\frac{M}{C} + k_n\right)V^2, \tag{4.159}$$

where again k_1, k_2 or k_3 equals 1/3, 1/2 or 3/5, respectively, for planar, cylindrical or spherical geometries. The function $U(v, v_o)$ is the mechanical potential energy of the explosive products and is a function of the equilibrium volume expansion v/v_o. For the system configurations

$$\frac{v}{v_o} = \left(\frac{R}{X_o}\right)^n, \tag{4.160}$$

where the coordinate $R \geqslant X_o$ provides the position of the plate ($n = 1$) or the radius of the expanding cylinder or sphere ($n = 2$ or 3).

Although any model for the potential energy can be used, a gamma-law gas is appropriate within the approximations of the Gurney theory. The gamma-law potential is written

$$U(R/X_o) = E\left(\frac{v}{v_o}\right)^{-(\gamma-1)} = E\left(\frac{R}{X_o}\right)^{-n(\gamma-1)}. \tag{4.161}$$

As before, $n = 1$, 2 or 3 corresponds to planar, cylindrical or spherical symmetry, respectively. Equations (4.159) and (4.161) combine to provide the dependence of the velocity on the position R,

$$V(R) = V_g\left[1 - (R/X_o)^{-n(\gamma-1)}\right]^{1/2}. \tag{4.162}$$

The time to expansion is provided by the additional quadrature

$$\frac{dR}{dt} = V(R). \tag{4.163}$$

The solution yields

$$t(R) = t_g f(R/X_o), \tag{4.164}$$

where the characteristic time $t_g = X_o/V_g$ and

$$f(R/X_o) = \int_{X_o/R}^{1} \frac{dy}{y^2(1 - y^{n(\gamma-1)})^{1/2}}. \tag{4.165}$$

The expansion velocity as a function of expansion radius and expansion time is illustrated in the plots of figure 4.45 for a gamma-law gas description of the

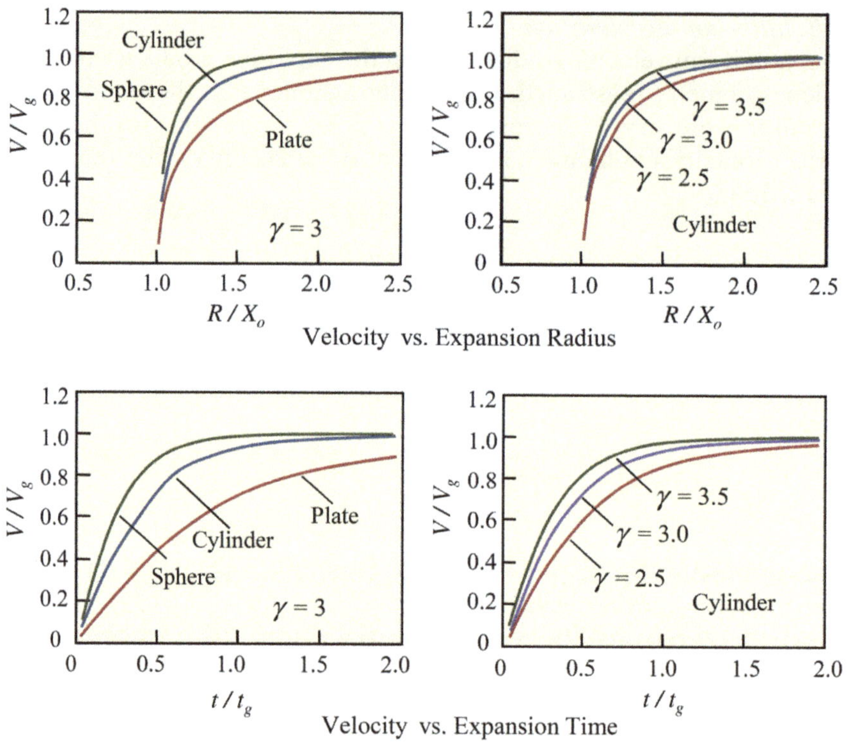

Figure 4.45. Gurney acceleration history of a plate, and a cylindrical and spherical shell, under gamma-law gas description of explosive products. Velocity versus radius in the upper plots and velocity versus time in the lower plots. Several values of gamma for cylinder expansion are illustrated on the right.

explosive products. Expansion velocity for several values of gamma is illustrated on the right. Note for a sphere or cylinder that near terminal velocity is achieved at an expansion of about one radius with that of the sphere somewhat sooner.

4.4.3.2 McGlaun observation

McGlaun (1990) at Sandia Laboratories (also see Jones *et al* 1980) pointed out the difficulty with the equilibrium acceleration solution. The fundamental assumptions of pressure equilibration throughout the explosive products, and that of a linear velocity gradient during acceleration, are not consistent with the continuum laws of mechanics. Writing the continuum equation of conservation of momentum in planar one-dimensional flow as

$$\frac{\partial u}{\partial t} + \frac{\partial p}{\partial m} = 0, \tag{4.166}$$

and assuming velocity linearity $u = (m/C)V$, yields the differential equation

$$\frac{\partial p}{\partial m} = -\frac{m}{C}\frac{dV}{dt}. \tag{4.167}$$

Application is to the one-dimensional plate acceleration on the left in figure 4.38. The velocity and pressure through the explosive products are $u(m)$ and $p(m)$ with $0 \leqslant u(m) \leqslant V$ as the coordinate spans $0 \leqslant m \leqslant C$. Integration of equation (4.167) yields a quadratic dependence of the pressure through the explosive products

$$p(m) = p(C) + \frac{1}{2}C\frac{dV}{dt}\left(1 - \frac{m^2}{C^2}\right). \tag{4.168}$$

Pressure equilibration is not achieved until plate acceleration approaches zero. An analytic model based on McGlaun's observation of position dependence of the pressure and mechanical potential can be performed but is probably not warranted considering other simplifications of the Gurney model. Use of the equilibrium acceleration model is still appropriate recognizing that some average measure of the accelerating pressure and mechanical potential is implied.

4.4.4 Gurney solutions in application

In the present subsection several examples of explosive-driven systems are explored in demonstrating application of the Gurney model to selected engineering problems. The examples pursued are chosen to demonstrate the coupling of Gurney solutions with other shock wave EOS and shock failure applications. First is that of a system of concentric metal cylinders with an intermediate explosive charge such that both inward and outward acceleration of the metal masses is produced. The inward accelerated metal collapses onto the central axis causing significant reshock, shock dissipation and subsequent rebound. The equivalent Hugoniot model is applied to assess the shock states and residual thermal states of the imploded metal mass. An assessment of the dynamic fragmentation of the rebounding metal completes the analysis. Consequences of a similar explosion experiment are illustrated in figure 3.32. The second example applies the Gurney acceleration solution analysis to the application of an explosion-driven metal plate that is impeded by a porous environment. Applications could include the coupling of explosive energy into a porous concrete, or the expansion of an exploding shell imbedded in an earth environment.

4.4.4.1 Explosion acceleration of concentric metal cylinders

The geometry of the Gurney calculation, illustrated on the left in figure 4.46, is that of a system of two concentric metal cylinders enclosing an explosive charge. Detonation of the explosive charge leads to inward acceleration of the inner metal to a velocity V_A and outward acceleration of the outer shell to a velocity V_B. Imparted motions are one-dimensional cylindrical. The Gurney solution methods and approximations that are applied are discussed in section 4.4.2.3.

For this analysis, material properties for iron (density 8920 kg m^{-3}) and TNT (density 1630 kg m^{-3} and Gurney energy $E = 2810$ kJ kg^{-1}) are used. Both the inner

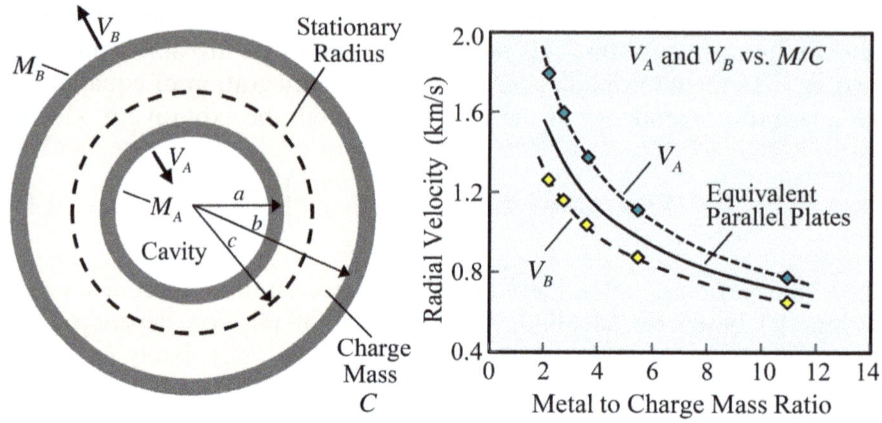

Figure 4.46. Geometry for Gurney analysis of concentric nested cylinders. The metal wall thicknesses are both 5 mm. The radius a of the inner shell is 5 cm. Radius b is, respectively, 5.5, 6.0, 6.5, 7.0 and 7.5 cm. Calculated inward and outward Gurney velocities are plotted on the right as a function of total metal mass to total explosive mass ratio. A plot of equivalent parallel plate velocities is shown for comparison.

and outer shell thicknesses are 5 mm. The outer radius of the inner shell is 5 cm. Explosive thicknesses of 5, 10, 15, 20 and 25 mm are used in the calculation. The inward and outward momentums in cylindrical coordinates corresponding to equations (4.157) and (4.158) are

$$I_A = \left(M_A + \frac{1}{3}\left(\frac{2 + c/a}{1 + c/a}\right) C_A \right) V_A, \quad I_B = \left(M_B + \frac{1}{3}\left(\frac{2 + c/b}{1 + c/b}\right) C_B \right) V_B. \tag{4.169}$$

Explosive masses C_A and C_B reside inside and outside of the stationary cylinder radius c, respectively. Corresponding Gurney velocities in cylindrical geometry are

$$V_A = \frac{\sqrt{2E}}{(M_A/C_A + 1/2)^{1/2}}, \quad V_B = \frac{\sqrt{2E}}{(M_B/C_B + 1/2)^{1/2}}. \tag{4.170}$$

The Gurney velocities are calculated and the stationary cylinder radius c is adjusted until momenta of the respective regions in equation (4.169) are equal. Calculated Gurney velocities are shown on the right in figure 4.46 for the equal thickness (5 mm) metal shells and increasing amounts of explosive mass. Specific dimensions are provided in the caption.

The inward accelerated metal mass converges on the cylindrical geometry centerline, experiences reshock and subsequent rebound. The state of the imploded metal, including shock dissipation and rebound velocity, is reasonably approximated through the equivalent Hugoniot theory described earlier in this chapter. In particular, the analysis leading to equation (4.67) is appropriate to the calculation. Assuming a linear shock-velocity versus particle-velocity Hugoniot response of the iron ($C_o = 3.57$ km s^{-1}, $S = 1.92$, $\Gamma_o = 1.7$), the equivalent Hugoniot theory provides a sensible estimate of the reduction in rebound kinetic energy. The difference of

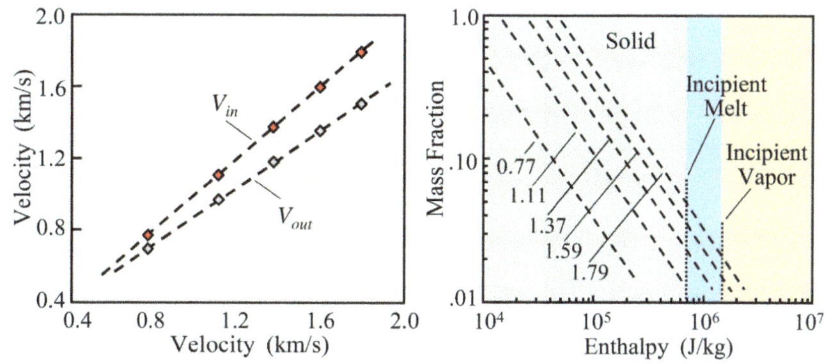

Figure 4.47. Left: comparison of the implosion velocity of iron metal from the Gurney calculation with the corresponding rebound velocity computed from the equivalent Hugoniot model. Right: the distribution of shock dissipation energy over metal mass of the rebound material. Numbers identify the initial implosion velocity.

course is a consequence of shock dissipation and increased thermal state of the rebound metal.

The initial inward specific kinetic energy is $E_{\text{in}} = V_{\text{in}}^2/2$ with $V_{\text{in}} = V_A$. This kinetic energy is equated to the equivalent Hugoniot energy, and equation (4.67) provides the dissipation energy incurred in the implosion shock event. The rebound kinetic energy for the five calculations is illustrated on the left in figure 4.47 in terms of a rebound velocity measure $V_{\text{out}} = \sqrt{2E_{\text{out}}}$. The rebound velocity is reduced by approximately 10% at the lowest-energy event, and by about 15% at the upper. Corresponding equivalent Hugoniot pressures for the lower and upper calculations are 30 and 100 GPa, respectively.

The Hugoniot dissipation energy, when assumed to be distributed over the rebound material according to equation (4.76), results in the plots shown on the right in figure 4.47. Most material remains in the solid state, but at elevated temperatures. At the highest implosion velocities, the percentage levels of liquid and vapor metal are predicted through the equivalent Hugoniot model.

4.4.4.2 Explosion acceleration of metal plate into porous media.

A further example of the Gurney solution method is provided by the explosive acceleration of a metal plate into a dissipative medium that subsequently decelerates and arrests the plate. The analysis makes use of the Gurney acceleration solution developed in section 4.4.3.1. The impeding medium is assumed to be of initial density ρ_o that compresses to a density $\rho_s > \rho_o$ on passage of the shock wave driven into the medium by the accelerating plate. The material model, sometimes referred to as a locking-solid or snow-plow model, approximates compaction of a porous medium. The model, for example, provides in a geological context a rough approximation to the resistance of a soil or porous rock environment. The example would be more interesting in a cylindrical or spherical geometry. Although doable, the analysis is less transparent due to complexities of the mathematics. Here the calculation is limited to a planar geometry. Extensions of the solution are discussed

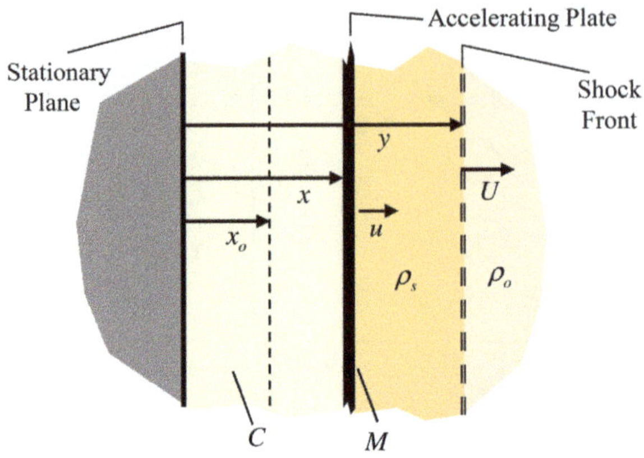

Figure 4.48. Gurney solution application in which a plate is explosively accelerated into a dissipative medium modeled as a locking solid. A shock wave propagates into the medium at a velocity U compressing the material from initial density ρ_o to compacted density ρ_s.

at the end of the section and addressed in part in the chapter on shock waves in porous solids.

The metal plate of mass M per unit area is initially at position x_o and with $x > x_o$ as the solution advances in time. The plate and all of the medium mass between the plate and the shock wave is at a velocity u when the plate is at position x. The problem geometry, and pertinent properties and parameters are illustrated in figure 4.48.

The kinetic plus potential energy of the system when the plate is at a position x is

$$W(x) = CE\left(\frac{x}{x_o}\right)^{-(\gamma-1)} + \frac{1}{2}\left(M + \frac{C}{3}\right)u^2 + \frac{1}{2}\rho_o(y - x_o)u^2. \tag{4.171}$$

The second and third terms on the right are the kinetic energies of the plate metal and explosive products, and the medium mass behind the shock front at current position y. A gamma-law potential in the first term characterizes the current potential energy of the explosive products. A compression factor, identified as the porosity, is provided by $p = 1 - \rho_o/\rho_s = (x - x_o)/(y - x_o)$. The position y of the shock front is eliminated from equation (4.171) through the porosity expression, yielding

$$W(x) = CE\left(\frac{x}{x_o}\right)^{-(\gamma-1)} + \left[\left(M + \frac{C}{3}\right) + \frac{\rho_o}{p}(x - x_o)\right]T, \tag{4.172}$$

where the system specific kinetic energy $T = u^2/2$ is carried as the dependent solution variable. The energy of the system is lost through passage of the shock wave through the dissipative medium. Namely,

$$dW = -\rho_o\frac{u^2}{2}dy = -\frac{\rho_o}{p}Tdx. \tag{4.173}$$

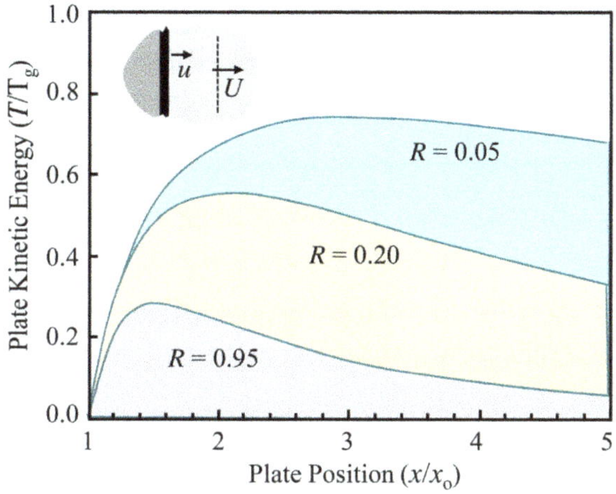

Figure 4.49. Plate kinetic energy as a function of position for selected values of medium resistance parameter R.

Equating the energy differential to the dissipation loss yields

$$dW = -(\gamma - 1)CE\left(\frac{x}{x_o}\right)^{-\gamma} + \left[\left(M + \frac{C}{3}\right) + \frac{\rho_o}{p}(x - x_o)\right]dT + \frac{\rho_o}{p}Tdx = -\frac{\rho_o}{p}Tdx. \tag{4.174}$$

Equation (4.174) reduces to the linear first-order inhomogeneous differential equation,

$$[1 + R(\xi - 1)]\frac{dT/T_g}{d\xi} + 2R(T/T_g) = (\gamma - 1)\xi^{-\gamma}, \tag{4.175}$$

where $\xi = x/x_o$ and $T_g = E/(M/C + 1/3)$ is the limiting Gurney kinetic energy if no impeding medium is present. The velocity history of the plate, where $u(\xi)/V_g = \sqrt{T/T_g}$, depends on a single system parameter,

$$R = \frac{\rho_o x_o}{p(M + C/3)}. \tag{4.176}$$

Equation (4.175) is readily solved through analytic methods. Solutions are illustrated for the velocity versus position history for selected values of the parameter R in figure 4.49.

There is value in outlining the solution approach in other Gurney geometries. As an example, the detonation of a long cylinder metal case of mass M per unit length and initial radius x_o enclosing an explosive of mass C per unit length leads to an expression for the energy corresponding to equation (4.171),

$$W(x) = CE\left(\frac{x}{x_o}\right)^{-\frac{1}{2}(\gamma-1)} + \frac{1}{2}\left(M + \frac{C}{2}\right)u^2 + \frac{\pi\rho_o}{1 - p}x^2u^2 \ln(y(x)/x), \tag{4.177}$$

where u is the radial velocity of the case when at radial position $x > x_o$. Only minor changes are noted in the first two terms. Increased complexity is introduced in the expression for the kinetic energy of the external media subsumed by the shock wave. This kinetic energy is first written

$$T = \frac{1}{2}\int_x^y u(m)^2 dm = \pi\rho_s \int_x^y u(r)^2 r dr = \pi\rho_s y^2 u(y)^2 \int_x^y \frac{dr}{r}. \tag{4.178}$$

Incompressibility of the shock compressed media relates velocities to the current velocity at the shock front through $ru(r) = yu(y)$ and accounts for the last step in equation (4.178). As in the planar calculation, the current position of the case and the shock wave are related through the porosity,

$$p = \frac{x^2 - x_o^2}{y^2 - x_o^2} = 1 - \frac{\rho_o}{\rho_s}. \tag{4.179}$$

Completing the integral and introducing the porosity results in the kinetic energy expression for the external medium in equation (4.177).

Dissipation at the shock front accounting for energy loss also entails added complexity,

$$dW = -\frac{u(y)^2}{2} dm = -\frac{u(y)^2}{2}\left(\rho_o \pi y dy\right), \tag{4.180}$$

and again using the porosity and incompressibility relations yields

$$dW = -\frac{\pi\rho_o}{2p}\frac{x^2}{y(x)^2} u(x)^2 x dx. \tag{4.181}$$

As in the planar solution, the differential of the energy in equation (4.177) is equated to the dissipation in equation (4.181) and the solution sought.

In the limit as the radius x_o of the case approaches zero, the equations take on a quite simple form and yield an analytic self-similar solution. Such solutions are described in the book of Zeldovich and Raizer (1967) and discussed in a later chapter on shock waves in porous media. For the present problem, solutions must be found through more laborious methods.

A final point is noted. After passage of the shock wave the subsumed medium continues in motion; namely, the kinetic energy term in the energy relation above. The divergent motion also entails distortion and some additional level of dissipation is expected from the shearing deformation. This dissipation if warranted can be added to the expression in equation (4.181).

References

Ang J A 1990 Impact flash jet initiation phenomenology *Int. J. Impact Eng.* **10** 23–33

Backman M E and Goldsmith W 1978 The mechanics of penetration of projectiles and targets *Int. J. Eng. Sci.* **16** 1–99

Bell R L, Baer M R, Brannon R M, Elrick M G, Hertel E S, Silling S A and Taylor P A 2000 CTH user's manual and input instructions, version 4.00 *Sandia National Laboratories Technical Report* April

Chou P C, Carleone J, Hirsh E, Flis W J and Ciccarelli R D 1981 Improved formulas for velocity, acceleration and projection angle of explosive driven liners *Proc. 6th Int. Symp. on Ballistics (Orlando, FL)*

Chou P C, Carleone J, Hirsh E, Flis W J and Ciccarelli R D 1983 Improved formulas for velocity, acceleration and projection angle of explosive driven liners *Propellants, Explosives Pyrotechnics* **8** 175–83

Getis A and Jackson P H 1971 The expected proportion of a region polluted by k sources *Geograph.Anal.* 3 256–61

Goldstein H 1965 *Classical Mechanics* (Reading, MA (also London): Addison-Wesley)

Grady D E 2001 Modeling and analysis tools for fragmenting munitions *Applied Research Associates Technical Report for Naval Surface Warfare Center, Dahlgren Division* May

Grady D E 2006a *Fragmentation of Rings and Shells* (Berlin: Springer)

Grady D E 2006b Comparison of hypervelocity fragmentation and spall experiments with Tuler–Butcher spall and fragment size criteria *Int. J. Impact Eng.* **33** 305–15

Grady D E 2010 Length scales and size distributions in dynamic fragmentation *Int. J. Fracture* **163** 85–99

Grady D E, Bergstresser T K and Taylor 1985 Impact fragmentation of lead and uranium plates *Sandia National Laboratories Technical Report* SAND84-1545, November, 1985

Grady D E and Passman S L 1990 Stability and fragmentation of ejecta in hypervelocity impact *Int. J. Impact Eng.* **10** 197–212

Grady D E and Kipp M E 1993 Dynamic fracture and fragmentation *High-pressure Shock Compression of Solids* ed J R Asay and M Shahinpoor (New York: Springer) pp 265–322

Grady D E and Kipp M E 1994 Experimental and computational simulation of the high velocity impact of copper spheres on steel plates *Int. J. Impact Eng.* **15** 645–60

Grady D E and Kipp M E 1995 Experimental measurement of dynamic failure and fragmentation properties of metals *Int. J. Solids Struct.* **32** 2779–91

Grady D E and Kipp M E 1997 Fragmentation properties of metals *Int. J. Impact Eng.* **20** 293–308

Grady D E, Mitlehner A and Potter K 1999 Characterization and analysis of a tungsten heavy alloy *Applied Research Associates Technical Report for Naval Surface Warfare Center, Dahlgren Division* TDMDG0199, March

Grady D E, Parsons D A and Winfree N A 2000 Physics-based models and analysis tools for debris formation in meteoroid and orbital debris impact *Applied Research Associates Report for National Aeronautics and Space Administration* March

Grady D E and Winfree N A 2000 Characterization of impact fragmentation, in emerging technologies in fluids, structures, and fluid/structure interactions *Proc. of ASME 2000 Pressure Vessel and Piping Conference, (Seattle, WA, 23–27 July)* pp 327–34

Grady D E and Winfree N A 2001 Impact fragmentation of high-velocity compact projectiles on thin plates: a physical and statistical characterization of fragment debris *J. Impact Eng.* **26** 249–62

Gurney R W 1943 The initial velocity of fragments from bombs, shells and grenades *Army Ballistic Research Laboratory Report BRL* 405

Johnson W A and Mehl R F 1939 Reaction kinetics in processes of nucleation and growth *Trans. AIMME* 135 414–58

Jones G E, Kennedy J E and Bertholf L D 1980 Ballistic calculations of R W Gurney *Am. J. Phys.* **48** 264–69

Kanel G I, Razorenov S V, Bogatch A, Utkin A V, Fortov V F and Grady D E 1996 Spall fracture properties of aluminum and magnesium at high temperatures *J. Appl. Phys.* **79** 8310–17

Kennedy J E 1997 The Gurney model for explosive output for driving metal *Explosive Effects and Applications* ed J A Zukas and W P Walters (Berlin: Springer) pp 221–58

Kipp M E, Grady D E and Swegle J W 1993 Experimental and numerical studies of high-velocity impact fragmentation Sandia National Laboratories Technical Report SAND93-0773, August

Kipp M E and Grady D E 1995 Experimental and numerical studies of high-velocity fragmentation *High-pressure Shock Compression of Solids II* ed L Davison, D E Grady and M Shahinpoor (New York: Springer) pp 282–339

McGlaun J M 1990 unpublished

McGlaun J M, Thompson S L and Elrick M G 1990 CTH: a three dimensional shock wave physics code *Int. J. Impact Eng.* **10** 351–60

Mott N F 1943 Fragmentation of HE shells: a theoretical formula for the distribution of weights of fragments *United Kingdom Ministry of Supply Report* AC3642, March

Piekutowski A J 1996 Formation and description of debris clouds produced by hypervelocity impact *NASA Contractor Report* 4704, February

Piekutowski A J 1997 Effects of scale on debris cloud formation *Int. J. Impact Eng.* **20** 639–50

Recht R F 1990 Analytic modeling of plate penetration dynamics *High Velocity Impact Dynamics* ed J A Zukas p 443 (New York: Wiley)

Recht R F and Ipson T W 1963 Ballistic perforation dynamics *J. Appl. Mech.* **30** 384–90

Recht R F and Yatteau J D 1983 Multiple plate penetration modeling for deforming and yawing ballistic penetrators which lose mass *Proc. Seventh Int. Symp. on Ballistics* (April) S5.4

Richardson A J 1969 Theoretical penetration mechanics of multisheet structures based on discrete debris particle modeling *AIAA Hypervelocity Impact Conference Cincinnati, OH (April–May 1969*) Paper No. 69-371

Romero L A 1991 The stability of stretching and accelerating plastic sheets. I *J. Appl. Phys.* **69** 7474

Romero L A 1991 The stability of stretching and accelerating plastic sheets. II *J. Appl. Phys.* **69** 7487

Swift H F, Bamford R and Chen R 1982 Designing dual-plate meteoroid shields—a new analysis *California Institute of Technology, Jet Propulsion Laboratory, NASA/JPL Publication* **8239** (March)

Swift H F, Bamford R and Chen R 1983 Designing space vehicle shields for meteoroid protection: a new analysis *Adv. Space Res.* **2** 219–34

Taylor G I 1941 Analysis of the explosion of a long cylindrical bomb detonated at one end *Scientific Papers of G. I. Taylor* Vol 3 ed G K Batchelor (Cambridge: Cambridge University Press) 1963 pp 277–86

Zeldovich Y B and Raizer Y P 1967 *Physics of Shock Waves and High-temperature Hydrodynamic Phenomena* Vol 2 (New York (also London): Academic)

IOP Publishing

Physics of Shock and Impact
Volume 1: Fundamentals and dynamic failure
Dennis Grady

Chapter 5

Thermoplastic deformation localization in solids

In the impact and explosive-driven deformation of solids, and particularly where strong gradients in the motions are imparted, the stability of the governing thermoplastic constitutive response is essential to uniform and homogenous deformation of the material. When instabilities in the relations governing inelastic deformation occur, deformation localization can follow. There is particular interest in dynamic thermoplastic instabilities where adiabatic trapping of dissipated heat and concomitant thermal softening governs the localization process. Deformation localization and heterogeneous shear play a key role in the front of a structured shock wave, and is a principal mechanism contributing to the shock failure and to the transmutation of solids. Adiabatic shear localization is central to failure and fragmentation in the ballistic impact environment. The present chapter extends and applies, from a physical perspective, models of the high-rate thermoplastic localization of shear deformation. Applications of linear stability analysis to thermoplastic shear localization are addressed first. A more extensive investigation pursues the dynamic localization process through a two-temperature adiabatic shear-band model. The models are used to explore the length, time and energy scales that govern the adiabatic shear-band width, spacing, localization time and dissipation features associated with the fabric of heterogeneous adiabatic shear deformation. Dimensional analysis of the properties determining thermoplastic shear localization is addressed, and dimensionless ratios of the governing properties are formulated and related to the propensity of materials to undergo failure through adiabatic shear localization. An extensive set of metals data is use to benchmark the analytic models. Extended topics include concepts of adiabatic shear-band toughness and its possible relationship to fracture toughness in ductile metals, some observations on turbulence in intense shear-band deformation, statistics of multiple shear bands, contributions of elastic strain energy in the shear-localization event, and issues of adiabatic shear localization in brittle solids.

doi:10.1088/978-0-7503-1254-7ch5

5.1 Introduction

A number of issues are addressed that relate to adiabatic shear deformation localization in solids subject to impact and high-strain-rate loading. The largest share of attention is given to a physics-based two-temperature model of the shear localization process (Grady and Kipp 1985, 1987, Grady 1992, 1994, 2010) that relies on plastic wave solutions originating from the early seminal work of Mott on the fracture and fragmentation of explosively driven cylinders (Mott 1943, 1947, Grady 2006).

Other theoretical approaches have been pursued for investigating thermoplastic deformation localization in solids. Perturbation analysis, assessing the stability of the initially homogeneous deformation solution of the governing system of thermoplastic equations, has been widely pursued over the past several decades. Applications of stability analysis to dynamic adiabatic shear localization in solid materials was pioneered in the late 1970s (e.g. Grady 1977, 1980, Clifton 1980, Bai 1981, 1982, Burns *et al* 1982), and has achieved a mature state in the extensive study of Wright (2002). These efforts in turn received their stimulus from even earlier studies, such as the seminal paper of Zener and Holloman (1944) and the theoretical efforts of Gruntfest (1963). Stability analysis methods have provided considerable insight into the mechanics of adiabatic shear.

It is interesting to note that the problem originally pursued by Mott—namely, the uniform plastic expansion, onset of fracture and fragmentation of a metal ring—can also be investigated through linear stability analysis. Later workers have pursued this approach. Mott, although aware of the applicability of stability analyses methods, chose instead to pursue an analysis that encompasses the more extensive nonlinear physics of the fracture problem (Mott 1943, 1947).

Linear stability analysis of thermoplastic shear is briefly addressed in the present chapter. This analysis closely parallels the earlier stability analysis of the author on adiabatic shear localization in solids (Grady 1977, 1980). The stability analysis incorporates the same thermomechanical material properties governing adiabatic shear, as does the more extensive physics-based two-temperature model addressed later in this chapter. Comparison of the results from applications of the two modeling methods lends insights into the shear localization process not afforded by either model alone.

The physics-based theory of thermoplastic shear localization principally pursued in the present chapter has similarities to the theory of thermal explosions introduced by Semenov (1928) and expanded on by Frank-Kamenetskii (1938). In the theoretical modeling of thermal explosions, temperature perturbations are introduced into discrete regions of the body and the competing processes of heat generation and thermal conduction are explored through a two-temperature model of the process to quantify the propensity for thermal runaway. In the present theory of thermoplastic shear a model comparable to the two-temperature theory of Semenov (1928) is provided dynamics through the concepts of plastic wave propagation incorporated in the fracture model of Mott (1943, 1947).

Thermoplastic shear localization is prevalent in a wide range of practical high-rate deformation applications. Shear localization is particularly important in terminal ballistics events where impact induces strong deformation gradients that can rapidly transform to a network of adiabatic shear bands (e.g. Zurek and Meyers 1996). Heterogeneous shear deformation plays a primary role in the high-rate failure and transformation of solids within the front of structured shock waves (e.g. Grady 1984). Adiabatic shear can also dominate the early deformation leading to breakup and fragmentation of exploding metal munitions (e.g. Hoggatt and Recht 1968). There are, of course, a number of metal forming and machining operations where adiabatic shear can play an important role (e.g. Dodd and Bai 1987).

Adiabatic shear localization has also been attributed to features of dynamic deformation occurring in brittle solids. In the 1960s and early 1970s microscopic deformation features were observed in quartz and feldspar silicate rock (e.g. Stoffler 1972, 1984). Microscopic planar lamella of amorphous silica were observed in crystalline silicate rock subjected to shock impact. These microscopic signatures of shock wave deformation were observed in rock recovered from meteor impact craters (e.g. Bunch 1968, Chao 1968), and concurrently in laboratory test specimens (DeCarli and Milton 1965, DeCarli 1968). More recently, comparable laminar features, suggestive of an adiabatic shear deformation cause, have been observed in boron carbide ceramic subjected to ballistic impact (Chen *et al* 2003). Adiabatic shear may also play a key role in the failure of bulk metallic glass (e.g. Trexler and Thadhani 2010).

Thermoplastic shear localization can affect strength and deformation in the strong shock environment in both metals and brittle solids. Polymorphic phase transformation under shock compression in silicate minerals has been related to heterogeneous thermoplastic shear within the shock wave (Grady *et al* 1974, Grady 1980). The dynamic yield and post-shock strength of metals appears to exhibit effects relating to similar thermoplastic shear localization (Grady and Asay 1982, Grady *et al* 1983, Swegle and Grady 1985).

Section 4.2 begins with discussion of the physical properties governing adiabatic shear localization. Continuum thermomechanical properties are selected that are deemed first-order relevant to the thermoplastic shear localization process. The same properties are used in the stability analysis model and the two-temperature model of adiabatic shear. Dimensional analysis of the governing properties is pursued in section 4.3 for purposes of determining functional constraints on shear-band features, key length, time and energy scales, and non-dimensional variables.

In section 4.4 stability analysis of the governing thermoplastic equations is addressed. Length scales emerging from the linear stability analysis are compared with the experimental study of adiabatic shear in stainless steel by Xue *et al* (2003, 2004, 2006).

A description of the development of the physics-based two-temperature model of adiabatic shear localization is provided in section 4.5. This effort primarily parallels earlier work (Grady and Kipp 1985, 1987), although several modifications are introduced. In particular, thermal softening during adiabatic shear is reinterpreted in terms of the material melt temperature. This modification of the model is deemed more appropriate for full thermoplastic localization. Analytic relations for shear-band width, spacing, localization time, dissipation and toughness are derived from the model.

In section 4.6 features of adiabatic shear emerging from the shear-band models are calculated for some common metals and compared with experimental observation. The metals selected for this effort have previously been provided appropriate thermoplastic constitutive models (Johnson and Cook 1983, Johnson 1993) and the additional thermophysical properties necessary for the model comparisons are available in the literature. Additionally, the substantial investigations of the adiabatic shear characteristics of stainless steel (Xue *et al* 2003, 2004, 2006) provide key experimental data for comparisons with the model predictions. A concept of shear-band toughness of metals has been introduced in earlier work (Grady 1992, 1994). Shear-band toughness is calculated for the Johnson–Cook metals and its relationship to other failure properties is explored. Two dimensionless numbers emerge from the dimensional analysis of the thermomechanical properties of metals. Values are calculated for these dimensionless numbers for the Johnson–Cook metals and then examined as possible indicators for adiabatic shear localization. The models of thermoplastic shear localization outlined in the present chapter and applied to the suite of Johnson–Cook metals specifically exclude elasticity and elastic strain energy. A method for including elastic strain energy in the two-temperature model is described, and an assessment is made of the effect of ignoring elasticity (Grady 2011). Adiabatic shear in brittle solids is deliberated in the context of the model including elastic strain energy. In particular, adiabatic shear is explored as a possible cause for deformation features observed in boron carbide resulting from ballistic impact (Chen *et al* 2003). Additionally, experimental and computational observations of the statistics of multiple shear-band frequency are compared with theoretical predictions of the shear localization models. Section 4.6 closes with observations on the onset of turbulence within the adiabatic shear band.

The present two-temperature theory of adiabatic shear localization provides an analytic prediction of energy dissipation in a propagating shear band based on commonly available thermal and mechanical properties. When couched in units of fracture toughness, dissipation predictions based on this relation are surprisingly close (within a factor of two or three) to measured quasi-static fracture toughness properties for the same metals. An explanation for this commonality is put forth in section 4.7 (Grady 2009a). The assumption is made and justification presented that static fracture advance is stochastically dynamic. It is then shown through application of the Dugdale model of crack-tip plasticity (Dugdale 1960) that the plasticity elements within the crack tip achieve length scales in which both local heat transport and momentum transport on the microscale play a balancing role in determining plastic dissipation in the crack advance process. The same physics governs dissipation in the propagation of adiabatic shear bands.

5.2 Physics of thermoplastic shear localization

Adiabatic shear in solids brought about by intense impulsive loading is a multiscale phenomenon. The interplay of physics on the microscale, the mesoscale and the continuum scale come together to create the display of localized deformation exhibited in the failure of many materials in response to such loads. Most of the present chapter explores adiabatic shear at the continuum level. Nonetheless, underlying the continuum deformation reside cooperative deformations on finer length scales that are key to the adiabatic shear phenomena. This section offers discourse on some of the physics underlying the adiabatic shear deformation event.

5.2.1 Deformations leading to thermoplastic shear localization

The deformations preceding adiabatic shear localization in metals are necessarily dynamic, and of moderately large strain, such as might occur in the impact breach of a metal plate or the explosive expansion of a fragmenting metal shell. The latter criterion of large strain is not necessary under certain circumstances, however. Global shear strain in the planar shock event that accommodates thermoplastic heterogeneous shear is quite modest. In the latter application, strains only need to be large within shear-band regions while such regions may only accommodate a reasonably small mass fraction of the total.

The dynamic deformations leading to thermoplastic shear localization are invariably non-homogeneous, however. The applications that lead to shear localization, such as impact plugging, high-rate machining, explosive welding, munitions fragmentation or strong shock compression, all involve large gradients in the deforming strains and strain rates. Inception of shear localization occurs at the specific points at which the localization criteria are first achieved. Shear localization features (adiabatic shear bands) then propagate away from such inception sites at speeds determined by both the rate of expansion of the regions achieving localization criticality, as well as by the intrinsic limiting velocity of propagating shear bands. Depending on the intensity of the shearing field fueling the localization process, propagating shear bands can multiply through branching to numbers adequate to accommodate the necessary dissipation of the driving field; or arrest, if the driving field becomes insufficient to support the shear-band system. It is reasonable to expect that under sufficiently intense loading the collective adiabatic shear process would achieve chaotic characteristics not unlike hydrodynamic turbulence or catastrophic brittle fracture. In common with these related phenomena, the asymptotic behavior of heterogeneous adiabatic shear such as shear-band numbers and shear-band spacings would exhibit universality features independent of the detailed dynamics of the deformation event.

These latter details of the thermoplastic shear localization process are not the primary purpose of the present chapter. Rather, attention is principally focused on the growth and culmination of adiabatic shear bands. It is assumed that inception of the shear band through some criterion has been achieved. Explored in some detail is the evolution of shear-band growth from the point of inception to the terminal steady state of a fully evolved shear band. Much of the analysis focuses on the physics of the shear band at the tip of the propagating shear band. This region is identified as the shear-band process zone. The propagation of a shear band and the two- or three-dimensional geometry of a propagating shear-band process zone are largely ignored.

It is assumed that the growth processes within a shear-band process zone and the movement of plastic release waves away from the process zone are sufficiently slow relative to the rate of advance of the propagating shear band, that a one-dimensional planar model of shear-band evolution is sensibly applicable. It is important; however, to bear in mind that the predictions based on such models are subject to the limitation of these simplifying assumptions and may not be more broadly applicable.

5.2.2 Onset of thermoplastic shear localization

Although the onset of thermoplastic shear localization is not an issue that is emphasized in the present chapter, it is nonetheless an essential feature, and a substantial amount of study has been devoted to this aspect of adiabatic shear over the past several decades. The onset of shear localization is of paramount importance to a number of practical applications where considerable effort is expended in avoiding deformation localization.

With the possible exception of brittle solids, most materials exhibit some level of plastic deformation before localization onset. This observation suggests that for many materials, plastic deformation is self-stabilizing at continuum length scales, and distributes uniformly over the neighborhood of observation. A criterion for localization onset is then to determine the set of conditions at which deformation ceases to be self-stabilizing and deformation localization ensues.

Shear deformation is always heterogeneous on some scale (usually microscopic and/or mesoscopic). Shear localization occurs when a synergy of such chaotic deformation leads to a localized runaway deformation in a quasi-continuum sense. Various localization criteria have been explored. These criteria rely on a sensible set of continuum equations governing material response under a prescribed set of initial and boundary conditions. Methods identified as load maximum, bifurcation and instability, have been associated with the various localization criteria (e.g. Dodd and Bai 1987). The different criteria are not, of course, independent.

Some clarification is necessary. Both boundary conditions and properties of the material play important roles in the onset of localization. For example, a body with a stress and deformation intensifying notch subjected to a load will have a greater propensity for shear localization than will the same body without the notch (e.g. Ritter *et al* 2008). A criterion for shear localization in the body is not the same as that for localization in the material itself, however. It is reasonable to consider intensified deformation in the vicinity of the notch and invoke a material criterion for localization onset. Although boundary conditions and local material conditions probably cannot be strictly separated in assessing localization onset, this is commonly done and is implicit in the following discussions.

The phenomena of thermoplastic shear localization are a competition among the physical processes of deformation dissipation, thermal softening and heat conduction. The various localization criteria commonly assume that the same physics also governs onset of localization. This may well be, but is not necessary. It is reasonable to contemplate onset of thermoplastic localization triggered by initially athermal localization effects (e.g. Luders bands, twinning).

Load maximum offers a straightforward criterion for onset of runaway thermoplastic shear localization. The criterion can be viewed as an application of Le Chatelier's principle, which states that for a system to be stable any deviation from equilibrium should bring about forces that would tend toward restoration of equilibrium (e.g. Reif 1965). A similar criterion was stated by Considère (1885), specialized to the onset of instability in the quasi-static deformation of plastic bodies. Onset of thermoplastic shear localization through a load-maximum criterion is commonly determined from a thermoplastic constitutive relation for the deformation of the form

$$\sigma = f(\varepsilon, \dot{\varepsilon}, T), \tag{5.1}$$

relating the effective stress to the plastic strain, strain rate and temperature. The localization criterion is determined from the stress–strain–temperature state, at which the differential of equation (5.1) is stationary,

$$d\sigma = f_{\varepsilon}(\varepsilon, \dot{\varepsilon}, T)d\varepsilon + f_{\dot{\varepsilon}}(\varepsilon, \dot{\varepsilon}, T)d\dot{\varepsilon} + f_{T}(\varepsilon, \dot{\varepsilon}, T)dT = 0. \tag{5.2}$$

Criteria for the onset of thermoplastic adiabatic shear also emerge from instability analysis of the governing equations. The application of instability analysis to the thermoplastic localization in the dynamic deformation of solids initiated in the late 1970s (Grady 1977, 1980, Clifton 1980, Bai 1981, 1982, Burns *et al* 1982). Instability analysis has played a strong role in revealing the physics of adiabatic shear over the intervening decades (e.g. Wright 2002). Differing levels of analytic sophistication can be employed in the implementation of instability analysis to thermoplastic localization. Fundamentally, a constitutive description of the material, as can be represented by equation (5.1), is joined with the governing equations of momentum and energy. A homogeneous spatially independent solution to the system of equations is found. A perturbation to the homogeneous solution is introduced and the governing equations are examined for the growth of some spectrum of the perturbation. If perturbation growth is determined, then the material is judged to be susceptible to thermoplastic shear localization under the applied kinematic conditions. The state of the material at which the solution transitions to unstable perturbation growth is identified as the condition for onset of deformation localization. Perturbation instability analysis also speaks further to the initial spatial fabric and rates of growth of thermoplastic localization.

Criteria for thermoplastic localization onset have also been determined from the principles of bifurcation theory. The aim of the bifurcation approach is to establish the state of the material at which a solution to the governing system of equations loses uniqueness. Subsequently, two or more deformation solutions satisfy the equations and boundary conditions. Localization criteria based on load maximum and bifurcation can differ for selected applications in plastic deformation (e.g. Needleman 1972).

5.2.3 Plastic heating and thermal softening

No feature of the physics of thermoplastic shear localization is more critical to the phenomenon than that of temperature softening of the flow stress. Thermal softening is intrinsic to the occurrence of a load maximum in the shear-stress versus shear-strain relation, to the onset of non-uniqueness and bifurcation of the defining shear-localization boundary value problem, or to the perturbation instability of the governing thermoplastic equations. Although thermal softening is clearly recognized in all of the numerous studies of thermoplastic shear localization and adiabatic shear, the underlying physics is seldom discussed and is not well understood.

Plastic deformation is a non-conservative process that leads to dissipative heating and defect production in the deforming material. The latter feature of defect production and associated configurational entropy falls outside of the scope of the present study. In the present model, plastic dissipation is assumed to convert to thermal energy and contributes to temperature rise in the affected material. In most cases energy in the damage-induced defect production is considered to be a second-order effect (of the order of 10% or less of the total dissipation). The detailed processes are probably significantly more complex. During early deformation hardening, energy stored in the strain field of defects can be quite appreciable. Much of this energy, however, is subsequently released and contributes to the thermal energy when temperatures in the shear band become large.

It is generally known that temperature enhances the ductility of most metals. Temperature assists dislocation plasticity in crystal systems through the thermally activated acceleration of the processes of dislocation multiplication and motion. Thermal softening is in turn limited in the shear localization process by the acceleration of the strain rate, and by the strain-rate sensitivity of the flow stress of most materials. Strain-rate dependence is commonly logarithmic at the lower strain rates and transitions to a stronger near-linear behavior at strain rates in the neighborhood of 10^3 s^{-1} and higher. Metals at quite high strain rates and temperature can exhibit near viscous-like flow-stress behavior.

When materials can be described by an analytic constitutive relation for the shear flow stress of the functional form

$$\tau(\gamma, \dot{\gamma}, T), \tag{5.3}$$

then a thermal softening coefficient is defined through

$$\alpha = -\frac{\partial \ln \tau}{\partial T}\bigg|_{\gamma,\dot{\gamma}} \tag{5.4}$$

It is noteworthy that when equation (5.3) is a multiplicative function of the form $\tau = A(\gamma)B(\dot{\gamma})C(T)$, as is common (the Johnson–Cook strength model (1983) is representative), then the thermal softening coefficient α is a function only of temperature, independent of the current strain and strain rate. In general such independence is not expected.

The constitutive relation for the material in equation (5.3), and through differentiation, the thermal softening coefficient of equation (5.4), are obtained through the curve fitting of theoretical or empirical relations to stable experimental stress versus strain data over appropriate ranges of the strain, strain rate and temperature. Stable stress–strain data imply that the load maximum is not achieved assuring nominally homogeneous deformation throughout the test sample. Hopkinson bar data are commonly exploited for this purpose, although sometimes supplemented with data from other test methods. The application of such constitutive relations to the thermoplastic shear localization process requires extrapolation through the point of load maximum to states outside the range of the experimental data. This extrapolation is probably appropriate when addressing issues of the onset of shear localization, or the early shear-band behavior, where only modest extrapolation is required. Extrapolation of the constitutive relation would not be appropriate to material response deep into the shear localization process, where strains of 10, or even 100, are achieved, strain-rates an order of magnitude or higher than the onset strain rate, and local shear-band temperatures approach melt.

There is no direct experimental evidence to guide constitutive modeling of the flow stress at such extremes of strain, strain rate and temperature confined within the limiting spatial length scale of a few, to a few tens, of micrometers shear-band widths. Constitutive modeling is usually guided by the intuitively reasonable observations that the flow stress at shear-band inception is provided by the stress at load maximum, or at the bifurcation state, determined from the experimental constitutive relation, and that the flow stress is negligible (effectively zero) when the temperature achieves melt. Again, the Johnson–Cook model (1983) is representative with the flow stress provided by

$$\tau = \tau_o(\gamma, \dot{\gamma})\left[1 - \left(\frac{T}{T_m}\right)^m\right], \tag{5.5}$$

where T and T_m are the temperature and temperature to melt, above some reference temperature. This relation captures the aforementioned intuitive argument. The added material exponent m in the Johnson–Cook relation is only applicable with any degree of confidence for deformations up to, and in the neighborhood of, the load maximum.

Accepting the reasonable premise that the flow stress degrades from the load-maximum value to near zero at the melt temperature, the thermal softening coefficient of equation (5.4), when averaged over the full range, is nominally

$$\alpha \simeq 1/T_m. \tag{5.6}$$

The proportionality is of order unity, while T_m is the temperature span from shear-band onset to melt. This approximation for the thermal softening property of solids governing the full adiabatic shear localization process provides for predictions of adiabatic shear based on a commonly available thermal–physical property; namely, the melt temperature.

5.2.4 Heat transfer

Heat transfer (or thermal conduction) also plays an important role in the evolution of an adiabatic shear band. Note that the common name for such zones of thermoplastic shear (adiabatic shear bands) contradicts the notion of thermal conduction. In most models of thermoplastic shear localization, the physical processes of heat transfer or thermal conduction provide the forces that limit how thin shear bands can become in the heterogeneous deformation process. All of the fields, including stress, strain, strain rate and temperature, have strong gradients transverse to the plane of the shear band in the evolution of thermoplastic shear. Gradient forces other than temperature may play a role in determining the spatial dimensions (thickness) of the shear band. Thermal gradients seem to be the most physically sensible, and are considered the responsible mechanism in most cases. This premise of thermally determined shear-band length scales is central to the analytic studies of thermoplastic shear localization pursued both here and elsewhere.

Less certain is the form that heat transfer takes in the environment of thermoplastic shear deformation. Heat transfer by radiation is unlikely to contribute significantly, as the photon (and phonon) mean free path in metals is small. Most micrographic evidence suggests that deformation in the adiabatic shear process is laminar, although the intensity of deformation raises concerns with possible onset of local turbulence in the shear flow. The underlying assumption of thermal-dependent deformation zones may still be appropriate within the extreme shock environment where length scales can approach the phonon mean free path. Supporting analysis follows from more detailed consideration of the statistical Boltzmann heat transport (Maasilta and Minnich 2014).

Turbulence in the shear localization process is suggested by the experimental work of Guduru *et al* (2001). Local turbulence offers the possibility of heat transfer through mass transport—a form of convective heat transfer. The more likely heat transfer mechanism is transport through Fourier heat conduction although, considering the intense gradients in temperature and deformation fields, linear thermal diffusion is probably only approximate.

In the present analytic model, heat transport through thermal diffusion is assumed to dominate the thermoplastic shear localization process. A two-temperature model is applied where a uniform temperature T is identified with the shear band, while a constant temperature T_o associated with the temperature and onset of instability is assigned to all material exterior to the shear band. A heat transfer coefficient is provided by the approximate relation $h_t \simeq \kappa / a$, where κ is the thermal conductivity of the material, and a is a measure of the width of the shear band. Temperature change in the shear band is then governed by the relation

$$\frac{dT}{dt} = -\frac{2}{\rho c a} h_t (T - T_o), \tag{5.7}$$

where ρc is the volumetric heat capacity of the material. The two-temperature representation of the shear band pursued in the present model has a basis in the analytic models of thermal explosions in energetic materials developed by Semenov (1928) and by Frank-Kamenetskii (1938). (See also the review article by Lee 1998.) Their theories of hot-spot ignition of reactive materials pursue a similar two-temperature and heat transfer description of the transient thermal phenomenon.

5.2.5 Wave propagation

Stress wave propagation is fundamental to the adiabatic shear localization process. As thermal softening proceeds within the shear band, the reduction in shear stress is communicated to regions away from the shear band through stress waves. An adiabatic shear band is a propagating deformation feature with a process zone at the front of the shear band within which the principal physics of shear localization takes place. The shear stress is the load-maximum value at the tip of the process zone and relaxes to a low value (of order zero) at the tail of the process zone. Thus, a stress release wave originates at the tip of the process zone and propagates outward much like a Mach wave encompassing a body traveling at supersonic speed through air. In a plastic material, this diffusion-like wave propagates into a region that is actively undergoing shear deformation. Such plastic diffusion waves have been named Mott waves (Lee 1967, Grady 1992).

5.3 Dimensional characteristics of thermoplastic shear localization

Dimensional issues are examined that relate to the properties of the materials that govern the onset and growth of thermoplastic shear localization. Preliminary dimensional considerations are generally useful to many problems of physics. This approach is particularly true for adiabatic shear localization, where the first-order properties governing the physics are relatively few, and worthwhile observations are made through dimensional arguments alone. The first objective is to identify the key kinematic and material properties central to the shear localization process. The identification of such properties requires judgments based on a best understanding of the underlying physics governing adiabatic shear as was explored in the previous section. The intent of this analysis is to identify the key continuum physical parameters that control the essential first-order features of unstable thermoplastic shear in solids. Other higher-order properties that may influence the shear-band process are not included in this initial assessment.

5.3.1 Properties governing adiabatic shear localization

The stress, deformation, thermal and kinematic fields in the material within which the adiabatic shear process takes place, and that will provide the backdrop for the present modeling and analysis, must first be characterized. The criterion used for onset of shear localization in the present study is that of achieving load maximum in the adiabatic stress versus strain response of the material. Other criteria can, of course, be used. Consequently, in the neighborhood of the nascent shear band the material is on the plastic yield surface at the load-maximum shear stress τ_o as well as at the load-maximum temperature T_o and plastic shear strain γ_o. Further, the material in the same neighborhood is considered to be undergoing plastic deformation at an initial shearing rate of strain $\dot{\gamma}_o$. In application it is evident that at the same point in the material there would be appreciable gradients in the stress, strain, strain rate and temperature field that would additionally influence the onset and growth of adiabatic shear. These complications are not considered here.

Accordingly, consider a region of material at density ρ undergoing uniform shear deformation parallel to the y–z plane at a shearing strain rate $\dot{\gamma}_o$. Initial conditions require the material to be flowing at a uniform initial load-maximum shear stress τ_o and at a uniform temperature T_o. Stress is assumed to be at the point of hardening saturation such that subsequent deformation is governed by some rule of flow-stress softening according to $\alpha = -(1/\tau)d\tau/dT$. In earlier studies, the softening coefficient α is identified as a material constant determined from the flow-stress relation (Grady and Kipp 1985, 1987), or from some description of the solid viscosity (Grady 1977, 1980).

Issues of thermal softening are discussed in the previous section. The Johnson–Cook model,

$$\tau = \tau_o(\gamma, \dot{\gamma})\left[1 - \left(\frac{T}{T_m}\right)^m\right], \tag{5.8}$$

is representative of phenomenological constitutive descriptions of metal deformation plasticity that account for thermal softening. In equation (5.8) T and T_m are the temperature and temperature to melt in excess of the load-maximum temperature T_o. Assuming the flow stress degrades from the load-maximum value to near zero at the melt temperature, the thermal softening coefficient α, when averaged over the full temperature range, is approximately given by $\alpha = 1/T_m$. In the subsequent exploration of thermoplastic shear localization the temperature rise to material melt T_m, rather than the softening constant α, is used to characterize the shear-band softening propensity of the material. Thermal and heat transfer properties are assumed to be adequately described through a constant specific heat c and a thermal diffusion coefficient χ.

Following perturbation of the homogeneous thermoplastic state of the material in the one-dimensional model of adiabatic shear centered about the y-plane at $x = 0$, properties of the material will subsequently be functions only of the time t and the one spatial coordinate x, namely, the temperature $T(x, t)$, the transverse velocity $u(x, t)$ and the shear stress $\tau(x, t)$. In metals that exhibit a proclivity for thermoplastic shear localization, the phenomenon observed is that of a rapid transition of deformation to thin elevated-temperature adiabatic shear bands. Shear bands are typically a few, to a few tens, of micrometers in width, as observed from optical or SEM imaging.

Dimensional analysis of the shear-localization event can be used for the purpose of identifying functional relations of the governing properties associated with observable features of the localization process including the widths and spacings of shear bands, characteristic localization strains, as well as times and energies to localization. These features relate here to the fully developed shear band. Time evolution of such features is not explored. Accordingly, such shear-band features should have the following functional dependence, based on the material and kinematic parameters deemed key to the shear-localization process,

$$f(\dot{\gamma}_o, \tau_o, \rho, c, \chi, T_m). \tag{5.9}$$

Several points should be noted. The elastic modulus is not included among the parameters considered in this analysis. The elastic response is judged to be less important to the shear-localization process, and the material response off the plastic yield surface is assumed rigid. This simplification is reasonable for many of the metals that will be considered, as the contribution of elastic strain energy to the shear-localization process is sensibly second-order. This assumption is not appropriate for adiabatic shear in brittle solids and is not necessarily appropriate for all of the metals of interest. Elasticity in the shear-localization process is addressed later in the chapter. In addition, a dependence on the confining pressure over the range of pressures experienced in the problems of interest is not included. This omission is probably reasonable and is consistent with the preponderance of literature on adiabatic shear localization. The study of Hanina *et al* (2007), however, shows that modest hydrostatic pressure delays the onset of adiabatic shear in metals and modifies early localization, presumably a consequence of the effect of pressure on the initial damage micro-mechanisms responsible for shear-band onset.

The thermal softening coefficient is probably a function of temperature. It is considered a constant here, and is assumed to be a representative average over the localization process, and captured by the temperature rise to material melt, as discussed earlier. Further, it is assumed that the late time values of features such as shear-band width and localization time are not dependent on details of the initial perturbations leading to localization. The event will be considered to initiate at the same initial ambient temperature T_o, and T_o will not be included in the list of independent parameters. Lastly, the many possible microstructure properties expected to influence adiabatic shear are not included, other than through their effect on the continuum properties selected.

5.3.2 Reduction of the functional dependence

An inspection of the dimensions of the selected properties governing adiabatic shear reveals that among the six variables identified, three independent length scales can be formed. An appropriate set is

$$L_1 = \sqrt{\tau_o/\rho\dot{\gamma}_o^{\,2}}, \quad L_2 = \sqrt{\chi/\dot{\gamma}_o}, \quad L_3 = \sqrt{cT_m/\dot{\gamma}_o^{\,2}}. \tag{5.10}$$

The reduced functional dependence on the dimensionless parameters is then

$$f(L_1, L_2, L_3). \tag{5.11}$$

Of course, any number of length scales can be formed from the independent length scales above. For example $L_4 = (L_1 L_2 L_3)^{1/3}$ provides a length scale mixing all of the selected material properties.

The functional dependence can be reduced by one further variable. Any normalized (non-dimensional) feature (such as the shear-band width a divided by one of the above length scales, say a/L_1, or the localization time multiplied by the shear strain rate $t_c\dot{\gamma}_o$) can be written as a function of two independent non-dimensional representations of the governing parameters. This dependence can be written without loss of generality as the dimensionally homogeneous function,

$$f(L_1/L_2,\ L_3/L_1). \tag{5.12}$$

Dimensional analysis does not proceed much further. Nothing can be said about the functional dependence on the two dimensionless ratios of any specific feature of the shear-localization process without further supporting modeling of the phenomena. These terms will appear in subsequent physical modeling of the shear-localization phenomena, and some initial familiarity is provided here.

The first dimensionless ratio in equation (5.12) will be written as

$$P = (L_1/L_2)^2 = \frac{\tau_o}{\rho\chi\dot{\gamma}_o}. \tag{5.13}$$

Noting that $\tau_o/\rho\dot{\gamma}_o$ has the dimensions of kinematic viscosity, the dimensionless number P is recognized as the solid-state equivalent of the Prandtl number in the physics of fluids. This solid-state Prandtl number provides a measure of the ratio of the momentum-diffusion to the thermal-diffusion properties of the material. The solid Prandtl number has emerged from earlier stability analysis of thermoplastic shear deformation (Grady 1980).

The second dimensionless number is written

$$Q = (L_3/L_1)^2 = \frac{\rho c T_m}{\tau_o}. \tag{5.14}$$

This number is independent of the rate of shear deformation and is strictly a property of the material. The dimensionless parameter Q is a measure of the stored heat content relative to the plastic dissipation intensity of the material. A low value of Q will be found to favor thermoplastic shear localization and is shown to be proportional to the localization strain in the adiabatic shear band.

A third dimensionless number is found from the ratio of L_2 and L_3,

$$(L_3/L_2)^2 = \frac{c T_m}{\chi\dot{\gamma}_o}. \tag{5.15}$$

This number provides a measure of the heat content relative to the heat diffusion property of the material. This number is not, however, an additional independent parameter but is derived from the product PQ. In fact any number of dimensionless combinations can be obtained through functional variations of P and Q. One combination,

$$R = \sqrt{Q^2/P} = \sqrt{\frac{\rho^3 c^2 \chi T_m^{\,2} \dot{\gamma}_o}{\tau_o^{\,3}}}, \tag{5.16}$$

provides a measure of the material Reynolds number within the shear band. This particular Reynolds number potentially relates to onset of turbulence in the adiabatic shear flow (e.g. Guduru *et al* 2001); a topic that is discussed later in the chapter.

5.4 Stability analysis of thermoplastic shear localization

Perturbation and stability analysis of the thermoplastic localization process in solids has provided considerable insight into the shear-band phenomena over the past several decades. The topic has been pursued in detail by many (e.g. Wright 2002, and references therein) and is not addressed in depth here. The essential results of stability analysis methods are nonetheless useful in contrasting with the alternative two-temperature theory of adiabatic shear that is explored in more detail in the present chapter. Consequently, a reasonably elementary linear stability analysis is described that encompasses the same thermoplastic material constants identified in the previous dimensional assessment, and utilized later in the two-temperature theory. The complementary nature of the two approaches for modeling adiabatic shear localization contributes insights not accessible from either model independently. The length scales predicted from stability analysis are in remarkably good accord with the spatial fabric of adiabatic shear in stainless steel (Xue 2001, Xue *et al* 2003, 2004, 2006).

5.4.1 Stability analysis

The stability analysis is one performed by the author in earlier attempts to shed light on adiabatic shear localization observed in the impact shock environment (Grady 1977, 1980). Focus was on the then recently recognized effects of deformation localization, and very likely thermal localization, in the strong shock compression of natural quartz and other quartz-like (silicate) minerals (Grady *et al* 1974, 1975, Grady 1980) and metals (Grady and Asay 1982, Grady *et al* 1983). Accordingly, consider a solid material deforming homogeneously in shear with particle motion in the y direction and described by a uniform velocity field $u_o(x)$. The initial temperature T_o and shear stress τ_o are correspondingly uniform through the deformation field. The initial strain rate is $\dot{\gamma}_o = du_o(x)/dx$, and is also initially constant. All subsequent functional dependence is only on the x coordinate orthogonal to the motion.

Under the specified deformation conditions the governing equations follow from conservation of momentum,

$$\frac{\partial u}{\partial t} - \frac{1}{\rho}\frac{\partial \tau}{\partial x} = 0, \tag{5.17}$$

and conservation of energy,

$$\frac{\partial T}{\partial t} - \chi\frac{\partial^2 T}{\partial x^2} = \frac{\tau}{\rho c}\frac{\partial u}{\partial x}. \tag{5.18}$$

The time and spatial coordinate are t and x, respectively, u is the particle velocity in the y direction, and ρ is the constant density of the material. T and c are the temperature and specific heat. The thermal diffusivity is $\chi = \kappa/\rho c$ where κ is the thermal conductivity. Fourier heat conduction is assumed. Non-thermal dissipation in the deformation process is ignored. A quite basic constitutive model for the shear deformation is assumed of the Newtonian viscous form

$$\tau = \eta(T)\frac{\partial u}{\partial x}, \tag{5.19}$$

where a temperature-dependent solid viscosity,

$$\eta(T) = \eta_o e^{-T/T_m}, \tag{5.20}$$

is presumed to account for the deformation-induced thermal softening during the dynamic process. In equation (5.20) T and T_m are, respectively, the current temperature and the melt temperature in excess of the initial temperature T_o. This simple constitutive model leads to a tractable analysis. Constitutive models of increasing fidelity can be pursued, of course, with the attendant increase in analytic complexity.

A non-dimensional representation of the governing and constitutive equations leads to a simplification of the perturbation analysis. The following characteristic scales are identified in the problem,

$$t^* = \chi/cT_m, \quad x^* = \chi/\sqrt{cT_m}, \quad u^* = \sqrt{cT_m}, \quad \tau^* = \rho cT_m, \quad T^* = T_m. \tag{5.21}$$

Each dependent and independent variable in equations (5.17)–(5.20) is rendered dimensionless by dividing by the respective characteristic scale. For example, time t becomes the dimensionless time t/t^* and so forth. The set of equations now become

$$\frac{\partial u}{\partial t} - \frac{\partial \tau}{\partial x} = 0, \tag{5.22}$$

$$\frac{\partial T}{\partial t} - \frac{\partial^2 T}{\partial x^2} = \tau\frac{\partial u}{\partial x}, \tag{5.23}$$

$$\tau = Pe^{-T}\frac{\partial u}{\partial x}, \tag{5.24}$$

where the variables t, x, u, τ, T, as noted above, are now considered dimensionless. The parameter P in equation (5.24) is the dimensionless solid Prandtl number $P = \eta_o/\rho\chi$ and, as noted in the previous section, provides a measure of the momentum diffusion to the thermal diffusion properties of the material. In the previous section, the solid Prandtl number P is identified as one of two independent dimensionless numbers derived from the material and kinematic properties deemed to govern the thermoplastic shear-localization process.

Introduction of small variations from the homogeneous field, $\tau_o + \bar{\tau}$, $u_o + \bar{u}$ and $T_o + \bar{T}$, leads to the following linear system of equations in the variation,

$$\frac{\partial \bar{u}}{\partial t} - \frac{\partial \bar{\tau}}{\partial x} = 0, \tag{5.25}$$

$$P\dot{\Gamma}_o\frac{\partial \bar{u}}{\partial x} + \dot{\Gamma}_o\bar{\tau} - \frac{\partial \bar{T}}{\partial t} + \frac{\partial^2 \bar{T}}{\partial x^2} = 0, \tag{5.26}$$

$$P\frac{\partial \bar{u}}{\partial x} - \bar{\tau} - P\dot{\Gamma}_o\bar{T} = 0. \tag{5.27}$$

In equations (5.26) and (5.27) a dimensionless uniform strain rate is $\dot{\Gamma}_o = \dot{\gamma}_o/\dot{\gamma}^*$ where $\dot{\gamma}^*=cT_m/\chi$ provides a characteristic strain-rate scale. This dimensionless strain rate is recognized as the inverse of the dimensionless ratio provided in equation (5.15).

A perturbation solution of the form

$$e^{\alpha t}e^{i\beta x} \tag{5.28}$$

is assumed for the three field variables $\bar{\tau}$, $\bar{u}$ and $\bar{T}$, where α is the growth factor and β is the wave number of the assumed oscillatory solution. The sign of the growth factor α determines the stability or instability of the perturbed system. Substitution of the perturbation solution into the above system of governing equations results in a system of three algebraic equations,

$$\alpha\bar{u} - i\beta\bar{\tau} = 0, \tag{5.29}$$

$$i\beta P\dot{\Gamma}_o\bar{u} + \dot{\Gamma}_o\bar{\tau} - \alpha\bar{T} - \beta^2\bar{T} = 0, \tag{5.30}$$

$$i\beta P\bar{u} - \bar{\tau} - P\dot{\Gamma}_o\bar{T} = 0. \tag{5.31}$$

A non-trivial solution requires that the determinant of coefficients be identically zero. The determinant expression, when solved for the growth factor, yields

$$\alpha = \frac{1}{2}\left(P\dot{\Gamma}_o^{\,2} + (1 + P)\beta^2\right)\left(\pm\sqrt{1 - \frac{4\left(\beta^4 P - \dot{\Gamma}_o^{\,2}\beta^2 P^2\right)}{\left(P\dot{\Gamma}_o^{\,2} + (1 + P)\beta^2\right)^2}} - 1\right). \tag{5.32}$$

The solution of equation (5.32) with a positive root provides a positive growth factor and implies unstable growth of thermoplastic deformation over the respective span of wavelengths. The growth factor $\alpha(\beta, P, \dot{\Gamma}_o)$ depends only on the wave number (or wavelength), the Prandtl number, and the initial shearing strain rate.

The dimensionless growth factor from equation (5.32) is plotted as a function of wavelength $\lambda = 2\pi/\beta$ in figure 5.1 for selected values of the dimensionless shearing strain rate $\dot{\Gamma}_o$ and Prandtl number P. The general trend of any curve is to exhibit no growth below a cut-off wavelength λ_c, an increasing rate of growth to some optimal wavelength λ_m at which growth is a maximum, and subsequent reduction in the growth rate with further increase in wavelength. At a specified strain rate an increase in Prandtl number (increasing viscosity η_o or decreasing thermal diffusivity χ) leads to an increase in the growth rate. Alternatively, an increase in the strain rate $\dot{\Gamma}_o$ for the same Prandtl number leads to a reduction of the maximum wavelength. Both trends are in agreement with common perceptions of the shear localization phenomena.

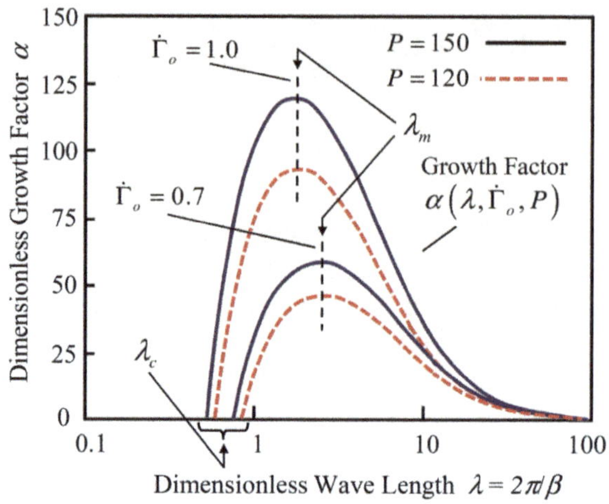

Figure 5.1. Plots of the dimensionless growth factor as a function of wavelength resulting from a sinusoidal perturbation of the linearized equations governing dynamic thermoplastic shear (Grady 1980). Selected variations of the material Prandtl number and the shear strain rate are illustrated. Both the wavelength at maximum growth rate and the cut-off wavelength are identified for the selected plots.

In the neighborhood of the maximum growth rate equation (5.32) is approximated by

$$\alpha \cong \frac{\beta^2 P\left(\dot{\Gamma}_o^2 P - \beta^2\right)}{\beta^2 + \left(\dot{\Gamma}_o^2 + \beta^2\right)P}. \tag{5.33}$$

The functional dependence of the wave number corresponding to the maximum growth rate can be determined from equation (5.33), and for reasonably large Prandtl number is provided by

$$\beta_m \cong \dot{\Gamma}_o P^{1/4}. \tag{5.34}$$

The Prandtl number P for the metal U3/4Ti is, for example, of order 10^4. The corresponding growth rate at the maximum wavelength is

$$\alpha_m \cong \dot{\Gamma}_o^2 P. \tag{5.35}$$

When dimensional variables are reintroduced the wavelength at maximum growth $\lambda_m = 2\pi/\beta_m$ is given by

$$\lambda_m \cong \frac{2\pi}{\dot{\gamma}_o}\left(\frac{\rho\chi c^2 T_m^2}{\eta_o}\right)^{1/4}, \tag{5.36}$$

with growth rate

$$\alpha_m \cong \dot{\gamma}_o^2 \frac{\eta_o}{\rho c T_m}. \tag{5.37}$$

Other authors (e.g. Birkhoff 1962) suggest a physical interpretation of λ_m as a statistical measure of the spacing of shear-deformation features. If the material is subjected to random inhomogeneous deformation stimulation, then those deformation perturbations favorably spaced (near λ_m) will grow most rapidly and dominate the adiabatic shear deformation fabric.

The Newtonian viscous constitutive shear-deformation model provided in equation (5.19) was assumed in earlier analysis (Grady 1977, 1980) and precludes a quantitative evaluation of the analysis with available material data. An appropriate viscosity coefficient η_o for the solid material is not known and, further, the strong rate sensitivity is probably not appropriate for most materials. The analysis can be repeated using more sensible constitutive models. This, of course, provided the underlying motivation for numerous later studies (e.g. Bai 1982, Burns *et al* 1982, Wright 2002) over the intervening several decades following the earlier efforts described here.

An expedient alternative is to substitute for the initial viscosity coefficient η_o the plastic shear stress at which adiabatic shear localization occurs through $\tau_o = \eta_o \dot{\gamma}_o$. This shear-stress amplitude is known, or is sensibly calculated, for a number of the metals that exhibit adiabatic shear localization. The Prandtl number then becomes $P = \tau_o / \rho \chi \dot{\gamma}_o$ as emerged from dimension considerations in the previous section.

The wavelength at the maximum growth rate provided by equation (5.36) is then

$$\lambda_m = 2\pi \left(\frac{\rho \chi c^2 T_m^2}{\tau_o \dot{\gamma}_o^{\,3}} \right)^{1/4}. \tag{5.38}$$

Later analysis will show that the shear-band spacing relation provided by equation (5.38) is functionally the same and quantitatively comparable to the characteristic spacing calculated from the two-temperature model of thermoplastic shear localization.

5.4.2 Application to adiabatic shear in metals

In the experimental microscopic observations of the adiabatic shear-localization, two length scales stand out starkly in characterizing the failure phenomenon. The first relates to the multiplicity of shear bands necessary to accommodate the dynamic deformation. The second characterizes the extent of active material of a single shear band participating in the deformation. In the present one-dimensional analysis of adiabatic shear, the first length scale relates to the typical spacing of shear bands, while the second relates to the thickness or width of a single shear band.

These features of adiabatic shear deformation are strikingly demonstrated by the experimental studies of Xue *et al* (Xue 2001, Xue *et al* 2003, 2004, 2006). In figure 5.2 results are shown for the explosive inelastic compression of open cylinders of stainless steel. The overall fabric of shear bands necessary to accommodate the deformation is depicted on the left. Details of the higher density of shear bands near the inner radius of the cylinder, and the early arrest of many from the active population, are seen in the center photograph. The graph on the right quantifies the

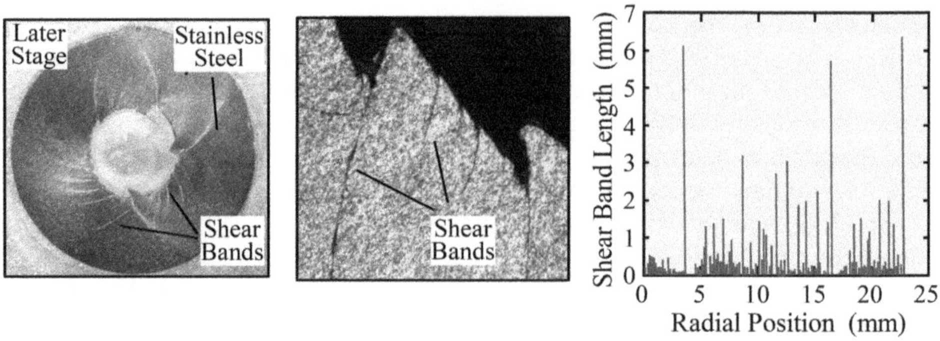

Figure 5.2. Adiabatic shear-band failure of explosively compressed cylinders of stainless steel (Xue *et al* 2003, 2004). Inner surface shear-band details are illustrated in the center image. A plot of shear-band length against radial position around the inner circumference is shown on the right.

number, location and distance of propagation of shear bands relative to their initial position on the inner cylinder circumference.

In figure 5.3 further experiments of Xue *et al* (2006) on stainless steel are shown in which top-hat geometry samples subjected to split-Hopkinson (Kolsky) bar loading experience forced shear and adiabatic shear deformation. The characteristic widths of shear bands in two stainless steels are illustrated in the photographs. The time durations shown in the figure assist in quantifying the extent of deformation in the dynamic event preceding imaging.

The present stability analysis of thermoplastic shear yields a length scale λ_m that corresponds to the maximum rate of shear-band growth. The functional relation for this length scale is provide by equation (5.38), as well as by earlier versions of the same equation (equations (5.36) and (5.34)). Physical arguments suggest that this characteristic length corresponds to the nominal spacing of fully developed adiabatic shear bands. The rate of growth at this optimum wavelength from equation (5.37) is, in dimensional units,

$$\alpha_m \cong \frac{\dot{\gamma}_o \tau_o}{\rho c T_m}. \tag{5.39}$$

This initial growth rate calculated from the stability analysis is found to be markedly higher than the average rate of shear-band growth determined from the two-temperature shear-band model described later in the chapter. The two growth rates differ by a power of the Prandtl number $P^{1/2}$.

The present stability analysis provides a second length scale through the wavelength λ_c. This wavelength corresponds to a cut-off below which unstable thermoplastic growth does not occur. From equation (5.32) the dimensionless form for the cut-off wave number is given by

$$\beta_c = \dot{\Gamma}_o P^{1/2}. \tag{5.40}$$

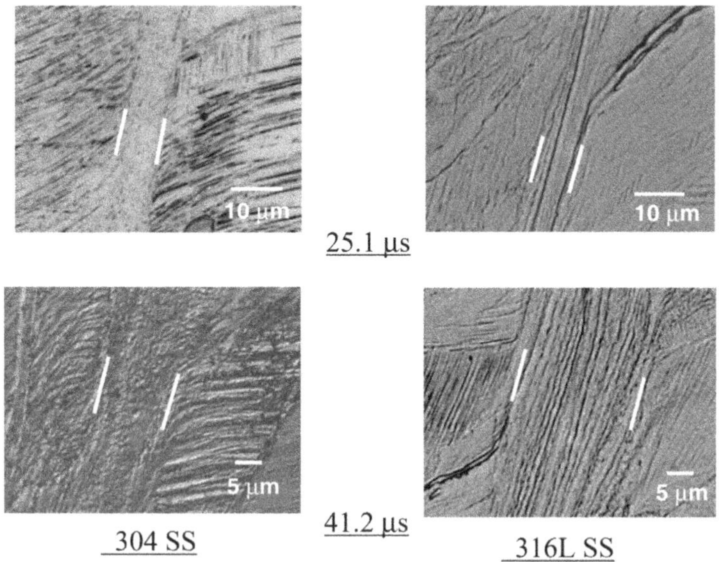

Figure 5.3. Adiabatic shear-band morphology in dynamic shear-deformation experiments for several stainless steel metals (Xue *et al* 2006). Copyright 2006., AIP Publishing LLC.

Introducing dimensional variables provides a cut-off wavelength

$$\lambda_c = 2\pi\left(\frac{\rho\chi c T_m}{\tau_o\dot{\gamma}_o}\right)^{1/2}. \tag{5.41}$$

The physical significance of this length scale is less readily interpreted.

Both the wavelength at the maximum rate of growth and the cut-off wavelength calculated from the stability analysis are plotted in figure 5.4, using the material properties for stainless steel, over the range of strain rates of interest to impact and explosive loading applications (approximately 6×10^4 s^{-1} for the cylinder tests of Xue *et al* (2003, 2004)). Values for the maximum growth wavelength are in sensible agreement with the experimental observations of Xue *et al* for the spacing of shear bands subsequent to the early self-organization and quench of the preponderance of activated incipient shear bands at the inner circumference (of order one millimeter and higher in figure 5.2).

The cut-off wavelength is about one order of magnitude smaller than the maximum growth wavelength. Although lacking physical motivation, the calculated cut-off wavelength can be compared with the observed widths of single shear bands in stainless steel. The cut-off wavelength predicts length scales in the neighborhood of several hundred micrometers over the experimental strain-rate range of interest (about 6×10^4 s^{-1}). This length scale is about one order of magnitude larger than the observed shear-band widths, which are typically of order of tens of micrometers, as indicated by the images in figure 5.3.

It is reasonable to suggest that the cut-off wavelength provides a measure of precisely what the analysis predicts. Namely, it is a measure of the minimum

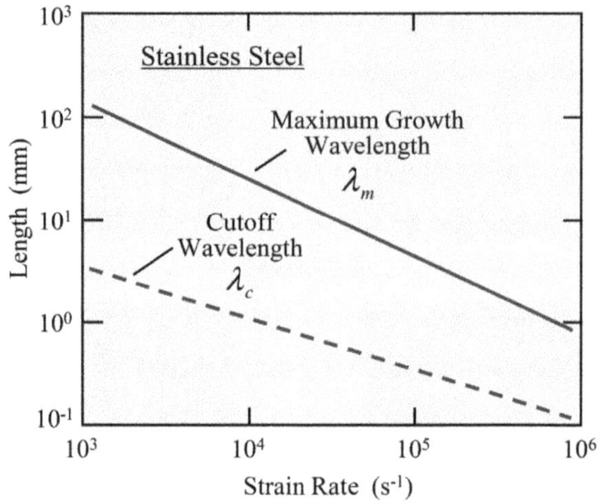

Figure 5.4. Maximum growth and cut-off wavelengths with properties for stainless steel based on stability analysis of the governing equations of thermoplastic shear (Grady 2010).

separation of shear bands that allows independent unstable incipient growth. This interpretation is consistent with the initial fine structure shear-band fabric observed in the tests of Xue *et al* (2003, 2004) preceding population reduction due to stress wave correlation and self-organization. Xue *et al* report shear-band spacing of this initial population in the neighborhood of ~150 μm. This explanation for the incipient shear-band population density is an alternative to the suggestion that material hardening or strain-rate sensitivity play a deciding role in establishing this initial spacing (Xue *et al* 2004).

5.5 Two-temperature theory of thermoplastic shear localization

The two-temperature thermoplastic shear-localization theory has been described in early papers (Grady and Kipp 1985, 1987), and has subsequently been applied with modifications in later research reports (Grady 2005, 2010). Features of the theory relevant to the present discussion of dynamic adiabatic shear are provided in the present section. Objectives of the theory are the determination of physical length, time and energy scales appropriate to the adiabatic shear-band phenomena. As noted previously, the theoretical development derives inspiration from the seminal theories of thermal explosions introduced by Semenov (1928) and from the plastic wave dynamics of Mott (1943, 1947).

The analysis focuses on one-dimensional thermoplastic shear localization. It is important to emphasize that this is a simplification of a profoundly more complex deformation process. Non-uniform deformation gradients, and differing time histories of the spatial variations of the deformation, are much more common in applications involving failure through thermoplastic shear localization. Further, shear localization is observed to initiate at free surfaces or interior interface discontinuities, as occurs for example in the onset of breakup of explosively expanding metal shells, or the impact breach of a metal barrier. In addition, in contrast to the assumed one-dimensionality of the localization process in the present model, the shear-band manifestations of thermoplastic localization propagate much like mode II, or perhaps mode III, cracks. Shear bands are observed to arrest at interior points. Single shear bands can branch into two or more multiple-shear bands.

Various features crucial to material response are not addressed in the present model of thermoplastic shear localization. Microstructure issues concerning the materials of interest are largely ignored. Practical concerns are not raised regarding grain size and structure, second phases, alloy precipitates, or twin and dislocation densities. Whether or not the nature of shear localization depends on these microstructure properties is not pursued. The governing continuum properties, such as flow stress, heat transport and thermal softening, are assumed to be some sensible average over the microstructure and are, to some extent, expected to reflect this microstructure. For example, as grain size is reduced, flow stress commonly increases and in turn is predicted to accelerate the rate of shear localization.

Also ignored are elastic properties of the body and their potential influence on the adiabatic shear-localization process. Early judgments suggested that dynamic plasticity largely dominates the localization physics in metals. Elasticity and elastic strain energy in the shear-localization process, along with possible adiabatic shear effects in more brittle solids, are explored in a later section of this chapter.

As noted earlier, analysis of the uniform one-dimensional thermoplastic shear localization of metals has been treated extensively in past studies (e.g. Grady 1977, 1980, Clifton 1980, Bai 1981, 1982, Merzer 1982, Burns *et al* 1982, Molinari and Clifton 1987, Fressengeas and Molinari 1987, Coleman and Newman 1989, Wright 1990, Batra and Kim 1990). Most of these studies emphasized perturbation and stability analysis.

The present two-temperature boundary-layer model focuses on an alternative physics-based approach that was introduced in the mid-1980s (Grady and Kipp 1985, 1987, Grady 1992, 1994). The method has been found in some cases to complement stability analysis studies such as that of Wright and Ockendon (1996) and Molinari (1997) as shown by Xue *et al* (2004). Comparisons of the two-temperature adiabatic shear model with the examples of stability analysis described in the previous section are made where appropriate.

5.5.1 Boundary layer model

Consider a solid body undergoing uniform plastic shear at a rate $\dot{\gamma}_o$. The Lagrangian h coordinate is normal to the plane of shear. The parameter $\dot{\gamma}_o$ is a kinematic constant in the problem, and captures both the initial state of plastic shear flow, and the intensity of kinetic energy available to fuel thermoplastic shear localization.

The onset and growth of adiabatic shear localization occur at the plane $h = 0$. The width of the shear-localization region a (the width or thickness of the shear band) is assumed to be small compared to other principal length scales in the problem. Consequently, stress, displacement and temperature conditions within the shear band will provide the plastic wave boundary conditions on the plane $h = 0$. The adequacy of the boundary layer assumption is tested after completion of the solution. The initial conditions for the shear stress and the gradient in the transverse velocity u over the domain $h \geqslant 0$ are

$$\begin{aligned} \tau(h, 0) &= \tau_o, \\ \frac{\partial u}{\partial h}(h, 0) &= \dot{\gamma}_o \end{aligned} . \tag{5.42}$$

The same conditions continue to apply at later time at all points ahead of the leading plastic wave disturbance. Again, $\dot{\gamma}_o$ is the constant shearing rate, while the shear stress τ_o and temperature T_o are values at the load-maximum conditions at onset of shear localization.

Time-dependent conditions in the shear band at the plane $h = 0$ provide the boundary conditions governing the Mott plastic wave propagation. They are

$$\begin{aligned} \tau(0, t) &= \tau(T(t)), \\ \int u(0, t)dt &= \psi(t), \\ T(0, t) &= T(t), \end{aligned} \tag{5.43}$$

for the shear stress, shear displacement and temperature within the shear band, respectively. Further, the two-temperature field is of the form

$$T(h, t) = \begin{cases} T(t), & h = 0 \\ T_o, & h > 0 \end{cases} . \tag{5.44}$$

The thermal energy transport is governed by a heat transfer coefficient between the elevated temperature boundary shear band and the adjacent material.

Equation (5.44) emphasizes the two-temperature character of the present model. A single time-dependent temperature applies through the width a of the shear band at the boundary $h = 0$. Gradients in temperature through the thickness of the shear band, as well as gradients in stress, deformation and deformation rate, are not included in the model. T_o is the temperature throughout the field at onset of instability and shear localization, and remains the constant temperature at later time for all $h > 0$. Thermal energy transport from the shear band is assumed to be negligible relative to the heat content of the neighboring material.

5.5.2 Plastic wave propagation

Onset of thermoplastic shear localization at the boundary $h = 0$ leads to a temperature-dependent reduction in the boundary shear stress $\tau(T) < \tau_o$. The concomitant misbalance in stress results in the propagation of a plastic wave disturbance away for the boundary shear band and reduces the shear stress in both the positive and negative h directions. Wave motion along the positive h coordinate is illustrated in figure 5.5. Initial motion is chosen such that velocity at the origin is zero. An anti-symmetric image of the same motion will propagate in the negative h direction.

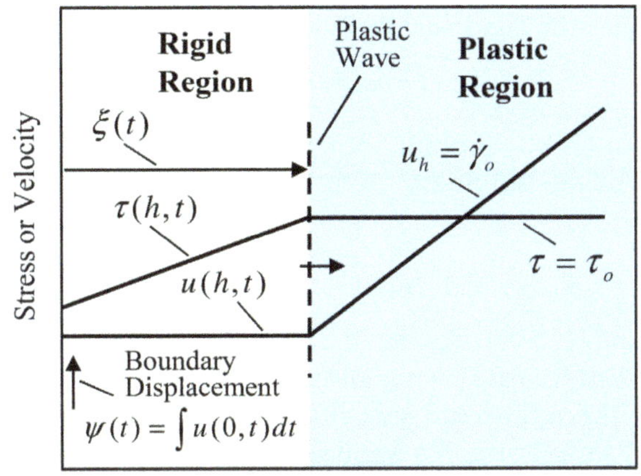

Figure 5.5. Illustrates the stress and velocity field for all $h > 0$ for Mott wave propagation at times $t > 0$ after shear-band activation at the plane $h = 0$. The Mott wave solution provides the time history of both the rigid–plastic interface and the boundary displacement.

The propagating disturbance in stress and velocity is identified as a Mott wave. Solution methods for this particular class of plastic wave propagation problems were first developed by Mott (1947), and subsequently pursued in further detail (Lee 1967, Grady and Kipp 1985, Kipp and Grady 1985, Grady 2006, 2010).

Structure and evolution of the plastic Mott wave is fully characterized by two time-dependent parameters, $\xi(t)$ and $\psi(t)$. The nature of this plastic wave at any time $t > 0$ is shown in figure 5.5. The position $\xi(t)$ is the front of the propagating plastic wave disturbance. Ahead of the plastic wave front, $h > \xi(t)$, the material is undisturbed and continues to shear uniformly at the initial shear deformation rate $\dot{\gamma}_o$ and flow stress τ_o. Behind this interface, $h < \xi(t)$, the material is off the plastic yield surface and assumed rigid. The material velocity is constant over the domain $0 < h < \xi(t)$, consistent with rigid motion, and the stress is linear from the reduced shear-band stress of $\tau(T)$ at $h = 0$ to the flow stress τ_o at $h = \xi(t)$. Momentum balance applied to the region encompassed by the Mott wave, $0 < h < \xi(t)$, yields a relation that governs the motion of the plastic wave front $\xi(t)$,

$$\rho\dot{\gamma}\xi\frac{d\xi}{dt} = \tau_o - \tau(T). \tag{5.45}$$

The parameter $\psi(t)$ provides the time-dependent shear displacement of the material at the boundary $h = 0$ and is given by

$$\psi(t) = \int_0^t u(0, t)dt, \tag{5.46}$$

or

$$\frac{d\psi}{dt} = u(0, t). \tag{5.47}$$

A constant velocity in the rigid region, $0 < h < \xi(t)$, yields

$$u(0,\, t) = u(\xi,\, t) = \dot{\gamma}_o \xi. \tag{5.48}$$

Equations (5.47) and (5.48) combine to provide the differential equation governing shear displacement at the boundary

$$\frac{d\psi}{dt} = \dot{\gamma}\xi. \tag{5.49}$$

Equations (5.45) and (5.49) provide the pair of ordinary differential equations that govern the Mott plastic wave motion. Since temperature appears in the boundary stress on the right-hand side of equation (5.45), one further constitutive equation is required for closure. An appropriate energy equation is developed of the form

$$\frac{dT}{dt} = f(\xi,\, \psi,\, T). \tag{5.50}$$

5.5.3 Shear-band model and the energy equation

In development of the energy equation governing the boundary conditions at $h = 0$, a shear band of width a is assumed centered about the origin, as illustrated in figure 5.6. Again, the width a is assumed sufficiently small relative to other pertinent length scales. When the time-dependent stress, displacement and temperature in the shear band are determined, they apply on the $h = 0$ boundary, and provide boundary conditions for the plastic wave boundary value problem developed in the previous subsection. Within the shear band the shear stress $\tau(t)$, the local shearing rate $\dot{\gamma}_l(t)$ and the temperature $T(t)$ are assumed to be uniform over the width a of the shear band.

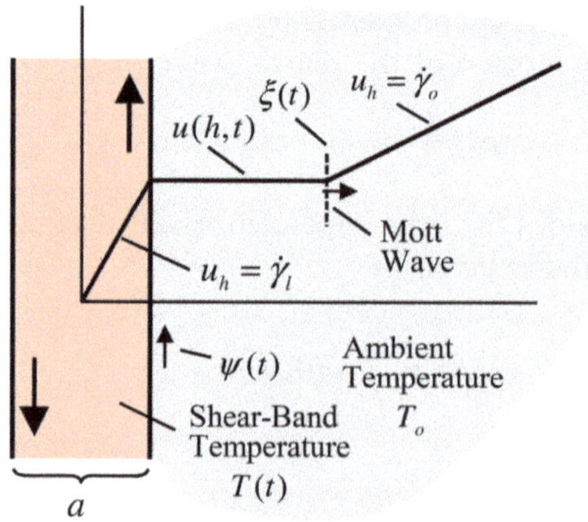

Figure 5.6. Details of the two-temperature boundary-layer adiabatic shear band providing boundary conditions for the Mott plastic wave propagation solution.

Temperature rise in the shear band due to plastic dissipation is provided by

$$\dot{T} = \frac{1}{\rho c}\tau\dot{\gamma}_l = \frac{2}{\rho c a}\tau\dot{\psi} = \frac{2}{\rho c a}\tau\xi\dot{\gamma}_o. \tag{5.51}$$

The density and specific heat in the shear band are ρ and c, respectively. All plastic dissipation is assumed to convert to thermal energy. The second equality follows from the local deformation rate in the shear band $\dot{\gamma}_l = 2\dot{\psi}/a$, where the transverse displacement at the shear-band boundary is ψ.

Temperature in the shear band is reduced by heat transport to the media adjacent to the shear band. Thermal transport is captured in the model through a heat transfer coefficient. Temperature decrease in the shear band is determined through the relations

$$\dot{T} = -\frac{2}{\rho c a}h_t T = -\frac{2}{\rho c a^2}\kappa T = -\frac{2\chi}{a^2}T. \tag{5.52}$$

In the first equality $\rho c a/2$ is the specific heat per unit area of a half-width of the shear band, and h_t is the heat transfer coefficient between the elevated temperature shear band and the neighboring material at temperature T_o. In equation (5.52), as noted previously, T is regarded as the temperature in excess of the temperature T_o at onset of load-maximum localization. This convenience saves carrying temperature differences though the train of equations.

In the second equality of equation (5.52) the heat transfer coefficient is approximated by $h_t = \kappa/a$ where κ is the thermal conductivity of the material. In the third equality thermal diffusivity is introduced through the relation $\chi = \kappa/\rho c$, and is the thermal parameter that will be carried through the subsequent analysis.

Equations (5.51) and (5.52) join to provide the energy equation governing the temperature in the shear band, and the $h = 0$ boundary constraint on the plastic wave boundary value problem. Repeating equations (5.45) and (5.49), the system of governing equations becomes

$$\rho\dot{\gamma}_o\xi\frac{d\xi}{dt} = \tau_o - \tau(T), \tag{5.53}$$

$$\frac{d\psi}{dt} = \dot{\gamma}_o\xi, \tag{5.54}$$

$$\frac{dT}{dt} = \frac{2}{\rho c a}\tau\xi\dot{\gamma}_o - \frac{2\chi}{a^2}T. \tag{5.55}$$

The temperature equation follows from equations (5.51) and (5.52). Solutions of this system of equations provide the motion of the plastic wave rigid–plastic interface $\xi(t)$, the accelerating growth of the shear band through the displacement $\psi(t)$, and the temperature rise $T(t)$ in the shear band.

Closure of the system of equations requires only specification of a shear-band thermal-softening relation through $\tau(T)$. Although not amenable in general to analytic solution, this system of ordinary differential equations is readily solved through numerical methods. Grady and Kipp (1987) have previously performed a numerical study of comparable equations.

5.5.4 Thermal softening in the shear-localization process

Difficulties in ascertaining the thermal softening properties of solids relevant to the adiabatic shear-banding phenomena are discussed in some detail in the opening section. Again, in their study of metal strength, Johnson and Cook (1983) used the functional form

$$\tau(T) = \tau_o\left(1 - \left(\frac{T}{T_m}\right)^m\right), \tag{5.56}$$

as a representation of thermal softening. Here, the relation is specialized to the present shear-localization problem. T and T_m are temperatures in excess of some initial reference temperature T_o at onset of localization, while τ_o is the corresponding shear flow stress. T_m is the temperate rise to material melt. Johnson and Cook determined the thermal softening parameter m from available elevated temperature experimental data. Whether or not equation (5.56) is appropriate to the deep thermal softening occurring in an adiabatic shear-localization process is arguable. Johnson–Cook constants that are available for the thermal softening relation are based on thermal-mechanical deformation data, unlike that experienced by material participating in the shear band.

For the present analysis a thermal softening function that is a modest variation of the Johnson and Cook (1983) relation,

$$\tau(T) = \tau_o(1 - T/T_m)^{1/2}, \tag{5.57}$$

is assumed for thermal softening through the growth of the adiabatic shear band. The shear stress is assigned the load-maximum value τ_o at initiation of shear-band growth and approaches zero as temperature rise above the load-maximum temperature approaches the melt temperature. In this softening relation temperature rise to melt is the sole material parameter governing thermal softening in the shear band. Equation (5.57) reasonably approximates softening curves of Johnson and Cook for at least some of the metals investigated. More importantly however, for present illustrative purposes, this specific thermal softening relation leads to a fully tractable analytic solution of the system of plastic wave equations provided in equations (5.53)–(5.55).

5.5.5 The adiabatic shear band

A solution to the system of equations (5.53)–(5.55) is first pursued for the simpler case in which the shear band at the boundary is assumed to be adiabatic (no heat loss through thermal conduction) during the shear-band growth process. That is, plastic dissipation in the shear band goes solely into temperature rise, and heat transfer to neighboring material does not occur. In the adiabatic approximation, the energy equation (5.55) reduces to

$$\frac{dT}{dt} = \frac{2}{\rho ca}\tau\xi\dot{\gamma}_o, \tag{5.58}$$

or, in differential form,

$$\rho ca\, dT = 2\tau\xi\dot{\gamma}_o\, dt = 2\tau\, d\psi, \tag{5.59}$$

with the second equality following from equation (5.54). The thermal softening relation provided by equation (5.57) yields

$$dT = -\frac{2T_m}{\tau_o^2}\tau\, d\tau. \tag{5.60}$$

Combining equations (5.59) and (5.60) gives

$$\frac{T_m}{\tau_o^2}d\tau = -\frac{1}{\rho ca}d\psi, \tag{5.61}$$

which integrates to

$$\tau_o - \tau = \frac{\tau_o^2}{\rho ca T_m}\psi. \tag{5.62}$$

The solution in equation (5.62) yields a stress τ that reduces linearly from τ_o to zero as displacement of the shear-band boundary increases from zero to a critical displacement $\psi_c = \rho ca T_m/\tau_o$. The plastic work dissipated per unit area of shear band is

$$W = \int_0^{t_c} \tau(\psi)\dot{\gamma}_l a\, dt = \int_0^{\psi_c} 2\tau_o(1 - \psi/\psi_c)d\psi, \tag{5.63}$$

where substitutions of equation (5.62) and the previous relation for the local shear strain rate in the shear band of $\dot{\gamma}_l = 2\dot{\psi}/a$ have been used. Carrying out the integral yields for the energy dissipated in the shear band,

$$W = \tau_o\psi_c = \rho ca T_m. \tag{5.64}$$

Substitution of equation (5.62) into equation (5.53) decouples temperature from the two plastic wave equations now written as

$$\rho\dot{\gamma}_o\xi\frac{d\xi}{dt} = \frac{\tau_o^2}{W}\psi, \tag{5.65}$$

$$\frac{d\psi}{dt} = \dot{\gamma}_o\xi. \tag{5.66}$$

Equations (5.65) and (5.66) are readily solved, yielding

$$\xi(t) = \xi_c(t/t_c)^2, \qquad \psi(t) = \psi_c(t/t_c)^3, \tag{5.67}$$

where

$$\psi_c = \frac{W}{\tau_o}, \qquad \xi_c = \left(\frac{3W}{2\rho\dot{\gamma}_o^{\,2}}\right)^{1/3}, \qquad t_c = \left(\frac{18\rho W^2}{\tau_o^{\,3}\dot{\gamma}_o}\right)^{1/3}. \tag{5.68}$$

From this solution the time-dependent reduction in shear stress in the adiabatic shear band is provided by

$$\tau(t) = \tau_o[1 - (t/t_c)^3], \tag{5.69}$$

while the temperature in the shear band, either through integration of the governing equation, or through direct substitution into the thermal softening relation from equation (5.57), is

$$T(t) = T_m\left[2(t/t_c)^3 - (t/t_c)^6\right]. \tag{5.70}$$

5.5.6 The shear band with thermal conduction

Numerical simulations of the three ordinary differential equations governing shear-band growth and provided in equations (5.53)–(5.55) (e.g. Grady and Kipp 1987) reveal three phases in the shear-band growth process. The first is an inception period immediately following onset of shear localization, when deformation in the shear-band region is slowly accelerating, and minimal shear-stress reduction and temperature rise occur. The second is a period of very rapid stress reduction and accompanying temperature rise. Lastly, there is a leveling off to a post-growth quasi-steady state of the shear band at a markedly reduced stress and elevated temperature.

The objective, as in the earlier paper of Grady and Kipp (1987), is obtaining approximate analytic solutions to the system of shear-band growth, equations (5.53)–(5.55) that adequately account for thermal conduction over the period of shear-band growth. The energy equation (5.55),

$$\frac{dT}{dt} = \frac{2}{\rho c a}\tau\frac{d\psi}{dt} - \frac{2\chi}{a^2}T, \tag{5.71}$$

when written in the differential form

$$\rho c a\left[1 + \frac{2\chi}{a^2}\frac{T}{dT/dt}\right]dT = 2\tau\, d\psi, \tag{5.72}$$

reveals that heat loss by thermal conduction during the shear-band growth process is determined by the second term within the brackets in equation (5.72). Compare equation (5.72) with equation (5.59) for the fully adiabatic case. The present development does not try to account for temporal details of conductive heat loss.

Rather, a sensible accounting of the total heat loss due to conduction over the effective shear-band growth period t_c is sufficient for the present purpose. In equation (5.72) the time varying expression $T/(dT/dt)$ has dimensions of time. Its average value over the time t_c is proportional to t_c, and a sensible average value is $t_c/2$. Equation (5.72) then simplifies to

$$\rho ca\left[1 + \frac{\chi}{a^2}t_c\right]dT = 2\tau\, d\psi. \tag{5.73}$$

Application of equation (5.73) to the solution for the thermal conducting shear band is then straightforward. If the thermal diffusivity χ is zero, then equation (5.73) reduces to that of equation (5.59) for the adiabatic shear band (no heat conduction), with ρca the heat capacity per unit area of the adiabatic shear band. Thus, for the thermal conducting shear band,

$$\rho ca\left[1 + \frac{\chi}{a^2}t_c\right] = \rho\bar{c}a \tag{5.74}$$

can be regarded as an effective area heat capacity which accounts for conductive heat loss from the band over the shear-band formation time t_c. Writing equation (5.73) in the form

$$\rho\bar{c}adT = 2\tau d\psi, \tag{5.75}$$

where $\bar{c} = c(1 + \chi t_c/a^2)$, allows the previous solution for the adiabatic shear band to apply. Introducing $\bar{c}$ into equation (5.64), the solution for the work dissipated in the shear band yields

$$W = \rho\bar{c}aT_m = \rho caT_m\left(1 + \frac{\chi t_c}{a^2}\right). \tag{5.76}$$

5.5.7 Optimum shear-band width and localization time

The plastic wave solution provided by equation (5.68) gives for the localization time

$$t_c = \left(\frac{18\rho W^2}{\tau_o{}^3\dot{\gamma}_o}\right)^{1/3}. \tag{5.77}$$

The shear-band energy W can be eliminated by combining equations (5.76) and (5.77), providing a functional relation between the width of the shear band a and the localization time $t_c(a)$. This relation is

$$\frac{t_c(a)}{t_{co}} = \frac{1}{2^{2/3}}\left[a/a_o + \frac{t_c(a)/t_{co}}{a/a_o}\right]^{2/3}, \tag{5.78}$$

where the normalizing shear-band width a_o and localization time t_{co} are, respectively,

$$a_o = \left(\frac{72\rho^3c^2\chi^3T_m{}^2}{\tau_o{}^3\dot{\gamma}_o}\right)^{1/4}, \tag{5.79}$$

$$t_{co} = \left(\frac{72\rho^3 c^2 \chi T_m^{\ 2}}{\tau_o^{\ 3} \dot{\gamma}_o} \right)^{1/2}. \tag{5.80}$$

It is readily shown that $t_{co} = a_o^2/\chi$. The numerical differences of a_o and t_{co} from the relations provided by Grady and Kipp (1987) reflect the modest differences in assumptions governing averaging of the thermal conduction over shear-band formation. The numerical values in the relations provided in Grady and Kipp (1987) can be recovered by replacing χ by $\chi/2$ in the present equations.

The shear-band localization time $t_c(a)$ from equation (5.78) is plotted as a function of shear-band width in figure 5.7. A comparison is provided with that of the strictly adiabatic shear-band solution. The curve shows that a fairly narrow band of shear-band widths occurs at or near the minimum growth time t_{co}. It is reasonable to propose that in the shear localization failure process shear-band widths in the neighborhood of a_o would be favored. Physical arguments supporting such a proposition follow the same logic outlined by Birkhoff (1962) regarding linear stability analysis.

The physical picture is reasonably clear. For shear bands that are significantly thinner than a_o, excessive thermal diffusion limits the rate of growth. Conversely, for bands substantially thicker than a_o, growth is limited by the rate at which momentum can be diffused to the shear-band site through plastic wave propagation. Bands of a width in the neighborhood of a_o achieve an optimal balance between thermal and momentum diffusion and exhibit the most rapid rate of growth.

The time-dependent solutions for shear stress and temperature in the shear band, based on the present model and provided by equations (5.69) and (5.70), are shown in figure 5.8. The shorter time duration solution corresponds to the strictly adiabatic shear band (no heat conduction) at the optimum width a_o. The second solution,

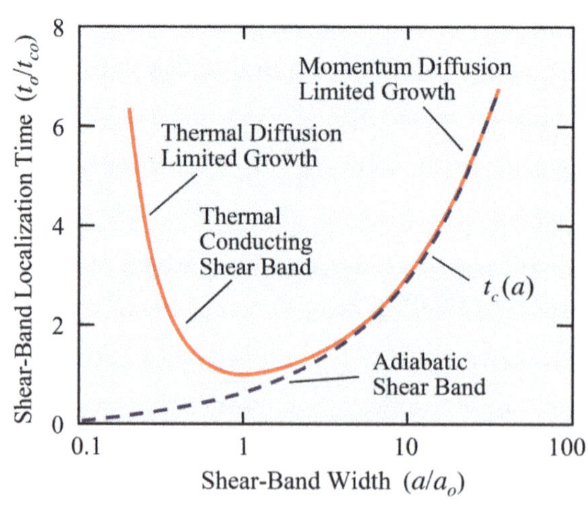

Figure 5.7. Shear-band localization time as a function of shear-band width. The thermal conducting solution is compared with that of the non-conducting (adiabatic) shear-band solution. Normalizing length and time scales are provide in the text.

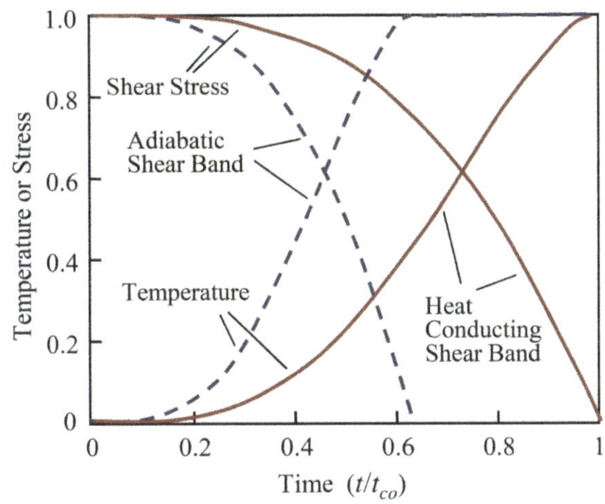

Figure 5.8. Solution for the time-dependent shear stress and temperature in the shear band assuming (1) strictly adiabatic behavior and (2) heat transfer from the shear band over the shear-band formation time.

at the same shear-band width, accounts for heat conduction. The relative delay in shear-band growth is a consequence of the added dissipation needed to accommodate thermal conduction.

5.5.8 Dissipation in the shear-band formation process

Alternatively, the localization time t_c can be eliminated between equations (5.76) and (5.77), and an expression for the plastic work or dissipation W in the shear band within the shear-band growth time can be obtained. This relation is written

$$\frac{W(a)}{W_o} = \frac{1}{2}\left[a/a_o + \frac{(W(a)/W_o)^{2/3}}{a/a_o}\right], \tag{5.81}$$

where the normalizing energy is

$$W_o = 2\rho c T_m a_o = 2\rho c T_m \left(\frac{72\rho^3 c^2 \chi^3 T_m{}^2}{\tau_o{}^3 \dot{\gamma}_o}\right)^{1/4}. \tag{5.82}$$

Equation (5.81) is plotted in figure 5.9 and provides the functional dependence of the shear-band dissipation $W(a)$ on the shear-band width a over the formation time t_c. The behavior is similar to that of the localization time $t_c(a)$. Dissipation is a minimum within a fairly narrow span of shear-band widths about the same optimum width a_o. Dissipation increases markedly for both narrower bands (limited by thermal diffusion) and for wider bands (limited by momentum diffusion). The minimum dissipation W_o in the shear localization process has the same dependence on the strain rate $\dot{\gamma}_o$ as has the optimum shear-band width a_o.

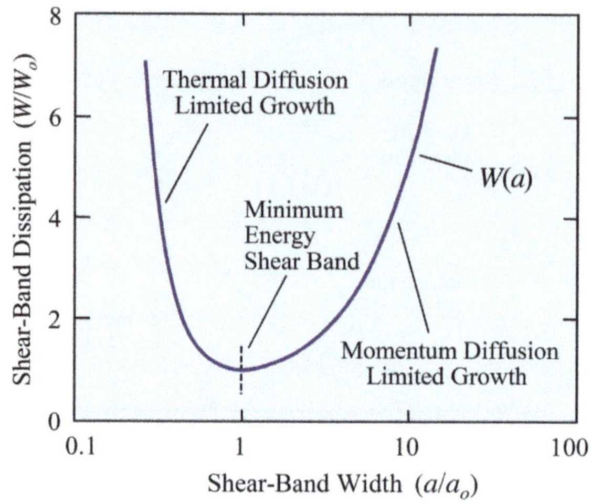

Figure 5.9. Plot of energy dissipation in the shear band over the formation time as a function of the shear-band width. Both energy and width are normalized by their value at the minimum.

5.5.9 Spacing of adiabatic shear bands

When adiabatic shear plays a dominant role in the dynamic failure of metals the frequency, or nominal spacing, of the multiple adiabatic shear-bands necessary to accommodate the deformation event is also of interest. Much like failure through ductile or brittle fracture, upon completion of the failure process, the body is partitioned into a number of regions (fragments) separated by a fabric of adiabatic shear bands. Commonly, some subset of the shear bands serves as surfaces of subsequent fracture in the dynamic failure event. Characteristic spacing of adiabatic shear bands in the dynamic localization process is also estimated from the present analysis.

On a plane of adiabatic shear, the kinetic energy of dynamic shear deformation in the immediate neighborhood of the shear band fuels the localization process. Energy is transported to the plane of localization through the Mott plastic wave propagation. It was shown from the earlier plastic wave solution that if the shear-localization process dissipates an energy W in the shear band, then a plastic wave propagates outward from the shear band a distance $\xi_c = (3W/2\rho\dot{\gamma}_o^{\,2})^{1/3}$ to supply the necessary shear-band energy (see equation (5.68)). Thus, for two adjacent shear bands to acquire sufficient energy to each independently achieve growth completion, they would have to be separated by a distance of at least $b = 2\xi_c$. It is sensible to suggest that

$$b(a) = 2\xi_c = \left(\frac{12W(a)}{\rho\dot{\gamma}_o^{\,2}}\right)^{1/3}, \tag{5.83}$$

provides a reasonable measure of shear-band spacing. At the minimum energy W_o provided by equation (5.82), and corresponding to the optimum shear-band width a_o, a shear-band spacing is calculated to be

$$b_o = 2\left(\frac{18\rho c^2\chi T_m^{\,2}}{\tau_o\dot{\gamma}_o^{\,3}}\right)^{1/4}. \tag{5.84}$$

Again, the flatness of the energy function in figure 5.9 suggests $b_o/2 \leqslant b \leqslant 2b_o$ as a reasonable range of shear-band spacing about b_o at a strain rate of $\dot{\gamma}_o$. Experimental and computational observations of the statistics of shear-band spacing are discussed in a later section. A relation between the shear-band spacing and the localization time, $b_o^2 = (2\tau_o/\rho\dot{\gamma})t_{co}$, can be derived. The ratio $\tau_o/\rho\dot{\gamma}_o$ is identified as a measure of the plastic viscosity, revealing that shear-band spacing is governed by momentum diffusion during the growth of the shear band in contrast to the shear-band width, which is determined by thermal diffusion (see equation (5.80) and the paragraph following).

It is notable that the stability analysis explored in an earlier section yields a maximum growth rate at a wavelength provided in equation (5.38) that is functionally the same, and numerically similar, to the present expression, equation (5.84), for the shear-band spacing resulting from the two-temperature model.

5.6 Application to adiabatic shear in metals

The two-temperature boundary layer model for adiabatic shear provides quantitative analytic expressions for the key length, time and energy scales associated with the adiabatic shear localization process. These calculated features include the characteristic width and spacing of adiabatic shear bands, the time from onset to full localization, and the energy per unit area dissipated in the formation of a shear band. Further features provided by the adiabatic shear model are introduced in the present section. These shear-band features are calculated from commonly available thermomechanical properties of the materials. In the present section, these calculations are tested for sensibility on common metals for which the necessary material property data are available. The metals examined are those for which reasonable metal plasticity models have been developed and published in the open literature. Johnson and Cook (1983) and later Johnson (1993) provided a practical plasticity model for metals, and additionally, undertook the effort necessary to generate the model constants for a selected set of metals—here referred to as the Johnson–Cook metals. These mechanical properties joined with handbook thermophysical properties for the same metals make the Johnson–Cook metals suitable for comparisons with the present shear-localization model.

The section starts by restricting the Johnson–Cook plasticity model to an adiabatic deformation path and establishing the state of load-maximum stress, strain and temperature. The load-maximum state is identified here as the criterion for onset of adiabatic shear localization. The systematics among the load-maximum states are examined for the Johnson–Cook metals. Shear-localization features are calculated for the metals from the two-temperature model, including characteristic width and spacing of the shear band, shear-localization time and localization strain, and energy dissipation in the shear band. The calculated shear-band features are examined for trends in the selected set of metals.

The solid state Prandtl number P and the heating number Q, previously identified through dimensional analysis of the shear-localization properties, are calculated for the Johnson–Cook metals and inspected for correlations with shear-localization propensity of the metals. The concept of shear-band toughness, proposed in earlier studies (Grady 1992, 1994), is also calculated for the Johnson–Cook metals. This section continues with a comparison of the two-temperature model and the earlier stability analysis with the base of experimental data for stainless steel provided by Xue *et al* (2003, 2004, 2006). Lastly, the section concludes with an assessment of the import of elasticity and elastic strain energy in the adiabatic shear localization of metals, some observations of spatial statistics of multiple shear bands, and some remarks on turbulence in the shear-band deformation process.

5.6.1 Johnson–Cook strength model

Johnson and Cook (1983) developed a constitutive plasticity model for metals that is appropriate for testing the present adiabatic shear-localization theory. The constitutive model of Johnson and Cook has seen wide application by the computational community over the intervening years. Through several papers (Johnson and Cook 1983, Johnson 1993) material constants for a suite of common metals have been made available.

The Johnson and Cook strength model takes the analytic form

$$\sigma(\varepsilon, \dot{\varepsilon}, T) = (A + B\varepsilon^n)(1 + C \ln \dot{\varepsilon}^*)\left(1 - T^{*^m}\right). \quad (5.85)$$

The starred parameters in the equation are $\dot{\varepsilon}^* = \dot{\varepsilon}/\dot{\varepsilon}_o$ with $\dot{\varepsilon}_o = 1.0\ \text{s}^{-1}$ a reference strain rate, and $T^* = (T - T_{\text{ref}})/(T_{\text{melt}} - T_{\text{ref}})$, where T_{melt} and T_{ref} are the absolute melt and reference (room) temperature, respectively. The constitutive equation equivalent stress and plastic strain are σ and ε, respectively. Equation (5.85) with $T^* = 0$ provides the ambient (room temperature) isothermal stress versus strain response of the metal.

The Johnson–Cook properties (Johnson and Cook 1983, Johnson 1993) for 16 metals along with thermophysical properties for the same metals obtained from the *Handbook of Chemistry and Physics* (Haynes 2014) are provided in table 5.1. These properties are sufficient to explore predictions of the two-temperature shear-band model outlined in the previous section.

5.6.2 Johnson–Cook adiabatic load-maximum criterion

The Johnson–Cook relation in equation (5.85) also provides the adiabatic stress versus strain plastic response of the material. In calculation of the adiabatic load path, plastic dissipation is assumed to fully contribute to adiabatic temperature rise. The adiabatic stress versus strain path for OFHC copper is compared with the isothermal load curve in figure 5.10. The adiabatic temperature rise normalized by the corresponding rise to the melt temperature $(T - T_{\text{ref}})/(T_{\text{melt}} - T_{\text{ref}})$ is also plotted. The load maximum of the adiabatic stress versus strain curve is identified as the criterion for instability and onset of shear localization. The sensitivity of equation (5.85) to strain rate is quite modest and for the present purposes calculated instability states will be appropriate for strain rates within several decades of $10^4\ \text{s}^{-1}$.

Table 5.1. Thermal and mechanical properties for metals.

Material[a]	Density[a] (kg m^{-3})	Shear modulus[a] (GPa)	Specific heat[b] (J $kg^{-1} \cdot K$)	Thermal conductivity[b] (J $s^{-1} \cdot m \cdot K$)	Thermal diffusivity[b] ($m^2\ s^{-1} \times 10^6$)	Melt temperature[b] (K)
OFHC copper	8945	46.3	383.5	400	116.6	1356
Cartridge brass	8517	41.4	385.5	220	67.0	1189
Nickel 200	8902	74.5	446.3	91	22.9	1726
Armco iron	7887	80.2	452.6	73	20.5	1811
Carp. elec. steel	7887	80.2	452.6	100	28.0	1811
1006 steel	7887	80.2	452.6	100	28.0	1811
Aluminum 2024-T351	2768	26.2	875.6	240	99.0	775
Aluminum 7039	2768	26.2	875.6	240	99.0	878
4340 steel	7823	77.5	477.8	100	26.8	1793
S-7 tool steel	7737	77.5	477.8	100	27.1	1763
Tungsten (.07Ni,.03Fe)	16 980	124.1	134.5	173	75.8	1723
Uranium (.0075Ti)	18 560	66.2	117.2	30	13.8	1473
Tantalum (EB)	16 590	68.9	150.7	60	24.0	3268
Aluminum 1100	2704	25.9	904.6	240	98.1	923
Titanium (.06Al, .04V)	4424	42.8	586.5	20	7.7	1905
Aluminum 6061-T6	2704	27.6	896.2	240	99.0	925

Material[a]	Density (kg m^{-3})	A (MPa)	B (MPa)	n	C	m
OFHC copper	8945	89.6	291.6	0.31	0.025	1.09
Cartridge brass	8517	111.7	504.7	0.42	0.009	1.68
Nickel 200	8902	163.4	648.1	0.33	0.006	1.44
Armco iron	7887	175.1	379.9	0.32	0.060	0.55
Carp. elec. steel	7887	289.6	338.5	0.40	0.055	0.55
1006 steel	7887	350.2	275.1	0.36	0.022	1.00
Aluminum 2024-T351	2768	264.8	426.1	0.34	0.015	1.00
Aluminum 7039	2768	336.5	342.7	0.41	0.010	1.00
4340 steel	7823	792.2	509.5	0.26	0.014	1.03
S-7 tool steel	7737	1539.0	476.4	0.18	0.012	1.00
Tungsten (.07Ni,.03Fe)	16 980	1506.0	176.5	0.12	0.016	1.00
Uranium (.0075Ti)	18 560	1079.0	1120	0.25	0.007	1.00
Tantalum (EB)	16 590	268.9	275.8	0.46	0.077	0.33
Aluminum 1100	2704	115.8	68.9	0.58	0.016	1.13
Titanium (.06Al, .04V)	4424	861.8	330.9	0.34	0.012	0.80
Aluminum 6061-T6	2704	324.1	113.7	0.42	0.002	1.34

[a] Johnson–Cook and elasticity properties from Johnson and Cook (1983), Johnson (1993).
[b] Thermal properties from the *Handbook of Chemistry and Physics* (Haynes 2014).

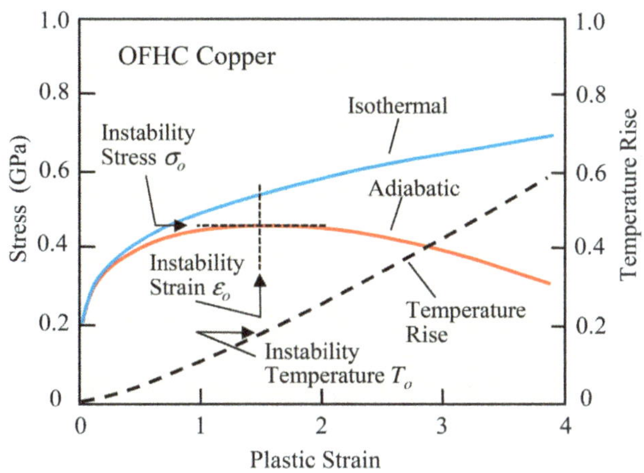

Figure 5.10. Isothermal and adiabatic stress versus plastic strain from the Johnson–Cook strength model for OFHC copper (Johnson and Cook 1983). The critical-state stress, strain and temperature are shown and identified with the onset of deformation instability. Temperature on the right-hand scale is the dimensionless term in the Johnson–Cook relation.

The critical stress, strain and temperature at instability for the Johnson–Cook metals based on the above analysis and identified as σ_o, ε_o and T_o, respectively, are provided in table 5.2. The shear stress and shear strain at criticality are calculated from $\tau_o = \sigma_o/\sqrt{3}$ and $\gamma_o = \varepsilon_o\sqrt{3}$. Again note that T_{melt} is the material absolute melt temperature, whereas T_m, which appears in the relations of the present analysis, is the temperature rise to melt from the critical temperature T_o through the expression $T_m = T_{\text{melt}} - T_o$. The critical stress, strain and temperature at instability calculated for the Johnson–Cook metals are plotted in figure 5.11. A trend is noted for the critical stress to reduce and the critical temperature to rise with increasing instability strain.

5.6.3 Predictions of adiabatic shear-band features

The previous section describes an analytic two-temperature boundary layer model that provides calculations of the thermoplastic evolution of an adiabatic shear band. The width a of the shear band is posed as a constant parameter in the model. Both the shear-band growth time $t_c(a)$ (also called the localization time) and the final energy dissipated in the shear band $W(a)$, illustrated in figure 5.7 and figure 5.9, respectively, are multi-valued functions of the shear-band width a within the model. A minimum occurs in both the time and the energy functions at a unique value a_o of the shear-band width. This optimum shear-band width provides for corresponding unique values for the localization time t_{co}, the dissipation energy W_o and the spacing of multiple shear bands b_o. Analytic relations for these optimum time, length and energy scales are provided by the previous equations (5.79), (5.80), (5.82) and (5.84).

Calculated values for this optimal shear-band width, spacing, localization time and dissipation energy for the 16 Johnson–Cook metals are tabulated in table 5.3. In addition to the thermomechanical properties provided in tables 5.1 and 5.2, a

Table 5.2. Adiabatic load-maximum instability conditions for metals.

Material	Critical strain ε_o	Critical stress (MPa) σ_o	Critical shear strain γ_o	Critical shear stress τ_o	Critical temperature (K) T_o	Melt temperature (K) T_{melt}	Temperature to melt (K) T_m
OFHC copper	1.49	457	2.58	264	409	1356	866
Cartridge brass	1.51	679	2.62	392	567	1189	622
Nickel 200	1.73	873	3.00	504	642	1726	1084
Armco iron	0.57	633	0.99	366	413	1811	1398
Carp. elec. steel	0.47	683	0.81	394	401	1811	1410
1006 steel	0.95	683	1.65	394	481	1811	1330
Aluminum 2024-T351	0.27	549	0.47	317	362	775	413
Aluminum 7039	0.32	538	0.55	311	369	878	509
4340 steel	0.34	1269	0.59	733	416	1793	1377
S-7 tool steel	0.07	2033	0.12	1174	337	1763	1426
Tungsten (.07Ni, .03Fe)	0.02	1893	0.03	1093	317	1723	1406
Uranium (.0075Ti)	0.12	1725	0.21	996	396	1473	1077
Tantalum (EB)	0.32	556	0.55	321	392	3268	2876
Aluminum 1100	1.46	202	2.53	116	418	923	505
Titanium (.06Al,.04V)	0.11	1089	0.19	629	348	1905	1557
Aluminum 6061-T6	0.46	392	0.80	226	375	925	550

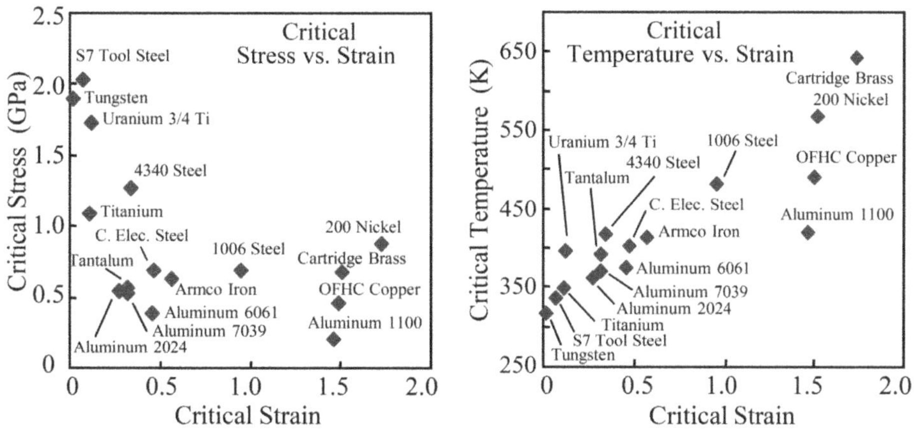

Figure 5.11. Critical stress, strain and temperature at the adiabatic load maximum for the Johnson–Cook metals. Load maximum is identified as a criterion for onset of adiabatic shear localization.

shearing strain-rate parameter $\dot{\gamma}_o$ must be specified to complete the calculation in each of the equations. A strain rate of $\dot{\gamma}_o = 10^4\,\text{s}^{-1}$ was chosen, as this rate is representative of the strain rate occurring in several applications in which adiabatic shear can be central to the failure process. One is the explosive expansion and fragmentation of metal shells. Another is the impact breach of metal plates. It is readily seen how variations in strain rate will affect the predicted properties from the functional dependence on strain rate in the respective equations.

5.6.4 Width of the adiabatic shear band

The widths of adiabatic shear bands based on the two-temperature boundary-layer model, and calculated from equation (5.79) for the present set of metals, are provided in table 5.3 and shown in figure 5.12. All material constants used in the calculations are provided in tables 5.1 and 5.2. The predicted shear-band widths for a deforming strain rate of $10^4\ \text{s}^{-1}$ range from several micrometers for metals such as titanium and uranium, to several tens of micrometers for soft aluminum and copper. The adiabatic shear-band width is reasonably insensitive to strain rate. The predicted width varies by less than a factor of two for one order of magnitude increase, or decrease, in the driving strain rate.

The present model does not address the evolution of the shear-band width during the transient thermal softening process. The model prediction relates more closely to the late time shear-band width. Also, the shear-band width is not provided by any of the fundamental length scales in equation (5.10) emerging from the dimensional analysis. Rather, the width is dependent on all three of the fundamental length scales, $a_o \sim L_1^{-3/2} L_2^{3/2} L_3$. Further, this shear-band width length scale does not surface from the stability analysis of the thermoplastic equations pursued in the earlier section. Calculations of shear-band width are in sensible agreement with the results of Xue *et al* (2006) shown in figure 5.3 for stainless steel.

Table 5.3. Calculated adiabatic shear-band properties for metals.

Material	Shear-band width (μm) a_o	Shear-band spacing (mm) b_o	Shear-band localization time (μs) t_{co}	Shear-band energy (kJ m^{-2}) W_o	Shear-band localization strain γ_o	Shear-band toughness (MPa m$^{1/2}$) K_s
OFHC copper	83.6	18.8	60.0	497	45.0	152
Cartridge brass	33.6	12.5	16.9	137	20.9	75
Nickel 200	18.3	12.8	14.6	157	34.2	108
Armco iron	22.3	15.0	24.3	222	54.6	134
Carp. elec. steel	26.8	16.0	25.6	270	51.1	147
1006 steel	26.0	15.5	24.2	247	48.1	141
Aluminum 2024-T351	27.9	13.4	7.9	56	12.6	38
Aluminum 7039	31.5	15.0	10.0	78	15.9	45
4340 steel	16.4	13.7	10.1	169	28.1	114
S-7 tool steel	11.7	12.4	5.1	124	18.0	98
Tungsten (.07Ni, .03Fe)	25.5	10.5	8.6	163	11.8	142
Uranium (.0075Ti)	6.6	5.9	3.2	31	9.4	45
Tantalum (EB)	40.1	16.1	66.9	576	89.5	199
Aluminum 1100	64.9	19.2	43.0	160	42.4	64
Titanium (.06Al, .04V)	5.6	10.7	4.0	45	25.7	44
Aluminum 6061-T6	41.3	17.0	17.2	110	23.6	55

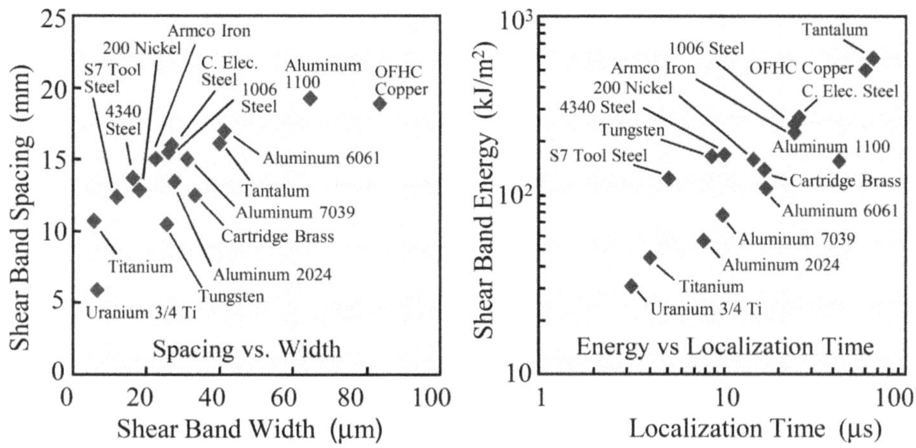

Figure 5.12. Adiabatic shear-band features, including shear-band width, spacing, dissipation energy and localization time, calculated for the Johnson–Cook metals from the two-temperature boundary layer model at a strain rate of 10^4 s^{-1}.

5.6.5 Spacing of the adiabatic shear bands

Dynamic failure of a metal through adiabatic shear is seldom accomplished through a single adiabatic shear band. More commonly, failure proceeds through a collection or network of shear bands in order to accommodate the transient dissipation required of the event. The situation is analogous to the dynamic tensile failure of a material through propagating fractures and the subsequent fragmentation of the body. Thus, the density, or population, of shear bands necessary to accommodate the adiabatic shear failure event is of interest. This feature of adiabatic shear is reasonably captured through calculation of the typical spacing of adiabatic shear bands in a region undergoing sensibly uniform dynamic shearing. This size measure is equivalent, for example, to the characteristic fragment size in a dynamic fragmentation event. A reasonably uniform dynamic shearing is produced experimentally in tests such as the imploding cylinder experiments conducted by Xue *et al* (2003, 2004) and illustrated earlier. The experimental photograph in figure 5.2 illustrates the collective adiabatic shear necessary to accommodate failure in these tests.

The typical spacing of shear bands for the Johnson–Cook metals, at a driving strain rate of 10^4 s^{-1}, is calculated from the model through equation (5.84). The calculated shear-band spacing is also tabulated in table 5.3 and plotted against shear-band width on the left in figure 5.12. Calculated values range from about 6 mm for uranium to about 19 mm for soft aluminum and copper. A reasonably strong dependence of shear-band spacing on strain rate to the inverse three-fourth power is predicted from the theory. This shear-band spacing is functionally the same relation assessed from the maximum growth rate from the stability analysis (Grady 1977, 1980) described in an earlier section. The same strain-rate dependence has been reported by Wright and Ockendon (1996), also obtained through stability analysis.

It is of interest to note that comparable strain-rate dependence is observed in the characteristic size of fragments resulting from failure through dynamic fracture in both experimental testing and theoretical analysis. The strain-rate dependence of fragment size for metals is discussed in chapters 2 and 3, and range from an inverse two-thirds to inverse first-power dependence. (See also Grady 2006, 2009b.)

Comparisons with measured experimental densities of adiabatic shear bands in stainless steel (Xue *et al* 2002, 2004) suggest that the present model over-predicts the typical spacing of initially activated shear bands by about one order of magnitude. The theory, however, is in sensible agreement with the spacing of the surviving, longer-lasting shear bands subsequent to failure self-correlation, and accommodating most of the inelastic deformation of the test samples. Application of the present models of adiabatic shear to the experimental work of Xue *et al* (2003, 2004, 2006) is pursued further later in the present section. Zhou *et al* (2006) have made a similar observation in the computational simulations of multiple adiabatic shear-band formation. Again, adiabatic shear-band spacing is not provided by any of the three fundamental length scales in equation (5.10). Instead, a dependence of $b_o \sim L_1^{-1/2} L_2^{1/2} L_3$ is determined.

5.6.6 Localization strain in the adiabatic shear band

A finite shear strain within the shear band is required to accommodate the adiabatic heating process and thermally soften the shear stress. Within the two-temperature model this plastic strain is achieved within the localization time t_{co}. Within this time span the band is heated through the temperature rise to melt T_m and shear stress in the band is reduced to zero. The localization strain is calculated from the model through the relation $\gamma_l = 2\psi_{co}/a_o$ where ψ_{co} is the transverse displacement provided from equation (5.68) by the relation $\psi_{co} = W_o/\tau_o$. The calculation yields for the localization strain

$$\gamma_l = 4\rho c T_m/\tau_o = 4Q, \tag{5.86}$$

where Q is the second of the two dimensionless shear localization numbers developed in an earlier section.

Calculated values for the localization strain are provided in table 5.3 and range from about $\gamma_l = 10$ for tungsten and uranium to as high as $\gamma_l = 90$ for tantalum. Shear strain will continue subsequent to the predicted localization strain until global deformations consistent with the failure event are accommodated. To the extent that adiabatic shear-band features calculated through the model can be associated with a shear-band process zone at the tip of a propagating shear band, the present localization strain identifies the shear-band plastic strain immediately following passage of the process zone.

5.6.7 Solid Prandtl number features of the adiabatic shear band

The ratio of the adiabatic shear-band width to the shear-band spacing, a_o/b_o, has several implications to the shear localization process. The ratio provides a measure of the fraction of material actively deforming following localization in the failure event. Further, this ratio should be sensibly small for validity of the boundary-layer assumption in the two-temperature model development. This ratio of length scales is dimensionless and consequently should be a function of the two dimensionless numbers P and Q provided by equations (5.13) and (5.14) in the earlier dimensional assessments. Joining the two relations for a_o and b_o yields a functional dependence only on the solid Prandtl number,

$$a_o/b_o = \sqrt{1/2P} = \sqrt{\rho\chi\dot{\gamma}_o/2\tau_o}. \tag{5.87}$$

The solid Prandtl number P calculated for a strain rate of $10^4\ \mathrm{s}^{-1}$ ranges in magnitude over about two decades for the metals provided in the tables. Values range from about 2.5×10^4 for OFHC copper to 180×10^4 for the titanium alloy. For all of the metals the ratio, a_o/b_o, again for a strain rate of $10^4\ \mathrm{s}^{-1}$, are slightly less than a few times 10^{-3}, indicating that in a fully formed adiabatic shear-localization process, deformation is accommodated by a very small fraction of the material.

The solid Prandtl number can be written as $P = \dot{\gamma}_p/\dot{\gamma}_o$ where $\dot{\gamma}_p = \tau_o/\rho\chi$ is a characteristic Prandtl strain rate. This characteristic rate is of order $10^9\ \mathrm{s}^{-1}$ to $10^{10}\ \mathrm{s}^{-1}$, indicating that the present boundary-layer model is appropriate over all strain rates of practical interest, including rates commensurate with structured shocks in metals. A later plot of the metal data will be provided in terms of the Prandtl time defined as $t_p = \dot{\gamma}_p^{-1} = \rho\chi/\tau_o$.

5.6.8 Adiabatic shear effectiveness of the metals

Dimensional analysis in an earlier section led to two dimensionless parameters that govern the thermoplastic localization within the restricted set of thermal and mechanical properties selected. First, was the Prandtl number $P = \tau_o/\rho\chi\dot{\gamma}$. The second was the material heating number $Q = \rho c T_m/\tau_o$. The latter is shown in the previous subsection to provide the extent of plastic strain accommodated by the shear band during the localization event. The Prandtl number determines the fraction of material participating in the shear-band deformation.

Intuition gleaned from the earlier analysis of adiabatic shear localization suggests that materials that have a higher Prandtl number and a lower heating number would be more effective in failure through adiabatic thermoplastic shear localization. There is always interest in attempting to order material according to their propensity for exhibiting a physical effect of significance. A plot of the Johnson–Cook metals in terms of the calculated Prandtl number P and heating number Q is provided in figure 5.13.

To eliminate a dependence on strain rate, the abscissa coordinate is the Prandtl time defined as $t_p = (\dot{\gamma}P)^{-1} = \rho\chi/\tau_o$. This plot for the metals is not found to sharply order the materials as to adiabatic shear effectiveness. Nonetheless, common experience indicates that the more effective shear-banding materials reside towards the lower left of the plot, whereas those metals towards the upper right are not commonly effective shear-banding materials. The tendency to fall on two curves in this plot is not understood.

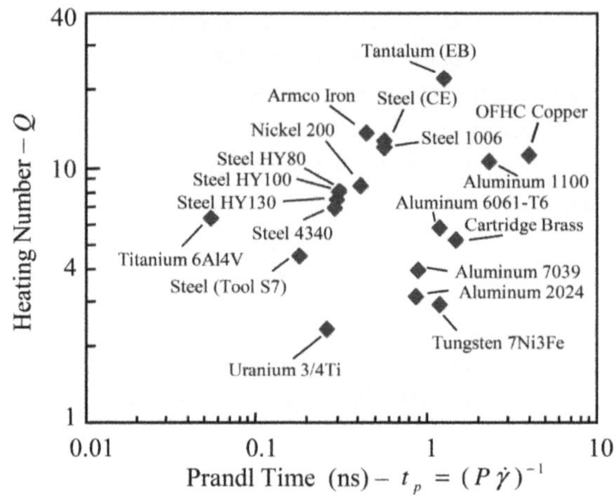

Figure 5.13. Ordering of the metals based on fundamental thermomechanical constants.

5.6.9 Localization time for the adiabatic shear band

The evolution to a fully developed adiabatic shear band subsequent to the onset of instability occurs over a characteristic time. When viewed as a propagating shear band this same characteristic time corresponds to the passage of the shear-band process zone past a material point. In the analytic formulation of the present two-temperature boundary-layer model that results in closed-form equations for the features of the shear band, this characteristic time is called the localization time, and is provided by equation (5.80).

That a finite time is required to develop an adiabatic shear band is a consequence of local inertia within the deforming region neighboring the shear band, and the concomitant time necessary to flux momentum to the shear band from the nearby regions. Under the assumption of a rigid–plastic model of the material at the stress–strain load maximum, momentum is transported to the shear band through stress diffusion. The characteristic shear-band spacing provides a measure of the length scale over which stress diffusion must take place. Momentum transport proceeds through propagation of a diffusive Mott wave. When addressed in more depth, elastic stress wave propagation also contributes to the momentum transport and should be accounted for in a more detailed model. Elasticity in the shear-localization process is considered later in the present section.

The localization time t_{co} for the Johnson–Cook metals calculated from equation (5.80) are provided in table 5.3 and are also plotted in figure 5.12 based on a shear strain rate of 10^4 s^{-1}. Localization times range over a few microseconds to a few tens of microseconds. The functional dependence on strain rate provided by equation (5.80) shows that localization time decreases as the loading strain rate increases.

5.6.10 Energy dissipation in the adiabatic shear band

In the present two-temperature model a thin layer or band of the metal experiences accelerated plastic deformation subsequent to onset of thermoplastic instability.

Optimum conditions for growth of adiabatic shear predicted by the model are achieved when the band has a characteristic width of a_o as provided by equation (5.79). At this optimum width both the localization time t_{co} and the energy W_o dissipated in the shear band achieve minimum values. Plots of the localization time and energy dissipation as a function of the shear-band width in figures 5.7 and 5.9 illustrate the functionality predicted by the model. The minimum energy W_o is identified as the shear-band dissipation energy and provides a measure of the resistance of the metal to failure through adiabatic shear localization. The shear-band energy has dimensions of energy per unit area and is a consequence of dissipation through plastic work within the shear band during the localization event. For the present two-temperature model this dissipated energy equally partitions into heating of the shear band and heat transfer to neighboring material through thermal conduction during the localization event. The source of this dissipated energy is from the transport of energy to the shear band from neighboring material on the microscale. Within constraints of the rigid–plastic material response assumed in the model, energy is transported from the nearby kinetic energy field through stress diffusion. More generally, both kinetic energy and elastic strain energy will contribute in fueling the shear-band dissipation. When viewed as a propagating shear band, the energy W_o is dissipated within the process zone at the front of the advancing shear band. Shear-band energies calculated from equation (5.82) for the Johnson–Cook metals are provided in table 5.3 and range over several tens to several hundreds of kJ m^{-2}. The same shear-band energies are plotted on the right in figure 5.12. The lowest shear-band energies are predicted for the metals known to have a proclivity for failure through adiabatic shear.

A further observation concerning the shear-band energy is of interest. When failure occurs through adiabatic shear, a network of shear bands commonly activates and cooperates in accommodating the failure event. The necessary density of adiabatic shear bands is captured through the model by the length scale b_o, calculated from equation (5.84), which provides a measure of the nominal spacing of shear bands in the region of the material undergoing failure through a network of shear bands. The specific energy (per unit mass) of dissipation due to collective adiabatic shear within the region is provided by

$$E_o = \frac{W_o}{\rho b_o} = cT_m\left(\frac{2\rho\chi\dot{\gamma}_o}{\tau_o}\right)^{1/2} = cT_m\sqrt{\frac{2}{P}}, \tag{5.88}$$

where again P is the solid Prandtl number considered previously. This specific shear-band energy increases with strain rate with the same dependence as the localization time t_{co} decreases with strain rate. Also recall that the thermal softening coefficient $\alpha \simeq 1/T_m$ within the constraints of the model.

The product of this specific energy and the localization time is independent of strain rate and is given by (Grady and Kipp 1987)

$$A_o = E_o t_{co} = 12\chi\left(\frac{\rho c T_m}{\tau_o}\right)^2 = 12\chi Q^2. \tag{5.89}$$

The property A_o has dimensions of specific action (e.g. Goldstein 1965). Similar invariance has been observed for other solid dynamics processes including pressure-induced reaction of energetic materials (Walker and Wasley 1969), impact-induced spall and dynamic fragmentation (Steverding and Lehnigk 1976, Grady and Lipkin 1980), and structured steady waves in the shock compression of metals (Grady 1981b, Swegle and Grady 1985). This fundamental property associated with the dynamic failure of solids is visited further in chapter 8.

5.6.11 Toughness of the adiabatic shear band

The similarities of adiabatic shear to dynamic fracture are apparent. The propagation of an adiabatic shear band clearly mimics the expected behavior of a mode II or a mode III fracture. It has become common practice to characterize the resistance of solid material failure through fracture by a fracture toughness property of the material. The fracture toughness of many materials has been measured, and the underlying theoretical fracture mechanics basis is sound. The mode I critical stress intensity factor K_{IC} (fracture toughness) and the corresponding strain energy release G_{IC} are related (within a function of Poisson's ratio of order unity) through $G_{IC} = K_{IC}^2/E$ with E the elastic modulus (e.g. Lawn and Wilshaw 1975). The fracture property G_{IC} provides a measure of the energy dissipated per unit area of the advancing fracture front.

A parallel correspondence can be proposed relating the shear-band energy W_o to a property identified as the shear-band toughness K_s through $W_o = K_s^2/G$, with G the elastic shear modulus (Grady 1992, 1994). Although there is no apparent theoretical justification for this exercise, the effort uncovers intriguing questions concerning the relationship between dynamic fracture and adiabatic shear in metals. At a minimum, the conversion couches the shear-band energy derived from the model in the more familiar units of fracture toughness. Repeating the specific relations used in deriving this measure of shear-band toughness, the shear-band energy developed from the two-temperature model resulting in equation (5.82) is

$$W_o = 2\rho c T_m a_o = 2\rho c T_m \left(\frac{72\rho^3 c^2 \chi^3 T_m{}^2}{\tau_o{}^3 \dot{\gamma}} \right)^{1/4}, \tag{5.90}$$

and the shear-band toughness is then defined through

$$K_s = \sqrt{G W_o}. \tag{5.91}$$

Using the above two relations, and the properties for the present set of metals provided in the earlier tables, values for K_s are collected in the final column of table 5.3. It is observed that these toughness values are within a factor of two or three of, and even show a tendency to correlate with, the quasi-static mode I fracture toughness values for the same metals. It is worth pondering what relationship, if any, this unique collection of familiar continuum thermomechanical properties provided by equation (5.90) has with the fracture toughness of ductile and semi-ductile metals.

A sensible explanation has been suggested (Grady 2009a) and is described in the concluding section of this chapter.

5.6.12 Adiabatic shear failure in stainless steel

The experimental investigations of Xue *et al* (2001, 2003, 2004, 2006) are unique in providing constraints for models of adiabatic shear deformation in metals. Particularly encompassing is the collection of data on 304 stainless steel. The experimental results detail the activation and growth of a network of adiabatic shear bands necessary to accommodate failure in well-characterized dynamic loading experiments (Xue *et al* 2003, 2004). Further, detailed dynamics and morphology of individual shear bands have been examined on the same steel (Xue *et al* 2006). Variations in metallurgy are explored, as is the influences of surface conditions at the site of shear-band activation, and the effect of shock pre-stress. Material properties and appropriate constitutive models are provided for the steel. (A Johnson–Cook model for 304 stainless steel is available in Xue *et al* (2003).)

These seminal data provide an opportunity to test the analytic adiabatic shear-localization models described in the present chapter. Comparisons of the test data of Xue *et al* are made with results of the adiabatic shear failure stability analysis in an earlier section.

The Johnson–Cook constitutive model provided for 304 stainless steel yields an adiabatic load-maximum critical state of $\tau_o = 810$ GPa, $T_o = 585$ K and $\gamma_o = 1.4$. Using the appropriate thermal properties, equation (5.84) for the shear-band spacing provides a prediction of 3.2 mm at the reported strain rate of $6 \times 10^4\ \mathrm{s}^{-1}$ for the cylinder implosion tests. This value for the spacing is about 30% smaller than that provided by the functionally identical relation with a slightly different coefficient resulting from the stability analysis in equation (5.38).

Calculation of the shear-band spacing does not differ markedly from the same calculation performed by Xue *et al* (2004) using an earlier presentation of the two-temperature model (Grady and Kipp 1985). (The authors report a calculated shear-band spacing of 2.4 mm from the two-temperature model). As noted earlier, the predicted spacing appears to have relevance to the surviving late-time shear bands after self-organization arrests the preponderance of incipient shear bands activated at the inner free surface of the test cylinder. Zhou *et al* (2006) show comparable agreement and reach a similar conclusion from the computational simulations of multiple adiabatic shear-band formation. Nominal spacing of these incipient shear bands was found to be remarkably close to the cut-off wavelength predicted from the stability analysis discussed in section 5.4.

The driving strain rate is calculated to decrease with increasing radius from the inner to outer circumference of the test cylinder, but not markedly (a factor of 2 to 3). An increase in shear-band spacing should mirror this drop in strain rate. Experimental shear-band spacing shown in figure 5.2 supports this trend. Shear bands are not observed to propagate to the outer cylinder surface. Presumably, deformation load maximum and onset of shear-band instability are not achieved in the outer domain of the cylinder.

The two-temperature model also provides a measure of the width of a mature adiabatic shear band. The top-hat Hopkinson bar experiments of Xue *et al* (2006) on 304 stainless steel provide the shear-band morphology shown in figure 5.3. The governing strain rate representative of conditions leading to shear-band formation is less easily assessed in these tests. The deformation rate of the top-hat specimens in the Hopkinson bar tests is reported to be $2 \times 10^3\ s^{-1}$. The authors estimate the strain rate within the band following full formation to be approximately $10^5\ s^{-1}$. A sensible estimate of strain rate in the shear deforming material preceding onset of localization is about $10^4\ s^{-1}$. Prediction of shear-band width is not overly sensitive to this strain rate. At a strain rate of $10^4\ s^{-1}$ equation (5.79) predicts a shear-band width of approximately 15 μm, in reasonable accord with the micrograph images in figure 5.3. A decade change in strain rate results in less than a factor of two change in the predicted shear-band width. The two-temperature model does not capture the apparent growth in the shear-band width with increasing deformation suggested by the images.

The Johnson–Cook model for 304 stainless steel provided in the earlier paper of Xue *et al* (2003) would not be appropriate for the pre-shocked material tested in Xue *et al* (2006). Accounting for the pre-strain, the stress and stress hardening is shown by the authors to sensibly approximate the untreated material. Intuition suggests that the critical state stress τ_o will be somewhat higher and the critical state temperature T_o somewhat lower than that resulting from the Johnson–Cook model for the untreated stainless steel. The differences are unlikely to be substantial, but would result in modest differences in calculated shear-band features.

The stability analysis outlined in an early section and the subsequent two-temperature model of adiabatic shear localization are complementary in the sense that the same thermal and mechanical material properties governing the shear-band physics are applied in each of the analytic models. Two independent physical length scales emerge from each of the models of adiabatic shear. Dimensional analysis governing the material properties considered appropriate to the physics of thermoplastic adiabatic shear, and used in both models, reveals three independent length scales.

Stability analysis and the two-temperature model yield one length scale in common. Namely, the wavelength of maximum growth rate in the former, and the characteristic spacing of adiabatic shear bands in the latter. Both yield length scales in the 2–4 mm range under loading rates applicable to the experiments of Xue *et al* (2003, 2004) on stainless steel cylinders. This length scale appears consistent with the spacing of the mature, propagating and self-organized shear bands that optimally dissipate the energy imparted to the body in the dynamic event.

The two-temperature model yields a second length scale. This length scale provides the characteristic width of a fully developed adiabatic shear band. The analytic form for this length scale results from a minimum of the time to full localization or, equivalently, a minimum of the plastic work dissipated in the shear band. This length scale ranges from a few micrometers to a few tens of micrometers, for the metals investigated, and is consistent with observed shear-band widths.

A prediction of approximately 15 μm for stainless steel is in line with the experimental observations of Xue *et al* (2006).

A further length scale emerges from the stability analysis of adiabatic shear that does not appear in the two-temperature model. This length scale is a cut-off wavelength that establishes a boundary between stable and unstable thermoplastic deformation. This cut-off length scale resides in the hundreds of micrometers range for the present Johnson–Cook metals, and is intermediate between predicted shear-band widths and shear-band spacings.

Calculation of this cut-off length scale for stainless steel (see figure 5.4) falls remarkably close to the approximately 150 μm nominal spacing of the incipient shear bands that dominate the near-surface volume of material at the activating inner circumference of the stainless steel cylinders tested by Xue *et al* (2003, 2004). A preponderance of unstable incipient shear bands preceding adequate communication through stress diffusion is reasonable in the early formation phase of shear localization. Thus, the cut-off wavelength emerging from the linear stability analysis may provide a sensible spacing of the incipient shear bands preceding the stress wave correlation and self-organization phase leading to the late-time spacing of shear bands.

The shear-band toughness for stainless steel can be calculated from the relation derived from the two-temperature model of adiabatic shear provided through equations (5.90) and (5.91). A shear-band toughness of 75–100 MPa $m^{1/2}$ is calculated over the range of strain rates of the several experimental tests performed by Xue *et al* (2003, 2004, 2006) on stainless steel ($\sim 10^4$ s^{-1} to 10^5 s^{-1}). The values are again remarkably close to an approximately 100 MPa $m^{1/2}$ fracture toughness reported for stainless steel. Explanations for this thought-provoking commonality between the calculated shear-band toughness and measured fracture toughness are pursued in a later section.

The characteristic time required to develop a mature shear band is determined by the time required to transport momentum and energy to the shear band from nearby material of a magnitude sufficient to achieve full plastic softening. Within the two-temperature model, this transport is achieved through the propagation of diffusive plastic Mott waves. An analytic expression for the localization time t_{co} is provided by equation (5.80). A localization time t_{co} of 8–10 μs is calculated for strain rates comparable with the forced-shear localization experiments of Xue *et al* (2006) on stainless steel. Measurements in the tests of Xue *et al* (2006) reveal a load rise to peak, identified by the authors as onset of localization, followed by rapid load loss that is approximately exponential. Accepting a drop of one exponential power as a measure of the shear-band localization time, an experimental value of approximately 15 μs is in sensible agreement with the theoretical localization time.

5.6.13 Elasticity in the adiabatic shear-band process

In the analysis of adiabatic shear to this point, elasticity and the possible participation of elastic strain energy in the shear localization event has been ignored. In this subsection the relative importance of elasticity, and how elasticity would influence the relations that predict the fundamental length, time and energy scales intrinsic to the adiabatic shear failure event, are assessed (Grady 2011). Using physical insights that have been provided by the previous more detailed analysis of the two-temperature boundary layer model, a more terse analysis based on elementary energy and momentum arguments is pursued to provide order-of-magnitude solutions and functional forms for the shear-band features that include elasticity of the material. Previous relations for shear-band width and spacing based on a strictly rigid thermoplastic response are again derived, but with the inclusion of elastic properties of the material. The relative importance of elasticity in the adiabatic shear of materials is then assessed.

Consider first the energy required to create a fully formed planar shear band of thickness a that is elevated from the temperature at onset of localization to the melt temperature. The required thermal energy per unit area of the planar region is

$$\mathcal{E}_r = \rho c T_m a, \tag{5.92}$$

where ρ is the material density, c is the specific heat and T_m is the temperature change from localization onset to melt. In earlier analysis, within the framework of the two-temperature model of adiabatic shear, it is found that for a shear band near optimum width, twice the adiabatic dissipation of equation (5.90) is required to account for both temperature rise and conductive heat loss over the shear-band formation time. Thus, accounting for conductive heat loss, the energy in equation (5.92) applies to one-half of the shear band, as illustrated in the sketch on the left in figure 5.14.

The shear-band energy in equation (5.92) must be supplied by energy residing in the material near the shear band. The material is initially deforming in shear at a strain rate $\dot{\gamma}_o$ and is on the yield or failure surface at a shear stress τ_o. The strain rate $\dot{\gamma}_o$, shear stress τ_o and an elastic shear modulus G, are intrinsic to the deformation dynamics in the neighborhood of the shear band. Energy near the shear band resides

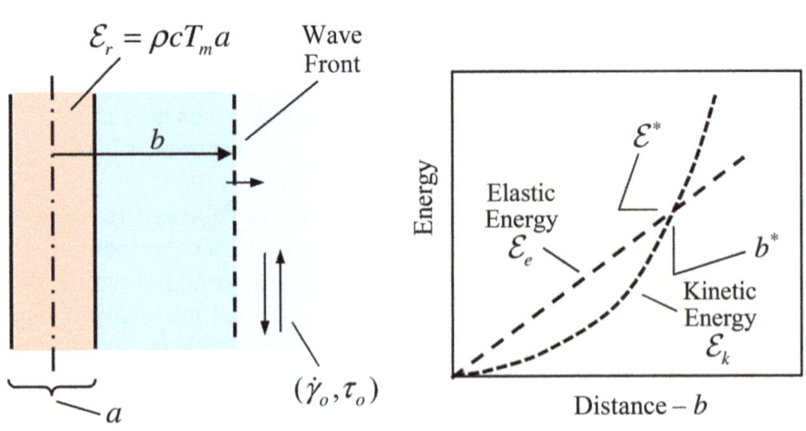

Figure 5.14. Left: a two-temperature shear-band model and the energy required to heat the shear band to the melt temperature. Right: on the right are plots of the elastic strain energy and kinetic energy residing in the material neighboring the shear band that will fuel shear-band growth.

as both kinetic energy and elastic strain energy. The cumulative energies of each contribution as a function of the normal distance b from the shear band are plotted on the right in figure 5.14.

The strain energy is (in units of energy per unit area)

$$\mathcal{E}_e = \frac{1}{2}\frac{\tau_o^2}{G}b, \tag{5.93}$$

while the kinetic energy is given by

$$\mathcal{E}_k = \frac{1}{6}\rho\dot{\gamma}_o^2 b^3. \tag{5.94}$$

The available strain energy increases linearly with distance b from the shear band. The kinetic energy has a cubic dependence on the distance b. The energies are equal at a distance

$$b^* = \sqrt{\frac{3\tau_o^2}{\rho G\dot{\gamma}_o^2}}, \tag{5.95}$$

as illustrated in the figure and the total energy within this distance to

$$2\mathcal{E}^* = \sqrt{\frac{3\tau_o^6}{\rho G^3\dot{\gamma}_o^2}}. \tag{5.96}$$

If the energy $\mathcal{E}_r = \rho c T_m a$ from equation (5.92), required for full adiabatic localization of the shear band, significantly exceeds $2\mathcal{E}^*$, then the shear band is fueled predominantly by the kinetic energy field. This is the premise of much of the earlier analysis, where strictly rigid plastic response of the material is assumed. If, on the other hand, the required shear-band energy is less than $2\mathcal{E}^*$, then the neighboring elastic strain energy is the larger and would be the principle energy fueling growth of the shear band. For most metals, the kinetic energy source appears to be the more important. This dichotomy, and transport of the requisite energy, is aptly accommodated through the Lee (1967) solution of the Mott wave considered in chapter 2.

Consider first the case where the kinetic energy from equation (5.94) is the dominant source fueling shear localization. If kinetic energy residing within a distance $b > b^*$ is solely required to fuel shear localization, then momentum diffusion requires that localization will occur over a time

$$t_f \simeq \rho\dot{\gamma}_o b^2/\tau_o. \tag{5.97}$$

Thermal diffusion necessitates that the width of the shear band relates to this localization time according to

$$a \simeq \sqrt{\chi t_f}, \tag{5.98}$$

where χ is the thermal diffusivity of the material. The characteristic time as written in equation (5.97) is provided through the solution of the Mott plastic diffusion equations. Alternatively, a straightforward application of Newton's second law provides the same result.

Eliminating the localization time between equations (5.97) and (5.98) provides a relation between the shear-band width a and the transport distance b,

$$a = \sqrt{\frac{\rho\chi\dot{\gamma}_o b^2}{\tau_o}}. \tag{5.99}$$

Equating the required shear-band energy from equation (5.92) to the kinetic energy in equation (5.94) provides a second relation between a and b. Solving the two equations relating a and b provides unique relations for both the shear-band width,

$$a_o \simeq \left(36\frac{\rho^3 c^2 \chi^3 T_m^2}{\tau_o^3 \dot{\gamma}_o}\right)^{1/4}, \tag{5.100}$$

and the transport distance,

$$b_o \simeq \left(36\frac{\rho c^2 \chi T_m^2}{\tau_o \dot{\gamma}^3}\right)^{1/4}, \tag{5.101}$$

where the subscripted a_o and b_o signify the optimum shear-band width and transport distance predicted from the theory. The transport distance of equation (5.101) can again be regarded as a measure of the spacing of multiple shear bands (actually twice this distance if one considers the transport distance of two interacting shear bands). The above elementary energy-based analysis again provides estimates of shear-band width and spacing that are functionally the same and remarkably close in magnitude to the more detailed analysis presented in the earlier sections.

A comparable analysis can be undertaken when the required shear-band energy $\rho c T_m a$ is less than $2\mathcal{E}^*$, and it is assumed that elastic strain energy predominantly fuels the shear-localization process. Energy is then transported to the shear band through elastic wave propagation and the localization time is of the order

$$t_f \simeq \frac{b}{\sqrt{G/\rho}}, \tag{5.102}$$

where $c_s = \sqrt{G/\rho}$ is the elastic shear wave speed. The same analysis leading to equations (5.100) and (5.101) now provides for the shear-band width

$$a_o \simeq \left(\frac{4G\rho^3 c^2 \chi^2 T_m^2}{\tau_o^4}\right)^{1/2}, \tag{5.103}$$

and the transport distance

$$b_o \simeq \frac{4G^{3/2}\rho^{5/2} c^2 \chi T_m^2}{\tau_o^4}. \tag{5.104}$$

A comparison of the two solutions independently assuming kinetic energy fueled shear localization on one hand, and elastic strain energy localization on the other, is illustrated for 4340 steel in figure 5.15.

The transport transition distance as a function of strain rate, at which strain energy and kinetic energy contribute equally, is provided by equation (5.95) and is shown as the dashed curve as a function of strain rate in the figure. The shear-band energy requirement and transport distance for 4340 steel exceeds the transition distance for all strain rates in excess of a point slightly above $10^3\ s^{-1}$. The solid curve transport distance above about $10^3\ s^{-1}$ is provided by equation (5.101). The strain-rate-independent portion of the solid curve below $10^3\ s^{-1}$ is given by equation (5.104). The same calculation using properties for 1100 aluminum identifies the corresponding transition strain rate at slightly more than one order of magnitude lower ($\sim 10^2\ s^{-1}$). Thus, at the higher strain rates, shear localization is fueled principally by the neighboring kinetic energy field and the spacing of multiple shear bands is expected to exhibit a strong dependence on strain rate as is predicted by equation (5.101). Conversely, at lower loading rates strain energy is predicted to play an increasingly important role in shear localization. The transition strain rate from kinetic-energy to strain-energy dominated shear localization for any material is calculated by joining equations (5.95) and (5.101),

$$\dot{\gamma}_t = \frac{1}{4}\frac{\tau_o^5}{\rho^3 c^2 T_m^2 G^2 \chi}. \tag{5.105}$$

At strain rates less than about $10^3\ s^{-1}$ elastic strain energy is predicted to dominate the shear localization event for 4340 steel. The transport distance and characteristic shear-band spacing is constant independent of strain rate in this range. It is also

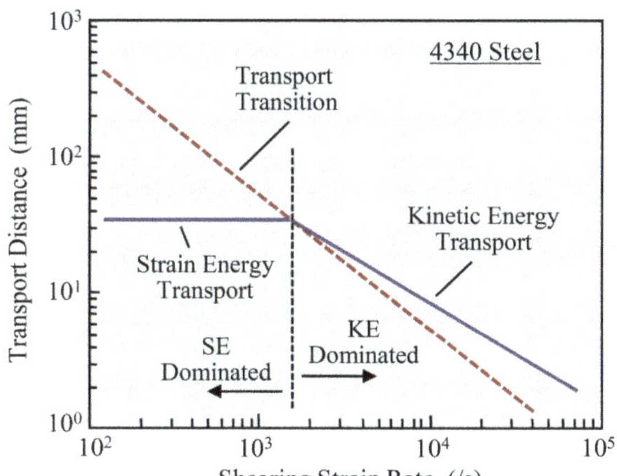

Figure 5.15. Distance of energy transport to the adiabatic shear band and transition curve from strain-energy to kinetic-energy dominated localization. Based on properties for 4340 steel. The constants used in the calculation for 4340 steel are $\rho = 7820\ \mathrm{kg\ m^{-3}}$, $c = 480\ \mathrm{J\ kg^{-1}} \cdot K$, $\chi = 27\ \mathrm{mm^2\ s^{-1}}$, $\tau_o = 730$ MPa and $T_m = 1370$ K. The corresponding properties for 1100 aluminum are $\rho = 2700\ \mathrm{kg\ m^{-3}}$, $c = 904\ \mathrm{J\ kg^{-1}} \cdot K$, $\chi = 98\ \mathrm{mm^2\ s^{-1}}$, $\tau_o = 120$ MPa and $T_m = 505$ K. The temperature rise T_m from onset of shear instability to melt is calculated from a Johnson and Cook (1983) plastic strength model.

apparent that in the neighborhood of a strain rate of $10^3\ s^{-1}$ both elastic and kinetic energy sources contribute comparably. At strain rates of $10^4\ s^{-1}$ to $10^5\ s^{-1}$, achieved in many impact events, the elastic strain energy contribution to shear localization for 4340 steel and other comparable metals is small but certainly not negligible. Further, even when kinetic energy dominates the full shear localization process, early growth of the shear band should reflect the larger component of nearby elastic strain energy. The strong dependence of the transition strain rate on the supporting shear stress in equation (5.105) is also noted.

The present analysis provides a method for contrasting adiabatic shear fueled by local kinetic energy versus that fueled by elastic strain energy. The analysis also points to a middle region at which kinetic and strain energy are comparable, and how this mid-region will depend on the dynamics of the loading that leads to shear localization. The analysis so far does not provide solutions for the adiabatic shear length scales when both kinetic and strain energy explicitly contribute.

This analysis exploits momentum conservation and energy balance to estimate shear-band width and spacing length scales. Key to the solution is the characteristic time required to flux momentum to the plane of the developing shear band. Equation (5.97) provides the momentum diffusion time when elasticity does not play a role. When elasticity dominates, momentum transport is governed by equation (5.102). Momentum transport when both elastic strain energy and kinetic energy participate in comparable proportions is suggested by a solution of the Mott problem provided by Lee (1967; see also Grady 2006, 2009b). The analysis of Lee was performed in 1943 in support of Mott's calculations for dynamic fragmentation (Mott 1943), but not openly published until several decades later. The Mott and Lee solutions in relation to dynamic fragmentation are discussed in detail in chapter 2.

Lee solves the problem in which an elastic–perfectly-plastic body with elastic modulus G is undergoing uniform one-dimensional deformation at a rate $\dot{\gamma}$ at a constant flow stress τ_o. At time zero, the stress at a point is reduced to zero (replicating instantaneous fracture). Concurrently, the solution provides the motion history of the fracture boundary—the requisite momentum transport. The Lee problem is the elastic–plastic equivalent of the rigid–plastic problem solved by Mott.

Mott and Lee were envisioning a body deforming in one-dimensional tension. Their solutions are equally applicable to comparable deformation in shear. The solution for the front of the propagating release wave is shown for both the Mott and the Lee solution on the left in figure 5.16. The parabolic wave front results from the rigid–plastic analysis of Mott. The elastic–plastic wave front from the Lee solution results from a more complex reverberation of elastic acceleration waves that is found to be enveloped by the Mott solution.

The key point for the present application is that early-time momentum transport is impeded by elasticity of the body. Late in time, plastic diffusion limits momentum transport. The Mott and Lee solutions suggest a characteristic momentum transport time as the sum of equations (5.97) and (5.102),

$$t_f \simeq \frac{1}{\sqrt{G/\rho}} b + \frac{\rho \dot{\gamma}_o}{\tau_o} b^2. \tag{5.106}$$

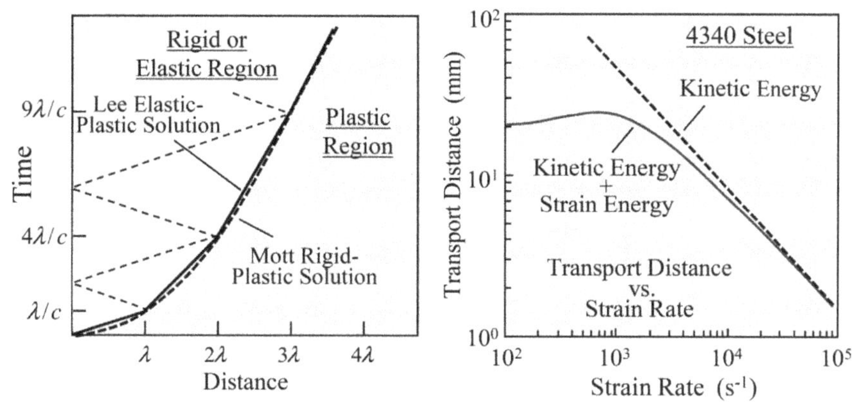

Figure 5.16. Left: wave fronts separating the plastic region and the rigid or elastic region from solutions of Mott (1943, 1947) and Lee (1967). Right: the two-temperature model predictions of shear-band transport distance (or shear-band spacing) based on properties for 4340 steel.

The use of some appropriate averaging of the two transport times does not significantly alter the analysis. The strain and kinetic energy available for shear localization within this transport time is then

$$\mathcal{E} = \frac{1}{2}\frac{\tau_o^2}{G}b + \frac{1}{6}\rho\dot{\gamma}_o^2 b^3. \tag{5.107}$$

The same algebra used in the previous analysis yields the following relation for a transport distance sufficient to fuel localization,

$$\frac{\tau_o^2}{2G\rho c T_m}b_o + \frac{\dot{\gamma}_o^2}{6cT_m}b_o^3 = \sqrt{\frac{\chi}{\sqrt{G/\rho}}b_o + \frac{\chi\rho\dot{\gamma}}{\tau_o}b_o^2}. \tag{5.108}$$

The transport distance (approximate shear-band spacing) from equation (5.108) is compared with that assuming strictly rigid–plastic, kinetic-energy-driven localization from equation (5.101) on the right in figure 5.16. The curves provide the dependence on shearing strain rate. Again, the properties are for 4340 steel. The two curves do not differ significantly in the 10^4 s^{-1} to 10^5 s^{-1} strain-rate range, as was inferred previously. The two curves differ markedly at strain rates of order 10^3 s^{-1} and lower.

The two-temperature model of adiabatic shear localization, based on elementary momentum and energy concepts, suggests that for 4340 steel, strain energy is increasingly important to the localization process below strain rates of a few times 10^3 s^{-1}. Further, the model also provides an approximate, but reasonable, estimate (~20 mm) of the low-strain-rate shear-band spacing assuming the body undergoing deformation accommodates regions of near-uniform shear of this spatial scale. Some test methods used to investigate adiabatic shear do not accommodate such length scales. Hopkinson bar experiments, for example, typically entail test specimens of sizes of order of tens of millimeters that are subjected to deformation at rates in the neighborhood of 10^3 s^{-1}. Failure through adiabatic shear in Hopkinson bar experiments (e.g. Dormeval and Ansart 1985, Stelly and Dormeval 1986) would, according to implications of the present model, exhibit structural failure that is largely elastic strain energy fueled. The model also supports the conjecture of Lemaire and Backofen (1972) that adiabatic shear in metal machining is governed primarily by the release of elastic strain energy. A comparable observation was made by Rogers (1974) that, when adiabatic shear failure occurs in compact specimens at modest loading rates, both sample strain energy and compliance of the loading apparatus play a role in supplying the requisite energy.

5.6.14 Adiabatic shear in brittle solids

Much less is known about the prevalence and the mechanisms of shear localization deformation in brittle solids in the impact environment. As noted in the introduction, the most renowned occurrence of shock-induced shear localization has been observed in naturally occurring silicate quartz and feldspar minerals. Considerable attention was focused on this discovery in the 1960s and early 1970s where petrographic studies revealed the distinctive laminar amorphous deformation features in rocks within meteor impact craters (e.g. Stoffler 1972). Such shear deformation features became a defining signature for the occurrence and intensity of impact craters (e.g. Bunch 1968, Chao 1968). Laboratory studies of shock deformation in silicate minerals (e.g. DeCarli 1968, Grady *et al* 1974, 1975) revealed the adiabatic nature of shock-induced shear localization and associated the effect with loss of strength on the shock Hugoniot in brittle solids. These observations provided the impetus for early explorations of linear stability analysis in assessing the most favorable fabric of adiabatic shear bands in the impact shock environment (Grady 1977, 1980).

More recently, observations of shear deformation localization have been made on boron carbide ceramic subjected to ballistic impact in excess of 20 GPa shock stress (Chen *et al* 2003). Subsequently, other laboratory methods have been used to investigate shear localization in boron carbide, including indentation testing (Yan *et al* 2006) and diamond anvil compression (Yan *et al* 2009). High-resolution electron microscopy reveals the signature amorphous lamella associated with regions of intense local shear deformation. Computational molecular dynamics (MD) analysis suggested that amorphization is a decompression recovery condition of a metastable distorted crystal state brought about by both pressurization and shear deformation experienced in the ballistic event (Yan *et al* 2009).

The static transition in quartz has been explained in terms of free-energy and kinetics arguments (Sikka 1992) in which the α-quartz crystalline state is driven by elevated pressure to a high free-energy state with respect to both the high-density crystalline and amorphous state, but transformation is prohibited by transition kinetics. When the mechanical stability limit is achieved the material follows the easiest kinetic path to an amorphous high-density state. MD simulations of the α-quartz transition have been satisfactory in reproducing the expected transition pressure and volume strain (Tse and Klug 1992, Chaplot and Sikka 1992).

Neither quasi-static high-pressure nor shock wave testing, however, has uncovered unambiguous evidence for phase transformation in boron carbide. Nonetheless, unusual features in the shock Hugoniot of boron carbide lend tentative support for phase transformation in the dynamic event (Grady 2010, 2015a). The previous studies on boron carbide (Chen *et al* 2003, Yan *et al* 2009) do not explore the possibility of adiabatic shear in the deformation processes leading to the observed amorphous deformation features.

The low-rate deformation tests on boron carbide (indentation and diamond anvil compression) do not preclude adiabatic shear and thermal softening in the shear localization event. Dynamic adiabatic shear localization can occur in such testing if the onset of local failure is unstable and catastrophic. Such localized catastrophic failure has not been established for boron carbide. Localized catastrophic failure has, however, been suggested as an explanation for the unique chaotic precursor wave structure observed in the shock compression of boron carbide (Grady 1994, Vogler *et al* 2004, Zhang *et al* 2006). It is a prerequisite to applicability of the adiabatic shear analysis as presented in the previous subsection.

The relations for adiabatic shear within an elastic solid developed in the previous subsection can be tested with the experimental results reported for boron carbide (Chen *et al* 2003, Yan *et al* 2009). The application of properties for boron carbide in equation (5.105) suggests a transition strain rate of order 10^9 s^{-1} to 10^{10} s^{-1}—achievable only in the rise time of quite intense shock waves. Thus, elastic strain energy, not kinetic energy, is expected to fuel a shear localization process in boron carbide. Equations (5.103) and (5.104) for the shear-band thickness and the spacing of multiple shear bands, if the theory leading to them is appropriate, would be applicable.

The thermomechanical properties available in the literature for boron carbide are used to assess the predicted shear-band features and provide comparisons with the recent observations. Reported melt temperatures for boron carbide provide a temperature rise to melt of approximately T_m = 2400 K. Boron carbide density and shear modulus are, respectively, $\rho = 2510$ kg m^{-3} and $G = 195$ GPa. The specific heat for boron carbide is c = 950 J kg^{-1} K. Reported values of the thermal conductivity of boron carbide range over 30–90 W m^{-1} K providing a thermal diffusivity in the range of $\chi = 25 \times 10^{-6}$ m^2 s^{-1}. The Hugoniot elastic limit of boron carbide ceramic is about 15–20 GPa. Shear-stress concentrations leading to localized failure in the shock impact event should be of the same order and $\tau_o = 20$ GPa is used in the calculation.

Application of these properties in equations (5.103) and (5.104) provides predictions for the shear-band width and shear-band spacing of about a_o = 16 nm and b_o = 90 nm, respectively (180 nm for $2b_o$). These predicted values for shear-band width and spacing are about three orders of magnitude smaller than corresponding predictions for impact-induced thermal deformation localization in metals. A predicted shear-band width of a_o = 16 nm is about a factor of four larger than the 3–5 nm width of amorphous deformation lamella observed by Chen *et al* (2003), and about an order of magnitude larger than the 1–2 nm width lamella reported for boron carbide by Yan *et al* (2009). Fracture steps observed in the images provided by Chen *et al* (2003) suggest a deformation band spacing of about 150 nm, not out of line with a prediction of b_o = 90–180 nm. Energy dissipation in the adiabatic shear band is estimated from equation (5.92) and provides a value for $W_o = 2\mathcal{E}_r$ of about 180 J m^{-2}. Expressed in units of fracture toughness results in a shear-band toughness of $K_s = \sqrt{GW_o}$, or about K_s = 6 MPa m$^{1/2}$—remarkably close to the measured fracture toughness values of K_f = 3–5 MPa m$^{1/2}$ reported for boron carbide ceramic. A sensible explanation for comparable values of calculated shear-band toughness and measured fracture toughness is discussed in the next section of this chapter (Grady 2009a).

The shear-localization model in an elastic solid culminating in equations (5.103) and (5.104) provides predictions of shear-band morphology (widths and spacings) based on the premise that shear localization is a dynamic and non-isothermal event. The theory has no application to processes of quasi-static and isothermal shear localization. Predictions of the theory for the widths and nominal spacings of adiabatic shear bands in boron carbide, based on independent commonly available thermo-elastic properties, are not too far out of line with observed amorphization morphology (e.g. Chen *et al* 2003). Such accord may be fortuitous. The comparative agreement between experiment and theory, on the other hand, may point to thermal-localization adiabatic shear as a deformation mechanism responsible for the observed amorphization.

5.6.15 Spatial statistics of multiple shear bands

Failure through adiabatic shear in metals commonly requires a multiplicity of shear bands. Many have explored the natural spacing of shear bands, and its dependence on material properties and the dynamics of deformation. The spacing of adiabatic shear bands is one of the paramount topics addressed in the present chapter. The frequency of adiabatic shear bands in the dynamic deformation event is, however, a statistical process. Investigations of the statistics of adiabatic shear are not common. In the present subsection, shear-band statistics are examined for two studies from the literature. The first is the experimental investigation of Xue *et al* (2004) on the explosive-driven adiabatic shear in stainless steel previously pursued in this chapter. The second is a computational study of adiabatic shear in metals by Zhou *et al* (2006).

Adiabatic shear experiments

The experimental study of Xue *et al* (2001, 2004) provides a reasonably clear description of the spatially evolving fabric of adiabatic shear deformation. From these data, at least a first-order assessment of spatial shear-band statistics can be extracted. The spatial plot of adiabatic shear for tests on as-received stainless steel from the study of Xue *et al* is shown in figure 5.17.

Cylinders of steel subjected to explosive deformation experienced extensive plastic strain and adiabatic shear. A photograph of this test was shown earlier in figure 5.2. The plot in figure 5.17 identifies the spatial position on the inner circumference of the dynamically compressed cylinder at which each adiabatic shear band

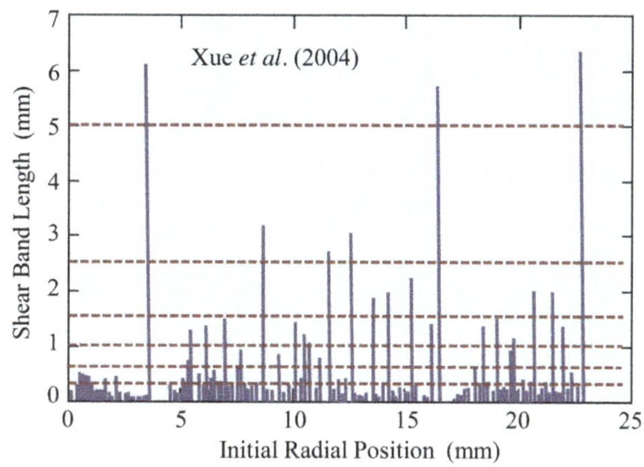

Figure 5.17. Spatial distribution and propagation lengths of adiabatic shear bands in explosively compressed cylinders of as-received stainless steel (Xue 2001, Xue *et al* 2004). The statistical spacing of shear bands is assessed at shear-band propagation distances of 0.3, 0.6, 1.0, 1.5, 2.5 and 5.0 mm.

accommodating the deformation is activated. The plot also provides the distance of propagation of each shear band. The location of each shear band on the inner circumference provides a reasonable assessment of the statistical position of shear bands, and this statistical distribution has been determined at six increasing distances of shear-band propagation as indicated in the figure.

From the plot in figure 5.17 the initial spacing between surviving shear bands at each of the six propagation distances is determined. These spacings are plotted for each successive distance as a cumulative number smaller in a semi-logarithmic graph in figure 5.18. This representation of the statistical fragment size data is commonly referred to as a Mott plot (e.g. Grady 2009c) in recognition of the seminal theoretical study of Mott (1943, 1947) on dynamic fragmentation. The statistical data at each of the shear-band propagation distances are observed to be reasonably well described by an exponential curve, and a best exponential fit to the data is shown in each plot.

It is shown (chapter 2, and also Grady 2009b, 2009c) that if breaks on a line curve are introduced at random with some constant frequency, then the distribution in spacings between breaks will be exponential. The shear-band data of Xue *et al* (2004) suggest that as shear bands drop out of the population as the propagation distance increases, the remaining active shear bands will distribute according to a statistically random spacing that is largely independent of the governing physics.

Exponential fits at each of the six shear-band lengths provide a characteristic size scale (inverse of the exponent constant in each exponential fit). This size scale is plotted as a function of shear-band propagation distance in figure 5.19. If, as a first approximation, material is assumed to be incompressible during plastic deformation, then the strain rate in the imploding cylindrical specimen would reduce with radial distance according to the proportionality $\dot{\gamma}_o \sim r^{-2}$. Shear-band spacing based on equation (5.84) depends on strain rate according to $b_o \sim \dot{\gamma}_o^{-3/4}$. Thus, it follows that shear-band spacing should increase with distance as $b_o \sim r^{3/2}$. This prediction of

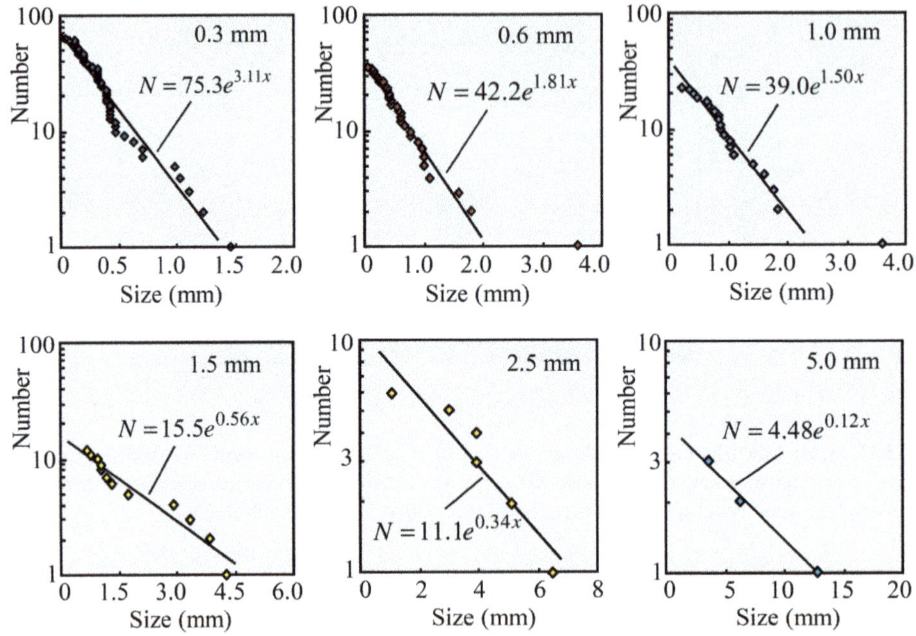

Figure 5.18. Semi-logarithmic Mott plots of the spacings of adiabatic shear bands from figure 5.17 at shear-band propagation distances of 0.3, 0.6, 1.0, 1.5, 2.5 and 5.0 mm (data from Xue *et al* 2004).

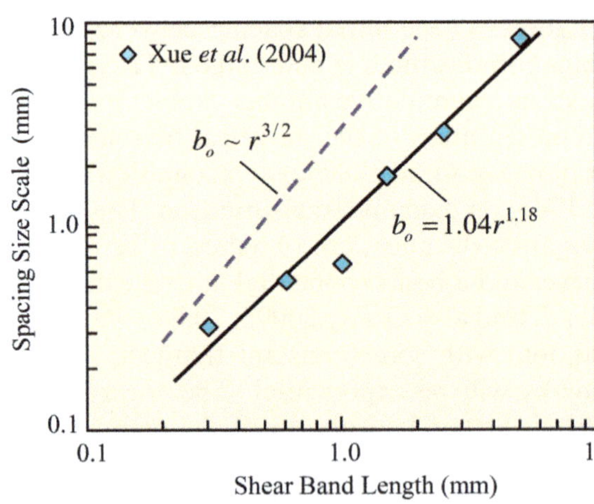

Figure 5.19. Plot of shear-band spacing size scales from statistical fit at increasing shear-band propagation distance. The data are compared with the functional dependence of shear-band spacing predicted by the two-temperature theory (data from Xue *et al* 2004).

shear-band spacing is in qualitative agreement with a power of about 1.2 exhibited by the data in figure 5.19. It is shown in the earlier section 5.5.12 that equation (5.84) provides a reasonable quantitative prediction of shear-band spacing for the present stainless steel data.

Adiabatic shear numerical calculations

Zhou *et al* (2006) have undertaken a numerical analysis of the one-dimensional shear deformation of a plastic solid, including thermal softening and rate-dependent hardening. Through a random introduction of perturbations, the numerical solutions lead to the statistical activation, growth and interaction of shear bands. Conclusions are extracted from the computational solution concerning the characteristic spacing of multiple adiabatic shear bands, and the dependence of this shear-band feature on thermomechanical properties of the material and the deformation strain rate of the event. Further, the authors have assessed the statistical distribution of shear-band spacing. Comparisons of the computational results of Zhou *et al* (2006) can be made with the previous experimental results of Xue *et al* (2004).

The cumulative distribution of shear-band spacings resulting from the computational simulations of adiabatic shear at two different rates of shearing (5×10^4 s^{-1} and 5×10^5 s^{-1}) are shown on the right in figure 5.20. Although there are clear differences in the conditions leading to adiabatic shear, the computational distributions nonetheless differ markedly from the near-linear exponential distributions observed in the experimental study of Xue *et al* (2004).

The computational model (Zhou *et al* 2006) of one-dimensional thermoplastic shear leading to onset of adiabatic shear localization at or near load maximum bears striking similarity to the statistical model posed by Mott (1943, 1947) to describe the tensile fracture of a uniformly stretching plastic body. That similar statistics should govern the present shear-localization event is not surprising. Mott envisioned a body undergoing plastic deformation at the load maximum. Fractures are randomly introduced in space and time until stress release through the diffusive propagation of plastic waves from sites of fracture subsumes the body. The solution to the statistical problem was first performed by Mott (1943, 1947) using graphical methods. An analytic solution was obtained later (Grady 1981a, 2006) for the statistical size

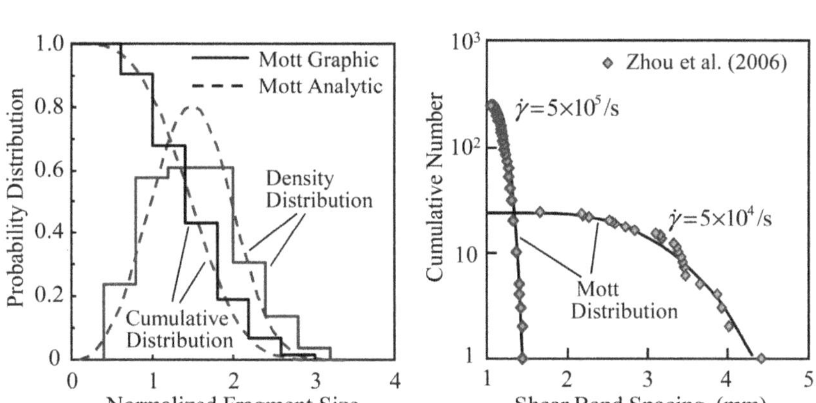

Figure 5.20. Left: the graphic probability density fragment size distributions resulting from the seminal model of Mott (1943, 1947) and the analytic solution of the same model (Grady 1981a). Right: distributions of shear-band spacing resulting from computational calculations at two strain rates (Zhou *et al* 2006) and compared with the Mott fragment size cumulative distribution.

distribution resulting from the same Mott model. The Mott, and later analysis, are described in chapter 2. The graphical size distribution derived by Mott and the subsequent analytic distribution are shown on the left in figure 5.20. The analytic representation of the cumulative Mott distribution is compared with the computational distributions of Zhou *et al* (2006) for the two loading rates on the right in figure 5.20. The Mott size distribution and the calculated computational distribution are in remarkably good agreement.

The Mott distribution of fragment (or shear-band) spacings differs from that of the exponential distribution resulting from random breakage of the same one-dimensional body, because regions of the body neighboring points of fracture (or shear-band activation) become inaccessible to further fracture due to the plastic release waves emanating from the sites of fracture. The smallest fracture spacings are rendered statistically less probable because of this release wave physics. There is evidence that the population of the smallest shear-band spacings are also under represented (relative to random fragmentation) in the experimental results of Xue *et al* (2004), but not to the extent observed in the computational study. The tendency for the experimental shear-band spacings to more closely replicate the natural statistical partitioning of a one-dimensional body is not fully understood, but may relate to physics of shear-band interaction not captured by the computational model.

5.6.16 Transition to turbulence in the adiabatic shear band

An interesting issue concerns late-time flow in the adiabatic shear band following passage of the shear-band process zone. If bounding deformation of the body requires continued shear deformation within the adiabatic shear band, then the high temperature and corresponding low shear stress must be sustained. In turn, however, the shear stress must be sufficient to generate the dissipation necessary to counter the cooling due to thermal conduction out of the shear band. If the stress level is not sufficient to account for the necessary dissipation through laminar flow, then a transition to turbulent flow deformation within the shear band may be needed to amplify dissipation to the necessary level. Experimental observations of adiabatic shear bands in steel with infrared diagnostics (Guduru *et al* 2001) provide evidence of temperature instabilities in the shear band, suggestive of deformation turbulence.

A transition from laminar to turbulent deformation within the adiabatic shear band suggests a rapid changing of the Reynolds number characterizing shear flow in the shear band. A sensible measure of the shear-band Reynolds number within the context of the present two-temperature boundary layer model is provided by

$$R = \frac{\rho \dot{\psi}^2/2}{\tau}, \tag{5.109}$$

which is a ratio of the dynamic stress at the boundary and the dissipative shear stress within the band. The time-dependent velocity of the boundary displacement $\dot{\psi}(t)$ and the stress $\tau(t)$ within the shear band are provided through the Mott wave solution. Using the solutions provided by equations (5.67)–(5.69), yields a time-dependent Reynolds number

$$R(t) = \left(\frac{9}{8}\frac{\rho W^2 \dot{\gamma}^2}{\tau_o^3}\right)^{1/3} \frac{(t/t_c)^4}{1 - (t/t_c)^3}. \tag{5.110}$$

The optimum shear-band width is not yet specified in equation (5.110). Using values of t_{co} and W_o for the optimum width provided in equations (5.80) and (5.82) yields for the Reynolds number

$$R(t) = \frac{3}{2}\left(\frac{\rho^3 c^2 \chi T_m^2 \dot{\gamma}}{\tau_o^3}\right)^{1/2} \frac{(t/t_{co})^4}{1 - (t/t_{co})^3} = R_o \frac{(t/t_{co})^4}{1 - (t/t_{co})^3}. \tag{5.111}$$

The dimensionless number $R_o = \sqrt{Q^2/P}$ was identified earlier in the dimensional analysis leading to equation (5.16). The constant R_o calculated at a strain rate of $10^4\ \mathrm{s}^{-1}$ yields values ranging over about 0.01–0.1 for the Johnson–Cook metals, with the larger values corresponding to the more ductile metals such as copper and tantalum and the smaller values to metals such as the hard steels, uranium and titanium. The time-dependent Reynolds number provided by equation (5.111) is plotted in figure 5.21. This number rapidly grows towards the end of the shear-band growth process and may correlate with onset of turbulence.

The present analysis relates the properties governing adiabatic shear to a parameter (the Reynolds number) that is familiar in the topic of turbulent flow. The analysis does not predict shear-band turbulence. If, however, turbulence in the late-time adiabatic shear deformation is a viable mechanism, then it is reasonable to suggest that the analysis provides a method for identifying the relative propensity of a material to experience turbulent deformation.

In an event favoring failure through adiabatic shear, two mechanisms stand out as available to the material to dissipate excess energy. First, the network of adiabatic shear bands can expand and refine. More adiabatic shear bands will dissipate more

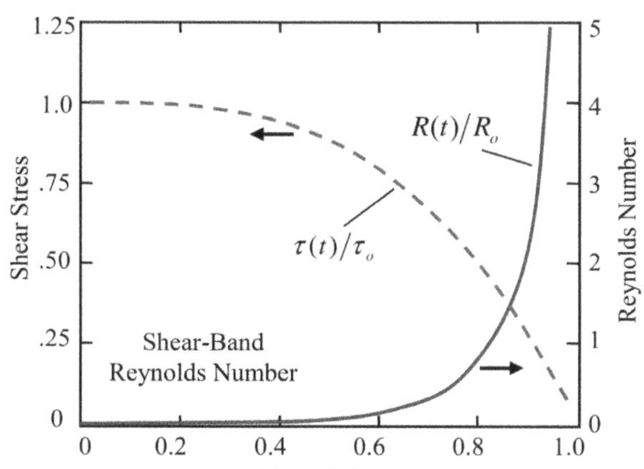

Figure 5.21. Time-dependent Reynolds number through the shear-band growth period.

energy. Impact breach of metal plates often reveals a profusion of shear bands that accommodate the required deformation and dissipation. Second, the onset of turbulent deformation in individual shear bands would further enhance dissipation. Plugging of metal plates by flat-nose projectiles commonly leads to failure through a single dominant shear band. Conditions for onset of turbulence would be enhanced in such geometries. The partitioning of energy dissipation among the available shear-band dissipation mechanisms is beyond the scope of our present understanding of shear-band physics.

5.7 Commonalities of adiabatic shear and fracture toughness

A final effort in the present chapter is to explore the proposition that the thermoplastic adiabatic shear and fracture toughness properties of metals have underlying physical commonalities (Grady 2009a). The arguments draw on physical principles inherent in the Dugdale model of crack-tip plasticity (Dugdale 1960). Namely, that an advancing crack tip in a ductile metal is not atomistically sharp. Rather, a layer of material with finite thickness normal to the crack plane undergoes plastic deformation during advance of the crack and accounts for the resisting dissipation in the fracture process.

Foremost in this consideration is the question of heat generated in the plastic fracture process, and the diffusion of this heat during fracture. There is little doubt that in unstable crack growth and fast fracture, near-adiabatic plastic deformation can occur in the fracture process zone. Near the fracture threshold, and in subsequent slow stable crack growth associated with toughness testing, however, quasi-static fracture and corresponding isothermal plasticity is usually assumed. If quasi-static fracture and isothermal plasticity occurs on all length scales, then there is no apparent relationship to the present physical model for adiabatic shear, and the similarities in resisting energies is an interesting, but fortuitous, observation.

On the other hand, if on some length scales fracture advance is not quasi-static, then the possibility of non-isothermal plasticity is present. Given the heterogeneous character of all engineering materials, and the fluctuating stress field corresponding to length scales of the heterogeneity, it is reasonable to propose that crack motion, even in stable crack growth, is stochastically dynamic. Crack advance on the microscopic and mesoscopic scale may proceed through erratic jumps and starts as energy barriers at pinning points are overcome. Then crack-tip plasticity may occur, on average, at relatively high deformation rates. This physical picture of the ductile fracture process would offer the possibility of non-isothermal plasticity in the fracture process zone.

There is ample evidence for stochastically dynamic fracture. Slow fracture in heterogeneous brittle solids proceeds through depinning transitions wherein fracture progression is through random jumps (Bonomy *et al* 2008). The sizes of stochastic fracture motions span a range of length scale that are power-law in character. Similar behavior is observed in the motion of ferromagnetic domain walls in response to a magnetic field (Bonomy *et al* 2008). It is reasonable to propose that slow fracture in ductile metals belongs to the same universality class and exhibits stochastically dynamic fracture.

5.7.1 Dugdale model and dynamic fracture

Accepting the proposition that fracture propagation on some length scale is dynamic, even for stable crack growth, a model is advanced that is based on near-adiabatic plasticity in the fracture of metals. A phenomenological theory based on the Dugdale ductile fracture model and developed by others (Lawn and Wilshaw 1975, Lawn 1993) is applied here and modified to account for the fracture dynamics. The more transparent presentation for the present purposes is provided in the earlier of these two references. Their model is illustrated on the left in figure 5.22 where, within the process zone of an advancing crack, transverse microcracks form isolating ligaments or membranes with initial length d that subsequently undergo extension plastic deformation until ductile rupture occurs. The stress load σ applied to the extending ligament at the boundary of the process zone ranges from a maximum of σ_Y to zero as the ligament accommodates an average plastic strain ε_c to failure. The critical crack opening displacement $u_c = \varepsilon_c d/2$ provides for a strain energy release,

$$G_c = \frac{1}{2}\sigma_Y(2u_c) = \frac{1}{2}\sigma_Y(\varepsilon_c d), \tag{5.112}$$

where a linear reduction of stress over the displacement $0 \leqslant u \leqslant u_c$ is assumed.

The accumulation of the plastic strain ε_c in the extending ligament is critical to the present discussion. Instability in the stretching ligament leads to localized necking. The largest portion of the crack opening displacement, and plastic strain, is accommodated by deformation in the necking region of the ligament.

Failure through ductile necking in a macroscopic metal tensile specimen is a familiar phenomenon. The extent a of the necking region is comparable to the width

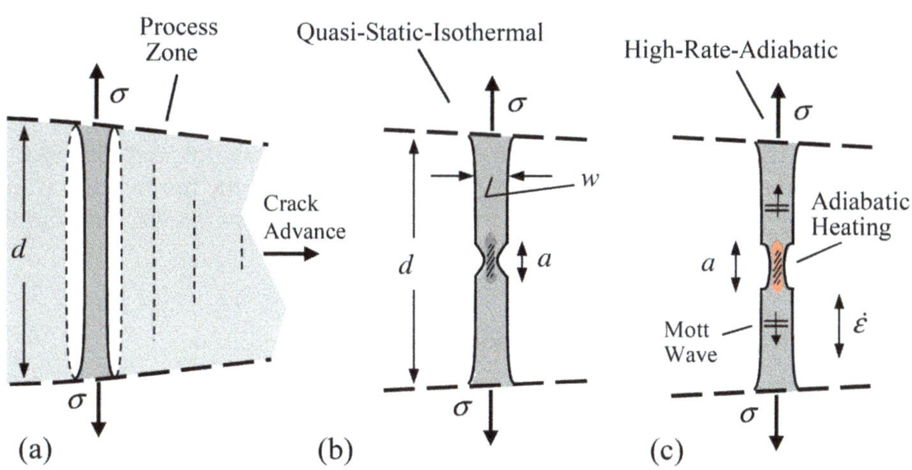

Figure 5.22. Model of advancing fracture dissipation in a ductile material (a), assuming quasi-static, isothermal and self-similar ductile plastic necking (b), in contrast to dynamic, near-adiabatic and non-self-similar necking (c) (Grady 2009a).

w of the test specimen. The geometry of the neck region, at least to first-order, is independent of the width. Necking deformation and failure is sensibly self-similar down to length scales comparable with the ligaments within the fracture process zone. This necking failure is illustrated in figure 5.22 by the sketch identified as quasi-static and isothermal.

5.7.2 Mott waves and thermal transport

Now consider a dynamic fracture process leading to plastic extension and onset of necking of the ligament at a high rate of strain as illustrated in sketch (c) in figure 5.22. Near-adiabatic plasticity in the neck region accommodating local heating and accompanying thermal softening may now play a role in the fracture physics at the neck.

Consider again the necking failure of a macroscopic tensile specimen but this time under dynamic conditions. The rapidly expanding ring experiment is representative. Failure in the neck region mirrors the static experiment to a large extent, except that a characteristic failure time is involved. Momentum and energy must be transported to the neck region through a propagating stress diffusion or Mott wave. This characteristic failure time is readily calculated (e.g. Kipp and Grady 1985), and is

$$t_f \sim \left(\frac{\rho a^2}{\sigma_Y \dot{\varepsilon}}\right)^{1/3}. \tag{5.113}$$

In this relation a is the extent of the neck region, and is comparable to the width of the tensile specimen, $a \sim w$. At onset of necking, the plastic yield stress is σ_Y and $\dot{\varepsilon}$ is the stretching rate.

A second characteristic time applicable to the present analysis is that determined by heat conduction over the dimension of the neck, and is

$$t_h \sim \frac{a^2}{\chi}, \tag{5.114}$$

where χ is the thermal diffusivity of the material. For macroscopic tensile specimens the heat transfer time t_h is orders of magnitude greater than the momentum transfer time t_f. Thus, plastic deformation in normal macroscopic dynamic necking failure is effectively adiabatic.

5.7.3 Ductile fracture and adiabatic shear commonalities

Consider tensile specimens as the characteristic size decreases. The two characteristic times from equations (5.113) and (5.114) are plotted in figure 5.23 for several metals as a function of the neck dimension a. The neck geometry is assumed to be self-similar so this is also a reasonable plot of the characteristic times versus the tensile specimen diameter w.

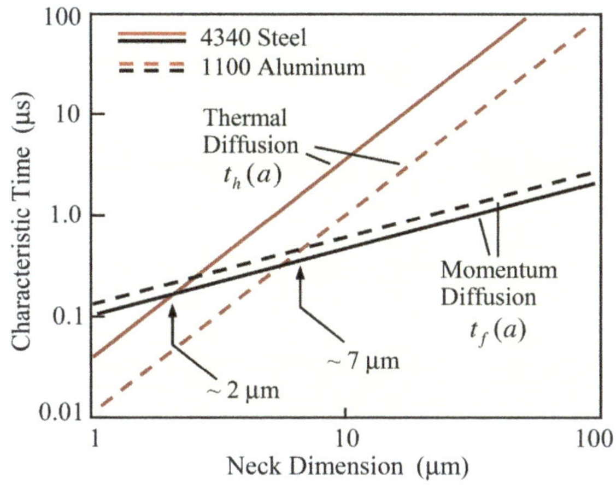

Figure 5.23. Characteristic thermal-diffusion and momentum-diffusion times in dynamic necking failure of a ductile metal. Properties for 4340 steel and 1100 aluminum are used (Grady 2009a).

As the specimen size and dimension of the neck region become increasingly smaller, the heat and momentum diffusion characteristic times become comparable. Equating the two characteristic times yield a length scale

$$a \sim \left(\frac{\rho\chi^3}{\sigma_Y\dot{\varepsilon}}\right)^{1/4}. \tag{5.115}$$

For the steel and aluminum, the time scales are comparable at size scales of approximately 2 μm and 7 μm, respectively, as illustrated in the figure. To account for the fracture dynamics, a strain rate of 10^4 s^{-1} was arbitrarily used in equations (5.113) and (5.115). At a strain rate of 10^5 s^{-1} this length scale would be about a factor of two smaller. The calculated length scales are not inconsistent with the observations of dimpling dimensions in the fracture of ductile metals. To maintain near-adiabatic conditions at this size scale and smaller, it becomes necessary for the necking region to elongate to counter heat conduction. Self-similarity of the necking geometry is lost at size scales where characteristic momentum diffusion and thermal diffusion times are comparable, as is illustrated in the right-hand sketch in figure 5.22. The competing physics of heat diffusion and momentum diffusion now govern the necking deformation and failure. Stress softening due to local heating at this size scale involves thermoplastic physics and plastic wave propagation comparable to the adiabatic shear localization process.

A mechanistic picture of the stochastically dynamic advance of fracture in a ductile metal is illustrated in figure 5.24. Cavitation, ligamentation and rupture occur on length scales and time scales at which both momentum and heat transport play comparable roles in establishing crack-tip plasticity. Momentum transport through diffusive Mott waves accommodates material displacement in the necking zone. Thermal transport establishes the material volume participating in plastic dissipation in the necking zone.

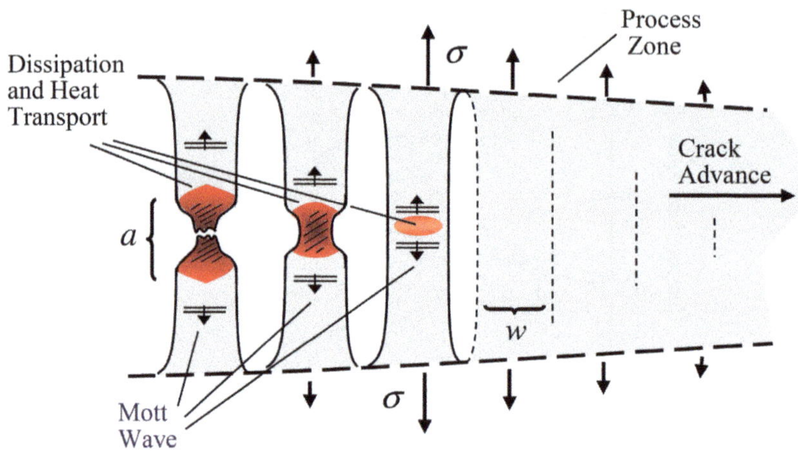

Figure 5.24. Mechanisms of momentum transport and heat transport in the resistance to dynamic ductile crack growth.

Although tentative, the physical ideas proposed offer a sensible explanation for the calculations of shear-band toughness that are found to be comparable to measured fracture toughness properties of the same ductile metals. Namely, stochastic dynamics in the ductile fracture of metals leads to processes of crack-tip plasticity in which the competition of microscopic momentum and thermal diffusion determines the extent of fracture dissipation. When such conditions are met, ductile fracture and adiabatic shear in metals exhibit comparable resistance.

To recap, a theory of thermoplastic shear localization based on a two-temperature model and Mott plastic wave dynamics leads to direct calculation of shear-band resistance of common metals using familiar thermal and mechanical properties. Calculated shear-band resistance is in remarkable accord with fracture resistance of the same metals based on static fracture toughness properties. An explanation for this commonality of shear-band toughness and fracture toughness is provided through application of the Dugdale (1960) model of crack-tip plasticity. The explanation assumes that static fracture is stochastically dynamic, and that the plasticity elements within the crack tip achieve length scales in which both local heat transport and momentum transport on the microscale play a balancing role in determining plastic dissipation in the crack advance process. The same physics governs dissipation in the propagation of adiabatic shear bands. Similar comparisons are made with deformation length scales achieved in the failure of metals under strong shock compression (Grady 2015b).

5.8 Summary and discussion

The present chapter explores the implications of applying two models of adiabatic shear localization to a suite of common metals for which the thermal and mechanical properties key to the shear-localization process are available. The suite includes 16 metals whose constitutive plastic response has been provided by Johnson and Cook (1983) and Johnson (1993) (here called the Johnson–Cook metals), and

304 stainless steel whose adiabatic shear properties have been extensively studied by Xue *et al* (2003, 2004, 2006). An exploratory excursion into application of the models to brittle solids is also undertaken. Boron carbide ceramic is the brittle solid examined.

The shear-localization models result from earlier efforts of the present author. The largest portion of this study focuses on a physics-based two-temperature boundary layer model of adiabatic shear (Grady and Kipp 1985, 1987); the development of which finds motivation from both the metal fragmentation plastic wave studies of Mott (1947), and the theory of thermal explosions of Semenov (1928). An early linear stability analysis of adiabatic shear localization (Grady 1977, 1980), utilizing the same material properties, provides additional insights through comparative calculations with the respective models.

A dimensional assessment is undertaken of the material properties governing the adiabatic shear-localization process in metal. It is argued, and ultimately assumed, that six material and kinematic properties are first-order constants governing features of adiabatic shear localization in regions of metals subjected to approximately uniform dynamic shear deformation. These properties are the material density, the shear stress at load maximum, the specific heat, the thermal diffusivity, the temperature rise to melt and the shearing deformation rate. The melt temperature is introduced in lieu of a thermal softening coefficient used in earlier presentations of the models. Notably excluded is material elasticity, which is considered later in the chapter.

Features of the shear-localization process (shear-band width, shear-band spacing, etc) are functionally dependent only on the six governing properties. Dimensional assessment reduces functional dependence first to three independent length scales, and then to two dimensionless numbers. Length scales emerging from both the two-temperature model and the stability analysis model are functionally related to the three independent length scales. The two dimensionless numbers, the solid Prandtl number P and the material heating number Q, similarly relate to key features of the shear-localization process.

Solutions from both the linear stability analysis and the two-temperature model of adiabatic shear localization each yield two physical length scales. The stability analysis yields a wavelength corresponding to the maximum unstable growth rate and a cut-off wavelength demarking the boundary between stable and unstable growth. The two-temperature model provides a length scale identified with the terminal width or thickness of an adiabatic shear band, and results from a minimization of the shear-band growth time or, equivalently, the shear-band dissipation energy. The second two-temperature model length scale is a local transport distance necessary to the energy and momentum requirements for completion of the shear-band growth process. The latter length scale is identified with the characteristic spacing of multiple shear bands. The latter shear-band spacing and the wavelength at maximum growth rate from the stability analysis are functionally the same and within a factor of two in magnitude. This length scale has been found to sensibly predict the typical spacing of fully mature adiabatic shear bands. The shear-band width from the two-temperature model ranges over several micrometers to several tens of micrometers for common metals—consistent with observation. The cut-off wavelength from the stability analysis has not previously been ascribed a physical interpretation. The cut-off wavelength for common metals is of the order of several hundred micrometers. It is suggested that the cut-off wavelength relates to the typical spacing of insipient shear bands early in the process of failure activation preceding the self-organization and down-select process leading to shear-band numbers sufficient to accommodate the failure event. The predicted spacing of shear bands based on the cut-off wavelength are in reasonable agreement with adiabatic shear-band measurements of Xue *et al* (2003, 2004) on stainless steel.

The models of adiabatic shear localization explored here assume that elasticity and elastic strain energy do not substantially contribute to the shear-localization phenomena in metals. Materials are regarded as rigid–plastic and the participation of material elasticity is ignored. An assessment of the contribution of elastic strain energy to adiabatic shear is undertaken. Analysis shows that higher strain rates favor kinetic energy dominated shear localization. All of the metals investigated in the 10^4 s^{-1} to 10^5 s^{-1} strain-rate range exhibit kinetic energy dominated shear localization and support the modeling assumption of rigid–plastic deformation. A transition strain rate is calculated below which elastic strain energy dominates the localization process. For 4340 steel and comparable metals, the transition strain rate is approximately 10^3 s^{-1}. Shear-band width and spacing based on elastic strain energy localization are analytically derived and found to be independent of the shearing strain rate. The calculations support the conjecture of Lemaire and Backofen (1972) that adiabatic shear in metal machining is governed primarily by the release of elastic strain energy. Rogers (1974) observed that, when adiabatic shear failure occurs in compact specimens at modest loading rates, both sample strain energy and compliance of the loading apparatus play a role in supplying the requisite energy.

Calculated shear-band width and spacing applied to adiabatic shear localization in boron carbide ceramic are found to be about three orders of magnitude smaller than comparable adiabatic shear in metals. Calculated characteristic lengths based on elastic strain energy are not out of line with observations of deformation localization observed in ballistic impact tests on boron carbide (Chen *et al* 2003).

The two-temperature model provides a prediction of the energy dissipated in the formation and growth of adiabatic shear bands. Experimental observations show that shear bands propagate, branch, arrest and otherwise exhibit many of the same features characteristic of dynamic metal fracture. It has been proposed (Grady 1992, 1994) that energy calculated from the model captures the dissipation in the process zone of a propagating adiabatic shear band. When this energy dissipation is couched in units of fracture toughness (a shear-band toughness has been suggested) the calculated values are remarkably close (within a factor of two or three) to reported fracture toughness for the same metals. This commonality has not previously been explained. An explanation is advanced. Critical to this explanation is that ductile metal fracture is stochastically dynamic even for quasi-static fracture. Given this premise, a modification of the Dugdale model of ductile fracture (Dugdale 1960) shows that adiabatic temperature rise, heat conduction and Mott momentum diffusion at the microscale, can underlie the processes of ductile fracture dissipation.

The same physical processes govern adiabatic shear deformation. The cavitation length scales suggested by the calculation are not inconsistent with observed dimpling in the fracture of ductile metals. Similar comparisons are made with deformation length scales achieved in the failure of metals under strong shock compression (Grady 2015b).

References

Bai Y L 1981 A criterion for the thermo-plastic shear instability ed M A Meyers and L E Murr *Shock Waves and High-strain-rate Phenomena in Metals* (New York: Plenum) pp 277–84

Bai Y L 1982 Thermo-plastic instability in simple shear *J. Mech. Phys. Solids* **30** 195–207

Batra R C and Kim C H 1990 Effect of viscoplastic flow rules on the initiation and growth of shear bands at high strain rates *J. Mech. Phys. Solids* **38** 859

Birkhoff G 1962 Helmholtz and Taylor instabilities *Proc. Symp. Appl. Math. Amer. Math. Soc.* **92** 55–76

Bonomy B, Santucci S and Ponson L 2008 Crackling dynamics in material failure as the signature of a self-organized dynamic phase transition *Phys. Rev. Lett.* **101** 045501

Bunch T E 1968 Some characteristics of selected minerals from craters *Shock Metamorphism of Natural Materials* ed B French and N Short (Baltimore, MD: Mono) p 509

Burns T J, Grady D E and Costin L S 1982 On a criterion for thermo-plastic shear instability shock waves *Shock Waves in Condensed Matter* ed W J Nellis, L Seaman and R A Graham *AIP Conf. Proc.* pp 372–75

Chao E C T 1968 Pressure and temperature histories of impact metamorphosed rock based on petrographic observations *Shock Metamorphism of Natural Materials* ed B French and N Short (Baltimore, MD: Mono) p 149

Chaplot S L and Sikka S K 1992 Molecular dynamics simulations of pressure induced crystal to amorphous transitions in some corner linked polyhedral compounds *Recent Trends in High Pressure Physics* (Oxford: Blackwell) pp 259–61

Chen M W, McCauley J W and Hemker K J 2003 Shock induced localized amorphization in boron carbide *Science* **299** 1563

Clifton R J 1980 Adiabatic shear banding *Material Response to Ultra-High Loading Rates* (Washington DC: National Academy of Sciences) pp NMAB-356, chapter 8

Coleman B D and Newman D C 1989 On adiabatic shear bands in rigid-plastic materials *Acta Mechanica* **78** 263

Considère A 1885 Mémoire sur l'emploi du fer et de l'acier dans les construction *Ann. Ponts Chaussees* 9 574–775

DeCarli P S and Milton D J 1965 Stishovite: synthesis by shock wave *Science* **147** 144–5

DeCarli P S 1968 Observations of the effects of explosive shock on crystalline solids *Shock Metamorphism of Natural Materials* ed B French and N Short (Baltimore, MD: Mono) p 129

Dodd B and Bai Y 1987 *Ductile Fracture and Ductility, With Applications to Metalworking* (New York: Academic)

Dormeval R and Ansart J P 1985 Cisaillement adiabatique—influence d'une predeformation *J. Phys. Colloq* **46** 299–306

Dugdale D S 1960 Yielding of steel sheets containing slits *J. Mech. Phys. Solids* **8** 100

Frank-Kamenetskii D A 1938 *Dokl. Nauk SSSR* **18** 413

Fressengeas C and Molinari A 1987 Instability and localization of plastic flow in shear at high strain rates *J. Mech. Phys. Solids* **35** 185

Goldstein H 1965 *Classical Mechanics* (Reading, MA: Addison-Wesley)

Grady D E 1977 Processes occurring in shock wave compression of rocks and minerals *High Pressure Research Applications in Geophysics* (New York: Academic) pp 389–438

Grady D E 1980 Shock deformation in brittle solids *J. Geophys. Res.* **85**(B2) 913–24

Grady D E 1981a Fragmentation of solids under impulsive stress loading *J. Geophys. Res.* **86** 1047–54

Grady D E 1981b Strain-rate dependence of effective viscosity under steady-wave shock compression *Appl. Phys. Lett.* **38** 825–6

Grady D E 1984 Microstructural effects on wave propagation in solids *Int. J. Eng. Sci.* **22**(8-10) 1181–6

Grady D E 1992 Properties of an adiabatic shear-band process zone *J. Mech. Phys. Solids* **40** 1197–215

Grady D E 1994 Dissipation in adiabatic shear bands *Mech. Materials* **17** 289–93

Grady D E 2005 Investigation of explosively driven fragmentation of metals—metal breach, enhanced ductility, adiabatic shear, shock structure and strength *Applied Research Associates Technical Report prepared for University of California, Lawrence Livermore National Laboratory* September

Grady D E 2006 *Fragmentation of Rings and Shells* (Berlin: Springer)

Grady D E 2009a Commonalities of adiabatic shear resistance and fracture toughness in metals *Proc. of the SEM Annual Conf. (Albuquerque, NM, 1–4 June)* pp 51–7

Grady D E 2009b Dynamic fragmentation of solids *Shock Wave Science Technology Reference Library* vol 3 ed Y Horie (Berlin: Springer) pp 169–276

Grady D E 2009c Length scales and size distributions in dynamic fragmentation *Int. J. Fracture* **163** 85–99

Grady D E 2010 Analysis of shock and high-rate data for ceramics: adiabatic shear localization and shock phase transformation in boron carbide in the ballistic environment *Applied Research Associates, Inc. Report for US Army* TACOM-TARDEC, June

Grady D E 2011 Adiabatic shear failure in brittle solids *Int. J. Impact Eng.* **38** 661–7

Grady D E 2015a Hugoniot equation of state and dynamic strength of boron carbide *J. Appl. Phys.* **117** 165904

Grady D E 2015b Unifying role of dissipative action in the dynamic failure of solids *J. Appl. Phys.* **117** 165905

Grady D E, Murri W J and Fowles G R 1974 Quartz to stishovite: wave propagation in the mixed phase region *J. Geophys. Res.* **79** 332–8

Grady D E, Murri W J and DeCarli P S 1975 Hugoniot sound velocities in two silicates *J. Geophys. Res.* **80** 4857–61

Grady D E and Asay J R 1982 Calculation of thermal trapping in shock deformation of aluminum *J. Appl. Phys.* **54** 7350–54

Grady D E, Asay J R, Rohde R W and Wise J L 1983 Microstructure and mechanical properties of precipitation hardened aluminum under high rate deformation *Material Behavior under High Stress and Ultrahigh Loading Rates* ed J Mescall and V Weiss (New York: Plenum) pp 81–100

Grady D E and Kipp M E 1985 The growth of inhomogeneous thermoplastic shear *J. Physique* **46** pp 291–8

Grady D E and Kipp M E 1987 The growth of unstable thermoplastic shear with application to steady-wave shock compression of solids *J. Mech. Phys. Solids* **35**(1) 95–118

Grady D E and Lipkin J 1980 Criteria for impulsive rock fracture *Geophys. Res. Lett.* **7** 255–8

Gruntfest I J 1963 Thermal feedback in liquid flow; plane shear at constant stress *Trans. Soc. Rheol.* **7** 195–207

Guduru P R, Rosakis A J and Ravichandran G 2001 Dynamic shear bands: an investigation using high speed optical and infrared diagnostics *Mech. Materials* **33** 371–402

Hanina E, Rittel D and Rosenberg Z 2007 Pressure sensitivity of adiabatic shear banding in metals *Appl. Phys. Lett.* **90** 021915

Haynes W M 2014 *Handbook of Chemistry and Physics* (Boca Raton, FL: CRC Press)

Hoggatt G R and Recht R F 1968 Fracture behavior of tubular bombs *J. Appl. Phys.* **39** 1856–62

Johnson G R 1993 Material characterization for warhead computation *Tactical Missile Warheads: Progress in Astronautics and Aeronautics* vol 155 ed J Carleone (Washington DC: American Institute of Astronautics and Aeronautics) pp 165–97

Johnson G R and Cook W H 1983 A constitutive model and data for metals subjected to large strains, high strain rates and high temperatures *Proc. 7th Int. Symp. on Ballistics, (The Netherlands)*, pp 541–548.

Kipp M E and Grady D E 1985 Dynamic fracture growth and interaction in one dimension *J. Mech. Phys. Solids* **33**(4) 399–415

Lawn B R 1993 *Fracture of Brittle Solids* 2nd edn (Cambridge: Cambridge University Press)

Lawn B R and Wilshaw T R 1975 *Fracture of Brittle Solids* (Cambridge: Cambridge University Press)

Lee E H 1967 The continuum mechanics aspect of material properties determination *Energetics III* ed W Mueller and M Shaw (New York: Gordon and Breach) pp 85–122

Lee P R 1998 Theories and techniques of initiation *Explosive Effects and Applications* ed J A Zukas and W P Walters (Berlin: Springer) pp 173–219

Lemaire J C and Backofen W A 1972 *Met. Trans.* **3** 477–13

Maasilta I and Minnich A J 2014 Heat under the microscope *Phys. Today* **67**(8) 27

Merzer A M 1982 Modeling of adiabatic shear band development from small imperfections *J. Mech. Phys. Solids* **30** 323

Molinari A 1997 Collective behavior and spacing of adiabatic shear bands *J. Mech. Phys. Solids* **45** 1551–75

Molinari A and Clifton R J 1987 Analytical characterization of shear localization in thermovisco-plastic materials *J. Appl. Mech.* **54** 806

Mott N F 1943 A theory of the fragmentation of shells and bombs *United Kingdom Ministry of Supply* AC4035, May

Mott N F 1947 *Fragmentation of shell cases Proc. R. Soc.* A***189*** *300–8*

Needleman A 1972 Void growth in an elastic-plastic medium *J. Appl. Mech.* **39** 964–70

Reif F 1965 Fundamentals of statistical and thermal physics (New York: McGraw Hill)

Ritter D, Landau P and Venkert A 2008 Dynamic recrystallization as a potential cause for adiabatic shear failure Phys. Rev. Lett. **101** 165501

Rogers H C 1974 Adiabatic shearing—a review *Drexel University Report*

Semenov N N 1928 Zur Theorie des Verbrennungsprozesses *Z. Phys. Chem.* **48** 571–82

Sikka S K 1992 Pressure induced glasses *Recent Trends in High Pressure Physics* (Oxford: Blackwell) pp 254–8

Stelly M and Dormeval R 1986 Adiabatic shearing *Metallurgical Applications of Shock Waves and High-strain-rate Phenomena* ed L E Murr, K P Staudhammer and M A Meyers (New York: Marcel Dekker) pp 607–32

Steverding B and Lehnigk S H 1976 The fracture penetration depth of stress pulses *Int. J. Rock Mech. Min. Sci.* **13** 75–80

Stoffler D 1972 Deformation and transformation of rock-forming minerals by natural and experimental shock processes *Fortschr. Min* **49** 50–113

Stoffler D 1984 Glasses formed by hypervelocity impact *J. Non-Cryst. Solids* **67** 465–502

Swegle J W and Grady D E 1985 Shock viscosity and the prediction of shock wave rise times *J. Appl. Phys.* **58**(2) 692–701

Trexler M M and Thadhani N N 2010 Mechanical properties of bulk metallic glasses *Prog. Mater. Sci* **55** 759

Tse J S and Klug D D 1992 High pressure phase transitions in silica and berlinite by molecular dynamics simulations *Recent Trends in High Pressure Physics* (Oxford: Blackwell) pp 274–9

Vogler T J, Reinhart W D and Chhabildas L C 2004 Dynamic behavior of boron carbide *J. Appl. Phys.* **95** 4173

Walker F E and Wasley R J 1969 Critical energy for shock initiation of heterogeneous explosives *Explosivstoffe* **17** 9

Wright T W 1990 Approximate analysis for the formation of adiabatic shear bands *J. Mech. Phys. Solids* **38** 515

Wright T W 2002 *The Physics and Mathematics of Adiabatic Shear Bands* (Cambridge: Cambridge University Press)

Wright T W and Ockendon H 1996 A scaling law for the effect of inertia on the formation of adiabatic shear bands *Int. J. Plasticity* **12** 927–34

Xue Q 2001 Spatial evolution of adiabatic shear localization in stainless steel, titanium, and Ti–6Al–4V alloy *PhD Thesis* (San Diego: University of California)

Xue Q, Meyers M A and Nesterenko V F 2003 Evaluation of the collapsing thick-walled cylinder technique for shear-band spacing *Int. J. Impact Eng.* **28** 257–80

Xue Q, Meyers M A and Nesterenko V F 2004 Self organization of shear bands in stainless steel *Mater. Sci. Eng* **384** 35–46

Xue Q, Cerreta E K and Gray G T 2006 Influence of explosive-driven shock prestraining on the microstructural evolution and shear localization of 304 and 316L stainless steels *Shock Waves in Condensed Matter* (New York: Plenum) pp 783–6

Yan X Q, Li W J, Goto T and Chen M W 2006 Raman spectroscopy of pressure-induced amorphous boron carbide *Appl. Phys. Letts* **88** 131905

Yan X Q, Tang Z, Zhang L, Guo J J, Jin C Q, Zhang Y, Goto T, McCauley J W and Chen M W 2009 Depressurization amorphization of single-crystal boron carbide *Phys. Rev. Letts.* **102** 075505

Zener C and Holloman J H 1944 Effect of strain rate upon plastic flow of steel *J. Appl. Phys.* **15** 22–32

Zhang Y, Fukuoka K, Kikuchi M, Kodama M, Shibata K and Mashimo T 2005 Shock compression behavior of titanium diboride *Int. J. Impact Eng.* **32** 643–9

Zhang Y, Mashimo T, Uemura Y, Uchino M, Kodama M, Shibata K, Fukuoka K, Kikuchi M, Kobayashi T and Sekine T 2006 Shock Compression Behaviors of Boron Carbide, *J. Appl. Phys.* **100** 113536

Zhou F, Wright T W and Ramesh K T 2006 The formation of multiple adiabatic shear bands *J. Mech. Phys. Sol.* **54** 1376–400

Zurek A K and Meyers M A 1996 Microstructural aspects of dynamic failure *High-pressure Shock Compression of Solids II—Dynamic Fracture and Fragmentation* ed L Davison, D Grady and M Shahinpoor (Berlin: Springer) pp 25–70

www.ingramcontent.com/pod-product-compliance
Ingram Content Group UK Ltd.
Pitfield, Milton Keynes, MK11 3LW, UK
UKHW051042120926
8556IPUK00030B/810

9 780750 312554